Radical Theory of Rings

PURE AND APPLIED MATHEMATICS

A Program of Monographs, Textbooks, and Lecture Notes

EXECUTIVE EDITORS

Earl J. Taft
Rutgers University
New Brunswick, New Jersey

Zuhair Nashed
University of Central Florida
Orlando, Florida

EDITORIAL BOARD

M. S. Baouendi
University of California,
San Diego

Jane Cronin
Rutgers University

Jack K. Hale
Georgia Institute of Technology

S. Kobayashi
University of California,
Berkeley

Marvin Marcus
University of California,
Santa Barbara

W. S. Massey
Yale University

Anil Nerode
Cornell University

Donald Passman
University of Wisconsin,
Madison

Fred S. Roberts
Rutgers University

David L. Russell
Virginia Polytechnic Institute
and State University

Walter Schempp
Universität Siegen

Mark Teply
University of Wisconsin,
Milwaukee

MONOGRAPHS AND TEXTBOOKS IN
PURE AND APPLIED MATHEMATICS

1. *K. Yano*, Integral Formulas in Riemannian Geometry (1970)
2. *S. Kobayashi*, Hyperbolic Manifolds and Holomorphic Mappings (1970)
3. *V. S. Vladimirov*, Equations of Mathematical Physics (A. Jeffrey, ed.; A. Littlewood, trans.) (1970)
4. *B. N. Pshenichnyi*, Necessary Conditions for an Extremum (L. Neustadt, translation ed.; K. Makowski, trans.) (1971)
5. *L. Narici et al.*, Functional Analysis and Valuation Theory (1971)
6. *S. S. Passman*, Infinite Group Rings (1971)
7. *L. Dornhoff*, Group Representation Theory. Part A: Ordinary Representation Theory. Part B: Modular Representation Theory (1971, 1972)
8. *W. Boothby and G. L. Weiss, eds.*, Symmetric Spaces (1972)
9. *Y. Matsushima*, Differentiable Manifolds (E. T. Kobayashi, trans.) (1972)
10. *L. E. Ward, Jr.*, Topology (1972)
11. *A. Babakhanian*, Cohomological Methods in Group Theory (1972)
12. *R. Gilmer*, Multiplicative Ideal Theory (1972)
13. *J. Yeh*, Stochastic Processes and the Wiener Integral (1973)
14. *J. Barros-Neto*, Introduction to the Theory of Distributions (1973)
15. *R. Larsen*, Functional Analysis (1973)
16. *K. Yano and S. Ishihara*, Tangent and Cotangent Bundles (1973)
17. *C. Procesi*, Rings with Polynomial Identities (1973)
18. *R. Hermann*, Geometry, Physics, and Systems (1973)
19. *N. R. Wallach*, Harmonic Analysis on Homogeneous Spaces (1973)
20. *J. Dieudonné*, Introduction to the Theory of Formal Groups (1973)
21. *I. Vaisman*, Cohomology and Differential Forms (1973)
22. *B.-Y. Chen*, Geometry of Submanifolds (1973)
23. *M. Marcus*, Finite Dimensional Multilinear Algebra (in two parts) (1973, 1975)
24. *R. Larsen*, Banach Algebras (1973)
25. *R. O. Kujala and A. L. Vitter, eds.*, Value Distribution Theory: Part A; Part B: Deficit and Bezout Estimates by Wilhelm Stoll (1973)
26. *K. B. Stolarsky*, Algebraic Numbers and Diophantine Approximation (1974)
27. *A. R. Magid*, The Separable Galois Theory of Commutative Rings (1974)
28. *B. R. McDonald*, Finite Rings with Identity (1974)
29. *J. Satake*, Linear Algebra (S. Koh et al., trans.) (1975)
30. *J. S. Golan*, Localization of Noncommutative Rings (1975)
31. *G. Klambauer*, Mathematical Analysis (1975)
32. *M. K. Agoston*, Algebraic Topology (1976)
33. *K. R. Goodearl*, Ring Theory (1976)
34. *L. E. Mansfield*, Linear Algebra with Geometric Applications (1976)
35. *N. J. Pullman*, Matrix Theory and Its Applications (1976)
36. *B. R. McDonald*, Geometric Algebra Over Local Rings (1976)
37. *C. W. Groetsch*, Generalized Inverses of Linear Operators (1977)
38. *J. E. Kuczkowski and J. L. Gersting*, Abstract Algebra (1977)
39. *C. O. Christenson and W. L. Voxman*, Aspects of Topology (1977)
40. *M. Nagata*, Field Theory (1977)
41. *R. L. Long*, Algebraic Number Theory (1977)
42. *W. F. Pfeffer*, Integrals and Measures (1977)
43. *R. L. Wheeden and A. Zygmund*, Measure and Integral (1977)
44. *J. H. Curtiss*, Introduction to Functions of a Complex Variable (1978)
45. *K. Hrbacek and T. Jech*, Introduction to Set Theory (1978)
46. *W. S. Massey*, Homology and Cohomology Theory (1978)
47. *M. Marcus*, Introduction to Modern Algebra (1978)
48. *E. C. Young*, Vector and Tensor Analysis (1978)
49. *S. B. Nadler, Jr.*, Hyperspaces of Sets (1978)
50. *S. K. Segal*, Topics in Group Kings (1978)
51. *A. C. M. van Rooij*, Non-Archimedean Functional Analysis (1978)
52. *L. Corwin and R. Szczarba*, Calculus in Vector Spaces (1979)
53. *C. Sadosky*, Interpolation of Operators and Singular Integrals (1979)
54. *J. Cronin*, Differential Equations (1980)
55. *C. W. Groetsch*, Elements of Applicable Functional Analysis (1980)

56. I. Vaisman, Foundations of Three-Dimensional Euclidean Geometry (1980)
57. H. I. Freedan, Deterministic Mathematical Models in Population Ecology (1980)
58. S. B. Chae, Lebesgue Integration (1980)
59. C. S. Rees et al., Theory and Applications of Fourier Analysis (1981)
60. L. Nachbin, Introduction to Functional Analysis (R. M. Aron, trans.) (1981)
61. G. Orzech and M. Orzech, Plane Algebraic Curves (1981)
62. R. Johnsonbaugh and W. E. Pfaffenberger, Foundations of Mathematical Analysis (1981)
63. W. L. Voxman and R. H. Goetschel, Advanced Calculus (1981)
64. L. J. Corwin and R. H. Szczarba, Multivariable Calculus (1982)
65. V. I. Istrățescu, Introduction to Linear Operator Theory (1981)
66. R. D. Järvinen, Finite and Infinite Dimensional Linear Spaces (1981)
67. J. K. Beem and P. E. Ehrlich, Global Lorentzian Geometry (1981)
68. D. L. Armacost, The Structure of Locally Compact Abelian Groups (1981)
69. J. W. Brewer and M. K. Smith, eds., Emmy Noether: A Tribute (1981)
70. K. H. Kim, Boolean Matrix Theory and Applications (1982)
71. T. W. Wieting, The Mathematical Theory of Chromatic Plane Ornaments (1982)
72. D. B. Gauld, Differential Topology (1982)
73. R. L. Faber, Foundations of Euclidean and Non-Euclidean Geometry (1983)
74. M. Carmeli, Statistical Theory and Random Matrices (1983)
75. J. H. Carruth et al., The Theory of Topological Semigroups (1983)
76. R. L. Faber, Differential Geometry and Relativity Theory (1983)
77. S. Barnett, Polynomials and Linear Control Systems (1983)
78. G. Karpilovsky, Commutative Group Algebras (1983)
79. F. Van Oystaeyen and A. Verschoren, Relative Invariants of Rings (1983)
80. I. Vaisman, A First Course in Differential Geometry (1984)
81. G. W. Swan, Applications of Optimal Control Theory in Biomedicine (1984)
82. T. Petrie and J. D. Randall, Transformation Groups on Manifolds (1984)
83. K. Goebel and S. Reich, Uniform Convexity, Hyperbolic Geometry, and Nonexpansive Mappings (1984)
84. T. Albu and C. Năstăsescu, Relative Finiteness in Module Theory (1984)
85. K. Hrbacek and T. Jech, Introduction to Set Theory: Second Edition (1984)
86. F. Van Oystaeyen and A. Verschoren, Relative Invariants of Rings (1984)
87. B. R. McDonald, Linear Algebra Over Commutative Rings (1984)
88. M. Namba, Geometry of Projective Algebraic Curves (1984)
89. G. F. Webb, Theory of Nonlinear Age-Dependent Population Dynamics (1985)
90. M. R. Bremner et al., Tables of Dominant Weight Multiplicities for Representations of Simple Lie Algebras (1985)
91. A. E. Fekete, Real Linear Algebra (1985)
92. S. B. Chae, Holomorphy and Calculus in Normed Spaces (1985)
93. A. J. Jerri, Introduction to Integral Equations with Applications (1985)
94. G. Karpilovsky, Projective Representations of Finite Groups (1985)
95. L. Narici and E. Beckenstein, Topological Vector Spaces (1985)
96. J. Weeks, The Shape of Space (1985)
97. P. R. Gribik and K. O. Kortanek, Extremal Methods of Operations Research (1985)
98. J.-A. Chao and W. A. Woyczynski, eds., Probability Theory and Harmonic Analysis (1986)
99. G. D. Crown et al., Abstract Algebra (1986)
100. J. H. Carruth et al., The Theory of Topological Semigroups, Volume 2 (1986)
101. R. S. Doran and V. A. Belfi, Characterizations of C*-Algebras (1986)
102. M. W. Jeter, Mathematical Programming (1986)
103. M. Altman, A Unified Theory of Nonlinear Operator and Evolution Equations with Applications (1986)
104. A. Verschoren, Relative Invariants of Sheaves (1987)
105. R. A. Usmani, Applied Linear Algebra (1987)
106. P. Blass and J. Lang, Zariski Surfaces and Differential Equations in Characteristic $p > 0$ (1987)
107. J. A. Reneke et al., Structured Hereditary Systems (1987)
108. H. Busemann and B. B. Phadke, Spaces with Distinguished Geodesics (1987)
109. R. Harte, Invertibility and Singularity for Bounded Linear Operators (1988)
110. G. S. Ladde et al., Oscillation Theory of Differential Equations with Deviating Arguments (1987)
111. L. Dudkin et al., Iterative Aggregation Theory (1987)
112. T. Okubo, Differential Geometry (1987)

113. D. L. Stancl and M. L. Stancl, Real Analysis with Point-Set Topology (1987)
114. T. C. Gard, Introduction to Stochastic Differential Equations (1988)
115. S. S. Abhyankar, Enumerative Combinatorics of Young Tableaux (1988)
116. H. Strade and R. Farnsteiner, Modular Lie Algebras and Their Representations (1988)
117. J. A. Huckaba, Commutative Rings with Zero Divisors (1988)
118. W. D. Wallis, Combinatorial Designs (1988)
119. W. Wiesław, Topological Fields (1988)
120. G. Karpilovsky, Field Theory (1988)
121. S. Caenepeel and F. Van Oystaeyen, Brauer Groups and the Cohomology of Graded Rings (1989)
122. W. Kozlowski, Modular Function Spaces (1988)
123. E. Lowen-Colebunders, Function Classes of Cauchy Continuous Maps (1989)
124. M. Pavel, Fundamentals of Pattern Recognition (1989)
125. V. Lakshmikantham et al., Stability Analysis of Nonlinear Systems (1989)
126. R. Sivaramakrishnan, The Classical Theory of Arithmetic Functions (1989)
127. N. A. Watson, Parabolic Equations on an Infinite Strip (1989)
128. K. J. Hastings, Introduction to the Mathematics of Operations Research (1989)
129. B. Fine, Algebraic Theory of the Bianchi Groups (1989)
130. D. N. Dikranjan et al., Topological Groups (1989)
131. J. C. Morgan II, Point Set Theory (1990)
132. P. Biler and A. Witkowski, Problems in Mathematical Analysis (1990)
133. H. J. Sussmann, Nonlinear Controllability and Optimal Control (1990)
134. J.-P. Florens et al., Elements of Bayesian Statistics (1990)
135. N. Shell, Topological Fields and Near Valuations (1990)
136. B. F. Doolin and C. F. Martin, Introduction to Differential Geometry for Engineers (1990)
137. S. S. Holland, Jr., Applied Analysis by the Hilbert Space Method (1990)
138. J. Oknínski, Semigroup Algebras (1990)
139. K. Zhu, Operator Theory in Function Spaces (1990)
140. G. B. Price, An Introduction to Multicomplex Spaces and Functions (1991)
141. R. B. Darst, Introduction to Linear Programming (1991)
142. P. L. Sachdev, Nonlinear Ordinary Differential Equations and Their Applications (1991)
143. T. Husain, Orthogonal Schauder Bases (1991)
144. J. Foran, Fundamentals of Real Analysis (1991)
145. W. C. Brown, Matrices and Vector Spaces (1991)
146. M. M. Rao and Z. D. Ren, Theory of Orlicz Spaces (1991)
147. J. S. Golan and T. Head, Modules and the Structures of Rings (1991)
148. C. Small, Arithmetic of Finite Fields (1991)
149. K. Yang, Complex Algebraic Geometry (1991)
150. D. G. Hoffman et al., Coding Theory (1991)
151. M. O. González, Classical Complex Analysis (1992)
152. M. O. González, Complex Analysis (1992)
153. L. W. Baggett, Functional Analysis (1992)
154. M. Sniedovich, Dynamic Programming (1992)
155. R. P. Agarwal, Difference Equations and Inequalities (1992)
156. C. Brezinski, Biorthogonality and Its Applications to Numerical Analysis (1992)
157. C. Swartz, An Introduction to Functional Analysis (1992)
158. S. B. Nadler, Jr., Continuum Theory (1992)
159. M. A. Al-Gwaiz, Theory of Distributions (1992)
160. E. Perry, Geometry: Axiomatic Developments with Problem Solving (1992)
161. E. Castillo and M. R. Ruiz-Cobo, Functional Equations and Modelling in Science and Engineering (1992)
162. A. J. Jerri, Integral and Discrete Transforms with Applications and Error Analysis (1992)
163. A. Charlier et al., Tensors and the Clifford Algebra (1992)
164. P. Biler and T. Nadzieja, Problems and Examples in Differential Equations (1992)
165. E. Hansen, Global Optimization Using Interval Analysis (1992)
166. S. Guerre-Delabrière, Classical Sequences in Banach Spaces (1992)
167. Y. C. Wong, Introductory Theory of Topological Vector Spaces (1992)
168. S. H. Kulkarni and B. V. Limaye, Real Function Algebras (1992)
169. W. C. Brown, Matrices Over Commutative Rings (1993)
170. J. Loustau and M. Dillon, Linear Geometry with Computer Graphics (1993)
171. W. V. Petryshyn, Approximation-Solvability of Nonlinear Functional and Differential Equations (1993)

172. E. C. Young, Vector and Tensor Analysis: Second Edition (1993)
173. T. A. Bick, Elementary Boundary Value Problems (1993)
174. M. Pavel, Fundamentals of Pattern Recognition: Second Edition (1993)
175. S. A. Albeverio et al., Noncommutative Distributions (1993)
176. W. Fulks, Complex Variables (1993)
177. M. M. Rao, Conditional Measures and Applications (1993)
178. A. Janicki and A. Weron, Simulation and Chaotic Behavior of α-Stable Stochastic Processes (1994)
179. P. Neittaanmäki and D. Tiba, Optimal Control of Nonlinear Parabolic Systems (1994)
180. J. Cronin, Differential Equations: Introduction and Qualitative Theory, Second Edition (1994)
181. S. Heikkilä and V. Lakshmikantham, Monotone Iterative Techniques for Discontinuous Nonlinear Differential Equations (1994)
182. X. Mao, Exponential Stability of Stochastic Differential Equations (1994)
183. B. S. Thomson, Symmetric Properties of Real Functions (1994)
184. J. E. Rubio, Optimization and Nonstandard Analysis (1994)
185. J. L. Bueso et al., Compatibility, Stability, and Sheaves (1995)
186. A. N. Michel and K. Wang, Qualitative Theory of Dynamical Systems (1995)
187. M. R. Darnel, Theory of Lattice-Ordered Groups (1995)
188. Z. Naniewicz and P. D. Panagiotopoulos, Mathematical Theory of Hemivariational Inequalities and Applications (1995)
189. L. J. Corwin and R. H. Szczarba, Calculus in Vector Spaces: Second Edition (1995)
190. L. H. Erbe et al., Oscillation Theory for Functional Differential Equations (1995)
191. S. Agaian et al., Binary Polynomial Transforms and Nonlinear Digital Filters (1995)
192. M. I. Gil', Norm Estimations for Operation-Valued Functions and Applications (1995)
193. P. A. Grillet, Semigroups: An Introduction to the Structure Theory (1995)
194. S. Kichenassamy, Nonlinear Wave Equations (1996)
195. V. F. Krotov, Global Methods in Optimal Control Theory (1996)
196. K. I. Beidar et al., Rings with Generalized Identities (1996)
197. V. I. Arnautov et al., Introduction to the Theory of Topological Rings and Modules (1996)
198. G. Sierksma, Linear and Integer Programming (1996)
199. R. Lasser, Introduction to Fourier Series (1996)
200. V. Sima, Algorithms for Linear-Quadratic Optimization (1996)
201. D. Redmond, Number Theory (1996)
202. J. K. Beem et al., Global Lorentzian Geometry: Second Edition (1996)
203. M. Fontana et al., Prüfer Domains (1997)
204. H. Tanabe, Functional Analytic Methods for Partial Differential Equations (1997)
205. C. Q. Zhang, Integer Flows and Cycle Covers of Graphs (1997)
206. E. Spiegel and C. J. O'Donnell, Incidence Algebras (1997)
207. B. Jakubczyk and W. Respondek, Geometry of Feedback and Optimal Control (1998)
208. T. W. Haynes et al., Fundamentals of Domination in Graphs (1998)
209. T. W. Haynes et al., eds., Domination in Graphs: Advanced Topics (1998)
210. L. A. D'Alotto et al., A Unified Signal Algebra Approach to Two-Dimensional Parallel Digital Signal Processing (1998)
211. F. Halter-Koch, Ideal Systems (1998)
212. N. K. Govil et al., eds., Approximation Theory (1998)
213. R. Cross, Multivalued Linear Operators (1998)
214. A. A. Martynyuk, Stability by Liapunov's Matrix Function Method with Applications (1998)
215. A. Favini and A. Yagi, Degenerate Differential Equations in Banach Spaces (1999)
216. A. Illanes and S. Nadler, Jr., Hyperspaces: Fundamentals and Recent Advances (1999)
217. G. Kato and D. Struppa, Fundamentals of Algebraic Microlocal Analysis (1999)
218. G. X.-Z. Yuan, KKM Theory and Applications in Nonlinear Analysis (1999)
219. D. Motreanu and N. H. Pavel, Tangency, Flow Invariance for Differential Equations, and Optimization Problems (1999)
220. K. Hrbacek and T. Jech, Introduction to Set Theory, Third Edition (1999)
221. G. E. Kolosov, Optimal Design of Control Systems (1999)
222. N. L. Johnson, Subplane Covered Nets (2000)
223. B. Fine and G. Rosenberger, Algebraic Generalizations of Discrete Groups (1999)
224. M. Väth, Volterra and Integral Equations of Vector Functions (2000)
225. S. S. Miller and P. T. Mocanu, Differential Subordinations (2000)

226. R. Li et al., Generalized Difference Methods for Differential Equations: Numerical Analysis of Finite Volume Methods (2000)
227. H. Li and F. Van Oystaeyen, A Primer of Algebraic Geometry (2000)
228. R. P. Agarwal, Difference Equations and Inequalities: Theory, Methods, and Applications, Second Edition (2000)
229. A. B. Kharazishvili, Strange Functions in Real Analysis (2000)
230. J. M. Appell et al., Partial Integral Operators and Integro-Differential Equations (2000)
231. A. I. Prilepko et al., Methods for Solving Inverse Problems in Mathematical Physics (2000)
232. F. Van Oystaeyen, Algebraic Geometry for Associative Algebras (2000)
233. D. L. Jagerman, Difference Equations with Applications to Queues (2000)
234. D. R. Hankerson et al., Coding Theory and Cryptography: The Essentials, Second Edition, Revised and Expanded (2000)
235. S. Dăscălescu et al., Hopf Algebras: An Introduction (2001)
236. R. Hagen et al., C*-Algebras and Numerical Analysis (2001)
237. Y. Talpaert, Differential Geometry: With Applications to Mechanics and Physics (2001)
238. R. H. Villarreal, Monomial Algebras (2001)
239. A. N. Michel et al., Qualitative Theory of Dynamical Systems: Second Edition (2001)
240. A. A. Samarskii, The Theory of Difference Schemes (2001)
241. J. Knopfmacher and W.-B. Zhang, Number Theory Arising from Finite Fields (2001)
242. S. Leader, The Kurzweil-Henstock Integral and Its Differentials (2001)
243. M. Biliotti et al., Foundations of Translation Planes (2001)
244. A. N. Kochubei, Pseudo-Differential Equations and Stochastics over Non-Archimedean Fields (2001)
245. G. Sierksma, Linear and Integer Programming: Second Edition (2002)
246. A. A. Martynyuk, Qualitative Methods in Nonlinear Dynamics: Novel Approaches to Liapunov's Matrix Functions (2002)
247. B. G. Pachpatte, Inequalities for Finite Difference Equations (2002)
248. A. N. Michel and D. Liu, Qualitative Analysis and Synthesis of Recurrent Neural Networks (2002)
249. J. R. Weeks, The Shape of Space: Second Edition (2002)
250. M. M. Rao and Z. D. Ren, Applications of Orlicz Spaces (2002)
251. V. Lakshmikantham and D. Trigiante, Theory of Difference Equations: Numerical Methods and Applications, Second Edition (2002)
252. T. Albu, Cogalois Theory (2003)
253. A. Bezdek, Discrete Geometry (2003)
254. M. J. Corless and A. E. Frazho, Linear Systems and Control: An Operator Perspective (2003)
255. I. Graham and G. Kohr, Geometric Function Theory in One and Higher Dimensions (2003)
256. G. V. Demidenko and S. V. Uspenskii, Partial Differential Equations and Systems Not Solvable with Respect to the Highest-Order Derivative (2003)
257. A. Kelarev, Graph Algebras and Automata (2003)
258. A. H. Siddiqi, Applied Functional Analysis: Numerical Methods, Wavelet Methods, and Image Processing (2004)
259. F. W. Steutel and K. van Harn, Infinite Divisibility of Probability Distributions on the Real Line (2004)
260. G. S. Ladde and M. Sambandham, Stochastic Versus Deterministic Systems of Differential Equations (2004)
261. B. J. Gardner and R. Wiegandt, Radical Theory of Rings (2004)
262. J. Haluška, The Mathematical Theory of Tone Systems (2004)

Additional Volumes in Preparation

E. Hansen and G. W. Walster, Global Optimization Using Interval Analysis: Second Edition, Revised and Expanded (2004)

C. Menini and F. Van Oystaeyen, Abstract Algebra: A Comprehensive Treatment (2004)

Radical Theory of Rings

B. J. Gardner
University of Tasmania
Hobart, Tasmania, Australia

R. Wiegandt
A. Rényi Institute of Mathematics
Hungarian Academy of Sciences
Budapest, Hungary

CRC Press
Taylor & Francis Group
Boca Raton London New York

CRC Press is an imprint of the
Taylor & Francis Group, an informa business

FIRST INDIAN REPRINT, 2015

Although great care has been taken to provide accurate and current information, neither the author(s) nor the publisher, nor anyone else associated with this publication, shall be liable for any loss, damage, or liability directly or indirectly caused or alleged to be caused by this book. The material contained herein is not intended to provide specific advice or recommendations for any specific situation.

Trademark notice: Product or corporate names may be trademarks or registered trademarks and are used only for identification and explanation without intent to infringe.

Library of Congress Cataloging-in-Publication Data
A catalog record for this book is available from the Library of Congress.

ISBN: 0-8247-5033-0

Headquarters
Marcel Dekker, Inc., 270 Madison Avenue, New York, NY 10016, U.S.A.
tel: 212-696-9000; fax: 212-685-4540

Distribution and Customer Service
Marcel Dekker, Inc., Cimarron Road, Monticello, New York 12701, U.S.A.
tel: 800-228-1160; fax: 845-796-1772

Eastern Hemisphere Distribution
Marcel Dekker AG, Hutgasse 4, Postfach 812, CH-4001 Basel, Switzerland
tel: 41-61-260-6300; fax: 41-61-260-6333

World Wide Web
http://www.dekker.com

The publisher offers discounts on this book when ordered in bulk quantities. For more information, write to Special Sales/Professional Marketing at the headquarters address above.

Copyright © 2004 by Marcel Dekker, Inc. All Rights Reserved.

Neither this book nor any part may be reproduced or transmitted in any form or by any means, electronic or mechanical, including photocopying, microfilming, and recording, or by any information storage and retrieval system, without permission in writing from the publisher.

Printed and bound in India by Bhavish Graphics.

FOR SALE IN SOUTH ASIA ONLY

To Louisa, Anna and Julia (B. J. G.)

To Peter and Thomas (R. W.)

Preface

So eine Arbeit wird eigentlich nie fertig. Man muß sie für fertig halten, wenn man nach Zeit und Umständen das Möglichste getan hat.

(J. W. Goethe)

This discontinued way of writing may have occasioned, besides others, two contrary faults, viz. that too little and too much may be said in it. If thou findest anything wanting, I shall be glad that what I have writ gives thee any desire that I should have gone further. If it seems too much to thee, thou must blame the subject.

(John Locke)

We started writing this book long ago, and now — as the first motto claims — we consider it ready for publication. Our aim was to give a systematic treatment of the radical theory of rings. This book, of course, does not contain *the* radical theory of rings: the theory is still in progress and it is not possible to define its borders. Besides the most important topics we discuss only some selected parts of the theory. At many places we give only references for more results and further directions of investigation. Some important branches of radical theory have been deliberately omitted but not neglected, for instance the study of lattices of radicals, of the behaviour of the (Jacobson) radical of group rings, of radicals of topological rings (see Arnautov [1] and [2]).

Although the basic idea of introducing a radical goes back to Wedderburn [1] (1908), we may say that the genesis of radical theory was in

1930 when Köthe introduced the nil radical in his fundamental paper [1]. In the next two decades prominent algebraists introduced several successful concrete radicals. Between 1952 and 1954 Amitsur [2], [3], [4] and Kurosh [1] defined the notion of general radicals and proved basic results concerning them. Further milestones of the theory were the papers of Andrunakievich [1] in 1958 and of Anderson, Divinsky and Suliński [1] in 1965.

The first book on radical theory of rings was written by Divinsky [2] in 1965. This was followed by Leavitt's lecture notes [5] in about 1970, then the books of Wiegandt [4] in 1974, Szász [6] in 1975 (German edition, 1975, English translation, 1981), and Andrunakievich and Ryabukhin [6] (in Russian) in 1979. The development of the radical theory in the quarter-century since these books appeared calls for an up-to-date account of the subject. Gardner [18] deals with radical theory for group-based structures; radical theory for rings is a special case which is not treated in an encyclopaedic way. Consequently, while some overlap between that book and the present one is unavoidable, it is not excessive.

The reader is assumed to be familiar with the basic notions, techniques and results of algebra, in particular of ring theory. We aimed to write a selfcontained exposition of radical theory which can introduce the reader to research work and which can become a handbook of the researcher. The book has also served as the basis for a graduate course.

It is natural that as a theory gets developed, the exposition of important and original results is not optimal as far as its natural place in the theory is concerned. We tried to do our best to present the theory of radicals in a natural and organic way. The proofs are not always the shortest, but possibly the easiest; we have endeavoured to keep them on an elementary level.

The bibliography is far from being complete. In Szász [6] and Wiegandt [4] it was fairly complete, but after a quarter of a century, seeking for completeness would have been meaningless. Nevertheless, at the end of the sections we give hints for more references, results of which are not discussed in this book.

We advise the reader to begin this book with Chapter 2 and go back to Chapter 1 whenever necessary. In Chapters 2 and 3 the general radical theory of rings is developed, and here concrete radicals are introduced gradually as examples of the general theory. So, arriving at Chapter 4, the reader has got a fair amount of knowledge on concrete radicals, and is well-prepared for the study of concrete radicals, and to prove structure theorems for rings. Finally, in Chapter 5 we sketch the basic features of radical theory in varieties of nonassociative rings, rings with involution and near-rings, respectively.

A pure mathematician is one who – in contrast to an applied mathematician – does not claim that his/her results are applicable. They are,

Preface

therefore, often charged with doing research mainly for fun, for prestige, or to a lesser extent for other purposes. This concerns also algebraists engaged with radical theory. We are often faced with the nasty question: what is the use of radical theory. An indisputable answer is that it does not harm, in sharp contrast with many expensive and subsidised human activities. Our opinion is that radical theory has contributed to the development of mathematics in the following five aspects (at least).

i) Living up to the original expectations, it provides structure theorems for rings which are semisimple with respect to certain radicals (cf. for instance, Sections 4.5, 4.7 and 4.10).

ii) Providing a context for studying and comparing properties, that is, classes of rings via closure operations (as in most of the sections of Chapter 3, in particular, in 3.1, 3.2, 3.9, 3.18 and 3.20).

iii) Constructing rings which distinguish given properties of rings, as minimally embeddable rings in 3.13, rings distinguishing nil radicals in 4.2, one-sided primitive rings and simple idempotent Jacobson radical rings in 4.6. Though the construction of such rings may ruin beautiful dreams, it definitely serves the better understanding of the structure of rings.

iv) Revealing hidden properties of rings which can be successfully used in various contexts of ring theory, for instance, the Andrunakievich Lemma 1.2.7, Krempa's Lemma 3.4.2, Stewart's Lemma 3.4.11 and Gardner's Lemma 3.19.17.

v) The infiltration of radical theory into other branches of mathematics has opened new dimensions for research, and enriched the arsenal of investigations. First in the mid-sixties a fast development of (hereditary) torsion theories took place in module categories and abelian categories. The interpretation of radical theory in general topology is known as the connectedness and disconnectedness theory which goes back to Preuß [1], Arhangel'skiĭ and Wiegandt [1], and has become a branch of categorical topology. The radical theory of graphs and abstract relational structures (a natural generalization of graphs and topological spaces) was developed by Fried and Wiegandt [1], [2], [3]. Radical theory has been applied also for Banach algebras (see Palmer [1]). Recent investigations have led to interesting results in the theory of incidence algebras and Petri nets (see Veldsman [16] and [18]).

The most general (Kurosh–Amitsur) radical theory was developed in the paper [1] of Márki, Mlitz and Wiegandt; all the so far known theories fit into its framework. Gardner [18] gives a unified treatment of the radical theory of group based structures, including (abelian) groups, modules, lattice-ordered groups (Martinez [1]), topological (abelian) groups. Also semifields (Weinert and Wiegandt [1]) and group automata (Fong, Huang and Wiegandt [1]) are group based structures. Other algebraic (but not group-based) stuctures for which a decent radical theory exists include semirings (see for instance Olson and Jenkins [2], Hebisch and Weinert [1], [2], [3], Morak [1]), and acts (Lex and Wiegandt [1]).

For purely categorical aspects of radical theory the reader is referred to the recent papers of Janelidze, Márki [1], [2] and Janelidze, Márki and Tholen [1].

These topics will not be touched in this book.

Some authors deal exclusively with rings with unity element. This assumption is all right and not restrictive, *if the ring is fixed*, as in module theory or group ring theory or sometimes investigating polynomial rings and power series rings (if the ring of coefficients does not possess a unity element, the indeterminate x is not a member of the polynomial ring). Dealing, however, simultaneously with several objects in a category of rings, demanding the existence of a unity element leads to a bizarre situation. Rings with unity element include among their fundamental operations the nullary operation $\mapsto 1$ assigning the unity element. Thus in the category of rings with unity element the morphisms, in particular the monomorphisms, have to preserve also this nullary operation: subrings (i.e. subobjects) have to contain the same unity element, and so a proper ideal with unity element is not a subring, although a ring and a direct summand; there are no infinite direct sums, no nil rings, no Jacobson radical rings, the finite valued linear transformations of an infinite dimensional vector space do not form a ring, etc. Thus, in many, maybe most, branches of ring theory the requirement of the existence of a unity element is not sensible, and therefore unacceptable. This applies also to radical theory, and so *in this book rings need not have a unity element*.

Sincere thanks are due to N. J. Divinsky for careful reading of and kind advice on Chapter 2 and the first 8 sections of Chapter 3 which are decisive in the exposition of radical theory. We are grateful to Zsuzsa Erő for preparing the TEX version of this book and to the Hungarian Research Grant OTKA #T034530 for financial support. Finally we would like to thank the staff of Marcel Dekker, Inc., for the cordial cooperation in the production of this book.

B. J. Gardner

R. Wiegandt

Contents

Preface   v

Interdependence Chart   xi

Chapter I. General Fundamentals   1
  1.1 Rudiments 1
  1.2 Some elementary ring theory 8
  1.3 Skew polynomial rings 14

Chapter II. The General Theory of Radicals   21
  2.1 Radical classes 21
  2.2 Radical constructions 28
  2.3 Semisimple classes 31

Chapter III. Radical Theory for Associative Rings   39
  3.1 Semisimple classes of associative rings 39
  3.2 Hereditary radicals and their semisimple classes 45
  3.3 Lower radical constructions 51
  3.4 The termination of the Kurosh radical construction 54
  3.5 The Suliński–Anderson–Divinsky problem 60
  3.6 Supernilpotent radicals and their semisimple classes 65
  3.7 Supernilpotent radicals and weakly special classes 73
  3.8 Special radicals 79
  3.9 Supplementing and dual radicals 89
  3.10 Subidempotent radicals 98
  3.11 Hypernilpotent and hypoidempotent radicals 101
  3.12 Partition of simple rings, unequivocal rings 102
  3.13 Minimally embeddable rings 108
  3.14 Modules and radicals 118
  3.15 Radicals defined by means of elements 131
  3.16 One-sided hereditary radicals and stable radicals 135

3.17 Strong radicals and strict radicals 142
3.18 Normal radicals 149
3.19 A-radicals 165
3.20 Radical semisimple classes 173

Chapter IV. Concrete Radicals and Structure Theorems 183

4.1 The principal nil radicals 183
4.2 Separation of the nil radicals 188
4.3 Coincidence of the nil radicals 196
4.4 The Jacobson radical 202
4.5 Structure theorems for Jacobson semisimple rings 208
4.6 One-sided primitivity and idempotent simple quasi-regular rings 218
4.7 Weakly primitive rings 235
4.8 The Brown–McCoy radical 253
4.9 Radicals of matrices and polynomials 256
4.10 Radicals on artinian rings 276
4.11 Concrete hypernilpotent radicals 284
4.12 Concrete hypoidempotent radicals 294

Chapter V. Special Features of the General Radical Theory 299

5.1 Degeneracy and pathology of nonassociative radical theory 299
5.2 Sufficient condition for a well-behaved radical theory:
    Terlikowska–Osłowska's approach 305
5.3 Sufficient condition for a well-behaved radical theory:
    Beidar's approach 309
5.4 On the radical theory of associative rings with involution 318
5.5 On the radical theory of near-rings 325

References 337

List of Symbols 367

List of Standard Conditions 373

Author Index 375

Subject Index 381

Interdependence Chart

Chapters II and III

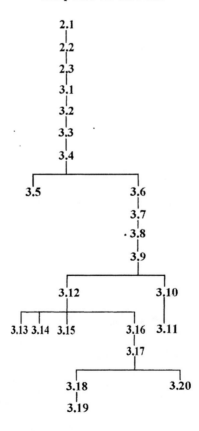

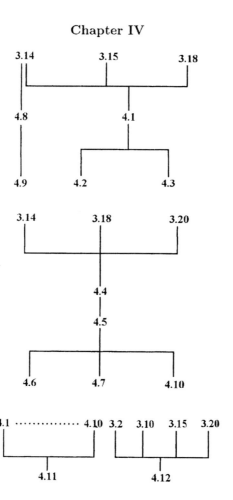

Chapter I
General Fundamentals

In this Chapter we fix the terminology, recall basic definitions and results of ring theory which will be needed later on. Therefore, we advise the reader to skip this Chapter but to look at it whenever necessary in understanding the subsequent Chapters.

1.1. Rudiments

Throughout the fact that B is a subset of the set A will be denoted by $B \subseteq A$, and $B \subset A$ will mean a proper subset, that is, $B \subseteq A$ and $B \neq A$.

From the set theory we shall make use of (infinite) cardinals and ordinals. As usual, the cardinality of a set S will be denoted by $|S|$. Ordinals will be denoted by lower case Greek letters, mostly by λ, and ω will stand for the first limit ordinal.

Among the many equivalent formulations of the Axiom of Choice, it is Zorn's Lemma [1] which is most elegantly applicable in algebra, in particular in ring theory. It is interesting to mention that this Lemma was discovered by Kuratowski [1] 13 years before Zorn.

ZORN'S LEMMA 1.1.1. *If S is a partially ordered set in which every chain has a least upper bound, then S has a maximal element.*

Note that the set S may have several maximal elements. Applying Zorn's Lemma it is important that *S is a set*.

A *ring* $A = (A, +, \cdot)$ is an algebraic strucutre with two binary operations $+$ and $\cdot$, called *addition* and *multiplication* such that $(A, +)$ is an abelian group, $(A, \cdot)$ is a semigroup and these operations are linked with the distributive laws

$$x(y+z) = xy + xz \qquad (y+z)x = yx + zx$$

for all $x, y, z \in A$, and we say that the addition is distributive with respect to the multiplication. In the definition of rings we do not demand — among

others — the existence of a nullary operation $\mapsto e$ designating an element $e \in A$ subject to
$$xe = x = ex \qquad \forall x \in A,$$
that is, *rings need not have a unity element*. The one-element ring 0 is called the *trivial ring*. Without fear of ambiguity we may denote the also the zero element of rings by 0. The additive group $(A, +)$ of a ring A will be denoted by A^+. Given an abelian group A^+, we may define always a *ring A^0 with zero-multiplication*, called also a *zero-ring* by the rule
$$xy = 0 \qquad \forall x, y \in A.$$

A ring A is *nilpotent*, if there exists an integer $n > 1$ such that $A^n = 0$. The smallest such exponent n is called the *degree of nilpotency* of A.

A ring A with commutative multiplication is called a *commutative ring*. Let K be a commutative ring with unity element 1, in particular a field. A *K-algebra* A is a ring A on which also a *scalar multiplication* is defined: to each $(k, x) \in K \times A$ an element $kx \in A$ is designated subject to the identities

$$1x = x$$
$$(k\ell)x = k(\ell x)$$
$$k(xy) = (kx)y = x(ky)$$
$$(k + \ell)x = kx + \ell x$$
$$k(x + y) = kx + ky$$

for all $k, \ell \in K$ and $x, y \in A$. Thus every ring is a $\mathbb{Z}$-algebra over the integers $\mathbb{Z}$.

A *subring* S is a subset of a ring A which is closed under the ring operations. A subring L of A is a *left ideal*, if $AL \subseteq L$. A *right ideal* is defined correspondingly. A subring I of A is called a *(two-sided) ideal*, if I is a left as well as a right ideal of A. If S is a subring, L is a left ideal, R is a right ideal and I is an ideal of A, then we shall write

$$S \subseteq A, \quad L \triangleleft_\ell A, \quad R \triangleleft_r A, \quad I \triangleleft A.$$

(Using the subset symbol $\subseteq$ for denoting subrings, will not cause ambiguity.) Every ring $A \neq 0$ has two *trivial* subrings (left ideals, right ideals, two-sided ideals, namely 0 and A. A ring A is said to be *simple*, if A has only trivial ideals. If A is a simple ring then either $A^2 = A$ (and A is a simple idempotent ring), or $A^2 = 0$ (and $A = A^0$ is a zero-ring on a cyclic additive group of prime order.

A subring, left ideal, right ideal and two-sided ideal generated by a single element $a \in A$ will be denoted by

$$\langle a \rangle, \ (a], \ [a), \ (a),$$

respectively.

A *homomorphism* f of a ring A into a ring B is a mapping $f : A \to B$ such that
$$f(x+y) = f(x) + f(y) \quad \text{and} \quad f(xy) = f(x)f(y)$$
for all $x, y \in A$. If f maps A onto B, we speak of a *surjection*, and write $A \twoheadrightarrow B$. A surjection $f : A \twoheadrightarrow B$ which is also a *bijection* (that is, a one-to-one mapping), is called an isomorphism, and we shall write $A \cong B$. A homomorphism $f : A \to B$ is an *embedding* or *injection*, if $A \twoheadrightarrow f(A)$ is an isomorphism. The set
$$\ker f = \{x \in A \mid f(x) = 0\}$$
of a homomorphism $f : A \to B$ is called the *kernel* of f. The kernel of any homomorphism is an ideal, and every ideal I of a ring A is the kernel of a homomorphism: consider the *factor ring* A/I in which the elements are the cosets $x + I$, $x \in A$, and
$$(x + I) + (y + I) = (x + y) + I$$
$$(x + I) \cdot (y + I) = xy + I$$
for all $x, y \in A$.

The precise connection between kernels and ideals is given in the

HOMOMORPHISM THEOREM 1.1.2. *$f : A \twoheadrightarrow B$ is a surjection if and only if $B \cong A/I$ where I is an ideal of A and $I = \ker f$.*

Beside the Homomorphism Theorem also the Isomorphism Theorems are of fundamental importance.

FIRST ISOMORPHISM THEOREM 1.1.3. *Let I and K be ideals of a ring A such that $I \subseteq K$. Then*
$$A/K \cong \frac{A/I}{K/I}.$$

The First Isomorphism Theorem tells us that the ideals of a factor ring A/I are of the form K/I where $I \subseteq K \triangleleft A$.

SECOND ISOMORPHISM THEOREM 1.1.4. *Let S be a subring of a ring A, and $I \triangleleft A$. Then*
$$S/(S \cap I) \cong (S + I)/I.$$

It is a consequence of the Second Isomorphism Theorem that the lattice of ideals of a ring A satisfies the *modularity law*: if I, J, K are ideals of a ring A such that $I \subseteq K$, then

$$(I + J) \cap K = I + (J \cap K).$$

The modularity law allows us to use, for instance, the following reasoning:

$$\frac{J \cap K}{I \cap K} = \frac{J \cap K}{I \cap J \cap K} \cong \frac{I + (J \cap K)}{I} = \frac{(I + J) \cap K}{I}.$$

Notice that also the lattice of all additive subgroups of a ring satisfies the modularity law, as it is easy to verify.

Given a set $\{A_\lambda \mid \lambda \in \Lambda\}$ of rings we may define the *direct product* (called also *complete direct sum*) $C = \prod_{\lambda \in \Lambda} A_\lambda$ as a ring C such that

i) to each $\lambda \in \Lambda$ there exists a surjection $\pi_\lambda : C \twoheadrightarrow A_\lambda$,

ii) if B is a ring and $f_\lambda : B \to A_\lambda$, $\lambda \in \Lambda$, system of homomorphisms then there exists a unique homomorphism $f : B \to C$ such that $\pi_\lambda f = f_\lambda$ for all $\lambda \in \Lambda$.

The direct product C of rings is determined uniquely up to isomorphisms, and C can be represented as the ring on the cartesian product of A_λ, $\lambda \in \Lambda$, with componentwise addition and multiplication. Thus the elements of the product can be viewed as (infinite) vectors $(\cdots, a_\lambda, \cdots)$, $a_\lambda \in A_\lambda$, and then the surjections π_λ are the projections onto the λ-th component.

The *direct sum* (called also *discrete direct sum* or *restricted direct sum*) $D = \bigoplus_{\lambda \in \Lambda} A_\lambda$ is the subring of the direct product C such that in every element $(\cdots, a_\lambda, \cdots) \in D$ only finitely many components a_λ differ from 0. Obviously $D \triangleleft C$, further, $D = C$ if and only if Λ is a finite set of indices.

A ring A is the direct sum $A = B \oplus C$ of the rings B and C if and only if B and C are ideals of A and $B \cap C = 0$.

A ring A is a *subdirect sum* $A = \sum_{\text{subdirect}} (A_\lambda \mid \lambda \in \Lambda)$, if

i) A is a subring of the direct product $C = \prod(A_\lambda \mid \lambda \in \Lambda)$;

ii) $\pi_\lambda(A) = A_\lambda$ for every projection $\pi_\lambda : C \twoheadrightarrow A_\lambda$, $\lambda \in \Lambda$.

THEOREM 1.1.5. *A ring A is a subdirect sum of the rings A_λ, $\lambda \in \Lambda$ if and only if there exist ideals I_λ, $\lambda \in \Lambda$, in A such that $\cap I_\lambda = 0$ and $A_\lambda = A/I_\lambda$.*

The whole scene of our investigations will take place in a *universal class* $\mathbb{A}$ of rings, that is, the class $\mathbb{A}$ is *hereditary* ($I \triangleleft A \in \mathbb{A}$ implies $I \in \mathbb{A}$) and *homomorphically closed* ($I \triangleleft A \in \mathbb{A}$ implies $A/I \in \mathbb{A}$). If the universal class $\mathbb{A}$ is closed also under forming arbitrary direct products and taking subrings, we speak of a *variety* of rings. In the sequel $\mathbb{A}$ will stand for the variety of all associative rings, unless we specify $\mathbb{A}$ as some other universal class or variety.

General Fundamentals

Given a set $\{\cdots, A_\lambda, \cdots\}$ of rings the variety generated by these rings (that is, the smallest variety containing these rings) can be obtained in the following way:

i) take all direct products of these rings A_λ,
ii) take all subrings of these direct products,
iii) take all homomorphic images of these subrings.

Varieties of rings are characterized by

BIRKHOFF'S THEOREM 1.1.6. *A class of rings is a variety if and only if it is defined by identities.*

For instance, commutative rings form a subvariety of associative rings which is defined by the additional identity $xy = yx$ for all $x, y \in A$. Another example is the class of Boolean rings which satisfy the identity $x^2 = x$ for all $x \in A$. This identity implies the identities

$$2x = 0 \quad \text{and} \quad xy = yx \quad \forall x, y \in A.$$

The *heart* $H(A)$ of a ring A is defined as

$$H(A) = \bigcap (I \triangleleft A \mid I \neq 0).$$

A ring A is said to be *subdirectly irreducible*, if $H(A) \neq 0$, that is, A has a unique minimal ideal.

The most general structure theorem is the following subdirect decomposition theorem.

BIRKHOFF'S THEOREM 1.1.7. *Every ring is a subdirect sum of subdirectly irreducible rings.*

A ring may have many substantially different subdirect decompositions into subdirectly irreducible rings. For instance, take the ring $\mathbb{Z}$ of integers and the sets of ideals

$$U = \{(p_i) \mid p_i \text{ ranges through the prime numbers}\}$$

and

$$V = \{(p^k) \mid p \text{ is a fixed prime and } k = 1, 2, \dots\}.$$

Then $\bigcap_{i=1}^{\infty}(p_i) = 0$ and $\bigcap_{i=1}^{\infty}(p^k) = 0$, and in view of Theorem 1.1.5 we have

$$\sum_{\text{subdirect}} (\mathbb{Z}/(p_i) \mid (p_i) \in U) \cong \mathbb{Z} \cong \sum_{\text{subdirect}} (\mathbb{Z}/(p^k) \mid (p^k) \in V).$$

Let us notice that all the notions occurring in Birkhoff's Theorems 1.1.6 and 1.1.7 make sense for universal algebras, and these theorems are valid in the generality of universal algebra.

A property $\mathcal{M}$ of rings can be given by the class $\mathbb{M}$ of all rings possessing this property $\mathcal{M}$, and vice versa. For instance, commutativity or nilpotency of rings can be given by the classes of commutative or nilpotent rings, respectively. We shall deal only with *abstract properties* (or *abstract classes*) of rings: if A has property $\mathcal{M}$ (or $A \in \mathbb{M}$) and $A \cong B$ then also B has property $\mathcal{M}$, (or $B \in \mathbb{M}$). We assume tacitly that all abstract classes of rings will contain the trivial ring 0.

For any set $X = \{x_\lambda \mid \lambda \in \Lambda\}$ the *free ring* $F\langle X \rangle$ generated by the symbols $x_\lambda \in X$ is the ring of all expressions $\sum_{\text{finite}} n_i x_{\lambda_{i_1}} \ldots x_{\lambda_{i_k}}$ ($n_i \in \mathbb{Z}$, and the indeterminates x_{λ_j} are not necessarily different), that is, $F\langle X \rangle$ is the ring of all polynomials with non-commuting indeterminates $x_\lambda \in X$ over the integers $\mathbb{Z}$ having no constant terms. We shall make use of the case $X = \{x, y\}$ in Section 4.2.

We shall deal, of course, also with matrices and matrix rings. The matrix ring of $n \times n$ matrices over a ring A is

$$M_n(A) = \{(a_{ij})_{n \times n} \mid a_{ij} \in A\}.$$

Let S_{ij}, $(i, j = 1, \ldots, n)$, be subsets of the ring A. We shall use the notation

$$(S_{ij})_{n \times n} = \{(s_{ij}) \in M_n(A) \mid s_{ij} \in S_{ij}\}.$$

This notation makes sense also for $m \times n$ matrices, $m \neq n$, in particular for vectors.

For any positive integer n the *standard polynomial of degree n* is defined by

$$s_n = s_n(x_1, \ldots, x_n) = \sum_\sigma (-1)^\sigma x_{\sigma(1)} \ldots x_{\sigma(n)}$$

where σ ranges to all permutations of $\{1, \ldots, n\}$ and

$$(-1)^\sigma = \begin{cases} 1 & \text{if } \sigma \text{ is an even permutation} \\ -1 & \text{if } \sigma \text{ is an odd permutation.} \end{cases}$$

AMITSUR–LEVITZKI THEOREM 1.1.8 ([1]). *If F is a field then the matrix ring $M_n(F)$ satisfies the standard polynomial identity s_{2n}, that is, $s_{2n} = 0$.*

We shall make use of the Amitsur–Levitzki Theorem in Example 4.3.8 in the special case $n = 2$.

A *domain* is a ring without zero divisors. A ring A is called a *division ring* (or a *skew field* according to the German terminology) if $(A \setminus 0, \cdot)$ is a group. If $(A \setminus 0, \cdot)$ is an abelian group, we speak of a *field*. A ring A is a division ring if and only if A has only trivial left ideals.

We shall need the following notions in Section 4.7. A ring Q is said to be the *ring of left quotients* of the ring A if $A \subseteq Q$, every *cancellable element* (that is, element which is not a zero divisor) of A has a two-sided inverse in Q, and every element $q \in Q$ is of the form $q = y^{-1}x$ where $x, y \in A$ and y^{-1} is the inverse of y in Q. As is well known, *a ring A with cancellable elements has a left ring of quotients Q if and only if for every element $a, b \in A$ with cancellable b there exist elements $x, y \in A$ with cancellable x such that $xa = yb$.* A *left Ore domain* A is a domain in which $Aa \cap Ab \neq 0$ for all nonzero elements $a, b \in A$. *For any left Ore domain A the left ring of quotients Q always exists and Q is a division ring.*

In Section 3.20 we shall need the following two theorems.

WEDDERBURN'S THEOREM 1.1.9. *Every finite division ring is a field.*

The next statement is commonly referred to as *Jacobson's Commutativity Theorem* [3], although it has been discovered independently also by Forsythe and McCoy [1].

THEOREM 1.1.10. *If in a ring A to every element $x \in A$ there exists an integer $n(x) > 1$ such that $x^{n(x)} = x$, then the ring A is commutative.*

In Sections 3.12 and 4.7 injectivity of modules will be needed. Next, we recall some definitions and basic facts.

Let A be a ring with unity element. An *A-module* M is an abelian group M on which a scalar multiplication $m \mapsto am$, $a \in A$, $m \in M$, is defined and subject to the identities

$$1m = m$$
$$(ab)m = a(bm)$$
$$(a+b)m = am + bm$$
$$a(m+n) = am + an$$

for all $a, b \in A$ and $m, n \in M$. In particular, abelian groups are $\mathbb{Z}$-modules. An A-module Q is said to be *M-injective* (with respect to the A-module M) if for each submodule K of M each homomorphism $f : K \longrightarrow Q$ can be extended to a homomorphism $g : M \longrightarrow Q$, that is, the diagram

$$\begin{array}{ccc} K & \longrightarrow & M \\ f \downarrow & \swarrow g & \\ Q & & \end{array}$$

where the horizontal arrow represents the inclusion, has a commutative completion g. An A-module Q is said to be *injective* if it is M-injective for all A-modules M. In the particular case $Q = M$ we speak of a *quasi-injective* module Q. It is well-known that *every module M is contained in a minimal injective extension Q which is unique up to isomorphism.* This extension Q

is called the *injective hull* of M. An injective module is a direct summand in each of its overmodules. The injective abelian groups (i.e. $\mathbb{Z}$-modules) are exactly the divisible groups (see Section 1.2).

For more ring theory the reader may consult the books Herstein [1], [2], [3], Jacobson [4], Jans [1], Kaplansky [2], Kertész [1], McCoy [2] and Rowen [1], for nonassociative rings we advise Zhevlakov, Slin'ko, Shestakov and Shirshov [1].

1.2. Some elementary ring theory

In this section we recollect some elementary ring theoretical facts which will be used later on and which are not necessarily covered in a course on ring theory.

The infinite cyclic group $C(\infty)$ and the zero-ring $Z(\infty)$ built on $C(\infty)$ have some remarkable properties:

i) *every nonzero subring (ideal) of $Z(\infty)$ is isomorphic to $Z(\infty)$*,

ii) *$Z(\infty)$ is infinite, but every homomorphic image $f(Z(\infty))$ with* $\ker f \neq 0$ *is finite*.

The *quasi-cyclic* or *Prüfer group* $C(p^\infty)$ and the zero-ring $Z(p^\infty)$ built on $C(p^\infty)$ features to some extent dual properties to the previous ones. $C(p^\infty)$ is the additive group isomorphic to the multiplicative group of all complex p^n-th roots of 1 for a prime p and $n = 1, 2, \ldots$.

i) *$Z(p^\infty)$ is infinite, but every proper subring (ideal) of $Z(p^\infty)$ is finite*,

ii) *every homomorphic image $f(Z(p^\infty))$ with $\ker f \neq Z(p^\infty)$ is isomorphic to $Z(p^\infty)$*.

The additive group A^+ of a ring A may have a strong effect on the structure of A. Next, we recall some results concerning the additive group A^+.

The set of additively torsion elements of a (not necessarily associative) ring A forms an ideal

$$T = \{a \in A \mid o(a) < \infty\},$$

which is called the *maximal torsion ideal* of A.

An element a of an abelian group G is said to be *divisible*, if the equation $nx = a$ is solvable in G for every nonzero integer n. G is *divisible* if each element $a \in G$ is divisible. Every abelian group has a largest divisible subgroup. The largest divisible subgroup of the additive group of a ring A is an ideal of A which we shall call *the maximal divisible ideal*. It is well known from the theory of abelian groups that *in every abelian group a divisible subgroup is always a direct summand*. Moreover, *every divisible abelian group is a direct sum of copies of the additive group $\mathbb{Q}^+$ of rationals and of copies of quasi-cyclic groups $C(p^\infty)$ for some primes p*.

A ring A is called a *p-ring*, if the additive order $o(a)$ of every element $a \in A$ is a power of the prime number p. In any ring A the subset

$$A_p = \{a \in A \mid o(a) \text{ is a power of } p\}$$

is an ideal. A ring A whose additive group is a torsion group, is called a *torsion ring* and A_p is its *p-component*. One can readily verify that *every torsion ring is a direct sum of its p-components*.

PROPOSITION 1.2.1. *If A is a p-ring and the equation $p^n x = a$ is solvable in A for every $n = 1, 2, \ldots$, then a is in the annihilator of A:*

$$a \in \operatorname{ann} A = \{x \in A \mid xA = Ax = 0\}.$$

The maximal divisible ideal D of a torsion ring A is in $\operatorname{ann} A$. In particular, the only ring multiplication on the quasi-cyclic group $C(p^\infty)$ is the zero-multiplication.

Proof: Let $b \in A$ be an arbitrary element, $o(b) = p^n$ and $x_0 \in A$ a solution of the equation $p^n x = a$. Then

$$ab = (p^n x_0)b = x_0(p^n b) = 0$$

and also $ba = 0$. Thus $a \in \operatorname{ann} A$.

The further statements are obvious consequences. □

A ring A is said to be *torsionfree*, if A^+ is a torsionfree group. In a torsionfree ring (or group) the equality $na = nb$ implies $a = b$, as one can readily check.

PROPOSITION 1.2.2. *Let F^+ be a divisible torsionfree subgroup of the additive group A^+ of a ring A, and T the maximal torsion ideal of A. Then $FT = TF = 0$.*

Proof: Let $f \in F$ and $t \in T$ be arbitrary elements, $o(t) = n$ and $x_0 \in F$ a solution of the equation $nx = f$. Then

$$ft = (nx_0)t = x_0(nt) = 0,$$

and similarly, $tf = 0$. □

PROPOSITION 1.2.3. *Let A be a ring such that its additive group decomposes into a direct sum $A^+ = D^+ \oplus B^+$, where D^+ is the maximal divisible subgroup of A^+ and B^+ is a torsion subgroup of A^+. Further, let T denote the maximal torsion ideal of A. Then*

(i) $A^+ = T^+ \oplus F^+$ *where F^+ is a divisible torsionfree subgroup.*
(ii) *every divisible torsion subgroup is in $\operatorname{ann} A$,*
(iii) $D/(T \cap D) \cong A/T$.

Proof: (i) D^+ decomposes as $D^+ = (D \cap T)^+ \oplus F^+$ with a torsionfree divisible subgroup F^+. Hence

$$A^+ = D^+ \oplus B^+ = F^+ \oplus (D \cap T)^+ \oplus B$$

holds, and clearly $T^+ = (D \cap T)^+ \oplus B^+$.

(ii) follows readily from Propositions 1.2.1 and 1.2.2.

(iii) By (i) we have

$$A/T \cong F \cong F/(F \cap (D \cap T)) \cong (F + (D \cap T))/(D \cap T) = D/(D \cap T). \quad \square$$

In view of Proposition 1.2.3 (ii) the zero-ring $Z(p^\infty)$ is quite often in the annihilator of an overring A, although it is not always so. For instance, *the endomorphism ring of $C(p^\infty)$ is the ring P of p-adic integers*. Consider the set

$$A = \{(a, x) \mid a \in P, x \in C(p^\infty)\}$$

and define addition on A componentwise and a multiplication by

$$(a, x)(b, y) = (ab, ay)$$

where ay means the image of y by the endomorphism a. As one can easily verify, A is a ring, the subset $(0, C(p^\infty)) \cong Z(p^\infty)$ is an ideal of A, but

$$0 \neq (0, C(p^\infty)) = (P, 0)(0, C(p^\infty)) = A(0, C(p^\infty)).$$

Note that A is the *split-null extension* of the P-bimodule $C(p^\infty)$ by P, where the elements of P act as endomorphisms from the left and $C(p^\infty)P = 0$ (split-null extension will be introduced and used in Section 3.6). For another example where $Z(p^\infty)$ is not a part of the annihilator, see the Dorroh extension below.

More on the additive groups of rings can be found in Feigelstock [1] and [2].

Rings with unity play a distinguished role in the variety of rings.

DORROH'S EXTENSION THEOREM 1.2.4 (Dorroh [1]). *Every ring A can be embedded as an ideal into a ring A^1 with unity element.*

The ring A^1 is referred to as *the Dorroh extension of A.*

Proof: On the set

$$A^1 = \{(a, n) \mid a \in A, n \in \mathbb{Z}\}$$

define addition componentwise and multiplication by

$$(a, n)(b, m) = (ab + ma + nb, nm).$$

A straightforward verification shows that A^1 is a ring with unity element $(0, 1)$ and

$$A \cong (A, 0) \triangleleft A^1. \quad \square$$

Rings with unity element possesses a characteristic property which corresponds to that of divisible groups in the variety of abelian groups. This was discovered by Szendrei [1], though one part is Lemma 2 in McCoy [1].

SZENDREI'S THEOREM 1.2.5. *A ring B has a unity element if and only if B is a direct summand in every extension A of B, that is, $B \triangleleft A$ implies $A = B \oplus C$ for some ideal C of A.*

Proof: Let $0 \neq B \triangleleft A$, and consider the left annihilator

$$C = \{x \in A \mid xB = 0\}$$

of B in A. It is straightforward to see that $C \triangleleft A$ and $B \cap C = 0$. Moreover, if e is the unity element of B then every element $a \in A$ can be written as

$$a = ae + (a - ae) \in B + C,$$

proving $A = B + C$.

Conversely, the Dorroh extension $A = B^1$ decomposes as $A = B \oplus C$, and so $B \cong A/C$ has a unity element. □

PROPOSITION 1.2.6. *A simple idempotent ring A has a unity element, if it has a left unity element.*

Proof: Let e be a left unity element of A. Then

$$(a - ae)A = (a - ae)eA = 0$$

for all elements $a \in A$. Since A is simple and idempotent, its left annihilator ideal is 0, and consequently $a = ae$ holds for all $a \in A$. □

ANDRUNAKIEVICH LEMMA 1.2.7 ([1]). *If $K \triangleleft I \triangleleft A$ and $\overline{K}$ denotes the ideal of A generated by K, then $\overline{K}^3 \subseteq K$.*

Proof:

$$\overline{K}^3 \subseteq I(K + AK + KA + AKA)I \subseteq IKI + IAKI + IKAI + IAKAI \subseteq K. \quad \square$$

PROPOSITION 1.2.8. *The heart H of a subdirectly irreducible ring A is either a simple idempotent ring or a ring with zero-multiplication.*

Proof: Since $H^2 \triangleleft A$ we have $H^2 = 0$ or $H \subseteq H^2$, that is, $H^2 = 0$ or $H^2 = H$. In the latter case, if $0 \neq I \triangleleft H$, then $\overline{I} \subseteq H$ but $\overline{I} \neq 0$, so $H \subseteq \overline{I}$. But by the Andrunakievich Lemma,

$$H = H^2 = \overline{I}^3 \subseteq I \subseteq H,$$

so $I = H$ and H is simple. □

The notion of a prime ideal (p) in the ring $\mathbb{Z}$ of integers can be easily extended to arbitrary rings. An ideal P of a ring A is called a *prime ideal*, if $IK \subseteq P$ implies $I \subseteq P$ or $K \subseteq P$ whenever I and K are ideals of A. A ring A is said to be a *prime ring*, if 0 is a prime ideal of A, that is, $IK = 0$ implies $I = 0$ or $K = 0$ for any ideals I, K of A. Thus P is a prime ideal of a ring A if and only if A/P is a prime ring.

LEMMA 1.2.9. *A ring A is prime if and only if $xAy = 0$ implies $x = 0$ or $y = 0$.*

Proof: If A is a prime ring and $xAy = 0$, then

$$(AxA)(AyA) \subseteq A(xAy)A = 0,$$

and by the primeness of A we conclude that $AxA = 0$ or $AyA = 0$. Suppose that $AxA = 0$. Then Ax is in the left annihilator of A, and again by the primeness of A we get $Ax = 0$. Since x is in the right annihilator of the prime ring A, it follows $x = 0$.

Conversely, assume that A is not a prime ring. Then there exist nonzero ideals I and K of A such that $IK = 0$. Selecting nonzero elements $x \in I$ and $y \in K$ are have $xAy \subseteq IK = 0$. □

A natural generalization of primeness is the notion of semiprimeness. A ring A is said to be *semiprime*, if A has no nonzero ideal I with $I^2 = 0$. An equivalent definition is that *the ring A has no nonzero nilpotent ideal*. A similar proof to that of Lemma 1.2.9 gives

LEMMA 1.2.10. *A ring A is semiprime if and only if $xAx = 0$ implies $x = 0$.* □

We shall say that I is a *semiprime ideal* of A, if A/I is a semiprime ring.

Finally, as a useful example, we discuss some properties of the ring of linear transformations on an infinite dimensional vector space. Linear transformations are *endomorphisms*, so we shall denote the ring of linear transformations of a vector space V by $\operatorname{End} V$ (which can be denoted also as $\operatorname{Hom}(V, V)$).

EXAMPLE 1.2.11. Let $T = \operatorname{End} V$ be the ring of all linear transformations on a countably infinite dimensional vector space V over a division ring D. A linear transformation $t \in T$ is said to be finite valued, if tV is a finite dimensional subspace in V.

(i) *The ring T is not simple*, because

$$H = \{t \in T \mid t \text{ is finite valued}\}$$

is an ideal in T and $0 \neq H \neq T$, as one can readily verify.

(ii) *If $t \in H$ is any nonzero element, then to every finite dimensional subspace W of V there exists an element t^* in the principal ideal (t) which*

General Fundamentals

is a projection of V onto W. For the proof we apply induction. Suppose first that $\dim W = 1$ and $W = Dw$. Moreover, let $\{tv_1, \ldots, tv_n\}$ be a basis of the image space tV. Now there exist elements $r, s \in T$ such that

$$rw = v_1, \; rz = 0 \text{ for any } z \in V \text{ being independent from } w$$

and

$$s(tv_1) = w, \; sz = 0 \text{ for any } z \in V \text{ being independent from } tv_1.$$

By definition $r, s \in H$. Furthermore, we have

$$(str)w = w \quad \text{and} \quad (str)V = Dw = W,$$

which shows that the element $t^* = str$ is a projection of V onto W and that $t^* \in (t)$. Assume now that $\dim W = m > 1$ and that the statement is true for every subspace of dimension less than m. Let W_1 be any $m-1$-dimensional subspace of W. By the hypothesis there is a projection $e \in (t)$ of V onto W_1. Denote e^* the restriction of e to W. Obviously $\ker e^*$ is a 1-dimensional subspace of W and $W = W_1 \oplus W_2$ with $W_2 = \ker e^*$. By the previously established result for 1-dimensional subspaces there exists a projection $f \in (t)$ of V onto W_2. We set

$$t^* = e + f - fe$$

and claim that t^* is a projection of V onto W. We have

$$t^*V = eV + fV + f(eV) \subseteq W.$$

Moreover, if $x_0 = x_1 + x_2$ with $x_1 \in W_1$ and $x_2 \in W_2$, then we get

$$t^*x_0 = (e + f - fe)(x_1 + x_2) = x_1 + fx_1 - fx_1 + x_2 = x_0.$$

Thus t^* is indeed a projection of V onto W and $t^* \in (t)$.

(iii) *H is a simple ring*. We show that for every nonzero element $t \in H$ the principal ideal (t) coincides with H. Let $f \in H$ be an arbitrary nonzero element. By statement (ii) there exists a projection $e \in (t)$ from V onto fV. Then

$$(ef)x = e(fx) = fx$$

holds for all $x \in V$, therefore $f = ef \in (t)$. Thus $A = (t)$.

(iv) *T is subdirectly irreducible with heart H*. Let t be any nonzero element of an ideal I of T, and take a nonzero vector $u = t(v)$ from the subspace $t(V)$. For the projection $s \in T$ onto the 1-dimensional subspace Du of V we have

$$0 \neq u = s(u) = s(t(v)) = st(v),$$

and so
$$0 \neq st \in H \cap I$$
follows. Thus the simplicity of H yields $H \subseteq I$, proving the assertion.

(v) Let us consider the linear transformation $t \in T$ given by $te_i = e_{i+1}$ if i is odd and $te_i = 0$ if i is even, where $\{e_1, \ldots, e_n, \ldots\}$ is a basis for V. Then *the subring $S = \langle t, H \rangle$ is clearly subdirectly irreducible with heart H, S cannot be mapped homomorphically onto a simple prime ring, and S/H as well as its homomorphic images have no nonzero idempotents.*

1.3. Skew polynomial rings

We shall give here a generalization of the polynomial ring construction in terms of an endomorphism and a derivation, and discuss properties of the constucted rings which will be important in some sections, for instance in 3.13, and 4.6.

Let A be a ring with unity element 1, x an indeterminate (or variable), and σ an endomorphism of A with ker $\sigma = 0$. An additive homomorphism $\delta: A^+ \longrightarrow A^+$ is called a σ-*derivation*, if it satisfies

$$\delta(ab) = \sigma(a)\delta(b) + \delta(a)b \qquad \forall a, b \in A.$$

For the identical automorphism σ, a σ-derivation is merely a derivation. On the additive group $(A[x])^+$ of polynomials over A a multiplication is defined by the rule

$$xa = \sigma(a)x + \delta(a).$$

Observe that this multiplication extends to the basis $1, x, x^2, \ldots$ of the $\mathbb{Z}$-module (that is, abelian group) $(A[x])^+$, and thus a new multiplication has been defined on the set $A[x]$ of polynomials over A. We shall denote this structure with the usual addition and new multiplication by $A[x; \sigma, \delta]$.

The degree of a polynomial f will be denoted by $\partial(f)$ throughout this book.

THEOREM 1.3.1. *$A[x; \sigma, \delta]$ is a ring with unity element 1. If A is a domain then*

$$\partial(fg) = \partial(f) + \partial(g) \qquad \forall f, g \in A[x; \sigma, \delta],$$

and also $A[x; \sigma, \delta]$ is a domain. Moreover, if A is a division ring then $A[x; \sigma, \delta]$ is a principal left ideal domain, and if σ is an automorphism then $A[x; \sigma, \delta]$ is also a principal right ideal domain. If f and g are polynomials and $(A[x; \sigma, \delta])f = (A[x; \sigma, \delta])g$ then $f = qg$ with some $q \in A$, $\partial(f) = \partial(g)$ and the polynomials f and g have the same number of nonzero terms. If $I \triangleleft A[x; \sigma, \delta]$ then

$$(A[x; \sigma, \delta])g = I = g(A[x; \sigma, \delta])$$

for any polynomial $g \in I$ of least degree. 1 is the only nonzero idempotent element in $A[x;\sigma,\delta]$.

$A[x;\sigma,\delta]$ is called the *skew* or *twisted polynomial ring with endomorphism σ and derivation δ*.

Proof: Instead of a tedious verification we shall prove that $A[x;\sigma,\delta]$ is isomorphic to a subring S of the endomorphism ring $\text{End}_{\mathbb{Z}}(A[x])$. Define $\bar{x} \in \text{End}_{\mathbb{Z}}(A[x])$ by

$$\bar{x}\left(\sum a_i x^i\right) = \sum (\sigma(a_i) x^{i+1} + \delta(a_i) x^i).$$

The left multiplication by an element $a \in A$ is an endomorphism $\bar{a} \in \text{End}_{\mathbb{Z}}(A[x])$. Now, define the subset

$$S = \left\{ \sum_{i=0}^{n} \bar{a}_i \bar{x}^i \mid a_i \in A; n = 0, 1, 2, \ldots \right\}$$

of $\text{End}_{\mathbb{Z}}(A[x])$ which is clearly an additive subgroup. If $\sum \bar{a}_i \bar{x}^i = 0$ then

$$0 = \left(\sum \bar{a}_i \bar{x}^i\right) x = \sum a_i x^{i+1},$$

so each $a_i = 0$. Hence $\{\bar{1}, \bar{x}, \bar{x}^2, \ldots\}$ is a basis of S as a module over $\overline{A} = \{\bar{a} \mid a \in A\}$. We have

$$\bar{x}\bar{a}\left(\sum a_i x^i\right) = \bar{x}\left(\sum a a_i x^i\right) = \sum \left(\sigma(aa_i) x^{i+1} + \delta(aa_i) x^i\right) =$$
$$= \sum \left(\sigma(a)\sigma(a_i) x^{i+1} + (\sigma(a)\delta(a_i) + \delta(a)a_i) x^i\right) =$$
$$= \overline{\sigma(a)} \sum (\sigma(a_i) x^{i+1} + \delta(a_i) x^i) + \overline{\delta(a)} \sum a_i x^i =$$
$$= \left(\overline{\sigma(a)}\bar{x} + \overline{\delta(a)}\right) \sum a_i x^i,$$

implying

$$\bar{x}\bar{a} = \overline{\sigma(a)}\bar{x} + \overline{\delta(a)}.$$

By induction, a similar calculation shows that

$$\bar{x}^i = (\bar{x})^i \quad i = 2, 3, \ldots.$$

Thus S is isomorphic to a subring of $\text{End}_{\mathbb{Z}}(A[x])$ with unity element $\bar{1} \in S$.

If A is a domain and $f, g \in A[x;\sigma,\delta]$ with leading coefficients a_n and b_m, respectively, then fg has degree $n+m$ with leading coefficient $a_n \sigma^n(b_m) \neq 0$.

Let A be a division ring and L a nonzero left ideal of $A[x;\sigma,\delta]$. Then L is generated by any polynomial $g \in L$ of lowest degree: if $f \in L$ we may write

$$f = qg + r, \qquad \partial(r) < \partial(g) \text{ or } r = 0,$$

and so only $r = 0$ is possible.

Let σ be an automorphism of A. A moment's reflection shows that
$$ax = x\sigma^{-1}(a) - \delta(\sigma^{-1}(a))$$
and so one sees easily by induction that each monomial ax^j can be written in the form $\sum_{i=0}^{j} x^i b_i$. Hence also the polynomials of $A[x;\sigma,\delta]$ can be written in the form $\sum x^i b_i$. Now a right-handed analogue of the previous considerations is applicable yielding that $A[x;\sigma,\delta]$ is also a principal right ideal domain.

The rest is now straightforward. □

More on skew polynomial rings can be found in Jacobson [6], McConnell and Robson [1] and Rowen [1].

EXAMPLE 1.3.2. We shall specialize the skew polynomial ring $A[x;\sigma,\delta]$ by choosing $A = \mathbb{Q}[y]$ the ring of polynomials over the rational numbers, σ the identical automorphism and δ the partial derivation $\delta = \partial/\partial y$. The multiplication is defined by the rule
$$xf = fx + \frac{\partial f}{\partial y} \qquad \forall f \in \mathbb{Q}[y],$$
in particular,
$$xy - yx = 1.$$
Thus the skew polynomial ring $\mathbb{Q}[y][x;\delta]$ can be viewed also as the factor ring
$$\mathbb{Q}\langle x,y\rangle/(xy - yx - 1)$$
of the ring $\mathbb{Q}\langle x,y\rangle$ of rational polynomials with non-commuting indeterminates x and y by the ideal generated by $xy - yx - 1$. The ring
$$W_1 = \mathbb{Q}[y][x;\delta] = \mathbb{Q}\langle x,y\rangle/(xy - yx - 1)$$
is called the *first Weyl algebra* with indeterminates x and y over the reals (the n-th Weyl algebra involves $2n$ indeterminates x_i and y_i).

We claim that W_1 *is a simple domain with unity element*. Let I be any nonzero ideal of W_1. If $I \cap \mathbb{Q}[y] = 0$ then I contains a polynomial $f(x)$ of minimal degree $n > 0$ with coefficients $a_i \in \mathbb{Q}[y]$. Without loss of generality we may assume that the leading coefficient $a_n = 1$. From the rule
$$xy = yx + 1$$
we get by induction
$$x^n y = yx^n + nx^{n-1}.$$
Hence
$$f(x)y - yf(x) \in I$$

is a polynomial of degree at most $n-1 < n$. Thus by the choice of f we get

$$f(x)y - yf(x) = 0$$

implying $n = 0$, a contradiction. Hence I must contain a nonzero polynomial $g \in \mathbb{Q}[y]$ of minimal degree k. Assume that $k > 0$. Then xg and gx are elements of the ideal I, so $xg = gx + g' \in I$. Hence $g' = xg - gx \in I$ follows, contradicting the minimality of the degree k. ∎ Thus $k = 0$ and $g \in \mathbb{Q}$. Consequently also $1 \in I$, and so $I = W_1$.

Notice that *the Weyl algebra $W_1 = \mathbb{Q}\langle x, y\rangle/(xy - yx - 1)$ is not a principal ideal domain*, though by Theorem 1.3.1 we know that $W_1 = \mathbb{Q}(y)[x; \delta]$, as a polynomial ring in one indeterminate x, is a principal left ideal domain. We show that *every proper left ideal $L = W_1 f(x)$ of W_1 is a simple domain without nonzero idempotents*. Suppose that $K \neq 0$ is an ideal of L. Then

$$W_1 f(x) K W_1 f(x) = LKL \subseteq K.$$

Moreover, $W_1 f(x) K W_1$, being an ideal in W_1, is either 0 or W_1. If it is W_1 then

$$K \supseteq LKL = W_1 f(x) = L$$

and so $K = L$. If $W_1 f(x) K W_1 = 0$, then $f(x) K = 0$, since W_1 has a unity element, and so $K = 0$. Thus L is a simple ring. Since the elements of L are polynomials of x of degree ≥ 1, L has no nonzero idempotents in view of Theorem 1.3.1.

EXAMPLE 1.3.3. The skew or twisted polynomial ring $A[x; \sigma, \delta]$ will be denoted simply by $A[x; \sigma]$, if $\delta = 0$. Now the multiplication is given by

$$xa = \sigma(a)x \quad \text{and} \quad x^n a = \sigma^n(a) x^n.$$

In the sequel we suppose that $A = F$ *is a field and σ is an automorphism of infinite order*. For example, F may be the field extension of the rationals $\mathbb{Q}$ by infinitely many transcendental elements $\ldots, y_{-1}, y_0, y_1, y_2, \ldots$, and σ may be the automorphism for which $\sigma(q) = q$ for all $q \in \mathbb{Q}$ and $\sigma(t_n) = t_{n+1}$, for all integers n.

We prove that the *center of $F[x; \sigma]$ is the subring*

$$F^\sigma = \{a \in F \mid \sigma(a) = a\},$$

called *the fixed subring of σ*. Obviously, the elements of F^σ commute with the elements of $F[x; \sigma]$, whence F^σ is contained in the center of $F[x; \sigma]$. Conversely, suppose that $f = \sum a_i x^i$ is in the center of $F[x; \sigma]$. For the element x we have

$$0 = xf - fx = \sum (\sigma(a_i) - a_i) x,$$

implying $\sigma(a_i) = a_i$, that is, $a_i \in F^\sigma$ for each i. Moreover, for any element $a \in F \setminus F^\sigma$ we have

$$0 = af - fa = \sum a_i(a - \sigma^i(a))x^i,$$

proving $a - \sigma^i(a) = 0$ whenever $a_i \neq 0$. Since σ is of infinite order, we get $a_i = 0$ for $i \geq 1$. Hence $f = a_0 \in F^\sigma$ follows.

Next, we prove that *every nonzero ideal I of $F[x;\sigma]$ has the form $I = F[x;\sigma]x^k$*. From Theorem 1.3.1 we know that $I = F[x;\sigma]g$ where g is a polynomial of minimal degree k in I, and that g may be replaced by a polynomial $f \in I$ of degree k and leading coefficient 1. Now if

$$f = \sum_{i=0}^{k-1} a_i x^i + x^k,$$

then

$$xf - fx = \sum_{i=0}^{k-1}(\sigma(a_i) - a_i)x^{i+1}$$

has fewer terms than f. Hence by Theorem 1.3.1 we conclude that $xf - fx = 0$, and so $a_i \in F^\sigma$ for each $i = 0, 1, \ldots, k-1$. Moreover, for every $c \in F$ we see that $\sigma^k(c)f - fc \in F[x;\sigma]$ has degree less than k, so it must be 0, implying

$$\sigma^k(c)a_i = a_i\sigma^i(c)$$

for each $i = 0, 1, \ldots, k-1$. Hence from $a_i \in F^\sigma$ it follows $\sigma^k(c) = \sigma^i(c)$, that is, $\sigma^{k-i}(c) = c$ whenever $a_i \neq 0$. Since σ is of infinite order, all $a_0, a_1, \ldots, a_{k-1}$ have to be 0, whence $f = x^k$ and $I = F[x;\sigma]x^k$.

Let X be a set of finitely or infinitely many symbols (indeterminates). The *polynomial ring $A[X]$ with commuting indeterminates over A* is defined as the set of all finite sums of finite products of powers of the indeterminates $x_i \in X$ with coefficients from A, and the addition and multiplication are the usual ones. We can speak also of the *ring of polynomials $A\langle X \rangle$ with non-commuting indeterminates*. Here we mean that the indeterminates commute with the elements of A but not with each other.

The *formal power series ring $A[[x]]$ over a ring A* consists of all formal expressions $\sum_{n=0}^{\infty} a_n x^n$, $a_n \in A$, where the addition and multiplication are defined as in the polynomial ring $A[x]$. This definition extends to formal power series rings $A[[X]]$ and $A\langle\langle X \rangle\rangle$ over A with commuting and non-commuting indeterminates from a set X of indeterminates. Formal power series rings will occur in sections 4.2, 4.6 and 4.9.

The *Laurent polynomial ring $A[x, x^{-1}]$ over a ring A* is the ring of all finite sums $\sum_{i=k}^{n} a_i x^i$, where $k \leq n$ and k, n are integers (that is, also negative exponents are admitted), with the usual addition and multiplication.

General Fundamentals

The definition of a twisted polynomial ring $A[x;\sigma]$ extends in a natural way to Laurent polynomials, getting the *twisted Laurent polynomial ring* $A[x,x^{-1};\sigma]$ with the multiplication given by

$$xa = \sigma(a)x \quad \text{and} \quad x^n = \sigma^n(a)x^n.$$

Twisted Laurent polynomial rings will occur in section 3.13.

Chapter II
The General Theory of Radicals

It was Amitsur [2], [3], [4] and Kurosh [1] who were the first to discover independently that the classical radicals all had certain common properties and they used these algebraic properties to axiomatically define abstract radical classes. This can be done at a category theoretical level (see e.g. Amitsur [4], Shulgeifer [1], Veldsman [1], [2], Márki, Mlitz and Wiegandt [1] and Gardner [18]) and interpreted in various algebraic and nonalgebraic specific categories like graphs or topological spaces. We plan to introduce and characterize radical classes, semisimple classes and torsion theories in a universal class $\mathbb{A}$ which may consist of associative rings, alternative rings, not necessarily associative rings, near-rings, associative rings with involution, algebras over a commutative ring with unity element, or Ω-groups, etc. We shall call members of $\mathbb{A}$ rings even though they might be much more general objects.

2.1. Radical classes

One of the central problems in ring theory is to determine the structure of rings in terms of linear transformations of vector spaces. This is quite difficult and in fact not always possible. For example a zero-ring cannot be isomorphic to a full matrix ring over a division ring. Wedderburn (1908) suggested and Köthe (1930) used an ingenious technique. They considered a "bad" property γ of rings, they discarded or ignored a certain "bad" ideal $\gamma(A)$, called the γ-radical of a ring A, such that the factor ring $A/\gamma(A)$ is "good". Here if something is "bad" then it has no homomorphic image which is "good" (e.g. a full matrix ring), whereas a "good" ring has no "bad" ideal. The "good" rings are expected to be described somehow by a decomposition theorem in which the components admit faithful representations in terms of linear transformations. Of course the larger the class of "bad" rings, the more one discards, the less there is left over. In the extreme case we can call every ring "bad" and then there is nothing "good". Or we can say everything is "good" but then we cannot get a description in terms of linear transformations.

To get down to cases, let γ be a class of rings such that
(a) γ is homomorphically closed: $A \in \gamma$ and $A \twoheadrightarrow B$ imply $B \in \gamma$;
(b) for every ring A, the sum $\gamma(A) = \sum(I \triangleleft A \mid I \in \gamma)$ is in γ,
(c) $\gamma(A/\gamma(A)) = 0$ for every ring A.

DEFINITION 2.1.1. A class γ of rings satisfying (a), (b) and (c) will be called a *radical class in the sense of Kurosh and Amitsur* (briefly only a *radical*). $\gamma(A)$ is called the γ-*radical* of A. A ring A is called a γ-*radical ring* if $A \in \gamma$, that is, $\gamma(A) = A$.

Conditions (a), (b) and (c) are excellent for motivating and defining radicals, but if we are given a class of rings and wish to discover whether it is a radical class, they are not always the conditions we can most easily test. Furthermore, they are not the most appropriate for dualizing the notion of a radical class in a given universal class or for defining radical classes in a universal class consisting of non-ringlike objects (e.g. universal algebras, in particular monoids, or graphs or topological spaces). We would, therefore, like to find conditions equivalent to (a), (b) and (c) which are often easier to test or more appropriate for dualization and generalization.

PROPOSITION 2.1.2. *Assuming conditions* (a) *and* (b) *on a class* γ *of rings, condition* (c) *is equivalent to*

($\bar{c}$) *If I is an ideal of the ring A and if both I and A/I are in γ, then A itself is in γ.*

When γ satisfies ($\bar{c}$) we say γ is *closed under extensions*.

Proof: First, suppose that (c) holds and that both I and A/I are in γ. Then $I \subseteq \gamma(A)$ by (b) and

$$A/\gamma(A) \cong \frac{A/I}{\gamma(A)/I}$$

is a homomorphic image of A/I. Then by (a) $A/\gamma(A)$ is in γ because A/I is in γ. But by (c) and (b)

$$0 = \gamma(A/\gamma(A)) = A/\gamma(A).$$

Therefore $A = \gamma(A)$ is in γ and we have ($\bar{c}$).

Conversely, suppose that ($\bar{c}$) holds. Assume that $\gamma(A/\gamma(A)) \neq 0$. Now $\gamma(A/\gamma(A)) = K/\gamma(A)$ for some ideal K of A. Since both $\gamma(A)$ and $K/\gamma(A)$ are in γ, ($\bar{c}$) tells us that K is in γ. Then $K \subseteq \gamma(A)$ and $K/\gamma(A) = 0$, a contradiction. Thus (c) holds and the Proposition is established. □

PROPOSITION 2.1.3. *Assuming conditions* (a) *and* ($\bar{c}$) *on a class* γ *of rings, condition* (b) *is equivalent to*

The General Theory of Radicals

(b̄) *if $I_1 \subseteq \ldots \subseteq I_\lambda \subseteq \ldots$ is an ascending chain of ideals of a ring A and if each I_λ is in γ, then $\cup I_\lambda$ is in γ.*

When γ satisfies (b̄) we say γ has the *inductive property*.

Proof: Suppose (b) holds and let $B = \cup I_\lambda$. Then by (b) each I_λ is contained in $\gamma(B)$. Therefore $B = \gamma(B)$ is in γ, and we have (b̄).

Conversely, suppose (b̄) holds. Then we can apply Zorn's Lemma and obtain a maximal γ-ideal B in A. If K is any γ-ideal of A, then $(B+K)/K \cong B/(B \cap K)$ and this is in γ by (a). Thus both K and $(B+K)/K$ are in γ and by (c̄) also $B+K$ is in γ. Since B is maximal with respect to this property, K must be in B and thus $\gamma(A) = B$ which is in γ. Thus (b) holds and the Proposition is proved. □

We then have

THEOREM 2.1.4. *A class γ of rings is a radical class if and only if*
(a) *γ is homomorphically closed,*
(b̄) *γ has the inductive property,*
(c̄) *γ is closed under extensions.* □

THEOREM 2.1.5. *For any class γ of rings, the following conditions are equivalent:*

I. *γ is a radical class;*

II. (R1) *if $A \in \gamma$ then for every $A \twoheadrightarrow B \neq 0$ there is a $C \triangleleft B$ such that $0 \neq C \in \gamma$;*
(R2) *if A is a ring of the universal class $\mathbb{A}$ and for every $A \twoheadrightarrow B \neq 0$ there is a $C \triangleleft B$ such that $0 \neq C \in \gamma$, then A is in γ;*

III. *γ satisfies condition (R1), has the inductive property and is closed under extensions.*

Proof: I $\implies$ III. Since every homomorphically closed class clearly satisfies (R1), Theorem 2.1.4 gives us the implication.

III $\implies$ II. We must prove (R2). Assume then that A is a ring such that for every $A \twoheadrightarrow B \neq 0$ there exists a $C \triangleleft B$ such that $0 \neq C \in \gamma$, and that A itself is not in γ. Because we have the inductive property, we can use Zorn's Lemma and obtain an ideal I of A, maximal with respect to being in γ. Since $A \notin \gamma$, $A/I \neq 0$ holds. Then there is an ideal $C/I \triangleleft A/I$ such that $0 \neq C/I \in \gamma$. Since γ is closed under extensions, C itself must be in γ. But this contradicts the maximality of I and thus we have (R2) and II.

II $\implies$ I. We will prove I by showing γ is homomorphically closed, has the inductive property and is closed under extensions.

Take A in γ, and let B be a given homomorphic image of A. Then any homomorphic image C of B is also an image of A. Then if $A \twoheadrightarrow B \twoheadrightarrow C \neq 0$, (R1) tells us that there is a $D \triangleleft C$ with $0 \neq D \in \gamma$. Then by (R2) B must be in γ, and therefore γ is homomorphically closed.

Next, let $I_1 \subseteq \ldots \subseteq I_\lambda \subseteq \ldots$ be an ascending chain of ideals of A, each in γ. We show that $\cup I_\lambda$ is in γ. To this end, let $(\cup I_\lambda)/K$ be any nonzero factor ring of $\cup I_\lambda$. There must exist an index λ such that $I_\lambda \not\subseteq K$, and thus

$$0 \neq (I_\lambda + K)/K \triangleleft (\cup I_\lambda)/K.$$

Now

$$(I_\lambda + K)/K \cong I_\lambda/(I_\lambda \cap K)$$

and this is in γ since I_λ is, and γ is homomorphically closed. Then (R2) tells us that $\cup I_\lambda$ must be in γ.

Finally, we take I and A/I both in γ, and want to show that A is in γ. To this end, let A/K be any nonzero factor ring of A. In the case when $I \subseteq K$, we have

$$0 \neq A/K \cong \frac{A/I}{K/I}$$

and this is in γ because A/I is and because γ is homomorphically closed. In the case when $I \not\subseteq K$, we have $0 \neq (I+K)/K \triangleleft A/K$ and $(I+K)/K \cong I/(I \cap K)$ is in γ because I is in γ and because γ is homomorphically closed. Thus in both cases A/K has a nonzero ideal in γ and (R2) guarantees that A itself is in γ. This ends the proof of the theorem. □

The conditions of Theorem 2.1.4 are the most transparent and the easiest to test. Conditions (R1) and (R2) are not particularly attractive but they are the least ring-like because they express logical connections between the relations $\triangleleft$ and $\twoheadrightarrow$. They are therefore the most appropriate for dualization and generalization. The conditions of Theorem 2.1.5. III seem to be the least demanding.

Let us now pause, to consider some examples of radical classes. *We shall assume in these examples that the rings considered are associative.*

EXAMPLE 2.1.6. *Köthe's nil radical.* This was, historically, the first radical (1930). The *nil radical class* is the class

$$\mathcal{N} = \{A \mid \forall a \in A \; \exists n \geq 1, n \text{ depending on } a, \text{ such that } a^n = 0\}$$

that is, $\mathcal{N}$ is the class of all nil rings.

It is clear that $\mathcal{N}$ is homomorphically closed and has the inductive property. To see that it is closed under extensions, take a ring A such that both I and A/I are in $\mathcal{N}$. Then for any $a \in A$, the coset $\bar{a} = a + I$ is nilpotent and thus a^n is in I for some $n \geq 1$. But a^n must then be nilpotent, so there is some $k \geq 1$ such that $(a^n)^k = 0$. Then $a^{nk} = 0$ and a is nilpotent, A is nil.

Köthe worked on the structure of $A/\mathcal{N}(A)$, a ring with no nonzero nil ideals. The elements of $\mathcal{N}(A)$ are, formally, roots of zero, and since the Latin name for root is *radix*, he called $\mathcal{N}(A)$ the radical of A.

EXAMPLE 2.1.7. *The Jacobson radical.* This radical is the most popular and certainly the most important as far as the structure of rings is concerned. In a ring $(A, +, \cdot)$ Perlis [1] introduced another operation, the *circle operation* $\circ$, by defining

$$a \circ b = a + b - ab \qquad \forall a, b \in A.$$

It is straightforward to show that this is associative. Also $0 \circ a = a \circ 0 = a$ for every $a \in A$. Thus $(A, \circ)$ is a monoid with unity element 0. The *Jacobson radical class* $\mathcal{J}$ is then defined as

$$\mathcal{J} = \{A \mid (A, \circ) \text{ is a group}\}.$$

It is clear that $\mathcal{J}$ is homomorphically closed and has the inductive property. To see that it is closed under extensions, take a ring A such that I and A/I are in $\mathcal{J}$. If $\bar{a} \in A/I$, then there exists an $\bar{x} \in A/I$ such that $\bar{x} \circ \bar{a} = \bar{0}$, that is, $x \circ a \in I$. Then there must exist an element $y \in I$ such that $y \circ (x \circ a) = (y \circ x) \circ a = 0$. Thus A is also in $\mathcal{J}$.

EXAMPLE 2.1.8. *The Levitzki radical.* A ring A is said to be *locally nilpotent*, if any finitely generated subring of A is nilpotent. Every nilpotent ring is locally nilpotent (the converse is not true, see Example 2.1.10), and every locally nilpotent ring is nil (the converse of this is also false, see Theorem 4.2.4). The *Levitzki radical class* $\mathcal{L}$ is the class of all locally nilpotent rings. It is clear that $\mathcal{L}$ is homomorphically closed and satisfies the inductive property. To see that it is closed under extensions, take a ring A such that both I and A/I are in $\mathcal{L}$. Let S be any subring of A generated by some finite set $\{a_1, \ldots, a_n\}$. Consider the finite set of cosets $\{a_1 + I, \ldots, a_n + I\}$, and let them generate a subring $\overline{S}$ of A/I. Then $\overline{S}$ is nilpotent because A/I is in $\mathcal{L}$. Thus $\overline{S}^k = 0$, or $S^k \subseteq I$ for some $k \geq 1$. Now S^k is finitely generated, namely by the set of all products $a_{i_1} \ldots a_{i_k}$ of k factors from the set $\{a_1, \ldots, a_n\}$. Then S^k is nilpotent because I is in $\mathcal{L}$, whence $(S^k)^l = S^{kl} = 0$ for some $l \geq 1$. Thus S is nilpotent and A itself is in $\mathcal{L}$.

These three radicals contain some "bad" classes of rings. They contain all zero-rings and (by the extension property) all nilpotent rings. However, not all radical classes contain "bad" rings, and we must not be disturbed to find radical classes all of whose rings are "well behaved".

EXAMPLE 2.1.9. A ring A is said to be *von Neumann regular*, if for every $a \in A$ we have $a \in aAa$. The class ν of all von Neumann regular rings is a radical class. It is clear that ν is homomorphically closed and satisfies the inductive property. For extensions, we take a ring A such that both I and A/I are in ν. For any $\bar{a} \in A/I$ we then know that $\bar{a}$ is in $(a + I)(A/I)(a + I) = (aAa + I)/I$. Thus $a \in aAa + I$. Then there exist an element $x \in A$ such that $a - axa \in I \in \nu$. Therefore

$$a - axa \in (a - axa)I(a - axa) \subseteq aAa.$$

Thus $a \in aAa$ and A itself is in ν.

The radical class ν does not contain any nil ring because if $0 \neq a = axa$, then
$$ax = (ax)^2 = \ldots = (ax)^n = \ldots$$
is a nonzero idempotent and cannot be nilpotent. On the other hand ν *contains every $n \times n$ matrix ring $M_n(D)$ over a division ring D*. To see this, take a nonzero element $a \in M_n(D)$. Then a is a linear transformation of an n-dimensional vector space V_n over D. Let $\{ay_1, \ldots, ay_m\}$ be a basis for the image of V_n by a. If $v \in V$, then av is a linear combination of $ay_1, ay_2, \ldots, ay_m$, whence it follows that V_n is spanned by $\ker(a) \cup \{ay_1, ay_2, \ldots, ay_m\}$. Since $\{ay_1, ay_2, \ldots, ay_m\}$ is a linearly independent set (and so extends to a basis of V_n), there is a linear transformation x such that $x \cdot ay_i = y_i$ for $i = 1, 2, \ldots, m$. Then $axay_i = ay_i$ for $i = 1, 2, \ldots, m$ and if $u \in \ker(a)$ then $axau = 0 = au$. Thus $axa = a$ and $M_n(D) \in \nu$.

Some classes of rings, that seem intuitively certain to be radical classes, may not be.

EXAMPLE 2.1.10. *The class of all nilpotent rings is not a radical class* in the universal class of all associative rings. Although it is clear that this class is homomorphically closed and closed under extensions, unfortunately it does not satisfy the inductive property. To see this, take the ring T_n of all upper triangular (rational) $n \times n$ matrices
$$T_n = \{(a_{ij}) \mid a_{ij} \in \mathbb{Q}, \ a_{ij} = 0 \ \text{for} \ i \geq j\}$$
for $n \geq 2$. T_n is a nilpotent ring of nilpotence degree n: $T_n^n = 0$ but $T_n^{n-1} \neq 0$. Let us consider the direct sum
$$A = \bigoplus_{n=2}^{\infty} T_n.$$
In the ring A there is an ascending chain
$$T_2 \subset T_2 \oplus T_3 \subset \cdots \subset \bigoplus_{n=2}^{k} T_n \subset \cdots$$
of ideals such that A is the union
$$A = \bigcup_{k=2}^{\infty} \left(\bigoplus_{n=2}^{k} T_n \right)$$
of the members of the chain, and each member of the chain is nilpotent:
$$\left(\bigoplus_{n=2}^{k} T_n \right)^k = 0.$$

The General Theory of Radicals

However, the ring A is not nilpotent. Moreover, every element $a \in A$ as well as every finite subset $\{a_1, \ldots, a_n\} \subset A$ is in a finite direct sum $\bigoplus_{n=2}^{k} T_n$ for some k. Thus A is locally nilpotent but not nilpotent.

Let us observe that in the universal class of all finite rings the class of all (finite) nilpotent rings is a radical class.

From an axiomatic point of view a radical γ may be defined as an assignment $\gamma \colon A \mapsto \gamma(A)$ designating a certain ideal $\gamma(A)$ to each ring A. Such an assignment γ is called a *Hoehnke radical*, if

i) $f(\gamma(A)) \subseteq \gamma(f(A))$ for every homomorphism $f \colon A \to f(A)$,
ii) $\gamma(A/\gamma(A)) = 0$

for every ring A. A Hoehnke radical γ may satisfy also the following conditions:

iii) γ is *complete*: if $I \triangleleft A$ and $\gamma(I) = I$ then $I \subseteq \gamma(A)$,
iv) γ is *idempotent*: $\gamma(\gamma(A)) = \gamma(A)$ for every rings A.

THEOREM 2.1.11. *If γ is a Kurosh–Amitsur radical then the assignment $A \mapsto \gamma(A)$ is a complete, idempotent Hoehnke radical. Conversely, if γ is a complete, idempotent Hoehnke radical, then there is a Kurosh–Amitsur radical ϱ such that $\gamma(A) = \varrho(A)$ for every ring A. Moreover, $\varrho = \{A \mid \gamma(A) = A\}$.*

Proof: Let γ be a Kurosh–Amitsur radical. $\gamma(A) \in \gamma$ by condition (b) and also $f(\gamma(A)) \in \gamma$ by condition (a). This proves i).

ii) is just condition (c).

iii) and iv) are immediate consequences of condition (b).

Assume that γ is a complete, idempotent Hoehnke radical, and define the class ϱ by
$$\varrho = \{A \mid \gamma(A) = A\}.$$

If $A \in \varrho$ and $f \colon A \to B$ is a surjective homomorphism, then by i)
$$B = f(A) = f(\gamma(A)) \subseteq \gamma(f(A)) = \gamma(B),$$
so $B \in \varrho$. This gives us (a) for ϱ.

For every ring A, we have
$$\varrho(A) = \sum(I \triangleleft A \mid I \in \varrho) = \sum(I \triangleleft A \mid \gamma(I) = I).$$

If $I \triangleleft A$ and $\gamma(I) = I$, then $I \subseteq \varrho(A)$ and by iii) $I \subseteq \gamma(\varrho(A))$, whence $\varrho(A) = \gamma(\varrho(A))$, that is, $\varrho(A) \in \varrho$. This gives us (b) for ϱ.

It also follows from the equality $\varrho(A) = \gamma(\varrho(A))$ and completeness iii) that $\varrho(A) \subseteq \gamma(A)$. But by idempotence iv), $\gamma(A) \in \varrho$, so $\gamma(A) \subseteq \varrho(A)$. Thus $\gamma(A) = \varrho(A)$ for all rings A. Now ii) for γ gives us (c) for ϱ. □

A radical assignment $\gamma\colon A \mapsto \gamma(A)$ is called a *Plotkin radical*, if it satisfies conditions i), iii) and iv). From a theoretical point of view radical assignment is fundamental: it determines its radical class and its semisimple class. There is a one-to-one correspondence between Hoehnke radicals and their semisimple classes, and between Plotkin radicals and their radical classes. In this book we deal nearly exclusively with Kurosh–Amitsur radicals, so we do not need to distinguish between radical assignment (i.e. radical) and radical class (i.e. radical property).

More on Hoehnke and Plotkin radicals can be found in the papers of Hoehnke [1], Mlitz [1], Mlitz and Veldsman [1], Plotkin [1], [2], [3], de la Rosa, van Niekerk and Wiegandt [1], de la Rosa, Veldsman and Wiegandt [1].

2.2. Radical constructions

We wish to study two basic constructions, the so called lower and upper radicals, and these will supply us with many different and interesting radicals.

Let us begin with an arbitrary class δ of rings. In general such a class is far from being a radical. However, there is always the smallest radical class containing δ. To see this, we can use Theorem 2.1.4 and take the intersection of all radical classes that contain δ. This intersection $\mathcal{L}\delta$ is a radical class and is clearly the smallest radical containing δ. $\mathcal{L}\delta$ is called the *lower radical* determined by the class δ. There are several ways of actually constructing $\mathcal{L}\delta$ and we shall follow the *Tangeman–Kreiling lower radical construction* (see Tangeman and Kreiling [1]).

Starting with δ we define

$$\delta_1 = \{A \mid A \text{ is a homomorphic image of a ring } B \text{ in } \delta\},$$

the so called *homomorphic closure* of δ. Proceeding inductively, if δ_μ has been defined for all ordinals $\mu < \lambda$, we define

$$\delta_\lambda = \{A \mid \exists I \triangleleft A \text{ such that } I \in \delta_{\lambda-1} \text{ and } A/I \in \delta_{\lambda-1}\}$$

when $\lambda - 1$ exists. When λ is a limit ordinal, we define

$$\delta_\lambda = \left\{ A \;\middle|\; \begin{array}{l} A \text{ is the union of an ascending chain of} \\ \text{ideals, each one in some } \delta_\mu \text{ for } \mu < \lambda \end{array} \right\}.$$

THEOREM 2.2.1. $\mathcal{L}\delta = \cup \delta_\lambda$ *where the union extends over all ordinals* λ.

It is customary to call $\mathcal{L}$ the *lower radical operator**.

---

* No confusion should result from our use of the same letter $\mathcal{L}$ for the Levitzki radical and the lower radical operator.

Proof: First we want to show that if $\mu < \lambda$ then $\delta_\mu \subseteq \delta_\lambda$. It is clear that $\delta \subseteq \delta_1$ and that $\delta_\mu \subseteq \delta_\lambda$ for any limit ordinal λ. To see that $\delta_\mu \subseteq \delta_{\mu+1}$ we first note that the ring 0 is in δ_1 and by induction also in every δ_λ for $\lambda \geq 1$. Then for $A \in \delta_\mu$ we take $I = 0 \in \delta_\mu$ and $A = A/I \in \delta_\mu$. Thus A is in $\delta_{\mu+1}$ by definition and the construction is monotonically increasing.

Now we want to show that $\cup \delta_\lambda$ is a radical class. First let us deal with homomorphic closure. Of course, δ_1 is homomorphically closed. Let us assume that the same is true for δ_μ for every $\mu < \lambda$. We take $A \in \delta_\lambda$ and consider any image A/I of A. When λ is a limit ordinal, there is a chain $\{K_\iota\}$ of ideals of A such that $\cup K_\iota = A$ and each K_ι belongs to a class δ_μ, $\mu < \lambda$. Then $\{(I + K_\iota)/I\}$ is a chain of ideals of A/I such that A/I is the union of this chain. Since

$$(I + K_\iota)/I \cong K_\iota/(I \cap K_\iota),$$

by the assumption on δ_μ, each of these ideals belongs to some δ_μ with $\mu < \lambda$. This means that $A/I \in \delta_\mu$. If $\lambda - 1$ exists, then A contains an ideal J such that J and A/J are in $\delta_{\lambda-1}$. Again by the hypothesis on $\delta_{\lambda-1}$ we have

$$(J + I)/I \cong J/(J \cap I) \in \delta_{\lambda-1}$$

and

$$\frac{A/I}{(J+I)/I} \cong A/(J+I) \in \delta_{\lambda-1},$$

which implies $A/I \in \delta_\lambda$. Thus by transfinite induction every δ_λ is homomorphically closed and so is $\cup \delta_\lambda$ too.

Let $I_1 \subset \cdots \subset I_\iota \subset \cdots$ be a strictly ascending chain of ideals of a ring A such that each I_ι is in $\cup \delta_\lambda$. Since A is a set and the construction is monotonic, there is an ordinal μ such that $I_\iota \in \delta_\mu$ for each ι. If we then take any limit ordinal ν bigger than μ, we have $\cup I_\iota \in \delta_\nu \subseteq \cup \delta_\lambda$. Thus gives us the inductive property.

To prove that $\cup \delta_\lambda$ is closed under extensions, assume that both I and A/I are in $\cup \delta_\lambda$. Then there must exist an ordinal λ such that both I and A/I are in δ_λ. So A is in $\delta_{\lambda+1}$ by definition and thus A is in $\cup \delta_\lambda$.

Therefore $\cup \delta_\lambda$ is a radical class by Theorem 2.1.4.

If γ is any radical class that contains δ, then it must also contain δ_1. Furthermore, if all δ_μ are contained in γ for $\mu < \lambda$, then δ_λ must also be contained in γ by one of the properties ($\bar{\text{b}}$) or ($\bar{\text{c}}$) of radicals. Thus by transfinite induction $\cup \delta_\lambda \subseteq \gamma$, proving $\cup \delta_\lambda = \mathcal{L}\delta$. □

EXAMPLE 2.2.2. We have already seen that in the universal class of all associative rings the class of all nilpotent rings is not a radical class (Example 2.1.10). We can now use the lower radical construction and Theorem 2.2.1 to obtain the lower radical class β of all nilpotent rings. The class β is called the *Baer (or prime) radical*.

Another nonradical class of awkward or "bad" rings is the class $\mathcal{Z}$ of all zero-rings, that is rings with zero-multiplication. We can construct the lower radical $\mathcal{L}\mathcal{Z}$ and it turns out that

$$\mathcal{L}\mathcal{Z} = \beta.$$

To see this we note that if A is nilpotent with $A^3 = 0$ and $A^2 \neq 0$, then both A^2 and A/A^2 are zero-rings. Thus putting $\mathcal{Z} = \delta_1$, we have that $A \in \delta_2$. By induction we show that $A^m = 0$ implies $A \in \delta_{m-1}$ for $m > 3$. If also $A^{m-1} = 0$, then $A \in \delta_{m-2} \subseteq \delta_{m-1}$. Assume that $A^m = 0$ and $A^{m-1} \neq 0$. Then $A^{m-1} \in \mathcal{Z} = \delta_1 \subseteq \delta_{m-2}$. Further, by $(A/A^{m-1})^{m-1} = 0$ we have $A/A^{m-1} \in \delta_{m-2}$. Hence $A \in \delta_{m-1}$. Thus $\mathcal{L}\mathcal{Z}$ contains all nilpotent rings and therefore all rings of β.

When we work with a class of awkward or "bad" rings, we want the lower radical determined by this class, so that we can get a good grip on them in order to set them aside. However, when we meet a pleasant and familiar class ϱ of rings, like division rings or matrix rings, we certainly do not want to discard them. In fact, what we want is to find a radical class that has no pleasant rings in it. We shall seek the largest radical class γ such that $\gamma \cap \varrho = \{0\}$.

To this end we shall say that a class ϱ is *regular*, if

(S1) for every ring $A \in \varrho$, every nonzero ideal of A has a nonzero homomorphic image in ϱ.

There are many regular classes of rings. In particular, *every hereditary class is clearly regular*.

THEOREM 2.2.3. *If ϱ is a regular class of rings, then the class*

$$\mathcal{U}\varrho = \{A \mid A \text{ has no nonzero homomorphic image in } \varrho\}$$

is a radical class, $\varrho \cap \mathcal{U}\varrho = \{0\}$ and $\mathcal{U}\varrho$ is the largest radical having zero intersection with ϱ.

Proof: We shall use Theorem 2.1.5 and establish conditions (R1) and (R2) for $\mathcal{U}\varrho$. To get (R1) we will prove the contrapositive, i.e. if A has a nonzero homomorphic image B such that B has no nonzero ideal in $U\varrho$, then A cannot be in $\mathcal{U}\varrho$. If such a B exists, then B itself cannot be in $\mathcal{U}\varrho$ and thus B must have a nonzero homomorphic image D in ϱ. Then $A \to B \to D$ gives us a nonzero homomorphic image of A which is in ϱ. Therefore, A is not in $\mathcal{U}\varrho$ which proves (R1).

To establish (R2) we again use the contrapositive and assume that A is not in $\mathcal{U}\varrho$. Then A has a nonzero homomorphic image D in ϱ. Since ϱ is regular, every nonzero ideal of D has a nonzero homomorphic image in ϱ. This contradicts the first half of (R2) and establishes (R2).

Therefore $\mathcal{U}\varrho$ is a radical class.

Finally suppose that γ is a radical and $\gamma \cap \varrho = \{0\}$. If $\gamma \not\subseteq \mathcal{U}\varrho$, then there would exist a ring A in γ but not in $\mathcal{U}\varrho$. Then A has a nonzero homomorphic image in ϱ. But this image must also be in γ, a contradiction. Therefore $\gamma \subseteq \mathcal{U}\varrho$ and $\mathcal{U}\varrho$ is in fact the largest radical such that $\varrho \cap \mathcal{U}\varrho = \{0\}$. □

The operator $\mathcal{U}$ is called the *upper radical operator* and $\mathcal{U}\varrho$ is called the *upper radical of the class ϱ*.

Regularity of the class ϱ is not entirely necessary for $\mathcal{U}\varrho$ to be a radical class. A necessary and sufficient condition for that was given by Enersen and Leavitt [1]. If we begin with an arbitrary class ϱ, maybe we cannot find a unique minimal regular class that contains ϱ because the intersection of two regular classes may not be regular (cf. Proposition 5.1.10).

The upper radical construction does provide us with many radical classes because every hereditary class and in particular every class of simple rings is regular and can be used in the construction.

EXAMPLE 2.2.4. Let $\mathcal{M}$ be the class of all simple rings with unity. The upper radical $\mathcal{G} = \mathcal{U}\mathcal{M}$ is called the *Brown–McCoy radical class*.

2.3. Semisimple classes

We shall define semisimple classes by dualizing our definition of radical classes and for that purpose we shall concentrate on the characterizing conditions (R1) and (R2) of Theorem 2.1.5.

DEFINITION 2.3.1. A class σ of rings is said to be a *semisimple class* in the sense of Kurosh and Amitsur, if σ satisfies the following conditions:

(S1) σ is regular: if $A \in \sigma$, then for every nonzero $B \triangleleft A$ there exists a $B \twoheadrightarrow C \neq 0$ such that $C \in \sigma$,

(S2) if A is a ring of the universal class $\mathbb{A}$ and for every nonzero $B \triangleleft A$ there exists a $B \twoheadrightarrow C \neq 0$ such that $C \in \sigma$, then $A \in \sigma$.

Conditions (S1) and (S2) are dual to (R1) and (R2) where the relations $\twoheadrightarrow$ and $\triangleleft$ are interchanged. Since the relation $\twoheadrightarrow$ is transitive, we were able to show that every radical class is homomorphically closed (see the proof of II $\implies$ I in Theorem 2.1.5). However, the relation $\triangleleft$ is not transitive and thus it is more difficult to describe semisimple classes. We cannot prove that every semisimple class is hereditary (the dual of being homomorphically closed) because it is in general not true (for instance, in the variety of all not necessarily associative rings — see Section 5.1, in particular Theorem 5.1.6).

First let us connect radical and semisimple classes.

PROPOSITION 2.3.2. *If γ is a radical class, then*

$$\mathcal{S}\gamma = \{A \mid \gamma(A) = 0\}$$

is a semisimple class.

Proof: Take $A \in \mathcal{S}\gamma$, let B be a nonzero ideal of A, and assume B has no nonzero homomorphic image in $\mathcal{S}\gamma$. Since γ is a radical, we know that

$\gamma(B/\gamma(B)) = 0$ and therefore $B/\gamma(B)$ is in $\mathcal{S}\gamma$. Then $B/\gamma(B)$ must be 0 and $B = \gamma(B) \in \gamma$. Then $0 \neq B \subseteq \gamma(A)$ which contradicts the fact that $\gamma(A) = 0$. Thus (S1) is established.

If $A \notin \mathcal{S}\gamma$, then $\gamma(A) \neq 0$. Now no nonzero homomorphic image of $\gamma(A)$ can be in $\mathcal{S}\gamma$ because γ is homomorphically closed. This proves the contrapositive version of (S2) and proves the Proposition. □

The operator $\mathcal{S}$ is called the *semisimple operator*.

Now we obtain a Galois correspondence between radical and semisimple classes. It turns out that every semisimple class is the semisimple class of its upper radical and every radical class is the upper radical of its semisimple class. Thus it is consistent to call A γ-*semisimple* if $\gamma(A) = 0$.

THEOREM 2.3.3. *For any semisimple class σ and radical class γ we have*
$$\mathcal{S}\mathcal{U}\sigma = \sigma \quad \text{and} \quad \mathcal{U}\mathcal{S}\gamma = \gamma.$$

Proof: Take a ring $A \in \sigma$. By (S1) every nonzero ideal of A has a nonzero homomorphic image in σ. Hence A can have no nonzero ideals in $\mathcal{U}\sigma$, and thus $A \in \mathcal{S}\mathcal{U}\sigma$. Using similar reasoning with (S2) we get $\mathcal{S}\mathcal{U}\sigma \subseteq \sigma$, and thus they are equal.

Similarly, we use (R1) to get $\gamma \subseteq \mathcal{U}\mathcal{S}\gamma$ and (R2) to get $\mathcal{U}\mathcal{S}\gamma \subseteq \gamma$. □

In view of Theorem 2.3.3 we shall say that γ and σ are *corresponding radical and semisimple classes*, if $\gamma = \mathcal{U}\sigma$ and $\sigma = \mathcal{S}\gamma$. Theorem 2.3.3 tells us that if either one of these equalities holds, then so does the other.

When σ and γ are such a corresponding pair of classes, then $A/\gamma(A) \in \sigma$. Since we hope to discard "bad" rings, we would like $A/\gamma(A)$ to be well behaved. Nil (or Jacobson) semisimple artinian rings are direct sums of finitely many simple rings (cf. Corollary 4.5.9) and in fact this is why they were originally called semisimple.

We shall say that a class ϱ of rings is *closed under subdirect sums*, if $A_\lambda \in \varrho$, $\lambda \in \Lambda$, implies that also $\sum_{\text{subdirect}} A_\lambda \in \varrho$.

PROPOSITION 2.3.4. *Every semisimple class σ is closed under subdirect sums.*

Proof: Let B be any nonzero ideal of $\sum_{\text{subdirect}} A_\lambda$. Then there must exist an index λ such that the restriction of the homomorphism $\pi_\lambda : \sum_{\text{subdirect}} A_\lambda \twoheadrightarrow A_\lambda$ to B is nonzero, that is, $\pi_\lambda(B) \neq 0$. Now $\pi_\lambda(B) \triangleleft A_\lambda \in \sigma$, and by the regularity of σ, $\pi_\lambda(B)$ must have a nonzero homomorphic image in σ. But this is also a homomorphic image of B. We can apply (S2) to conclude that $\sum_{\text{subdirect}} A_\lambda$ must be in σ. □

The General Theory of Radicals

Radical classes satisfy the inductive property. The dual notion is the coinductive property: a class ϱ has the *coinductive property*, if for any descending chain of ideals $I_1 \supseteq \cdots \supseteq I_\lambda \supseteq \cdots$ of a ring A, such that A/I_λ is in ϱ for every λ, the ring $A/\cap I_\lambda$ is also in ϱ.

PROPOSITION 2.3.5. *Every semisimple class σ has the coinductive property.*

Proof: Since $A/\cap I_\lambda = \sum_{\text{subdirect}} (A/I_\lambda)$, the assertion follows immediately from Proposition 2.3.4. □

Notice that every class closed under subdirect sums has also the coinductive property. The converse is not true, for instance, any class of simple rings has trivially the coinductive property but it is obviously not closed under subdirect sums.

PROPOSITION 2.3.6. *Every semisimple class σ is closed under extensions.*

Proof: We take I and A/I in σ and we want to show that A is in σ. First we note that

$$(\mathcal{U}\sigma(A) + I)/I \cong \mathcal{U}\sigma(A)/(\mathcal{U}\sigma(A) \cap I) \in \mathcal{U}\sigma.$$

It is also clear that

$$(\mathcal{U}\sigma(A) + I)/I \triangleleft A/I \in \sigma = \mathcal{S}\mathcal{U}\sigma.$$

Therefore $(\mathcal{U}\sigma(A) + I)/I$ must be 0 and so $\mathcal{U}\sigma(A) \subseteq I$. Now by $\mathcal{U}\sigma(A) \triangleleft A$ also $\mathcal{U}\sigma(A) \triangleleft I$, and since $\mathcal{U}\sigma(A) \in \mathcal{U}\sigma$, we get $\mathcal{U}\sigma(A) \subseteq \mathcal{U}\sigma(I) = 0$. Thus $A \in \mathcal{S}\mathcal{U}\sigma = \sigma$.

Recall that the radical $\gamma(A)$ is a union of ideals. We dualize this and define

$$(A)\sigma = \cap(I_\lambda \triangleleft A \mid A/I_\lambda \in \sigma).$$

PROPOSITION 2.3.7. *If σ is any semisimple class, then $\mathcal{U}\sigma(A) = (A)\sigma$ for any ring A.*

Proof: Since $\mathcal{U}\sigma(A/\mathcal{U}\sigma(A)) = 0$, we have $A/\mathcal{U}\sigma(A) \in \mathcal{S}\mathcal{U}\sigma = \sigma$. Thus $(A)\sigma \subseteq \mathcal{U}\sigma(A)$. For the converse we consider

$$A/(A)\sigma = A/\cap I_\lambda \cong \sum_{\text{subdirect}} (A/I_\lambda).$$

Since each A/I_λ is in σ, Proposition 2.3.4 tells us that $A/(A)\sigma \in \sigma$. If $\mathcal{U}\sigma(A)$ is in fact bigger than $(A)\sigma$, then $\mathcal{U}\sigma(A)/(A)\sigma$ is a nonzero ideal of the ring $A/(A)\sigma \in \sigma$. But by $\mathcal{U}\sigma(A)/(A)\sigma \in \mathcal{U}\sigma$ we have

$$\mathcal{U}\sigma(A)/(A)\sigma \subseteq \mathcal{U}\sigma(A/(A)\sigma) = 0,$$

a contradiction. □

PROPOSITION 2.3.8. *If σ is a regular class, closed under subdirect sums and extensions, then the following conditions are equivalent:*
  (i) $\mathcal{U}\sigma(A) = (A)\sigma$ *for every ring A,*
  (ii) $((A)\sigma)\sigma = (A)\sigma$ *for every ring A,*
  (iii) $((A)\sigma)\sigma \triangleleft A$ *for every ring A.*

Proof: Let us begin by assuming that (i) holds. Since σ is regular, $\mathcal{U}\sigma$ is a radical class. Then by (i)

$$((A)\sigma)\sigma = \mathcal{U}\sigma((A)\sigma) = \mathcal{U}\sigma(\mathcal{U}\sigma(A)) = \mathcal{U}\sigma(A) = (A)\sigma$$

holds. Thus we have got (ii).

Now (ii) trivially implies (iii).

To complete the proof we assume (iii) and try to establish (i). Since $((A)\sigma)\sigma \triangleleft A$, we can talk about $A/((A)\sigma)\sigma$. Now $\sum_{\text{subdirect}} (A/I_\lambda)$ with each A/I_λ in σ is isomorphic to $A/\cap I_\lambda = A/(A)\sigma$ and this is in σ because σ is assumed to be closed under subdirect sums. Furthermore,

$$\frac{A/((A)\sigma)\sigma}{(A)\sigma/((A)\sigma)\sigma} \cong A/(A)\sigma \in \sigma.$$

Similarly, the bottom factor $(A)\sigma/((A)\sigma)\sigma$ is in σ. Then, since σ is also assumed to be closed under extensions, we have $A/((A)\sigma)\sigma \in \sigma$. Thus $((A)\sigma)\sigma$ is one of the I_λ in the definition of $(A)\sigma$ and we have

$$(A)\sigma \subseteq ((A)\sigma)\sigma \subseteq (A)\sigma.$$

This means $(A)\sigma \in \mathcal{U}\sigma$ and thus $(A)\sigma \subseteq \mathcal{U}\sigma(A)$. Then as in the proof of Proposition 2.3.7, second half, $\mathcal{U}\sigma(A) = (A)\sigma$ must hold, and we have (i). □

THEOREM 2.3.9 (van Leeuwen and Wiegandt [1], [2], and Mlitz [2]). *The following three conditions are equivalent:*
  I. σ *is a semisimple class.*
  II. σ *is regular, closed under subdirect sums and extensions, and $((A)\sigma)\sigma \triangleleft A$ for every ring A.*
  III. σ *is regular, closed under subdirect sums and $((A)\sigma)\sigma = (A)\sigma$ for every ring A.*
  IV. σ *is regular, closed under extensions, has the coinductive property and satisfies*
  (∗) *If $I \triangleleft A$ and I is minimal with respect to $A/I \in \sigma$, and if $K \triangleleft I$ and K is minimal with respect to $I/K \in \sigma$, then $K \triangleleft A$.*

Proof: Propositions 2.3.4, 2.3.6, 2.3.7 and 2.3.8 make it clear that I $\Longrightarrow$ II.

The implication II $\Longrightarrow$ III follows trivially from Proposition 2.3.8.

To go from III to IV, firstly we prove that σ is closed under extensions. Assume that $B \triangleleft A$, $B \in \sigma$ and $A/B \in \sigma$. By definition we have $(A)\sigma \subseteq B$. If $(A)\sigma = 0$, then we are done because σ is closed under subdirect sums. Therefore, suppose that $(A)\sigma \neq 0$. Since $B \in \sigma$ and σ is regular, the ideal $(A)\sigma$ of B has a nonzero homomorphic image $(A)\sigma/K$ in σ. Hence $((A)\sigma)\sigma \subseteq K \neq (A)\sigma$ contradicting the assumption. Thus σ is closed under extensions.

The coinductive property follows from subdirect sum closure of σ.

We still have to prove condition $(*)$. Since σ is closed under subdirect sums, the ideal $I = (A)\sigma$ is the unique ideal of A minimal with respect to $A/I \in \sigma$, and $K = (I)\sigma$ is the unique ideal of I minimal with respect to $I/K \in \sigma$. Then by III we have

$$K = (I)\sigma = ((A)\sigma)\sigma = (A)\sigma \triangleleft A,$$

and the validity of $(*)$ has been established.

Finally to go from IV to I we need only establish (S2). So we take a ring A such that every nonzero ideal of A has a nonzero homomorphic image in σ. Since σ has the coinductive property, we can use Zorn's Lemma to find an ideal I of A which is minimal with respect to $A/I \in \sigma$. If I is nonzero, then I must have a nonzero homomorphic image I/K in σ. With another application of Zorn's Lemma we can choose K so that it is minimal with respect to $I/K \in \sigma$. Then by $(*)$, $K \triangleleft A$ and

$$\frac{A/K}{I/K} \cong A/I \in \sigma.$$

By the extension closure of σ we find that A/K is in σ. But K is properly contained in I and this contradicts the minimality of I. Thus $I = 0$ must hold, and so $A \in \sigma$. This gives us (S2) and the Theorem is established. $\square$

THEOREM 2.3.10 (Mlitz [2]). *The classes γ and σ are corresponding radical and semisimple classes if and only if*

(i) $A \in \gamma$ and $A \twoheadrightarrow B \neq 0$ imply $B \notin \sigma$, that is, $\gamma \subseteq \mathcal{U}\sigma$,

(ii) $A \in \sigma$ and $0 \neq B \triangleleft A$ imply $B \notin \gamma$, that is, $\sigma \subseteq \mathcal{S}\gamma$.

(iii) *every ring A of the universal class $\mathbb{A}$ has an ideal B such that $B \in \gamma$ and $A/B \in \sigma$.*

Proof: If γ and σ are corresponding radical and semisimple classes, then the three conditions are clear (to get (iii) just take $B = \gamma(A)$).

Conversely, suppose we have classes γ and σ satisfying the three conditions. Let us consider a ring $A \in \mathcal{U}\sigma$. By (iii) A has an ideal $B \in \gamma$ such that $A/B \in \sigma$. Hence by $A \in \mathcal{U}\sigma$ we conclude that $A/B = 0$, and so $A = B \in \gamma$ holds, proving $\mathcal{U}\sigma \subseteq \gamma$. This and (i) gives $\gamma = \mathcal{U}\sigma$. A similar reasoning yields that $\sigma = \mathcal{S}\gamma$. Since $\sigma = \mathcal{S}\gamma = \mathcal{S}\mathcal{U}\sigma$, also $\sigma \subseteq \mathcal{S}\mathcal{U}\sigma$ holds and this is nothing but the regularity of the class σ. Hence $\gamma = \mathcal{U}\sigma$ is a radical class and $\sigma = \mathcal{S}\mathcal{U}\sigma = \mathcal{S}\gamma$ the corresponding semisimple class. □

Joubert and Schoeman [1] posed the question as whether the following conditions characterize corresponding radical and semisimple classes γ and σ:

(i) $\gamma \cap \sigma = \{0\}$,
(ii) γ satisfies (R1), that is, $\gamma \subseteq \mathcal{U}\mathcal{S}\gamma$,
(iii) σ satisfies (S1), that is, $\sigma \subseteq \mathcal{S}\mathcal{U}\sigma$ (regularity),
(iv) every ring A has an ideal B such that $B \in \gamma$ and $A/B \in \sigma$.

This problem was answered by Leavitt [12] in the negative: even *for associative rings* the above conditions do not characterize correspondig radical and semisimple classes.

Now it is perhaps good time to mention the notion of torsion theory. A pair (τ, φ) of classes of rings is called a *torsion theory*, if

(i) $\tau \cap \varphi = \{0\}$,
(ii) τ is homomorphically closed,
(iii) φ is hereditary,
(iv) every ring A has an ideal B such that $B \in \tau$ and $A/B \in \varphi$.

The class τ is called a *torsion class* and φ a *torsionfree class*. This idea was defined and investigated by Dickson [1] in abelian categories. In view of Mlitz's Theorem 2.3.10, in every torsion theory (τ, φ), τ and φ are corresponding radical and semisimple classes, and every torsionfree class is a hereditary semisimple class.

PROPOSITION 2.3.11. *A semisimple class σ is hereditary if and only if the corresponding radical class $\gamma = \mathcal{U}\sigma$ satisfies*

$(**)$ $\quad \gamma(I) \subseteq \gamma(A) \quad$ *for every* $\quad I \triangleleft A \in \mathbb{A}$.

Proof: If we have $(**)$, then for any $A \in \sigma$ and $I \triangleleft A$ we have $\gamma(I) \subseteq \gamma(A) = 0$, and so $I \in \sigma$. Thus σ is hereditary.

Conversely, suppose σ is hereditary. Then for $I \triangleleft A$ we have

$$((\gamma(I) + \gamma(A))/\gamma(A) \triangleleft (I + \gamma(A))/\gamma(A) \triangleleft A/\gamma(A) \in \sigma.$$

Hence

$$(I + \gamma(A))/\gamma(A) \in \sigma \quad \text{and} \quad ((\gamma(I) + \gamma(A))/\gamma(A) \in \sigma$$

because σ is hereditary. But this gives us

$$\gamma(I)/(\gamma(I) \cap \gamma(A)) \cong ((\gamma(I) + \gamma(A))/\gamma(A) \in \gamma \cap \sigma = \{0\}.$$

Thus $\gamma(I) \subseteq \gamma(A)$ as claimed. □

Let us mention that the hereditariness of σ does not imply in general the stronger condition

The General Theory of Radicals

$\gamma(I) \triangleleft A$ for every $I \triangleleft A$.

Such an example was constructed by Krempa [5] in a certain universal class of Lie algebras (cf. Gardner [18] 3.4.14 Example on p. 118).

Using Proposition 2.3.11 we get immediately

COROLLARY 2.3.12 (Leavitt and Wiegandt [1]). *The following are equivalent:*
  I. (τ, φ) *is a torsion theory,*
  II. τ *is a radical class with* (**) *and* $\varphi = \mathcal{S}\tau$,
  III. φ *is a hereditary semisimple class and* $\tau = \mathcal{U}\sigma$. □

Soon we shall see in Corollary 3.1.4 that *for associative rings* every semisimple class is hereditary, and so radical theory is the same as torsion theory. For not necessarily associative rings torsion theory will be discussed in Section 5.1.

To close this section, let us define, for every regular class ϱ, the class

$$\bar{\varrho} = \left\{ A \;\middle|\; \begin{array}{l} \text{every nonzero ideal of } A \text{ has a} \\ \text{nonzero homomorphic image in } \varrho \end{array} \right\}$$

PROPOSITION 2.3.13. *The class $\bar{\varrho}$ is the smallest semisimple class containing the regular class ϱ. Moreover, $\mathcal{U}\varrho = \mathcal{U}\bar{\varrho}$ and $\bar{\varrho} = \mathcal{S}\mathcal{U}\varrho$.*

The class $\bar{\varrho}$ is called the *semisimple closure* of ϱ.

Proof: Since ϱ is regular, $\varrho \subseteq \bar{\varrho}$. It is also clear that $\bar{\varrho}$ is itself regular. Now, if every nonzero ideal B of A has a nonzero homomorphic image C in $\bar{\varrho}$, then C has a nonzero homomorphic image D in ϱ, and thus also B has a nonzero homomorphic image, namely D, in ϱ. Thus A is in $\bar{\varrho}$, and hence $\bar{\varrho}$ satisfies also (S2), proving that $\bar{\varrho}$ is a semisimple class.

Let σ be any semisimple class that contains ϱ. If $A \in \bar{\varrho}$, then every nonzero ideal of A has a nonzero homomorphic image in $\varrho \subseteq \sigma$. By (S2) for σ, we have $A \in \sigma$. Thus $\bar{\varrho} \subseteq \sigma$ and $\bar{\varrho}$ is the smallest semisimple class containing ϱ.

From $\varrho \subseteq \bar{\varrho}$ we have $\mathcal{U}\bar{\varrho} \subseteq \mathcal{U}\varrho$. Conversely, take a ring $A \notin \mathcal{U}\bar{\varrho}$. Then A has a nonzero homomorphic image B in $\bar{\varrho}$, and then B has a nonzero homomorphic image C in ϱ. Thus also A has a nonzero homomorphic image in ϱ and so A cannot be in $\mathcal{U}\varrho$. Hence $\mathcal{U}\varrho \subseteq \mathcal{U}\bar{\varrho}$ follows and so they are then equal.

Finally, $\bar{\varrho} = \mathcal{S}\mathcal{U}\bar{\varrho} = \mathcal{S}\mathcal{U}\varrho$. □

Chapter III
Radical Theory for Associative Rings

We can of course do much more, if we restrict ourselves to associative rings. The purpose of this chapter is to develop the general radical theory for asscociative rings. We shall introduce and study hereditary, supernilpotent, special, supplementing, subidempotent, normal and A-radicals. Doing so, we get acquainted with many more important and interesting radicals as motivating examples for the general theory. We shall carefully examine lower radical constructions and semisimple classes, and we shall obtain many results that are true for associative rings, but not true in general.

3.1. Semisimple classes of associative rings

We shall now restrict our universal class to be the class of all associative rings. This will allow us to get deeper characterizations of semisimple classes and thus improve on Theorem 2.3.9. It will also allow us to overcome the problem of the relation $\triangleleft$ not being transitive. We will then obtain a complete duality between the characterizations of semisimple and radical classes.

LEMMA 3.1.1. *Assume that $K \triangleleft I \triangleleft A$ and $a \in A$. Then*
(i) $aK + K \triangleleft I$,
(ii) $(aK + K)^2 \subseteq K$,
(iii) *the mapping $f \colon K \to (aK + K)/K$ defined by $f(x) = ax + K$ for any $x \in K$ is a surjective homomorphism,*
(iv) $\ker f \triangleleft I$.

Proof: A straightforward calculation establishes (i) and (ii). As for f, it is obvious that it preserves addition. As for multiplication, we note that for any $x, y \in K$

$$f(xy) = axy + K \in (aKK + K)/K \subseteq (IK + K)/K = 0.$$

On the other hand

$$f(x)f(y) = axay + K \in (aKaK + K)/K \subseteq (IK + K)/K = 0,$$

and so also the multiplication is preserved. Thus f is a homomorphism. Clearly f is surjective and we have (iii).

To see (iv) we notice that

$$\ker f = \{x \in K \mid ax \in K\}.$$

Take any $z \in I$ and any $k \in \ker f$. Then

$$a(zk) = (az)k \in IK \subseteq K.$$

Also

$$a(kz) = (ak)z \in KI \subseteq K$$

because $k \in \ker f$. Hence $\ker f \triangleleft I$, and we have (iv). □

The next result was, historically, the first that overcame the obstacle of the relation $\triangleleft$ not being transitive.

ADS-THEOREM 3.1.2 (Anderson, Divinsky and Suliński [1]). *For any radical γ and any ring A, if $I \triangleleft A$, then $\gamma(I) \triangleleft A$.*

Proof: We know that $\gamma(I) \triangleleft I \triangleleft A$. Take any $a \in A$ and consider $a\gamma(I) + \gamma(I)$. Lemma 3.1.1 tells us that

$$f: \gamma(I) \to (a\gamma(I) + \gamma(I))/\gamma(I)$$

is a surjective homomorphism. Since γ is homomorphically closed and $\gamma(I) \in \gamma$, the image on the right hand side must be in γ. Since γ is closed under extensions, the ring $a\gamma(I) + \gamma(I)$ must be in γ. It is also an ideal of I by Lemma 3.1.1 and thus it must be in $\gamma(I)$. Therefore $a\gamma(I) \subseteq \gamma(I)$.

Similarly we can get $\gamma(I)a \subseteq \gamma(I)$ and thus $\gamma(I) \triangleleft A$. □

COROLLARY 3.1.3. *For any radical γ and any ring A, if $I \triangleleft A$, then $\gamma(I) \subseteq \gamma(A)$.* □

This Corollary together with 2.3.12 gives us

COROLLARY 3.1.4. *Every semisimple class is hereditary and every corresponding radical and semisimple class form a torsion theory.* □

The ADS-Theorem was a breakthrough in radical theory for associative rings, but it took eleven more years before a satisfactory description of semisimple classes was obtained by Sands [3].

LEMMA 3.1.5 (van Leeuwen, Roos and Wiegandt [1]). *If σ is a class which is hereditary, closed under subdirect sums and extensions, then $((A)\sigma)\sigma \triangleleft A$ for every ring A.*

Proof: Let us call $(A)\sigma = I$ and $((A)\sigma)\sigma = K$. Then we have $K \triangleleft I \triangleleft A$, and Lemma 3.1.1 tells us that

$$K/\ker f \cong (aK + K)/K \triangleleft I/K = I/(I)\sigma$$

for any $a \in A$. Since σ is closed under subdirect sums, it is clear that $I/(I)\sigma \in \sigma$. Since σ is hereditary, also $K/\ker f \in \sigma$. By Lemma 3.1.1 $\ker f \triangleleft I$, so we can work with $I/\ker f$ getting

$$\frac{I/\ker f}{K/\ker f} \cong I/K \in \sigma.$$

Hence, using that σ is closed under extensions, we conclude $I/\ker f \in \sigma$. But then $K = (I)\sigma \subseteq \ker f$, and thus $K = \ker f$. Hence $aK + K = K$, that is, $aK \subseteq K$ for any $a \in A$. Similarly $Ka \subseteq K$, and thus $((A)\sigma)\sigma = K \triangleleft A$.□

Lemma 3.1.5 together with Theorem 2.3.9 immediately gives us

THEOREM 3.1.6 (Sands [3] and van Leeuwen, Roos and Wiegandt [1]). *A class σ of rings is a semisimple class if and only if*
   (a) *σ is hereditary,*
   (b) *σ is closed under subdirect sums,*
   (c̄) *σ is closed under extensions.* □

This characterization is dual to the one for radical classes given in Proposition 2.1.2. Sands, by refining these methods, has suceeded in characterizing semisimple classes by weaker looking conditions, ones that are categorically dual to those given in Theorem 2.1.5.III for radical classes. This gives us a complete duality between radical and semisimple classes. To achieve this, we first need

LEMMA 3.1.7. *Let σ be a class of rings which is regular, has the coinductive property, and is closed under extensions. If $I \triangleleft A \in \sigma$ and $I^2 = 0$, then also $I \in \sigma$.*

Proof: First we shall prove the Lemma in the more special case when $A^2 = 0$. Since σ is coinductive, we can use Zorn's Lemma to find an ideal J of I for which $I/J \in \sigma$ and which is minimal for this property. We want to show that $J = 0$ and that $I \in \sigma$. Since we are assuming here that $A^2 = 0$, any additive subgroup of A is an ideal, and in particular, $J \triangleleft A$. If $J \neq 0$, then since $A \in \sigma$ and σ is regular, there is a nonzero homomorphic image J/K in σ, and $K \triangleleft A$. Then by

$$\frac{I/K}{J/K} \cong I/J \in \sigma$$

and by the extension closure of σ we get $I/K \in \sigma$. But $K \subseteq J$ and J is minimal. Thus $K = J$, a contradiction. Hence $J = 0$ and $I \in \sigma$.

Now we must tackle the case when $A^2 \neq 0$. As above we choose an ideal J of I minimal with respect to $I/J \in \sigma$. We want to show $J = 0$. If J happens to be an ideal of A, then the above argument holds (note that $K \triangleleft I$ because $I^2 = 0$) and $I \in \sigma$. The difficult part occurs when J is not an ideal of A. Assume then that there is an element $a \in A$ such that $aJ \not\subseteq J$.

We shall now use Lemma 3.1.1 where $f(x) = ax + J$ for all $x \in J$. We have then
$$J/\ker f \cong (aJ + J)/J \triangleleft I/J \in \sigma.$$
Since $(I/J)^2 = 0$, we can apply our special case (with $A^2 = 0$) to this situation and conclude that
$$J/\ker f \cong (aJ + J)/J \in \sigma.$$
Applying Lemma 3.1.1 we see that $\ker f \triangleleft I$, so it follows that
$$\frac{I/\ker f}{J/\ker f} \cong I/J \in \sigma.$$
Since σ is closed under extensions, we get $I/\ker f \in \sigma$. Since $\ker f \subseteq J$ and J is minimal, we must have $\ker f = J$, and $(aJ + J)/J = 0$, that is, $aJ \subseteq J$, a contradiction. Similarly we can prove that $Ja \subseteq J$. Thus $J \triangleleft A$ and in all cases $I \in \sigma$. □

THEOREM 3.1.8 (Sands [5]). *A class σ of rings is a semisimple class if and only if*
(S1)   *σ is regular,*
($\bar{\text{b}}$)   *σ has the coinductive property,*
($\bar{\text{c}}$)   *σ is closed under extensions.*

Remark. It is interesting to compare Theorems 3.1.6 and 3.1.8. Since being hereditary is stronger than just being regular, and since being closed under subdirect sums is stronger than just having the coinductive property, it certainly seems that (S1), ($\bar{\text{b}}$) and ($\bar{\text{c}}$) are weaker than the conditions of Theorem 3.1.6. And yet taken together, they do give us a full semisimple class.

Proof: If σ is a semisimple class, then by Theorem 3.1.6 we certainly have (S1), ($\bar{\text{b}}$) and ($\bar{\text{c}}$). Conversely, if we have (S1), ($\bar{\text{b}}$) and ($\bar{\text{c}}$) and we can prove that σ satisfies also condition (∗) of Theorem 2.3.9, then we will know that σ is a semisimple class. To this end we assume that $I \triangleleft A$, $A/I \in \sigma$ and I is minimal for this property, and that $K \triangleleft I$, $I/K \in \sigma$ and K is minimal for this property. We must show that $K \triangleleft A$. Suppose that there is an element $a \in A$ such that $aK \not\subseteq K$. Then as before, by Lemma 3.1.1 we have
$$K/\ker f \cong (aK + K)/K \triangleleft I/K \in \sigma$$
where $f(x) = ax + K$ for every $x \in K$. Since $(aK + K)^2 \subseteq K$ by Lemma 3.1.1, we may apply Lemma 3.1.7 for $(aK + K)/K \triangleleft I/K \in \sigma$, yielding $(aK + K)/K \in \sigma$. Again by Lemma 3.1.1 we know that $\ker f \triangleleft I$, and so
$$\frac{I/\ker f}{K/\ker f} \cong I/K \in \sigma.$$

Since σ is closed under extensions, we conclude that $I/\ker f \in \sigma$. Since $\ker f \subseteq K$ and K is minimal, we must have $\ker f = K$ and thus $(aK + K)/K = 0$ and $aK \subseteq K$. Similarly we get $Ka \subseteq K$. This proves $K \triangleleft A$. □

Certainly much ingenuity is involved in the proof of Theorem 3.1.8. Theorem 3.1.8 implies trivially Theorem 3.1.6, and by Theorem 2.3.9 also Lemma 3.1.5.

A subring B of a ring A is said to be an *accessible* subring of A and is denoted by $B \succ\!\!-A$, if there exists a finite sequence $C_1, \ldots, C_n$ of subrings of A such that
$$B = C_1 \triangleleft C_2 \triangleleft \cdots \triangleleft C_n = A$$
where $C_i \triangleleft C_{i+1}$ but C_i need not be an ideal of C_{i+2} or of A. In this case we shall say that B is an *n-accessible subring* of A. Ideals of A will, in this notation, be 2-accessible subrings of A.

We can now improve conditions (R1), (R2), (S1) and (S2) using accessible subrings.

THEOREM 3.1.9 (Divinsky [3]). γ *is a radical class if and only if*

(R1°) *if $A \in \gamma$, then every nonzero homomorphic image of A has a nonzero accessible subring in γ,*

(R2°) *if A is a ring and every nonzero homomorphic image of A has a nonzero accessible subring in γ, then also A is in γ.*

Proof: If γ is a radical class, then (R1) of Theorem 2.1.5 readily implies (R1°). To get (R2°), take A such that every nonzero homomorphic image of A has a nonzero accessible subring in γ. If A is not in γ, then $A/\gamma(A) \neq 0$. Now it must have a nonzero accessible subring, say $C/\gamma(A)$ in γ. Since the semisimple class $\mathcal{S}\gamma$ is hereditary, $C/\gamma(A)$ must be in $\mathcal{S}\gamma$ by $A/\gamma(A) \in \mathcal{S}\gamma$. Thus $C/\gamma(A) = 0$, a contradiction. Therefore, we have (R2°).

Conversely, suppose γ satisfies (R1°) and (R2°). To show γ is a radical class, we will establish (R1) and (R2). Now (R2°) easily gives us (R2). Since the relation $\twoheadrightarrow$ is transitive, (R1°) and (R2°) tell us that γ is homomorphically closed (see II $\Longrightarrow$ I in the proof of Theorem 2.1.5), and thus (R1) holds, proving that γ is a radical class. □

THEOREM 3.1.10 (Divinsky [3]). σ *is a semisimple class if and only if*

(S1°) *if $A \in \sigma$, then every nonzero accessible subring of A has a nonzero homomorphic image in σ,*

(S2°) *if A is a ring and every nonzero accessible subring of A has a nonzero homomorphic image in σ, then A itself is in σ.*

Proof: If σ is a semisimple class, then the hereditariness of σ gives us (S1°). If (S2°) does not hold, then there is a ring $A \notin \sigma$ such that every nonzero accessible subring of A has a nonzero homomorphic image in σ. Then $\mathcal{U}\sigma(A)$ is a nonzero ideal of A and it must have a nonzero homomorphic image in σ. But this is impossible, and thus (S2°) holds.

Conversely, suppose that σ satisfies (S1°) and (S2°). It is clear that (S1°) gives (S1). Since σ has (S1), that is, σ is regular, we have $\sigma \subseteq \mathcal{SU}\sigma$ and $\mathcal{U}\sigma$ is a radical class. Let A be any ring in $\mathcal{SU}\sigma$, that is, $\mathcal{U}\sigma(A) = 0$. We want to show that A is in σ, and we shall use (S2°) for this purpose. Let B be any nonzero accessible subring of A, say

$$B = C_1 \triangleleft C_2 \triangleleft \cdots \triangleleft C_n = A.$$

Suppose that B has no nonzero homomorphic image in σ. Then $B \in \mathcal{U}\sigma$ and thus $B \subseteq \mathcal{U}\sigma(C_2)$. By induction we get

$$0 \neq B \subseteq \mathcal{U}\sigma(C_{n-1}) \subseteq \mathcal{U}\sigma(A) = 0,$$

a contradiction. Therefore, every nonzero accessible subring of A has a nonzero homomorphic image in σ. Then by (S2°) A must be in σ. Thus $\sigma = \mathcal{SU}\sigma$ and this is certainly a semisimple class. $\square$

Note that the hereditariness of semisimple classes was used only in proving the implications (R1), (R2) $\implies$ (R2°) and (S1), (S2) $\implies$ (S1°).

The characterizations that we now have of semisimple classes, allow us to construct semisimple classes by a procedure dual to the Tangeman–Kreiling construction for lower radicals. We begin with an arbitrary class ϱ of rings and define

$$\varrho_1 = \{A \mid A \text{ is an accessible subring of a ring in } \varrho\}.$$

Then we proceed inductively. If $\lambda > 1$ is not a limit ordinal, we define

$$\varrho_\lambda = \{A \mid A \text{ has an ideal } K \text{ such that } K \in \varrho_{\lambda-1} \text{ and } A/K \in \varrho_{\lambda-1}\}.$$

If λ is a limit ordinal, we define

$$\varrho_\lambda = \left\{ A \,\middle|\, \begin{array}{l} A \text{ has a descending chain } \{K_\iota\} \text{ of ideals with} \\ \cap K_\iota = 0 \text{ and with each } A/K_\iota \in \varrho_\mu \text{ for some } \mu < \lambda \end{array} \right\}.$$

THEOREM 3.1.11 (Sands [3]). *The class $\sigma = \cup \varrho_\lambda$ is the smallest semisimple class containing ϱ. Moreover $\sigma = \mathcal{SU}\varrho_1$.*

Proof: We will use Theorem 3.1.8 and show that σ is hereditary, is closed under extensions and has the coinductive property.

Certainly ϱ_1 is hereditary. We proceed by induction. Suppose λ is not a limit ordinal, and take $0 \neq I \triangleleft A \in \varrho_\lambda$. We want to show that also $I \in \varrho_\lambda$. Now A has an ideal K in $\varrho_{\lambda-1}$ such that $A/K \in \varrho_{\lambda-1}$. Since $I \cap K \triangleleft K$ and since by induction we are taking $\varrho_{\lambda-1}$ to be hereditary, it follows that $I \cap K \in \varrho_{\lambda-1}$. Furthermore, also

$$I/(I \cap K) \cong (I + K)/K \triangleleft A/K \in \varrho_{\lambda-1},$$

and therefore by definition, $I \in \varrho_\lambda$.

To complete the induction we must finally consider the case when λ is a limit ordinal. Then for any $A \in \varrho_\lambda$, A has a descending chain $\{K_\iota\}$ of ideals with $\cap K_\iota = 0$ and each A/K_ι in ϱ_μ for some $\mu < \lambda$. For each index ι there is an ordinal $\mu < \lambda$ such that

$$I/(I \cap K_\iota) \cong (I + K_\iota)/K_\iota \lhd A/K_\iota \in \varrho_\mu.$$

Then by induction, $I/(I \cap K_\iota) \in \varrho_\mu$. Then I has a descending chain $\{I \cap K_\iota\}$ of ideals with $\cap (I \cap K_\iota) = 0$ and each $I/(I \cap K_\iota) \in \varrho_\mu$ for some $\mu < \lambda$. Thus by definition $I \in \varrho_\lambda$. This completes the induction and all the ϱ_λ are hereditary. Thus also $\sigma = \cup \varrho_\lambda$ is hereditary.

For extension closure, suppose that both I and A/I are in σ. Then there must exist an ordinal λ such that both I and A/I are in ϱ_λ and thus $A \in \varrho_{\lambda+1} \subseteq \sigma$. Therefore σ is closed under extensions.

For the coinductive property, let $\{K_\iota\}$ be a descending chain of ideals of A with $\cap K_\iota = 0$ and each $A/K_\iota \in \sigma$. Then $A/K_\iota \in \varrho_{\mu_\iota}$ for some ordinal μ_ι. Since the indices ι form a set, there exists a limit ordinal $\lambda > \mu_\iota$ for all ι and thus $A \in \varrho_\lambda \subseteq \sigma$. Therefore σ has the coinductive property. Thus σ is a semisimple class and it contains ϱ.

It is clear that if ϱ is contained in any other semisimple class, then so are all the ϱ_λ. Therefore σ is the smallest semisimple class containing ϱ.

Finally, $\varrho_1 \subseteq \sigma$ implies $\mathcal{SU}\varrho_1 \subseteq \mathcal{SU}\sigma = \sigma$ and by the minimality of σ we must have equality. □

Sometimes, this construction is more useful than the one given in Proposition 2.3.13.

Semisimple classes were studied also by Majumdar [1].

The results of this section also hold for alternative rings. The proofs are of course more involved (Anderson, Divinsky and Suliński [1], van Leeuwen, Roos and Wiegandt [1], Anderson and Wiegandt [3]). An alternative approach to these problems which includes beside alternative rings also Jordan algebras and other algebraic structures, will be treated in Sections 5.2 and 5.3.

3.2. Hereditary radicals and their semisimple classes

We have been concerned with semisimple classes being hereditary. It often happens that radical classes are hereditary and we call such radicals *hereditary radicals*.

EXAMPLE 3.2.1. (i) It is clear that *Köthe's nil radical is hereditary*.

(ii) *The Jacobson radical is hereditary.* To see this, let A be a Jacobson radical ring and $I \lhd A$. For any $a \in I$, $a \in A$ too, and so there exists an $x \in A$ such that $a + x - ax = 0$. However, $x = -a + ax \in I$, and thus I itself is a Jacobson radical ring.

(iii) It is obvious that *the Levitzki radical is hereditary*.

(iv) *The von Neumann regular radical class is hereditary.* To see this, take a regular ring A and let $I \triangleleft A$. For any $a \in I$ also $a \in A$, and thus there exists an $x \in A$ such that $a = axa$. Then

$$a = axa = ax(axa) = a(xax)a.$$

Now $xax \in I$, and so I itself is regular.

In fact all the established radicals we have introduced up to now, are hereditary. But this is not true in general.

EXAMPLES 3.2.2. (i) *The subclass τ_D of the class $\mathcal{Z}$ of all zero-rings with underlying additive groups being divisible torsion groups, is a radical class which is not hereditary.* It is clear that τ_D is homomorphically closed, has the inductive property and is closed under extensions (because a divisible group is a direct summand in each of its overgroups). Therefore, by Theorem 2.1.4 τ_D is a radical class. Now the zero-ring $Z(p^\infty)$ on the additive quasicyclic group $C(p^\infty)$ is in τ_D but its ideal $Z(p)$ is not in τ_D. Thus τ_D is not a hereditary radical.

(ii) *The class $\mathcal{I}$ of all idempotent rings is a radical class which is not hereditary.* To see this, we begin with the class $\mathcal{Z}$ of all zero-rings. $\mathcal{Z}$ is hereditary, and so $\mathcal{UZ}$ is a radical class. It is clear that $A \in \mathcal{UZ}$ if and only if $A^2 = A$, that is, $A \in \mathcal{I}$. Now the ring $\mathbb{Z}$ of integers is in $\mathcal{I}$ but its ideal $2\mathbb{Z}$ is not in $\mathcal{I}$ (in fact, $2\mathbb{Z}/4\mathbb{Z} \in \mathcal{Z}$).

PROPOSITION 3.2.3. *A radical class γ is hereditary if and only if $I \cap \gamma(A) \subseteq \gamma(I)$ for every ideal I of every ring A.*

Proof: If $I \triangleleft A$ and if γ is hereditary, then

$$I \cap \gamma(A) \triangleleft \gamma(A) \in \gamma$$

yields $I \cap \gamma(A) \in \gamma$. Therefore, by $I \cap \gamma(A) \triangleleft I$ also $I \cap \gamma(A) \subseteq \gamma(I)$ holds.

Conversely, suppose the stated condition holds and that $I \triangleleft A \in \gamma$ and $I \cap \gamma(A) \subseteq \gamma(I)$. Then we have

$$I = I \cap A = I \cap \gamma(A) \subseteq \gamma(I) \subseteq I,$$

proving $I \in \gamma$. Thus γ is hereditary. $\square$

This result together with Corollary 3.1.3 gives us

COROLLARY 3.2.4. *A radical class γ is hereditary if and only if $\gamma(I) = I \cap \gamma(A)$ for every ideal I of every ring A.* $\square$

An ideal I of a ring A is said to be an *essential* (or *large*) *ideal* of A, if $I \cap K \neq 0$ for every nonzero ideal K of A. We shall denote such an ideal I by $I \triangleleft \cdot A$. Note that *in a subdirectly irreducible ring every nonzero ideal, in particular its heart, is an essential ideal.* On the other hand, a ring with unity is never essential in any proper overring.

LEMMA 3.2.5. *For any ideal I of a ring A, there exists an ideal K of A such that $I \cap K = 0$, K is maximal with respect to this property, and*

$$I \cong (I+K)/K \triangleleft \cdot A/K.$$

Proof: Zorn's Lemma tells us that a K being maximal with respect to having zero intersection with I exists, and of course

$$I \cong (I+K)/K \triangleleft A/K.$$

To see that $(I+K)/K$ is essential in A/K, let L/K be any nonzero ideal of A/K. Then L is bigger than K and by the maximality of K we know that $I \cap L \neq 0$, and so $I \cap L \not\subseteq K$. Hence we have

$$0 \neq (K + (I \cap L))/K = (K+I)/K \cap L/K$$

by the modularity law. Thus $(K+I)/K \triangleleft \cdot A/K$. □

We shall say that a class ϱ of rings is *closed under essential extensions*, if

(λ) $I \triangleleft \cdot A$ and $I \in \varrho$ imply $A \in \varrho$.

PROPOSITION 3.2.6 (Armendariz [1], Ryabukhin [2]). *Let γ and σ be corresponding radical and semisimple classes. Then γ is hereditary if and only if σ is closed under essential extensions.*

Proof: Assume first that γ is hereditary, and let $I \triangleleft \cdot A$ with $I \in \sigma$. We want to show that $A \in \sigma$. Proposition 3.2.3 tells us that $I \cap \gamma(A) \subseteq \gamma(I) = 0$. Then $\gamma(A) = 0$ must hold because I is essential in A. Thus $A \in \sigma$.

Conversely, suppose that γ is not hereditary. Then there is a ring $A \in \gamma$ and an ideal I of A which is not in γ. Hence $\gamma(I) \neq I$. By the ADS-Theorem 3.1.2 we can consider

$$0 \neq I/\gamma(I) \triangleleft A/\gamma(I).$$

Let us call $A/\gamma(I) = B$, and note that $B \in \gamma$. Let us call $I/\gamma(I) = L$, and note that $L \in \sigma$. Using Lemma 3.2.5 there exists an ideal K of B such that $L \cap K = 0$ and K is maximal with respect to this property. Furthermore,

$$L \cong (L+K)/K \triangleleft \cdot B/K \in \gamma.$$

Thus σ is not closed under essential extensions, and this proves the converse. □

THEOREM 3.2.7. *A class σ of rings is the semisimple class of a hereditary radical if and only if*

(S1) σ is regular,
(b) σ is closed under subdirect sums,
(λ) σ is closed under essential extensions.

Proof: If σ is the semisimple class of a hereditary radical, then Theorem 3.1.6 gives us (S1) and (b) and Proposition 3.2.6 gives us (λ).

Conversely, suppose σ satisfies (S1), (b), (λ). Then from Theorem 3.1.8, if we can show that σ is closed under extensions, we will know that σ is a semisimple class. To this end, take I, A/I both in σ. Using Lemma 3.2.5, we take an ideal K of A which is maximal with respect to having zero intersection with I. We also know that

$$I \cong (I+K)/K \triangleleft \cdot A/K.$$

Since $I \in \sigma$, (λ) tells us that also $A/K \in \sigma$. Since $I \cap K = 0$, A is a subdirect sum of A/I and A/K both in σ. By (b) A must be in σ, and thus σ is a semisimple class. Its corresponding radical is hereditary by Proposition 3.2.6 since σ has (λ). This ends the proof. □

COROLLARY 3.2.8. *A class σ of rings is the semisimple class of a hereditary radical if and only if σ is hereditary, and is closed under subdirect sums and essential extensions.* □

This Corollary used to be called *van Leeuwen's Theorem* (see van Leeuwen [2] and van Leeuwen, Roos and Wiegandt [1]).

Condition (b) in Theorem 3.2.7 cannot be weakened to the coinductive property. To see this, let $\mathcal{M}$ be the class of all simple rings, with unity element (with the trivial ring 0). Then $\mathcal{M}$ is hereditary, has the coinductive property, and $\mathcal{M}$ is closed under essential extensions (because there aren't any). However, $\mathcal{M}$ is not closed under subdirect sums because $A \oplus A$ is not in $\mathcal{M}$, if A is simple. Thus $\mathcal{M}$ is not the semisimple class of a hereditary radical; in fact, the smallest semisimple class containing $\mathcal{M}$ is $\mathcal{SUM}$, the semisimple class of the Brown–McCoy radical $\mathcal{G} = \mathcal{UM}$ (Example 2.2.4).

To get more hereditary radicals, we shall show that the lower radical of a hereditary class is again hereditary. A class δ of rings is said to be *left hereditary (right hereditary, strongly hereditary)*, if $A \in \delta$ implies $L \in \delta$ for every left ideal (right ideal, subring, respectively) L of A.

THEOREM 3.2.9 (Puczyłowski [4]). *If γ is a radical class, then the largest hereditary subclass*

$$\widehat{\gamma} = \{A \mid \text{every accessible subring of } A \text{ is in } \gamma\}$$

of γ is a radical class. Also the largest left hereditary, right hereditary and strongly hereditary subclasses of γ are radical classes.

Proof: First, let us observe that in an arbitrarily given class γ of rings there exists a largest hereditary (left here

Radical Theory for Associative Rings

and it is just the union of all hereditary (left hereditary, strongly hereditary, respectively) subclass of γ. In particular, $\widehat{\gamma}$ is the largest hereditary subclass of γ.

It is clear that $\widehat{\gamma}$ is contained in γ and is homomorphically closed. Thus, to show that $\widehat{\gamma}$ is a radical class, we need only show that it has the inductive property and is closed under extensions (Theorem 2.1.4). For the inductive property, let $I_1 \subseteq \ldots \subseteq I_\lambda \subseteq \ldots$ be an ascending chain of ideals of A such that $\cup I_\lambda = A$ and each I_λ in $\widehat{\gamma}$. Let $K \rhd\!\!-A$. Then $K \cap I_\lambda \in \widehat{\gamma}$ for every λ because $K \cap I_\lambda \rhd\!\!- I_\lambda \in \widehat{\gamma}$. Therefore $\cup(K \cap I_\lambda) \in \gamma$ because $\widehat{\gamma} \subseteq \gamma$ and γ is a radical class. But $\cup(K \cap I_\lambda) = K$. Thus $A \in \widehat{\gamma}$ holds proving that $\widehat{\gamma}$ has the inductive property.

For extension closure, take I and A/I in $\widehat{\gamma}$. Let K be any accessible subring of A. Then

$$K/(K \cap I) \cong (K+I)/I \rhd\!\!- A/I \in \widehat{\gamma}.$$

Therefore, $K/(K \cap I) \in \gamma$. Furthermore, $K \cap I \rhd\!\!- I \in \widehat{\gamma}$ implies $K \cap I \in \gamma$. Since γ is a radical class, it is closed under extensions, and so we conclude that $K \in \gamma$. Thus $A \in \widehat{\gamma}$ follows, proving that $\widehat{\gamma}$ is closed under extensions.

Using a similar reasoning we get the same result for the largest left hereditary, right hereditary and strongly hereditary subclasses of the radical class γ. □

COROLLARY 3.2.10 (Hoffman and Leavitt [1], Ryabukhin [2] and Watters [1]). *If δ is a hereditary class of rings, then the lower radical class $\gamma = \mathcal{L}\delta$ is hereditary.*

Proof: It is clear that

$$\delta = \widehat{\delta} \subseteq \widehat{\mathcal{L}\delta} \subseteq \mathcal{L}\delta.$$

Theorem 3.2.9 tells us that $\widehat{\mathcal{L}\delta}$ is a radical class, and since $\mathcal{L}\delta$ is the smallest radical class containing δ, we must have $\widehat{\mathcal{L}\delta} = \mathcal{L}\delta$, that is, $\mathcal{L}\delta$ is hereditary. □

Similarly we get (cf. Tangeman and Kreiling [1], Rossa [1], Rossa and Tangeman [1]):

COROLLARY 3.2.11. *If δ is a left, right or strongly hereditary class of rings, then so is the lower radical $\mathcal{L}\delta$.* □

Remark. We did not use associativity for Theorem 3.2.9 and its two Corollaries, and thus they hold true for not necessarily associative rings.

If we begin with any hereditary class, we can thus construct a hereditary radical, and this gives us a good supply of hereditary (as well as left-hereditary, right hereditary and strongly hereditary) radicals.

EXAMPLE 3.2.12. *The Baer radical is strongly hereditary*, and thus in particular, hereditary. This is obvious by Example 2.2.2 and Corollary 3.2.11.

EXAMPLES 3.2.13. (i) We do not need Corollary 3.2.11 to see that the *Levitzki and Köthe's nil radicals are strongly hereditary.*

(ii) *The Jacobson radical is left and right hereditary but not strongly hereditary.* The left and right hereditariness follows from the proof given in Example 3.2.1.(ii). Further, take the set

$$J = \left\{ \frac{2x}{2y+1} \;\middle|\; x, y \in \mathbb{Z} \text{ and } (2x, 2y+1) = 1 \right\},$$

that is, the set of all rationals of even numerator and odd denominator. J is obviously a ring. For any rational $a = \frac{2x}{2y+1} \in J$ the equation $a \circ b = a + b - ab = 0$ has a solution

$$b = \frac{a}{a-1} = \frac{2x}{2(x-y-1)+1} \in J.$$

Hence $(J, \circ)$ is a group, that is, J *is a Jacobson radical ring.* Nevertheless, the ring $2\mathbb{Z}$ of even integers is clearly a subring of J, and $2\mathbb{Z}$ is Jacobson semisimple because $b = \frac{a}{a-1} \in 2\mathbb{Z}$ if and only if $b = a = 0$ or 2, and the subset $\{0, 2\}$ is not an ideal of $2\mathbb{Z}$. It is important to observe that J has clearly no nonzero nilpotent element, and so J *is a nil semisimple ring.*

Finally we shall have a look at the hereditariness of the upper radical construction.

PROPOSITION 3.2.14 (Enersen and Leavitt [1] and Rashid and Wiegandt [1]). *Let ϱ be a regular class of rings. The upper radical $\mathcal{U}\varrho$ is hereditary if and only if ϱ satisfies condition*

(H) *if $0 \neq I \triangleleft A$ and there is a $I \twoheadrightarrow C$ such that $0 \neq C \in \varrho$, then there exists an $A \twoheadrightarrow B$ such that $0 \neq B \in \varrho$ (or $B \in \mathcal{SU}\varrho$).*

The proof is straightforward, and is left to the reader. □

Though condition (H) is a necessary and sufficient condition for the hereditariness of the upper radical, it is not easy to test. We could therefore use some other insight.

PROPOSITION 3.2.15. *If $K \triangleleft I \triangleleft A$ and I/K is a ring with unity element, then $K \triangleleft A$.*

Proof: Let $e + K$ be the unity of I/K. Take any $a \in A$ and any $k \in K$. Then we have $ak \in I$. Therefore

$$ak + K = (e + K)(ak + K) = e(ak) + K = (ea)k + K = K.$$

Hence $ak \in K$ for all $a \in A$ and $k \in K$. Similarly we get also $ka \in K$, and so $K \triangleleft A$. □

PROPOSITION 3.2.16. *If ϱ is any regular class of rings with unity element, then the upper radical $\gamma = \mathcal{U}\varrho$ is hereditary.*

Proof: We must show that ϱ satisfies (H) of Proposition 3.2.14. Suppose that A has an ideal I which has a nonzero homomorphic image $C = I/K \in \varrho$. Then I/K has a unity and by Proposition 3.2.15, K is an ideal of A. Since I/K has a unity, it must be a direct summand of A/K. Then A/K can be mapped homomorphically onto I/K and this is in ϱ and nonzero. Therefore condition (H) holds and $\mathcal{U}\varrho$ is hereditary. $\square$

EXAMPLE 3.2.17. *The Brown–McCoy radical is hereditary* by Proposition 3.2.16 (see Example 2.2.4).

EXAMPLE 3.2.18. The upper radical $\mathcal{F}$ of all *division rings is called the Thierrin radical* [2], *and is hereditary* by Proposition 3.2.16.

Finally, we characterize hereditary radicals in terms of Hoehnke radicals.

THEOREM 3.2.19. *A Hoehnke radical γ satisfies condition*

$$(*) \qquad \gamma(I) = I \cap \gamma(A) \quad \text{for all} \quad I \triangleleft A$$

if and only if γ is a hereditary Kurosh–Amitsur radical.

Proof: Let γ be a Hoehnke radical with $(*)$. In view of Theorem 2.1.11 and Corollary 3.2.4 it suffices to show that γ is complete and idempotent. If $I \triangleleft A$ and $\gamma(I) = I$, then $I = \gamma(I) = I \cap \gamma(A)$ holds implying that $I \subseteq \gamma(A)$. Thus γ is complete. Further, for $I = \gamma(A)$ we have $\gamma(\gamma(A)) = \gamma(A) \cap \gamma(A) = \gamma(A)$, and therefore γ is idempotent.

Conversely, a hereditary Kurosh–Amitsur radical γ is a Hoehnke radical by Theorem 2.1.11, and satisfies $(*)$ by Corollary 3.2.4. $\square$

3.3. Lower radical constructions

In addition to the Tangeman–Kreiling lower radical construction, there are others which are useful and interesting in various aspects of radical theory. The first lower radical construction was given by Kurosh [1], and we shall present it in the simplified version of Suliński, Anderson and Divinsky [1].

Let δ be an arbitrary class of rings, and define transfinitely the classes δ_λ for each ordinal λ as follows. Define δ_1 to be the homomorphic closure of δ,

$$\delta_1 = \{A \mid A \text{ is a homomorphic image of a ring in } \delta\}.$$

Assuming δ_μ has been defined for every ordinal $\mu < \lambda$, we define

$$\delta_\lambda = \left\{ A \;\middle|\; \begin{array}{l} \text{every nonzero homomorphic image of } A \\ \text{has a nonzero ideal in } \delta_\mu \text{ for some } \mu < \lambda \end{array} \right\}.$$

THEOREM 3.3.1. *The lower radical $\mathcal{L}\delta$ is the union of all the δ_λ.*

Proof: Since δ_1 is homomorphically closed, it is clear that so is each δ_λ and hence $\delta_\mu \subseteq \delta_\lambda$ for every $\mu < \lambda$. Thus also $\mathcal{L}\delta$ is homomorphically closed.

To show that $\cup\delta_\lambda$ is a radical class we thus need only prove (R2) by Theorem 2.1.5.II. To this end, suppose that every nonzero homomorphic image A/I has a nonzero ideal in $\cup\delta_\lambda$, and therefore in some $\delta_{\lambda(I)}$ where the index $\lambda(I)$ depends on the ideal I of A. All the ideals I of A form a set, and therefore there must exist an ordinal ζ which is greater than every ordinal $\lambda(I)$, and so also $\delta_{\lambda(I)} \subseteq \delta_\zeta$. Thus every nonzero homomorphic image A/I of A has a nonzero ideal in δ_ζ. Hence $A \in \delta_{\zeta+1}$ by definition, and so $A \in \cup U\delta_\lambda$. This gives us (R2) and proves that $\cup\delta_\lambda$ is a radical class.

To see that $\cup\delta_\lambda$ is in fact equal to $\mathcal{L}\delta$, let γ be any radical class that contains δ. Then it is clear that $\delta_1 \subseteq \gamma$. Proceeding by induction, assume that $\delta_\mu \subseteq \gamma$ for every $\mu < \lambda$. If $A \in \delta_\lambda$, then every nonzero homomorphic image of A has a nonzero ideal in δ_μ for some $\mu < \lambda$. Thus this ideal is in $\delta_\mu \subseteq \gamma$. Since γ is a radical class, it has (R2), and then we know A must be in γ. Therefore $\delta_\lambda \subseteq \gamma$ for all λ, and so $\cup\delta_\lambda \subseteq \gamma$. Thus $\cup\delta_\lambda$ is the smallest radical class containing δ, and hence $\mathcal{L}\delta = \cup\delta_\lambda$. □

This Kurosh construction is dual to the notion of semisimple closure given in Theorem 2.3.13. The semisimple closure $\bar{\varrho}$ of a class ϱ is achieved in just one step, whereas it is not clear precisely when the Kurosh construction terminates. In general, δ_2 in the Kurosh construction, is not a full radical class. We shall discuss the termination of the Kurosh construction later on in this chapter.

If we start with a homomorphically closed class δ and if, in the Kurosh construction, we use accessible subrings instead of ideals, then we do reach the full radical class in just one step. This is the *Yu-lee Lee radical construction* (see Yu-lee Lee [1], and Leavitt and Yu-lee Lee [1]).

THEOREM 3.3.2. *If δ is a homomorphically closed class of rings, then the class*

$$\mathcal{Y}\delta = \left\{ A \ \middle| \ \begin{array}{l} \text{every nonzero homomorphic image of } A \\ \text{has a nonzero accessible subring in } \delta \end{array} \right\}$$

is a radical class.

Proof: It is clear that $\mathcal{Y}\delta$ is homomorphically closed and thus satisfies the weaker condition (R1).

To get (R2), suppose that every nonzero homomorphic image B of a ring A has a nonzero ideal I in $\mathcal{Y}\delta$. Then I itself must have a nonzero accessible subring J in δ. Thus B has a nonzero accessible subring, namely J, in δ and therefore $A \in \mathcal{Y}\delta$. □

In this section we have not yet used associativity. However, to prove that $\mathcal{Y}\delta$ is the lower radical determined by δ we need semisimple classes to be hereditary. It is at this point that we will use associativity which certainly guarantees the hereditariness of semisimple classes.

PROPOSITION 3.3.3 (Yu-lee Lee [1]). *If γ is a radical class, then $\mathcal{Y}\gamma = \gamma$.*

Proof: Clearly $\gamma \subseteq \mathcal{Y}\gamma$. Suppose that there is a ring $A \in \mathcal{Y}\gamma$ which is not in γ. Then $A/\gamma(A)$ is not zero and is in $\mathcal{S}\gamma$. Since $\mathcal{S}\gamma$ is hereditary by Corollary 3.1.4, every accessible subring of $A/\gamma(A)$ is also in $\mathcal{S}\gamma$. Therefore $A/\gamma(A)$ cannot have a nonzero accessible subring in γ. Thus A cannot be in $\mathcal{Y}\gamma$, a contradiction. Thus $\mathcal{Y}\gamma = \gamma$. □

COROLLARY 3.3.4. *$\mathcal{Y}\delta = \mathcal{L}\delta$ for any homomorphically closed class δ.*

Proof: Clearly $\mathcal{L}\delta \subseteq \mathcal{Y}\delta$ since $\mathcal{Y}\delta$ is a radical class containing δ. On the other hand by Proposition 3.3.3 we have

$$\mathcal{Y}\delta \subseteq \mathcal{Y}\mathcal{L}\delta = \mathcal{L}\delta,$$

and therefore $\mathcal{L}\delta = \mathcal{Y}\delta$. □

There is one more construction that we wish to consider. *Watters' construction* [2] gives us the lower radical $\mathcal{L}\delta(A)$ of a ring A where δ is any homomorphically closed class. We define an ideal $\mathcal{W}(A)$ transfinitely as follows:

$\mathcal{W}_1(A)$ is the ideal of A generated by all accessible δ-subrings of A,

$\mathcal{W}_\lambda(A)$ is the ideal of A with the property that $\mathcal{W}_\lambda(A)/\mathcal{W}_{\lambda-1}(A)$ is the ideal of $A/\mathcal{W}_{\lambda-1}$ generated by all accessible δ-subrings, for any ordinal λ such that $\lambda - 1$ exists,

$$\mathcal{W}_\lambda(A) = \bigcup_{\mu < \lambda} \mathcal{W}_\mu(A)$$

when λ is a limit ordinal.

For any given ring A this process must terminate at some ordinal ζ (depending on A). We shall call this $\mathcal{W}_\zeta(A)$, simply $\mathcal{W}(A)$.

THEOREM 3.3.5 (Leavitt and Yu-lee Lee [1]). *$\mathcal{W}(A) = \mathcal{L}\delta(A)$ for every ring A.*

Proof: By Corollary 3.3.4 it is enough to show that $\mathcal{W}(A) = \mathcal{Y}\delta(A)$. If $\mathcal{Y}\delta(A) \not\subseteq \mathcal{W}(A)$, then

$$0 \neq (\mathcal{Y}\delta(A) + \mathcal{W}(A))/\mathcal{W}(A) \cong \mathcal{Y}\delta(A)/(\mathcal{Y}\delta(A) \cap \mathcal{W}(A)).$$

Since $\mathcal{Y}\delta$ is homomorphically closed, $(\mathcal{Y}\delta(A) + \mathcal{W}(A))/\mathcal{W}(A)$ is in $\mathcal{Y}\delta$, and therefore it has a nonzero accessible subring in δ which is also an accessible

δ-subring of $A/\mathcal{W}(A)$. But by Watters' construction $A/\mathcal{W}(A)$ cannot have such a subring. Thus $\mathcal{Y}\delta(A) \subseteq \mathcal{W}(A)$ for every ring A.

To finish the proof we must show that $\mathcal{W}(A) \subseteq \mathcal{Y}\delta(A)$. To this end let us define $\mathcal{W}_0(A) = 0$. Clearly $0 \subseteq \mathcal{Y}\delta(A)$. Let $\lambda \geq 0$ any ordinal and suppose that $\mathcal{W}_\lambda(A) \subseteq \mathcal{Y}\delta(A)$, and consider $\mathcal{W}_{\lambda+1}(A)/\mathcal{W}_\lambda(A)$ which is generated by the accessible δ-subrings $K/\mathcal{W}_\lambda(A)$ of $A/\mathcal{W}_\lambda(A)$. Since

$$\frac{K + \mathcal{Y}\delta(A)}{\mathcal{Y}\delta(A)} \cong \frac{(K + \mathcal{Y}\delta(A))/\mathcal{W}_\lambda(A)}{\mathcal{Y}\delta(A)/\mathcal{W}_\lambda(A)} \triangleright \frac{A/\mathcal{W}_\lambda(A)}{\mathcal{Y}\delta(A)/\mathcal{W}_\lambda(A)} \cong \frac{A}{\mathcal{Y}\delta(A)} \in \mathcal{S}\mathcal{Y}\delta,$$

the hereditariness of $\mathcal{S}\mathcal{Y}\delta$ yields

$$\frac{K}{K \cap \mathcal{Y}\delta(A)} \cong \frac{K + \mathcal{Y}\delta(A)}{\mathcal{Y}\delta(A)} \in \delta \cap \mathcal{S}\mathcal{Y}\delta = 0.$$

Thus $K \subseteq \mathcal{Y}\delta(A)$ holds implying $\mathcal{W}_{\lambda+1}(A) \subseteq \mathcal{Y}\delta(A)$. Since $\mathcal{W}_\lambda(A) \subseteq \mathcal{Y}\delta(A)$ is obviously true for limit ordinals λ, by transfinite induction we have $\mathcal{W}(A) \subseteq \mathcal{Y}\delta(A)$. □

Another method for constructing the lower radical class via accessible subrings was given by Gardner [5].

3.4. The termination of the Kurosh radical construction

There are three interesting problems connected with the Kurosh lower radical construction. First, when does the construction stop? That is, is there some ordinal λ such that for every homomorphically closed class δ we have $\mathcal{L}\delta = \delta_\lambda$? Second, are there particular classes δ where the construction stops at step 2 or 3? In other words, what conditions on the class δ guarantee that $\mathcal{L}\delta = \delta_2$ or $\mathcal{L}\delta = \delta_3$? Finally, given any positive integer n, can we find a homomorphically closed class δ such that $\mathcal{L}\delta = \delta_n$ (and not δ_{n-1} of course)? We shall deal with the first two questions now and come to grips with the third one in the next sections. In the process we shall develop methods and discover results about the ideal-structure of rings.

PROPOSITION 3.4.1. *Let* $K \triangleleft I \triangleleft A$, *and define* $K_1 = K$ *and*

$$K_{n+1} = \sum(J \triangleleft I \mid J/K_n \text{ is a homomorphic image of } K)$$

for $n \geq 1$. *Then*

$$K_\omega = \cup(K_n \mid n = 1, 2, \dots)$$

is an ideal of A. *In particular, if* I *is the ideal of* A *generated by* K, *then* $K_\omega = I$.

Proof: Since $K \twoheadrightarrow K_n/K_n$, clearly $K_n \subseteq K_{n+1}$. It will then suffice to show that AK_n and $K_n A$ are both in K_ω for every $n \geq 1$. We begin to show by induction that $AK_n \subseteq K_{n+1}$. Take any $a \in A$, and let f be the homomorphism $f\colon K \to (aK+K)/K$ defined by $f(x) = ax+K$, $x \in K$, as in Lemma 3.1.1. Then we know that $(aK+K) \triangleleft I$ and $K/\ker f \cong (aK+K)/K$. Therefore, by definition $aK + K \subseteq K_2$, and so $AK_1 \subseteq K_2$. Assume now that $AK_{n-1} \subseteq K_n$. We want to see that then $AK_n \subseteq K_{n+1}$. If $K_n = K_{n-1}$, then $AK_n = AK_{n-1} \subseteq K_n \subseteq K_{n+1}$. If, however, $K_{n-1} \neq K_n$, then there must be an ideal J of I with $K_{n-1} \subseteq J \subseteq K_n$ such that $J/K_{n-1} \neq 0$ is a homomorphic image of K. If g is the mapping defined by $g(x) = ax + K_n$, $x \in J$, for a fixed but arbitrary element $a \in A$, then g maps J homomorphically onto $(aJ + K_n)/K_n$,

$$J/\ker g \cong (aJ + K_n)/K_n \triangleleft I/K_n$$

and $\ker g \triangleleft I$. This is slightly different from the statements of Lemma 3.1.1 but it can be proved by a similar reasoning. Now $K_{n-1} \subseteq \ker g$ because we are assuming that $AK_{n-1} \subseteq K_n$, and of course $\ker g \subseteq J$. We then have

$$\frac{J/K_{n-1}}{\ker g/K_{n-1}} \cong J/\ker g \cong (aJ + K_n)/K_n.$$

Since J/K_{n-1} is a homomorphic image of K, so is also $(aJ + K_n)/K_n$. Therefore, $aJ + K_n \subseteq K_{n+1}$ by definition, and thus $aJ \subseteq K_{n+1}$ for every $a \in A$. This implies $AJ \subseteq K_{n+1}$. Similarly we can get $JA \subseteq K_{n+1}$. Thus

$$AK_\omega + K_\omega A = A \cdot \sum K_n + \sum K_n \cdot A \subseteq \sum K_n = K_\omega. \qquad \square$$

KREMPA'S LEMMA 3.4.2 (Krempa [4]). *Let $K \triangleleft I \triangleleft A$ and let I be the ideal of A generated by K. Then every nonzero homomorphic image of the ring I contains a nonzero ideal which is a homomorphic image of K.*

Proof: Let $I/L \neq 0$ be a homomorphic image of I. We know from Proposition 3.4.1 that $I = K_\omega$. Since $L \neq I$, there must exist a least integer $n \geq 1$ such that $K_n \not\subseteq L$. If $n = 1$, then

$$0 \neq K/(K \cap L) \cong (K + L)/L \triangleleft I/L$$

and so $(K + L)/L$ is a nonzero homomorphic image of K. If $n > 1$, then by the choice of n we have $K_{n-1} \subseteq L$, and there must exist an ideal J of I such that $J/K_{n-1} \neq 0$ is a homomorphic image of K with $J \not\subseteq L$. So for J we have

$$0 \neq (J + L)/L \triangleleft I/K$$

and

$$\frac{J/K_{n-1}}{(J \cap L)/K_{n-1}} \cong J/(J \cap L) \cong (J + L)/L.$$

Hence $(J+L)/L$ is a nonzero homomorphic image of K. This completes the proof. $\square$

Krempa's Lemma gives us some insight into the Kurosh lower radical construction. Let δ be a class of rings and let δ_λ be as defined just before Theorem 3.3.1. Then we have immediately

COROLLARY 3.4.3. *Let $K \triangleleft I \triangleleft A$ and I the ideal of A generated by K. If $K \in \delta_{n-1}$, then $I \in \delta_n$.*

The Yu-lee Lee construction $\mathcal{Y}\delta$ can also be considered as an infinite construction. If we put $\eta_1 = \delta$ and

$$\eta_n = \left\{ A \ \middle| \ \begin{array}{l} \text{every nonzero homomorphic image of } A \\ \text{has a nonzero } n\text{-accessible subring in } \delta \end{array} \right\}$$

for $n > 1$. Then obviously

$$\mathcal{Y}\delta = \cup(\eta_n \mid n = 1, 2, \dots).$$

The next assertion is a bit surprising.

PROPOSITION 3.4.4 (Heinicke [2]). *$\eta_n = \delta_n$ for every $n = 1, 2, \dots$.*

Proof: Take a nonzero ring $A \in \delta_n$. Then A has a nonzero ideal I_{n-1} in δ_{n-1} and I_{n-1} has a nonzero ideal in δ_{n-2}, etc. Thus A has a nonzero n-accessible subring in δ. We know that δ_n is homomorphically closed (see the proof of Theorem 3.3.1), and thus every nonzero homomorphic image of A is also in δ_n, and as above, has a nonzero n-accessible subring in δ. Therefore A is in η_n and $\delta_n \subseteq \eta_n$.

Conversely, if $A \in \eta_n$, then any nonzero homomorphic image B of A has a chain
$$0 \neq K_1 \triangleleft K_2 \triangleleft \cdots \triangleleft K_{n-1} \triangleleft K_n = B$$
with $K_1 \in \delta_1$. Call $K_1 = I_1$ and let I_2 be the ideal of K_3 generated by I_1. Then $I_2 \in \delta_2$ by Corollary 3.4.3. Carry this process on, letting I_k be the ideal of K_{k+1} generated by I_{k-1} for $k = 2, 3, \dots, n-1$. Then $I_{n-1} \in \delta_{n-1}$, and so B has a nonzero ideal in δ_{n-1}. Therefore $A \in \delta_n$ and thus $\delta_n = \eta_n$. For $n = 1, 2$ the equality $\delta_n = \eta_n$ is trivial, thus $\delta_n = \eta_n$ for every $n = 1, 2, \dots$. □

THEOREM 3.4.5 (Suliński, Anderson and Divinsky [1]). *The Kurosh lower radical construction terminates at the first limit ordinal ω, that is, $\mathcal{L}\delta = \delta_\omega$.*

Proof: Applying Corollary 3.3.4 and Proposition 3.4.4 we get

$$\mathcal{L}\delta = \mathcal{Y}\delta = \cup \eta_n = \cup \delta_n \subseteq \delta_\omega \subseteq \mathcal{L}\delta. \qquad \square$$

In some cases, of course, the Kurosh construction terminates earlier than ω. For example, if δ is a radical class, then $\mathcal{L}\delta = \delta_1$.

THEOREM 3.4.6 (Leavitt [1], Snider [2]). *If δ is a union of radical classes γ_λ, $\lambda \in \Lambda$, then $\mathcal{L}\delta = \delta_2$ and $\mathcal{L}\delta = \mathcal{U}(\cap \mathcal{S}\gamma_\lambda)$.*

Proof: Take any ring $A = \delta_3 = \eta_3$ (by Proposition 3.4.4). Then every nonzero homomorphic image of B of A has a nonzero 3-accessible subring C in $\delta = \cup \gamma_\lambda$. Thus
$$0 \neq C = C_1 \triangleleft C_2 \triangleleft C_3 = B.$$
Now $C \in \gamma_\lambda$ for some radical γ_λ. Then by the ADS-Theorem 3.1.2 we have
$$C \subseteq \gamma_\lambda(C_2) \triangleleft B.$$
Therefore B has a nonzero ideal in $\gamma_\lambda \subseteq \delta$. Thus by definition $A \in \delta_2$, proving $\delta_2 = \delta_3 = \mathcal{L}\delta$.

Obviously $\cup \gamma_\lambda \subseteq \mathcal{U}(\cap \mathcal{S}\gamma_\lambda)$ which implies $\delta_2 \subseteq \mathcal{U}(\cap \mathcal{S}\gamma_\lambda)$. If $A \in \mathcal{U}(\cap \mathcal{S}\gamma_\lambda)$, then no nonzero homomorphic image B of A is in $\cap \mathcal{S}\gamma_\lambda$, that is, B has a nonzero ideal in $\cup \gamma_\lambda = \delta$, that is, $A \in \delta_2$. Thus also $\mathcal{U}(\cap \mathcal{S}\gamma_\lambda) \subseteq \delta_2$ holds. $\square$

There are many more cases for classes δ with $\mathcal{L}\delta = \delta_2$.

THEOREM 3.4.7 (Hoffman and Leavitt [2]). *Let δ be a homomorphically closed class of rings satisfying condition*

(∗) *if I is a nonzero δ-ideal of a ring A and $A^3 \subseteq I$, then there exists an integer $k > 1$ such that $0 \neq A^k \in \delta$.*

Then $\mathcal{L}\delta = \delta_2$.

Proof: Since by Proposition 3.4.4, $\delta_3 = \eta_3$, it is enough to show that $\eta_3 \subseteq \delta_2$. Take a nonzero ring $A \in \eta_3$, and let B be any nonzero homomorphic image of A. Since $\eta_3 = \delta_3$ is homomorphically closed, also $B \in \eta_3$. Then there is a chain
$$0 \neq K \triangleleft I \triangleleft B$$
with $K \in \delta$. Let $\overline{K}$ be the ideal of B generated by K. Then by the Andrunakievich Lemma 1.2.7, $\overline{K}^3 \subseteq K$, and so by condition (∗) there is a k such that $0 \neq \overline{K}^k \in \delta$. Thus every B has a nonzero ideal, namely $\overline{K}^k$, in δ implying $A \in \delta_2$. This proves $\eta_3 \subseteq \delta_2$. $\square$

COROLLARY 3.4.8 (Suliński, Anderson and Divinsky [1]). *If δ is a hereditary class of rings containing all zero-rings, then $\mathcal{L}\delta = \delta_2$.*

Proof: The class δ_1 is homomorphically closed as well as hereditary. If we can show it satisfies condition (∗), then Theorem 3.4.7 tells us that $\mathcal{L}\delta = \delta_2$. Suppose then that I is a nonzero ideal of A, that $A^3 \subseteq I$ and that $I \in \delta_1$. Since δ_1 is hereditary, also $A^3 \in \delta$. If $A^3 = 0$, then either A^2 or A is a nonzero ring with zero-multiplication, and therefore in $\delta \subseteq \delta_1$. Thus (∗) is satisfied with $k = 1, 2$ or 3. Hence $\mathcal{L}\delta = \delta_2$. $\square$

COROLLARY 3.4.9 (Dickson [2]). *Let δ be a homomorphically closed class of rings satisfying condition*

(**) if $I \triangleleft A$ and $I \in \delta$ and $(A/I)^2 = 0$, then $A \in \delta$. Then $\mathcal{L}\delta = \delta_2$.

Proof: Every ring with zero-multiplication must be in δ (just take $I = 0$ in (**)). We will show that δ fulfils condition (*). To this end, let I be a nonzero δ-ideal of A with $A^3 \subseteq I$. Then $(A^2 + I)/I$ is a ring with zero-multiplication. Using condition (**), we get $A^2 + I \in \delta$. Furthermore, $A/(A^2 + I)$ is a ring with zero-multiplication. Thus again using condition (**) we get $A \in \delta$. Therefore (*) holds (with $k = 1$, in fact). Then Theorem 3.4.7 gives us $\mathcal{L}\delta = \delta_2$. $\square$

COROLLARY 3.4.10 (Hoffman and Leavitt [2]). *If δ is a class of idempotent rings, then $\mathcal{L}\delta = \delta_2$.*

Proof: Let $I \neq 0$ be an ideal of a ring A such that $A^3 \subseteq I$ and $I \in \delta$. Since $I \in \delta$, it must be idempotent, and thus $I = I^3 \subseteq A^3 \subseteq I$. Therefore $A^3 = I \in \delta$, and hence condition (*) of Theorem 3.4.7 holds true. This proves that $\mathcal{L}\delta = \delta_2$. $\square$

We must not get the idea that for most cases we have $\mathcal{L}\delta = \delta_2$. There are many classes of rings for which the Kurosh construction goes on far beyond 2.

STEWART'S LEMMA 3.4.11 (Stewart [3]). *Let δ be a homomorphically closed class of nilpotent rings. If $K \neq 0$ is an accessible δ-subring of A, then there exists a 3-accessible δ-subring $J \neq 0$ of A with $J^2 = 0$.*

Proof: Let $\overline{K}$ denote the ideal of A generated by K, and

$$0 \neq K = K_1 \triangleleft \cdots \triangleleft K_{n-2} \triangleleft K_{n-1} \triangleleft K_n = A.$$

Clearly $\overline{K} \subseteq K_{n-1}$, and so

$$0 \neq K \subseteq K_{n-2} \cap \overline{K} \triangleleft \overline{K} \triangleleft A.$$

Hence $\overline{K}$ is the ideal of A generated also by $K_{n-2} \cap \overline{K}$. Applying the Andrunakievich Lemma, we get

$$\overline{K}^3 \subseteq K_{n-2} \cap \overline{K} \subseteq K_{n-2}.$$

Iterating this procedure, an appropriate power (a power of 3) of $\overline{K}$ will be in $K \in \delta$. Hence $\overline{K}$ is nilpotent: $\overline{K}^r = 0$ and $\overline{K}^{r-1} \neq 0$ for some $r > 1$. Then if $r > 2$,

$$\overline{K}^{r-1} = (K + AK + KA + AKA)\overline{K}^{r-2} \neq 0.$$

Hence there are elements $a, b, c, d \in A$ and $k \in \overline{K}^{r-2}$ such that

$$(K + aK + Kb + cKd)k \neq 0.$$

Then
$$Kk \cup aKk \cup Kbk \cup cKdk \neq 0,$$
and thus at least one of the mappings
$$f(x) = xk, \quad f(x) = axk, \quad f(x) = xbk, \quad f(x) = cxdk$$
for $x \in K$, maps K homomorphically onto a nonzero ideal $J = f(K)$ of $\overline{K}^{r-1}$ because $(\overline{K}^{r-1})^2 = 0$. Since $K \in \delta$ and δ is homomorphically closed, J must be in δ. Then
$$0 \neq J \triangleleft \overline{K}^{r-1} \triangleleft A$$
and so $J^2 = 0$.

If $r = 2$, then $\overline{K} \neq 0$ and $\overline{K}^2 = 0$, then $0 \neq K \triangleleft \overline{K} \triangleleft A$ with $K \in \delta$.
Thus $A \in \delta$ has always a 3-accessible subring. □

THEOREM 3.4.12 (Stewart [3], see also Gardner [18]). *If δ is a homomorphically closed class of nilpotent rings, then $\mathcal{L}\delta = \delta_3$.*

Proof: Let A be any nonzero ring in $\mathcal{L}\delta$. Then by Corollary 3.3.4 every nonzero homomorphic image B of A has a nonzero accessible subring K in δ. Then by Stewart's Lemma 3.4.11, B has a 3-accessible subring, nonzero and in δ. Thus $A \in \delta_3$ by Proposition 3.4.4, and so $\mathcal{L}\delta = \delta_3$. □

THEOREM 3.4.13 (Armendariz and Leavitt [1], Watters [1]). *If δ is a hereditary and homomorphically closed class of rings, then $\mathcal{L}\delta = \delta_3$.*

Proof: If $\mathcal{L}\delta \neq \delta_3$, then there is a ring $A \in \delta_4 \setminus \delta_3$. Then by Proposition 3.4.4, A has a nonzero homomorphic image B which has no 3-accessible subring in δ. But $B \in \delta_4$, and thus B has a 4-accessible δ-subring $I \neq 0$, that is
$$0 \neq I = I_1 \triangleleft I_2 \triangleleft I_3 \triangleleft I_4 = B.$$
Call $K_1 = I_1$ and let K_i be the ideal of I_{i+1} generated by K_{i-1} for $i = 2, 3$. Then we have
$$0 \neq K_1 \triangleleft K_2 \triangleleft K_3 \triangleleft B.$$
Since $K_1 \in \delta$, Corollary 3.4.3 tells us that $K_2 \in \delta_2$ and $K_3 \in \delta_3$. Using the Andrunakievich Lemma twice, we get $K_3^9 \subseteq K_1 \in \delta$. Since δ is hereditary, also $K_3^9 \in \delta$. Now $K_3^9 \triangleleft B$, and since B does not have a nonzero 3-accessible subring in δ, K_3^9 must be 0. Hence $K_1^9 \subseteq K_3^9 = 0$, and so K_1 is nilpotent. Obviously the nilpotent rings of δ form a homomorphically closed subclass η in δ. Applying Stewart's Lemma 3.4.11 to η, there is a nonzero 3-accessible subring J of B with $J \in \eta \subseteq \delta$. This is not possible, and therefore $\mathcal{L}\delta = \delta_3$. □

The statements of Theorems 3.4.12 and 3.4.13 are sharp. In the following example we shall give a homomorphically closed class δ of rings which is hereditary, and consists of nilpotent rings, and $\mathcal{L}\delta \neq \delta_2$.

EXAMPLE 3.4.14. Let us consider the class δ consisting of all rings $Z(n)$ with zero-multiplication over the cyclic additive groups of n elements for $n = 1, 2, \ldots$. Then δ is homomorphically closed, hereditary, and consists of nilpotent rings, and so by Theorem 3.4.12 or by Theorem 3.4.13, $\mathcal{L}\delta = \delta_3$. To see that $\mathcal{L}\delta \neq \delta_2$, let us consider the *Zassenhaus ring* A which consists of all finite sums $\sum a_\lambda x_\lambda$ where each λ is a rational number with $0 < \lambda < 1$, each a_λ is an element of the two-element field, and each x_λ is an indeterminate subject to the multiplication rule

$$x_\lambda x_\mu = \begin{cases} x_{\lambda+\mu} & \text{if } \lambda + \mu < 1 \\ 0 & \text{otherwise.} \end{cases}$$

It is easy to check that A is a ring. Moreover, A has no nonzero ideals in δ, for if I is a nonzero δ-ideal, then it would have to be a ring with zero-multiplication, and so for every element $y = \sum a_\lambda x_\lambda \in I$ necessarily $\lambda \geq 1/2$ would hold. Since $I \in \delta$, I must be finite. However, AI has infinitely many elements and $AI \subseteq I$. This contradiction proves that $A \notin \delta_2$.

We prove that $A \in \delta_3$. Let J_n be the set of all elements $\sum a_\lambda x_\lambda$ with $\lambda \geq \frac{1}{n}$, $n = 2, 3, \ldots$. Then $0 \neq J_n \triangleleft A$, $J_n^n = 0$ for all n and $A = \cup J_n$. Hence every nonzero homomorphic image B of A contains as an ideal, a nonzero homomorphic image of some J_n, and taking an appropriate power of this we see that B has an ideal I with $I^2 = 0 \neq I$. If $0 \neq r \in I$ then $\{r, 0\} \triangleleft I \triangleleft B$, and $\{r, 0\} \in \delta$. As B is an arbitrary homomorphic image of A, we conclude that $A \in \delta_3$ by Proposition 3.4.4. □

Rossa [3] considered a class δ of rings satisfying the descending chain condition on two-sided ideals, and proved that $\mathcal{L}\delta = \delta_3$ with the proviso that $\mathcal{L}\delta$ is hereditary.

3.5. The Suliński–Anderson–Divinsky problem

One of the outstanding problems of radical theory was the Suliński–Anderson–Divinsky problem [1]. *Are there examples of the Kurosh radical construction which terminate in precisely* $4, 5, \ldots$ *steps?* Beidar [1] answered this question affirmatively. The examples of Beidar are developed from a rather detailed examination of the accessible subring structure of the ring G of Gaussian integers and considerable use is made of the arithmetic properties of G. We shall now begin working toward these examples (slightly modifying the presentation of Beidar [1]). Let us first mention some notation.

In what follows, $i = \sqrt{-1}$, $\mathbb{Q}(i)$ is the field obtained by adjoining i to the field $\mathbb{Q}$ of rationals, p a prime of the form $p = 4k + 3$, $A_0 = G = \{a + bi : a, b \in \mathbb{Z}\}$ the ring of Gaussian integers and

$$A_n = \{pa + p^n bi \mid a, b \in \mathbb{Z}\}$$

for $n = 1, 2, \ldots$.

LEMMA 3.5.1. *Let $0 \neq I \triangleleft A_n$ and $f\colon A_n \to \mathbb{Q}(i)$ be a ring homomorphism. Then*
   (i) *A_n/I is finite,*
   (ii) *$f(A_n) = 0$ or $f(A_n) = A_n$.*

Proof: (i) Let $0 \neq a + bi \in I$. Then $a + bi = pc + p^n di$ with suitable integers c and d, and from $pc \in A_n$ it follows that $a - bi = 2pc - (pc + p^n di) \in A_n$. Hence
$$a^2 + b^2 = (a+bi)(a-bi) \in I \cdot A_n \subseteq I,$$
and so $(a^2 + b^2)A_n \subseteq I$. Thus, as an abelian group, A/I is bounded and finitely generated, hence finite.

(ii) Since $\mathbb{Q}(i)$ has no finite nonzero subrings, by (i) either $f(A_n) = 0$ or f is injective, that is, $f(A_n) \cong A_n$. The ring of quotients of A_n is $\mathbb{Q}(i)$ and f extends to a homomorphism $\widehat{f}\colon \mathbb{Q}(i) \to \mathbb{Q}(i)$, also injective if f is. But $\mathbb{Q}(i)$ has only two injective endomorphisms, the identity and conjugation, and these fix A_n (not elementwise), so we have $\operatorname{im} f = A_n$. □

LEMMA 3.5.2. *If $A_{n+m} \triangleleft A_n$, then either $m = 0$ or $m = 1$.*

Proof: Since $p \in A_{n+m}$ and $p^n i \in A_n$, we have $p^{n+1} i \in A_{n+m}$ and therefore $p^{n+m} \mid p^{n+1}$. Hence either $m = 0$ or $m = 1$. □

LEMMA 3.5.3. *If R is a subring of $\mathbb{Q}(i)$ such that $A_n \triangleleft R$, then $R = A_{n-1}$ or $R = A_n$ or $1 \in R$.*

Proof: For $n = 0$ we have $1 \in G = A_0 \triangleleft R$ which implies $R = 1 \cdot R \subseteq G$. Thus $1 \in R$.

Let $n > 0$ and $a + bi \in R$ be an arbitrary element. Since $p \in A_n$, by $A_n \triangleleft R$ we have
$$pa + pbi = p(a+bi) \in A_n.$$
Hence both a and b must be integers and therefore $R \subseteq G$. Moreover, $b = p^{n-1} b'$ with an integer b' as $p^n \mid pb$. Thus every element $x \in R$ has the form
$$x = a + p^{n-1} bi$$
where a and b are integers.

Case $n = 1$. Here $A_1 = pG$ and
$$R/A_1 \subseteq G/pG \cong Z_p[i].$$
Since $p = 4k + 3$, $Z_p[i]$ is a field extension of Z_p. Since $Z_p[i]$ is a finite field, so is every nonzero subring. Thus either $R = A_1$, or $1 + A_1 \in R/A_1$. In the latter case $1 + A_1 \subseteq R$ and so $1 \in R$.

Case $n > 1$. We distinguish two subcases.

Subcase 1. $p \mid a$ for every $x = a + p^{n-1} bi \in R$. In this case $a = pa'$ with an integer a' and so $x \in A_{n-1}$ yielding $R \subseteq A_{n-1}$. One can readily check that $A_n \triangleleft A_{n-1}$ and that A_{n-1}/A_n consists of p elements. Hence $A_n \triangleleft R \subseteq A_{n-1}$ implies either $R = A_n$ or $R = A_{n-1}$.

Subcase 2. There exists an element $x = a + p^{n-1}bi \in R$ such that $p \nmid a$. Now

$$1 = pu + av$$

with suitable integers u and v. Multiplying x by v we get

$$xv = av + p^{n-1}bvi = 1 - pu + p^{n-1}bvi.$$

This implies

$$1 + p^{n-1}bvi = xv + pu \in R + A_n = R.$$

Putting $z = xv + pu$ we have

$$p^{n-1}bvi = z - 1.$$

Squaring both sides we get

$$-p^{2n-2}b^2v^2 = z^2 - 2z + 1.$$

Since $n > 1$ and $z \in R$, we arrive at

$$1 = -p(p^{2n-3}b^2v^2) - z^2 + 2z \in A_n + R + R = R.$$

Thus $1 \in R$ and the proof is complete. □

LEMMA 3.5.4. *Let*

$$A_n \cong B_m \triangleleft B_{m-1} \triangleleft \cdots \triangleleft B_1 \triangleleft B_0 \subseteq \mathbb{Q}(i)$$

be a finite chain without repetitions.

(i) *If* $1 \notin B_0$, *then*

$$A_n = B_m, \quad A_{n-1} = B_{m-1}, \ldots, A_{n-m+1} = B_1, \quad A_{n-m} = B_0.$$

(ii) *If* $1 \in B_0$, *then*

$$A_n = B_m, \quad A_{n-1} = B_{m-1}, \ldots, A_{n-m+1} = B_1.$$

Proof: Let $A_n \to B_m$ be an isomorphism. We can regard f as a homomorphism from A_n to $\mathbb{Q}(i)$ and since $f(A_n) \neq 0$, Lemma 3.5.1.(ii) says that $B_m = f(A_n) = A_n$.

If $1 \notin B_0$, then $1 \notin B_k$ for all $k = 1, 2, \ldots, m$. In this case Lemma 3.5.3 says that $B_{m-1} = A_n$ or $B_{m-1} = A_{n-1}$. But there are no repetitions, so $B_{n-1} = A_{n-1}$. Similarly $B_{m-2} = A_{n-2}$, etc. Finally $B_0 = B_{m-m} = A_{n-m}$.

If $1 \in B_0$, then as $1 \notin A_n = B_m$, there exists j such that $1 \in B_{j-1} \setminus B_j$. Then B_{j-1} is idempotent, so $B_{j-1} \triangleleft B_0$, and then as B_0 has no zero-divisors, $B_{j-1} = B_0$. Since there are no repetitions, we have $j - 1 = 0$, that is, $j = 1$ and $1 \notin B_k$ for $k \geq 1$. Therefore, as before

$$A_n = B_m, \ A_{n-1} = B_{m-1}, \ldots, \ A_{n-m+1} = B_1. \qquad \square$$

COROLLARY 3.5.5. *Let $n > 0$ be a fixed integer and δ be the class consisting of A_n and all rings with torsion additive groups. Then $A_0 \in \delta_{n+1} \setminus \delta_n$.*

Proof: By Lemma 3.5.1(i) the class δ is homomorphically closed and hence $\delta = \delta_1$. In view of Lemma 3.4.4 we have also

$$\delta_k = \left\{ A \ \middle| \ \begin{array}{l} \text{every nonzero homomorphic image of } A \\ \text{has a nonzero } k\text{-accessible subring in } \delta \end{array} \right\}.$$

If $0 \neq I \triangleleft A_0$, then by Lemma 3.5.1.(i) we have $A_0/I \in \delta \subseteq \delta_{n+1}$. Also $A_n \in \delta$ and A_n is an $(n+1)$-accessible subring of A_0. Hence every nonzero homomorphic image of A_0 has a nonzero $(n+1)$-accessible subring in δ and so $A_0 \in \delta_{n+1}$.

Suppose that $A_0 \in \delta_n$. Then in particular A_0 itself has a chain

$$0 \neq B_m \triangleleft B_{m-1} \triangleleft \cdots \triangleleft B_1 \triangleleft A_0$$

without repetitions where $B_m \in \delta$ and $m < n$. Since B_m is torsion-free, we have $B_m \cong A_n$. As $1 \in A_0$, Lemma 3.5.4.(ii) implies $A_n = B_m, \ldots, A_{n-m+1} = B_1 \triangleleft A_0$. By Lemma 3.5.1.(ii) it follows that $A_{n-m+1} = A_1$, that is, $m = n$, a contradiction. Hence $A_0 \notin \delta_n$. $\qquad \square$

LEMMA 3.5.6. *If R is any additively torsion-free ring with an accessible subring $B \cong A_n$, then B is an $(n+1)$-accessible subring of R.*

Proof: Let $A_n \cong B \triangleleft C \triangleleft \cdots \triangleleft R$. Since everything in sight is additively torsion-free, we have a series

$$\mathbb{Z}^{-1}B \triangleleft \mathbb{Z}^{-1}C \triangleleft \cdots \triangleleft \mathbb{Z}^{-1}R,$$

where $\mathbb{Z}^{-1}(\ )$ denotes the ring of quotients. But $\mathbb{Z}^{-1}B = \mathbb{Q}(i)$, so $\mathbb{Z}^{-1}B \triangleleft \mathbb{Z}^{-1}R$, whence

$$B \subseteq R \cap \mathbb{Z}^{-1}B \triangleleft R \cap \mathbb{Z}^{-1}R = R.$$

Hence $A_n \cong B$ is an accessible subring of $R \cap \mathbb{Z}^{-1}B = R \cap \mathbb{Q}(i) \subseteq \mathbb{Q}(i)$. By Lemma 3.5.4 there exists a series of one of the following kinds:

$$A_n \triangleleft A_{n-1} \triangleleft \cdots \triangleleft A_{n-m+1} \triangleleft A_{n-m} = R \cap \mathbb{Z}^{-1}A_n \triangleleft R,$$

if $1 \notin R \cap \mathbb{Z}^{-1} A_n$,

$$A_n \triangleleft A_{n-1} \triangleleft \cdots \triangleleft A_{n-m+1} \triangleleft R \cap \mathbb{Z}^{-1} A_n,$$

if $1 \in R \cap \mathbb{Z}^{-1} A_n$.

Now if $1 \notin R \cap \mathbb{Z}^{-1} A_n = A_{n-m}$, then $A_{n-m} \neq A_0 = G$, so $m < n$, while if $1 \in R \cap \mathbb{Z}^{-1} A_n$, then by Lemma 3.5.4 (ii) $n - m + 1 > 0$, that is, $m < n+1$. Thus in the first case A_n is $(m+2)$-accessible with $m < n$, and in the second A_n is $(m+1)$-accessible with $m < n+1$. In either case, A_n is $(n+1)$-accessible. □

This completes the preparation, we can now proceed to our termination result.

THEOREM 3.5.7 (Beidar [1]). *Let $n > 0$ be a fixed integer and δ be the class consisting of A_n and all rings with torsion additive groups. Then*

$$\mathcal{L}\delta = \delta_{n+1} \neq \delta_n.$$

Proof: By Corollary 3.5.5 $\delta_{n+1} \neq \delta_n$. What we need to prove is that $\mathcal{L}\delta \subseteq \delta_{n+1}$. Let R be in $\mathcal{L}\delta$, $R \neq 0$. If every nonzero homomorphic image of R has a nonzero ideal T with torsion additive group, then $T \in \delta_2 \subseteq \delta_{n+1}$. If some nonzero homomorphic image $\overline{R}$ of R is additively torsionfree, then $\overline{R}$ has a nonzero accessible subring $B \cong A_n$, so by Lemma 3.5.6 B is $(n+1)$-accessible in $\overline{R}$. Thus $\overline{R} \in \delta_{n+1}$ and so $\mathcal{L}\delta = \delta_{n+1}$. □

The story of the Kurosh radical construction would be incomplete without an answer to the question: are infinitely many steps ever necessary? The answer to this question is affirmative; this was shown by Heinicke [2] long before Beidar's result was obtained and from consideration of quite different rings. We shall demonstrate this by means of an adaptation of Beidar's examples.

THEOREM 3.5.8 (Heinicke [2]). *There exists a class δ of rings such that*

$$\mathcal{L}\delta = \delta_\omega \neq \delta_n$$

for all finite n.

Proof: For $n \geq 1$, let $\Sigma(n) = 1 + 2 + \cdots + n$, and let δ consist of all rings with torsion additive groups and all rings $A_{\Sigma(n)}$. For each n, the ring $A_{\Sigma(n)+1}$ is in $\mathcal{L}\delta$ because its homomorphic images are either finite (and therefore in δ) or isomorphic to $A_{\Sigma(n)+1}$ which has $A_{\Sigma(n+1)} \in \delta$ as an accessible subring. But $A_{\Sigma(n+1)}$ is $(n+1)$-accessible in $A_{\Sigma(n)+1}$ and by Lemma 3.5.4 not n-accessible, so $A_{\Sigma(n)+1} \in \delta_{n+1} \setminus \delta_n$. □

Also Guo Jin-yun [1] (see also Liu Shao-xue [1] and Liu et alia [1]) solved the Suliński–Anderson–Divinsky problem. In constructing the beginning class δ, he used the ring

$$A_n = \left\{ \sum_{i=1}^{n-1} a_i x^{in} + \sum_{j>0} b_j x^{n^2+j} \,\bigg|\, a_i, b_j \in Z_p \right\}$$

of polynomials over the prime field Z_p.

It is remarkable that both Beidar and Guo settled the Suliński–Anderson–Divinsky problem in the much slimmer variety of *commutative*, associative rings.

Andruszkiewicz and Puczyłowski [1] settled the Suliński–Anderson–Divinsky problem in a more general setting which included also the case of the left strong lower radical construction. In [2] the same authors have shown that starting from any commutative noetherian integrally closed domain with unity (the ring of Gaussian integers is such a one) classes δ can be constructed such that the lower radical construction terminates in exactly n or ω steps.

The behavior of accessible subrings is apparently decisive in answering questions concerning the lower radical construction. Andruszkiewicz [1] studied accessible subrings, and got surprising results, for example:

i) The positive solution of the Suliński–Anderson–Divinsky problem is equivalent to the question whether for every $n \geq 3$ there exists a homomorphically closed class δ of rings such that if a ring A has a nonzero n-accessible subring in δ, then A has a nonzero $n-1$-accessible subring in δ.

ii) For every $n \geq 4$, a ring A is in the class

$$\mathbb{I}_n = \left\{ A \,\bigg|\, \begin{array}{l} \text{if } S \cong A \text{ is an } n\text{-accessible subring of } B, \\ \text{then } S \text{ is an } n-1\text{-accessible subring of } B \end{array} \right\}$$

if and only if the additive group $(A/A^2)^+$ is divisible and torsion. Thus $\mathbb{I}_4 = \mathbb{I}_5 = \cdots$.

The Suliński–Anderson–Divinsky problem has a positive solution also for algebras over a commutative ring with unity (L'vov and Sidorov [1] and Watters [6]) and for semigroups (Sidorov [1]).

3.6. Supernilpotent radicals and their semisimple classes

Zero-rings are certainly "bad" and therefore should be contained in any successful radical class. Andrunakievich, therefore, called a radical γ *supernilpotent*, if γ contained the class of all zero-rings and if, in addition, γ was hereditary. Since radicals are closed under extensions, it is clear that all nilpotent rings are contained in every supernilpotent radical class.

EXAMPLES 3.6.1. (i) *The Baer radical β is the smallest supernilpotent radical* (see 2.2.2 and 3.2.12). If a ring A is not semiprime then A has a nonzero ideal I such that $I^2 = 0$, and so $\beta(A) \neq 0$. If $\beta(A) \neq 0$ then by Corollary 3.4.8 A has a nonzero ideal J such that $J^2 = 0$, and therefore A is not semiprime. Thus, *the class of all semiprme rings coincides with the semisimple class $\mathcal{S}\beta$, and is closed under essential extensions* (see Proposition 3.2.6 and Example 3.2.12).

(ii) *Köthe's nil radical* (see 2.1.6 and 3.2.1.(i)) *and the Levitzki radical* (see 2.1.8 and 3.2.1.(iii)) *are supernilpotent.*

(iii) Every zero-ring is Jacobson radical (see 2.1.7 and 3.2.1.(ii)) because $a \circ (-a) = 0$ for every element a in a zero-ring. Thus *the Jacobson radical is supernilpotent.*

(iv) *The Brown–McCoy and the Thierrin radicals* (see Examples 3.2.17 and 3.2.18) *are supernilpotent* as seen from their definition.

We do insist that a radical be hereditary before it can call itself supernilpotent, and thus not every radical class that contains all zero-rings is supernilpotent.

EXAMPLE 3.6.2. Let σ be the hereditary closure of the ring $2\mathbb{Z}$ of even integers, that is, σ contains $2\mathbb{Z}$ and all of its (accessible) subrings. Then σ is hereditary, and we can construct the upper radical $\mathcal{U}\sigma$. It is clear that all zero-rings are in $\mathcal{U}\sigma$, and it is also clear that the ring $\mathbb{Z}$ of integers is in $\mathcal{U}\sigma$, because no ring in σ has a nonzero idempotent. The ideal $2\mathbb{Z}$ of $\mathbb{Z}$ is certainly not in $\mathcal{U}\sigma$. Thus $\mathcal{U}\sigma$ is not hereditary, though it does contain all zero-rings.

PROPOSITION 3.6.3. *If a radical class γ contains the class $\mathcal{Z}$ of all zero-rings, then the semisimple class $\mathcal{S}\gamma$ consists entirely of semiprime rings. If γ is the upper radical of a class σ of semiprime rings, then $\mathcal{Z} \subseteq \gamma$.*

Proof: The first statement follows readily from the hereditariness of $\mathcal{S}\gamma$. The second statement is straightforward. □

PROPOSITION 3.6.4. *If $K \triangleleft I \triangleleft A$ and I/K is a semiprime ring, then $K \triangleleft A$.*

Proof: Let $\overline{K}$ be the ideal of A generated by K. The Andrunakievich Lemma tells us that $(\overline{K}/K)^3 = 0$. Since I/K is semiprime, we must have $\overline{K}/K = 0$ or $K = \overline{K} \triangleleft A$. □

We can put Propositions 3.6.4 and 3.2.15 together, and conclude that any ring S which either has a unity or is semiprime satisfies condition

(F)   if $K \triangleleft I \triangleleft A$ and $I/K \cong S$, then $K \triangleleft A$.

Condition (F) can be regarded as a substitute for the missing transitivity of the relation $\triangleleft$, and it will turn out to be a useful tool in the sequel. We say that *a class ϱ of rings satisfies condition* (F), if each ring $S \in \varrho$ satisfies (F).

PROPOSITIONS 3.6.5. *If the class ϱ of rings is regular, closed under essential extensions and has condition* (F), *then the upper radical $\mathcal{U}\varrho$ is hereditary.*

Proof: By Proposition 3.2.14 it is enough to show that ϱ satisfies condition (H). Suppose then that $I \triangleleft A$ and $0 \neq I/K \in \varrho$ for some $K \triangleleft I$. By (F) we know that $K \triangleleft A$ and thus $I/K \triangleleft A/K$. Using Lemma 3.2.5 we can find an ideal J/K of A/K such that $I/K \cap J/K = 0$ or $I \cap J = K$ and J/K is maximal for this, and then

$$0 \neq I/K \cong (I+J)/J \triangleleft A/J.$$

Now $I/K \in \varrho$ and ϱ is closed under essential extensions, therefore A/J must be in ϱ. Thus A has a nonzero homomorphic image in ϱ and condition (H) holds. □

PROPOSITION 3.6.6. *Any essential extension B of a subdirectly irreducible ring A is subdirectly irreducible, and the heart of B contains the heart of A. If A is also semiprime, then their hearts coincide.*

Proof: If $0 \neq I \triangleleft B$, then $I \cap A \neq 0$ because B is an essential extension of A. Since $I \cap A \triangleleft A$, the ideal $I \cap A$ must contain the heart H of A. The same applies to the intersection of all nonzero ideals of B proving the assertions in the first statement. The second statement follows from Proposition 1.2.8. □

PROPOSITION 3.6.7. *If I is an ideal of a semiprime subdirectly irreducible ring A, then also I is subdirectly irreducible and $H(I) = H(A)$ holds for their hearts.*

Proof: Straightforward in view of Proposition 1.2.8. □

EXAMPLES 3.6.8. (i) Let ϱ denote the class of all subdirectly irreducible rings with idempotent hearts. The rings in ϱ are semiprime, and by Proposition 3.6.7 the class ϱ is hereditary. Thus $\mathcal{U}\varrho$ is a radical class, which is the *antisimple radical* of Andrunakievich, and will be denoted by β_φ. In view of Propositions 3.6.3, 3.6.4, 3.6.5 and 3.6.6 *the antisimple radical is supernilpotent*. The reason for the name antisimple is that a ring in β_φ cannot be mapped homomorphically onto a nonzero simple idempotent ring.

(ii) Let η denote the class of all subdirectly irreducible rings, each having a nonzero idempotent in its heart. The class η is hereditary and closed under essential extensions by Propositions 3.6.6 and 3.6.7, moreover, η consists obviously of semiprime rings. Thus by Propositions 3.6.3 and 3.6.5, *the upper radical $\mathcal{B} = \mathcal{U}\eta$ is a supernilpotent* radical and it is called the *Behrens radical*.

From Theorem 3.2.7 and Proposition 3.6.3 it follows immediately

COROLLARY 3.6.9 (Anderson and Wiegandt [2]). *A class σ of rings is the semisimple class of a supernilpotent radical γ if and only if*
  (i) σ *is regular,*
  (ii) σ *is closed under subdirect sums,*

(iii) σ is closed under essential extensions,

(iv) σ consists of semiprime rings. □

A class σ is said to be *weakly homomorphically closed*, if $I \triangleleft A \in \sigma$ and $I^2 = 0$ imply $A/I \in \sigma$. Clearly any class of semiprime rings is weakly homomorphically closed because the only ideal in a ring of that class whose square is 0, is the ideal 0. Therefore all semisimple classes of supernilpotent radicals are weakly homomorphically closed. This insight will play an important role in studying such semisimple classes.

We now wish to introduce something called a split-null extension. To this end, let A be any ring and M an A-bimodule, that is, M is both a left and a right A-module such that $(am)b = a(mb)$ for all $a, b \in A$ and $m \in M$. Next, we consider the cartesian product $A \times M$, and define addition on $A \times M$ componentwise:

$$(a_1, m_1) + (a_2, m_2) = (a_1 + a_2, m_1 + m_2),$$

and multiplication by

$$(a_1, m_1)(a_2, m_2) = (a_1 a_2, a_1 m_2 + m_1 a_2)$$

for all $a_1, a_2 \in A$ and $m_1, m_2 \in M$. It is easy to see that this makes $A \times M$ into a ring. We shall denote this ring by $A * M$, and call it the *split-null extension of M by A*.

It is clear that the following properties hold for $A * M$:

(1) $\qquad M^0 = (0, M) \triangleleft A * M \quad \text{and} \quad (M^0)^2 = 0,$

(2) $\qquad A' = (A, 0) \text{ is a subring in } A * M \text{ and } A' \cong A,$

(3) $\qquad (A * M)/M^0 \cong A.$

It is of course possible to use A itself as the bimodule M, and to talk about $A * A$. In view of (1) we may denote the zero-ring on the additive group of A by A^0.

PROPOSITION 3.6.10. *Let γ be a hereditary radical such that is semisimple class $\sigma = S\gamma$ contains all zero-rings. Then we have for any ring A:*

(i) $\qquad \gamma(A * A)A^0 + A^0 \gamma(A * A) = 0$

and

(ii) $\qquad \gamma(A * A) \subseteq A'.$

Proof: Since γ is hereditary, by (1) we get

$$\gamma(A * A) \cap A^0 = \gamma(A^0).$$

Since A^0 is a zero-ring, it must be in σ and therefore $\gamma(A^0) = 0$. Hence

$$\gamma(A * A)A^0 + A^0 \gamma(A * A) \subseteq A^0 \cap \gamma(A * A) = \gamma(A^0) = 0,$$

and so (i) holds.

To establish (ii), we define a mapping

$$f\colon \gamma(A*A) \to A^0$$

by $f((a,b)) = (0,b)$ for all pairs $(a,b) \in \gamma(A*A)$. It is clear that f preserves addition. For multiplication we consider

$$f((a_1,b_1))f((a_2,b_2)) = (0,b_1)(0,b_2) = (0,0).$$

Further, by (i) we get

$$f((a_1,b_1)(a_2,b_2)) = f((a_1a_2, a_1b_2+b_1a_2)) = (0, a_1b_2+b_1a_2) =$$
$$= (a_1,b_1)(0,b_2) + (0,b_1)(a_2,b_2) \in \gamma(A*A)A^0 + A^0\gamma(A*A) = 0.$$

Therefore f is a homomorphism and its image is in γ, and of course in A^0. Thus

$$f(\gamma(A*A)) \subseteq \gamma(A^0)$$

holds. Since $A^0 \in \sigma$, it follows that $\gamma(A^0) = 0$, whence

$$f(\gamma(A*A)) = 0.$$

Hence the second component of each $(a,b) \in \gamma(A*A)$ must be 0, and so $\gamma(A*A) \subseteq A'$. □

THEOREM 3.6.11 (Anderson and Wiegandt [1]). *A class σ is the semisimple class of a supernilpotent radical if and only if*

(i)   *σ is regular,*
(ii) *σ is closed under subdirect sums,*
(iii) *σ is closed under essential extensions,*
(iv) *σ is weakly homomorphically closed,*
(v)   *σ is not the class of all rings.*

Proof: If σ is such a semisimple class, then Corollary 3.6.9 gives us (i), (ii) and (iii). Also the discussion right after Corollary 3.6.9 gives us (iv). Finally no zero-rings are in σ, and we have (v).

Conversely, we assume that σ has all five conditions. By Theorem 3.2.7 we know that σ is the semisimple class of a hereditary radical. If we can show that σ has only semiprime rings, then Proposition 3.6.3 would tell us that the corresponding radical class γ contains all zero-rings and is thus supernilpotent. We assume then that σ contains a nonzero ring A which is not semiprime. Then A has a nonzero ideal B which is a zero-ring. Since σ is hereditary (being a semisimple class), also B is in σ. Every zero-ring, in particular B, contains an ideal which is isomorphic to the zero-ring $Z(n)$ on the cyclic group of order $n > 1$ (possibly $n = \infty$). For a finite n the hereditariness of σ yields $Z(n) \in \sigma$ and also $Z(p) \in \sigma$ whenever p is a prime divisor of n. Since σ is a semisimple class, it is closed under extensions, and so σ contains also $Z(p^i)$ for each $i = 1, 2, \ldots$. Using (ii) we get

$$Z(\infty) \cong Z(\infty) \Big/ \bigcap_{i=1}^{\infty} (p^i) \cong \sum_{\text{subdirect}} Z(\infty) \Big/ (p^i) \cong \sum_{\text{subdirect}} Z(p^i) \in \sigma$$

where $(p^i) = \{kp^i \mid k \text{ ranges over the integers}\}$. Thus $Z(\infty) \in \sigma$ in any case. Since the class σ is weakly homomorphically closed and $Z(\infty) \in \sigma$, every homomorphic image of $Z(\infty)$, in particular every zero-ring $Z(p)$ on the cyclic group of prime order has to be in σ. The heart of a subdirectly irreducible zero-ring must be simple because every subring of such a ring is an ideal. Hence every subdirectly irreducible zero-ring is an essential extension of its heart, and so by $Z(p) \in \sigma$ and condition (iii) it follows that every subdirectly irreducible zero-ring is in the class σ. Since every zero-ring is a subdirect sum of subdirectly irreducible zero-rings each in σ, condition (ii) implies that every zero-ring must be in σ. Thus the class σ contains all zero-rings.

The last phase of the proof is to show that if σ contains all zero-rings, then it must be the class of all rings. This is of course a contradiction to (v), and will finally prove the Theorem. Here we shall make use of the split-null extension and of Proposition 3.6.10.

Suppose that A is any ring, and consider the split-null extension $A * A$. By Proposition 3.6.10.(ii) we know that $\gamma(A * A) \subseteq A'$, that is, the elements of $\gamma(A * A)$ are of the form $(a, 0)$. By Proposition 3.6.10.(i) we know that $\gamma(A * A)$ annihilates A^0. Thus for every $(0, b) \in A^0$ and $(a, 0) \in \gamma(A * A)$ we have

$$(0, ab) = (a, 0)(0, b) \in \gamma(A * A)A^0 = 0,$$

that is, $ab = 0$ for every $b \in A$ and $(a, 0) \in \gamma(A * A)$. Hence for any two elements $(a, 0)$ and $(b, 0)$ in $\gamma(A * A)$ we have

$$(a, 0)(b, 0) = (ab, 0) = (0, 0),$$

which means that $\gamma(A*A)$ is a ring with zero-multiplication. Thus $\gamma(A*A) \in \gamma \cap \sigma = 0$, that is, $A * A \in \sigma$ holds for every ring A. Now using properties (1) and (3) of the split-null extension and condition (iv) we conclude

$$A \cong A * A/A^0 \in \sigma,$$

and so σ is the class of all rings, contradicting (v). □

In Theorem 3.6.11 we assumed that σ is closed under essential extension and this made its radical $\gamma = \mathcal{U}\sigma$ hereditary. Without this we could not prove that γ contained all zero-rings.

EXAMPLE 3.6.12. *A (nonhereditary) radical class having weakly homomorphically closed semisimple class, need not contain all zero-rings.* The class $\mathcal{I}$ of all idempotent rings is a nonhereditary radical class (see Example

3.2.2.(ii)). Its semisimple class $\mathcal{SI}$ contains every zero-ring. We can show that $\mathcal{SI}$ is weakly homomorphically closed and thus four of the five conditions of Theorem 3.6.11 are fulfilled, but $\mathcal{I}$ is not supernilpotent. To see that $\mathcal{SI}$ is weakly homomorphically closed, suppose there is a ring $A \in \mathcal{SI}$ with an ideal I such that $I^2 = 0$ and $A/I \notin \mathcal{SI}$. Then there exists an ideal J of A such that
$$0 \neq \mathcal{I}(A/I) = J/I \triangleleft A/I.$$
Now $J/I \in \mathcal{I}$ and thus $J/I = (J/I)^2$, that is, $J = J^2 + I$. Hence
$$J^2 = (J^2 + I)(J^2 + I) \subseteq J^4 + (J^2 \cap I),$$
and
$$J^4 \subseteq (J^4 + (J^2 \cap I))(J^4 + (J^2 \cap I)) \subseteq$$
$$\subseteq J^8 + J^4(J^2 \cap I) + (J^2 \cap I)J^4 + (J^2 \cap I)^2.$$
Since $I^2 = 0$, also $(J^2 \cap I)^2 = 0$ and we get
$$J^4 \subseteq J^8 + J^6 + J^6 = J^6 \subseteq J^4.$$
Therefore J^4 is idempotent and thus in $\mathcal{I}$. Since $J^4 \triangleleft A \in \mathcal{SI}$, it follows that $J^4 \in \mathcal{I} \cap \mathcal{SI} = 0$. Hence $J/I \neq 0$ is nilpotent and so in $\mathcal{SI}$, a contradiction. This proves that $\mathcal{SI}$ is weakly homomorphically closed.

Weakly homomorphically closed classes containing all zero-rings were studied by Sands [6].

To construct upper radicals we usually begin with a regular class. One can easily see that any regular class of rings with unity must consist of semiprime rings. This remains true if we require only the validity of condition (F).

PROPOSITION 3.6.13 (Anderson, Kaarli and Wiegandt [1]). *A regular class ϱ of rings consists of semiprime rings if and only if ϱ satisfies condition* (F).

Proof: Proposition 3.6.4 makes it clear that if ϱ has only semiprime rings, then ϱ satisfies (F). Thus we assume that ϱ satisfies (F) and want to show that every ring in ϱ is semiprime. Suppose not. Then the regularity of ϱ implies the existence of a ring $J \neq 0$ in ϱ with $J^2 = 0$. Let J^1 be the Dorroh extension of J, and consider the matrix rings
$$A = M_2(J^1), \quad I = M_2(J), \quad \text{and} \quad K = \begin{pmatrix} 0 & J \\ J & J \end{pmatrix}.$$
Then I is an ideal of A. Also K is an ideal of I because $KI = IK = 0$, since $J^2 = 0$. Furthermore, $I/K \cong J \in \varrho$. However, K is definitely not an ideal of A because
$$\begin{pmatrix} 0 & b \\ 0 & 0 \end{pmatrix}\begin{pmatrix} 0 & 0 \\ 1 & 0 \end{pmatrix} = \begin{pmatrix} b & 0 \\ 0 & 0 \end{pmatrix}$$

and this is not in K if we take $b \neq 0$. Therefore condition (F) does not hold for the class ϱ. This contradiction proves that ϱ consists of semiprime rings. □

In Proposition 3.6.13 we cannot drop the regularity of the class ϱ. The largest class ϱ_0 of rings which satisfy condition (F) is definitely bigger than that of all semiprime rings as there are non-semiprime rings with unity element and these satisfy (F) by Proposition 3.2.15. As a supplement to this section we describe the class ϱ_0. Following de la Rosa [1], [2] we say that a ring A is *quasi-semiprime*, if $0 \neq I \triangleleft A$ implies $AIA \neq 0$. We say that I is a *quasi-semiprime ideal* of A, if A/I is a quasi-semiprime ring. Sands [7] defines the *middle annihilator* M of a ring A as the set

$$M = \{a \in A \mid AaA = 0\}.$$

The middle annihilator M is clearly an ideal of A. The next Proposition is obvious.

PROPOSITION 3.6.14. *The following assertions are equivalent:*
(i) A *is a quasi-semiprime ring,*
(ii) *the ring A has zero middle annihilator,*
(iii) *if $a \in A$ is a nonzero element, then $AaA \neq 0$.* □

THEOREM 3.6.15 (Sands [10]). *A ring S is quasi-semiprime if and only if condition* (F) *is satisfied with the ring S.*

Proof: Let us assume that $C \triangleleft B \triangleleft A$ and $B/C \cong S$ is a quasi-semiprime ring. For the ideal $\overline{C}$ of A generated by C we have

$$B\overline{C}B = BCB + BACB + BCAB + BACAB \subseteq C,$$

and therefore

$$(B/C)(\overline{C}/C)(B/C) = 0.$$

Since $B/C \cong S$ is quasi-semiprime, by definition we conclude that $\overline{C}/C = 0$, whence $C \triangleleft A$.

For proving the converse, let us consider a ring S which satisfies condition (F), and its left annihilator

$$K = \{s \in S \mid sS = 0\}.$$

We claim that $K = 0$. For this end let us consider the matrix rings

$$B = \begin{pmatrix} S & K \\ 0 & 0 \end{pmatrix} \quad \text{and} \quad C = \begin{pmatrix} 0 & K \\ 0 & 0 \end{pmatrix}.$$

Since $K \triangleleft S$, one easily sees that $C \triangleleft B$, furthermore, $B/C \cong S$. The mapping $f \colon B \to B$ defined by

$$f \begin{pmatrix} s & u \\ 0 & 0 \end{pmatrix} = \begin{pmatrix} u & 0 \\ 0 & 0 \end{pmatrix}$$

Radical Theory for Associative Rings

for all $s \in S$ and $u \in K$, is an endomorphism of the left B-module $_BB$, as one readily verifies; moreover, $f^2 = 0$. Thus the subring H generated by f in the endomorphism ring of $_BB$, consists of integer multiples of f,

$$H = \{nf \mid n \in \mathbb{Z}\}.$$

On the additive group $A^+ = B^+ \oplus H^+$, let us define a multiplication by the rule

$$(a, kf)(b, nf) = (ab + nfa, 0)$$

for all $a, b \in B$ and $k, n \in \mathbb{Z}$. A straightforward verification shows that in this way we have got a ring A. Identifying B with $\{(b, 0) \mid b \in B\} \subseteq A$ we have

$$C \triangleleft B \triangleleft A \quad \text{and} \quad B/C \cong S.$$

Applying condition (F), we conclude that $C \triangleleft A$. Hence identifying H with $\{(0, nf) \mid n \in \mathbb{Z}\}$ we have $Cf = C$. Rewriting to the original notation, we get

$$\begin{pmatrix} K & 0 \\ 0 & 0 \end{pmatrix} = f \begin{pmatrix} 0 & K \\ 0 & 0 \end{pmatrix} \subseteq \begin{pmatrix} 0 & K \\ 0 & 0 \end{pmatrix}$$

so that $K = 0$, as claimed.

A similar reasoning shows that also the right annihilator of S is zero.

Let us suppose that $SIS = 0$ for some ideal I of S. Since the left and right annihilators of S are zero, we get $SI = 0$ and $I = 0$. Thus S is quasi-semiprime. □

Quasi-semiprime rings will occur later in Sections 3.12, 4.12 and 5.1.

3.7. Supernilpotent radicals and weakly special classes

If we begin with a regular class of semiprime rings, then we know that its upper radical will contain all zero-rings. However, this upper radical need not be hereditary. Proposition 3.6.5 tells us that essential extensions play an important role in guaranteeing that this radical is hereditary and thus supernilpotent.

For any class ϱ of rings we define the *essential cover* $\mathcal{E}\varrho$ by

$$\mathcal{E}\varrho = \{A \mid \text{there is a ring } B \in \varrho \text{ such that } B \triangleleft \cdot A\},$$

that is, $\mathcal{E}\varrho$ is the class of all essential extensions of rings from ϱ.

PROPOSITION 3.7.1 (Anderson and Wiegandt [2]). *If ϱ is a regular class of semiprime rings, then so is its essential cover $\mathcal{E}\varrho$.*

Proof: In view of Example 3.6.1 (i) the class $\mathcal{E}\varrho$ consists of semiprime rings.

To show that $\mathcal{E}\varrho$ is regular, we again take a ring $A \in \mathcal{E}\varrho$. A is an essential extension of a ring $B \in \varrho$. Let I be any nonzero ideal of A. Then

$0 \neq B \cap I \triangleleft B$. Now $B \in \varrho$ and ϱ is regular. Therefore there exists a nonzero homomorphic image $(B \cap I)/K$ of $B \cap I$ in ϱ. Proposition 3.6.4 tells us that K is an ideal of I and $(B \cap I)/K \triangleleft I/K$. Let us choose an ideal J of I such that $(B \cap I) \cap J \subseteq K$ and such that J is maximal with respect to this property. Then by Lemma 3.2.5 we know that

$$(B \cap I)/K \cong ((B \cap I) + J)/J \triangleleft \cdot I/J.$$

Since $0 \neq (B \cap I)/K \in \varrho$, we have $I/J \in \varrho$. Thus $\mathcal{E}\varrho$ is regular. □

Here we mention that the essential cover $\mathcal{E}\varrho$ of a regular class ϱ of semiprime rings need not be closed under essential extensions (see Watters [5]).

PROPOSITION 3.7.2 (le Roux, Heyman and Jenkins [1]). *If ϱ is a regular class of semiprime rings, and $\mathcal{E}\varrho$ its essential cover, then the upper radical $\mathcal{UE}\varrho$ is hereditary and therefore supernilpotent.*

Proof: If $\mathcal{UE}\varrho$ is not hereditary, then there must be a nonzero ring $A \in \mathcal{UE}\varrho$ with an ideal I which is not in $\mathcal{UE}\varrho$. Thus I can be mapped homomorphically onto a nonzero ring $I/K \in \mathcal{E}\varrho$. Now $K \triangleleft I \triangleleft A$ and I/K is semiprime by Proposition 3.7.1. Hence by Proposition 3.6.4 we have $I/K \triangleleft A/K \in \mathcal{UE}\varrho$. Thus without loss of generality we may assume that $0 \neq I \triangleleft A \in \mathcal{UE}\varrho$, $I \in \mathcal{E}\varrho$, I is a semiprime ring and $I \triangleleft \cdot A$ by Proposition 3.2.5. Then there exists an essential ideal J of I such that $0 \neq J \in \varrho$. Considering the ideal $\overline{J}$ of A generated by J, from the Andrunakievich Lemma we know that $\overline{J}^3 \subseteq J$, further $\overline{J} \subseteq I$ and $\overline{J}^3 \triangleleft A$. Let K be any ideal of A such that $\overline{J}^3 \cap K = 0$. Then we have

$$(\overline{J} \cap K)^3 \subseteq \overline{J}^3 \cap K = 0.$$

Hence $\overline{J} \cap K$ is a nilpotent ideal of the semiprime ring I, and so $\overline{J} \cap K = 0$. We have

$$J \cap (I \cap K) \subseteq J \cap K \subseteq \overline{J} \cap K = 0,$$

and hence by $J \triangleleft \cdot I$ it follow that $I \cap K = 0$. Taking into consideration that $I \triangleleft \cdot A$, we conclude that $K = 0$. Thus $\overline{J}^3$ is an essential ideal of A. Since $0 \neq A \in \mathcal{UE}\varrho$ and $\overline{J}^3 \triangleleft \cdot A$, we conclude that $\overline{J}^3 \notin \varrho$. Nonetheless, by $\overline{J}^3 \triangleleft J \in \varrho$ and by the regularity of ϱ there exists a nonzero homomorphic image $\overline{J}^3/L$ of $\overline{J}^3$ in ϱ. Now $\overline{J}^3/L$ is semiprime, so by Proposition 3.6.4 we have $L \triangleleft A$, and hence $\overline{J}^3/L \triangleleft A/L \in \mathcal{UE}\varrho$. Choose an ideal M of A such that M is maximal with respect to $\overline{J}^3 \cap M \subseteq L$. Then by Lemma 3.2.5 we have

$$\overline{J}^3/L \cong (\overline{J}^3 + M)/M \triangleleft \cdot A/M \cong \frac{A/L}{M/L} \in \mathcal{UE}\varrho.$$

From this and from $\overline{J}^3/L \in \varrho$ we get $A/M \in \mathcal{E}\varrho \cap \mathcal{U}\mathcal{E}\varrho = 0$, that is, $A = M$. Thus we have got

$$L \neq \overline{J}^3 = \overline{J}^3 \cap A = \overline{J}^3 \cap M \subseteq L,$$

a contradiction. Therefore $\mathcal{U}\mathcal{E}\varrho$ is hereditary. □

We shall say that *the radical class γ has the intersection property relative to the class ϱ*, if $\gamma(A) = (A)\varrho$ for every ring A. Recall that

$$(A)\varrho = \cap(I \triangleleft A \mid A/I \in \varrho).$$

From Proposition 2.3.7 we know that every radical class has the intersection property relative to its own semisimple class. Moreover, as one readily sees, the intersection property of γ with respect to ϱ implies that $\varrho \subseteq \mathcal{S}\gamma$, ϱ is a regular class and $\gamma = \mathcal{U}\varrho$.

THEOREM 3.7.3 (Ryabukhin [1] and le Roux, Heyman and Jenkins [1]). *For a regular class ϱ of semiprime rings the following two conditions are equivalent:*

(i) $\mathcal{U}\varrho$ *is hereditary,*

(ii) $\mathcal{U}\varrho$ *has the intersection property relative to the essential cover $\mathcal{E}\varrho$ of ϱ.*

Proof: Suppose that $\gamma = \mathcal{U}\varrho$ is hereditary. Then Proposition 3.2.6 says that its semisimple class $\mathcal{S}\gamma$ is closed under essential extensions. Of course $\varrho \subseteq \mathcal{S}\gamma$, and therefore also $\mathcal{E}\varrho \subseteq \mathcal{S}\gamma$. Hence by Proposition 2.3.7 we have

$$\gamma(A) = (A)\mathcal{S}\gamma \subseteq (A)\mathcal{E}\varrho$$

for every ring A. Assume that $\gamma(A) \neq (A)\mathcal{E}\varrho$ for a ring A, and let us call $(A)\mathcal{E}\varrho = B$. Since γ is hereditary, we have

$$\gamma(B) = \gamma(A) \cap B = \gamma(A) \neq B.$$

Thus $B \notin \gamma = \mathcal{U}\varrho$ and there exists a homomorphic image B/K of B such that $0 \neq B/K \in \varrho$. Since ϱ consists of semiprime rings and $K \triangleleft B \triangleleft A$, Proposition 3.6.4 yields that $K \triangleleft A$. Hence $B/K \triangleleft A/K$. Let us choose an ideal J/K of A/K such that J/K is maximal with respect to $B/K \cap J/K = 0$, or equivalently, $B \cap J \subseteq K$. Applying Lemma 3.2.5 we get

$$B/K \cong (B+J)/J \triangleleft \cdot A/J.$$

Hence $B/K \in \varrho$ implies $A/J \in \mathcal{E}\varrho$ which means that $B = (A)\mathcal{E}\varrho \subseteq J$. Thus $B = B \cap J \subseteq K$ follows contradicting $B/K \neq 0$. Hence the radical $\gamma = \mathcal{U}\varrho$ has the intersection property relative to $\mathcal{E}\varrho$.

Conversely, assume that (ii) holds. Since $\varrho \subseteq \mathcal{E}\varrho$, we know that $\mathcal{U}\mathcal{E}\varrho \subseteq \mathcal{U}\varrho$. Take a ring $A \notin \mathcal{U}\mathcal{E}\varrho$. Then there exists an ideal I of A

such that $0 \neq A/I \in \mathcal{E}\varrho$, and so $(A)\mathcal{E}\varrho \subseteq I \neq A$. Using (ii) we get $\mathcal{U}\varrho(A) = (A)\mathcal{E}\varrho \neq A$. Hence $A \notin \mathcal{U}\varrho$, and so $\mathcal{U}\varrho \subseteq \mathcal{U}\mathcal{E}\varrho$. Thus $\mathcal{U}\varrho = \mathcal{U}\mathcal{E}\varrho$. Applying Proposition 3.7.2, we get that $\mathcal{U}\varrho = \mathcal{U}\mathcal{E}\varrho$ is hereditary. □

An immediate consequence of Theorem 3.7.3 is

COROLLARY 3.7.4. *Let ϱ be a regular class of semiprime rings and $\mathcal{E}\varrho$ its essential cover. If $\gamma = \mathcal{U}\varrho$ is a supernilpotent radical, then every γ-semisimple ring $A \in \mathcal{S}\gamma$ is a subdirect sum*

$$A = \sum_{\text{subdirect}} (A_\alpha \mid A_\alpha \in \mathcal{E}\varrho).$$

Proof: Obvious, since $(A)\mathcal{E}\varrho = \gamma(A) = 0$. □

This is an interesting structure theorem and it gives us most information when $\mathcal{E}\varrho \neq \mathcal{S}\mathcal{U}\varrho$. In many cases ϱ is a well filled out class and is even hereditary, but sometimes ϱ is sparse and thin and $\mathcal{E}\varrho$ is not subdirectly closed. It is then that the Corollary is most revealing.

Notice that the intersection property relative to ϱ readily implies the decomposition of $\mathcal{S}\mathcal{U}\varrho$-rings into a subdirect sum of ϱ-rings, but not conversely, as shown by Leavitt [7].

PROPOSITION 3.7.5 (Heyman and Roos [1]). *If ϱ is a hereditary class of semiprime rings, then its essential cover $\mathcal{E}\varrho$ is also a hereditary class of semiprime rings.*

Proof: We know that all rings in $\mathcal{E}\varrho$ are semiprime (Proposition 3.7.1). To show that the class $\mathcal{E}\varrho$ is hereditary, take a ring $A \in \mathcal{E}\varrho$ and let I be any nonzero ideal of A. The ring A must have an essential ideal K in ϱ. Then $0 \neq I \cap K \in \varrho$ because the class ϱ is hereditary. To show that I is in $\mathcal{E}\varrho$, we will show that $I \cap K$ is an essential ideal in I. To this end, let L be any ideal of I such that $(I \cap K) \cap L = 0$ so that $L \cap K = 0$. Let $\overline{L}$ be the ideal of A generated by L. Then by the Andrunakievich Lemma we have

$$(\overline{L} \cap K)^3 \subseteq \overline{L}^3 \cap K \subseteq L \cap K = 0.$$

Since A is semiprime and $(\overline{L} \cap K)^3 = 0$, also $\overline{L} \cap K = 0$. But K is essential in A, and so $L \subseteq \overline{L} = 0$. Thus $I \cap K \triangleleft \cdot I$, and $I \cap K \in \varrho$ implies $I \in \mathcal{E}\varrho$. □

PROPOSITION 3.7.6 (Heyman and Roos [1]). *If $0 \neq K \triangleleft \cdot I \triangleleft \cdot A$ and $\overline{K}$ is the ideal of A generated by K, then $\overline{K}$ is an essential ideal of A.*

Proof: Let L be any nonzero ideal of A. Since I is essential in A, we have $I \cap L \neq 0$. Since K is essential in I, we must have

$$0 \neq K \cap (I \cap L) \subseteq K \cap L \subseteq \overline{K} \cap L.$$

Thus $\overline{K}$ is essential in A. □

PROPOSITION 3.7.7 (Heyman and Roos [1]). *If $0 \neq I \triangleleft \cdot A$ and A is a semiprime ring, then $I^n \triangleleft \cdot A$ for every positive integer n.*

Proof: For any nonzero ideal L of A we have $I \cap L \neq 0$. Then for any n, $(I \cap L)^n \neq 0$ since A is semiprime. Thus

$$0 \neq (I \cap L)^n \subseteq I^n \cap L$$

and I^n is essential in A. □

PROPOSITION 3.7.8 (Heyman and Roos [1]). *If ϱ is a hereditary class of semiprime rings, then its essential cover $\mathcal{E}\varrho$ is closed under essential extensions.*

Proof: Suppose that $I \triangleleft \cdot A$ and $I \in \mathcal{E}\varrho$. Then in view of Example 3.6.1 (i), I is a semiprime ring and so is A as well. Now there is an essential ideal $K \in \varrho$ of I. By Propositions 3.7.6 and 3.7.7 we get that $\overline{K}$ as well as $\overline{K}^3$ are essential ideals of A. Since $\overline{K}^3 \triangleleft K \in \varrho$, and ϱ is hereditary, it follows that $\overline{K}^3 \in \varrho$. Thus $A \in \mathcal{E}\varrho$ and the Proposition has been proved. □

Now we can easily construct the semisimple class of the largest supernilpotent radical γ which excludes a given class ϱ of semiprime rings (cf. Sands' construction Theorem 3.1.11). We define the *hereditary closure operator* $\mathcal{H}$ by

$$\mathcal{H}\varrho = \{A \mid A \text{ is an accessible subring of a ring } B \in \varrho\}.$$

THEOREM 3.7.9. *Let ϱ be any class of semiprime rings, and let σ be the class of all subdirect sums of rings in $\mathcal{E}\mathcal{H}\varrho$. Then $\mathcal{U}\sigma = \mathcal{U}\mathcal{E}\mathcal{H}\varrho$ and this is a supernilpotent radical. Moreover, σ is the smallest semisimple class containing ϱ and defining a supernilpotent upper radical.*

Proof: It is clear that $\mathcal{H}\varrho$ consists of semiprime rings (for, if $K \triangleleft I \triangleleft A \in \varrho$ and $K^2 = 0$ and $\overline{K}$ is the ideal of A generated by K, then by the Andrunakievich Lemma $\overline{K}$ is a nilpotent ideal of $A \in \varrho$, so $\overline{K} = 0$ and I is semiprime). The class $\mathcal{E}\mathcal{H}\varrho$ is just the essential cover of $\mathcal{H}\varrho$ and by Proposition 3.7.5 it is also a hereditary class of semiprime rings. Proposition 3.7.8 tells us that $\mathcal{E}\mathcal{H}\varrho$ is closed under essential extensions. Hence Proposition 3.7.2 yields that $\gamma = \mathcal{U}\mathcal{E}\mathcal{H}\varrho$ is a supernilpotent radical, and from Corollary 3.7.4 we infer that every ring in $\mathcal{S}\gamma = \mathcal{S}\mathcal{U}\mathcal{E}\mathcal{H}\varrho$ is a subdirect sum of rings in $\mathcal{E}\mathcal{H}\varrho$. Our σ contains all such subdirect sums, and therefore $\mathcal{S}\mathcal{U}\mathcal{E}\mathcal{H}\varrho \subseteq \sigma$. Since $\mathcal{E}\mathcal{H}\varrho$ is hereditary, we have $\mathcal{E}\mathcal{H}\varrho \subseteq \mathcal{S}\mathcal{U}\mathcal{E}\mathcal{H}\varrho$, and since semisimple classes are closed under subdirect sums, we get $\sigma \subseteq \mathcal{S}\mathcal{U}\mathcal{E}\mathcal{H}\varrho$. Thus $\sigma = \mathcal{S}\mathcal{U}\mathcal{E}\mathcal{H}\varrho$ and $\mathcal{U}\sigma = \mathcal{U}\mathcal{E}\mathcal{H}\varrho$ have been established. By the construction of σ it is clear that σ is the smallest semisimple class containing ϱ and having a supernilpotent upper radical. □

If I is an ideal of a ring A, we define

$$I^* = \{x \in A \mid xI = Ix = 0\},$$

namely the set of all two-sided annihilators of I in A. Clearly $I^* \triangleleft A$.

PROPOSITION 3.7.10 (Heyman and Roos [1]). *The following conditions are equivalent for any class ϱ of semiprime rings:*
- (R) *if $I \triangleleft A$, $I \in \varrho$ and $I^* = 0$, then $A \in \varrho$,*
- (A) *if $I \triangleleft A$ and $I \in \varrho$, then $A/I^* \in \varrho$,*
- (λ) *ϱ is closed under essential extensions.*

Proof: (R) $\Rightarrow$ (λ). Suppose that A has an essential ideal I in ϱ. Consider the ideal I^*. Since I is semiprime, $I \cap I^* = 0$, and therefore $I^* = 0$. Hence condition (R) gives us $A \in \varrho$. This proves (λ).

(λ) $\Rightarrow$ (A) Suppose that A has an ideal I in ϱ, and suppose that L is an ideal of A such that $I \cap L = 0$. Then

$$IL + LI \subseteq I \cap L = 0,$$

and therefore $L \subseteq I^*$. Also $I \cap I^* = 0$ because I is semiprime. Therefore I^* is maximal with respect to having zero intersection with I. Then Lemma 3.2.5 tells us that

$$I \cong (I + I^*)/I^* \triangleleft \cdot A/I^*.$$

Since $I \in \varrho$, condition (λ) tells us that A/I^* must be in ϱ, and this is (A).

Finally (A) clearly implies (R). $\square$

Conditions (A), (R) and (λ) have been introduced by Andrunakievich [1], Ryabukhin [1] and Heyman and Roos [1], respectively. Following Ryabukhin [1] we define a *weakly special class* ϱ as one which is hereditary, which consists of semiprime rings, and which satisfy any of the conditions (R), (A) or (λ).

COROLLARY 3.7.11 (Heyman and Roos [1]). *If ϱ is a hereditary class of semiprime rings, then its essential cover $\mathcal{E}\varrho$ is the smallest weakly special class containing ϱ.*

Proof: Propositions 3.7.5, 3.7.8 and 3.7.10 tell us that $\mathcal{E}\varrho$ is a weakly special class, and it is clear that $\mathcal{E}\varrho$ is the smallest such class containing ϱ. $\square$

THEOREM 3.7.12 (Ryabukhin [1]). *If ϱ is a weakly special class, then its upper radical $\gamma = \mathcal{U}\varrho$ is supernilpotent and has the intersection property relative to ϱ. Every supernilpotent radical is the upper radical of a weakly special class.*

Proof: If ϱ is a weakly special class, then by Propositions 3.7.2 γ is supernilpotent, and by Theorem 3.7.3 γ has the intersection property relative to ϱ.

If γ is any supernilpotent radical, then its semisimple class $\mathcal{S}\gamma$ is hereditary (Corollary 3.1.4), has only semiprime rings (Proposition 3.6.3), is closed under essential extensions (Proposition 3.2.6), and therefore $\mathcal{S}\gamma$ is a weakly special class. □

EXAMPLE 3.7.13. *The class of all semiprime rings is a weakly special class.* By definition the semisimple class $\mathcal{S}\beta$ of the Baer radical β consists of semiprime rings. Further, if $A \notin \mathcal{S}\beta$, then $\beta(A) \neq 0$ and in view of Corollary 3.4.8 there exists an ideal I of $\beta(A)$ such that $I \neq 0$ and $I^2 = 0$. Hence by the Andrunakievich Lemma the ideal $\overline{I}$ of A generated by I is nilpotent and $\overline{I} \neq 0$. Thus A is not semiprime, and therefore $\mathcal{S}\beta$ coincides with the class of all semiprime rings, and the Baer radical β is the upper radical of that class. Since β is hereditary, $\mathcal{S}\beta$ is closed under essential extensions (Proposition 3.2.6). Moreover, $\mathcal{S}\beta$ is hereditary (Corollary 3.1.4), and so $\mathcal{S}\beta$ *is a weakly special class*, in fact *the largest weakly special class*.

3.8. Special radicals

We begin with a characteristic property of semiprime rings, and this will explain the origin of the terminology "semiprime".

THEOREM 3.8.1. *A ring A is semiprime if and only if the intersection Q of all the prime ideals of A is zero.*

Proof: If A is not semiprime, then it has a nonzero ideal I with $I^2 = 0$. Then I is contained in every prime ideal of A, and therefore $I \subseteq Q$. Thus $Q \neq 0$.

Conversely, suppose that A is semiprime. Take any nonzero element x_1 of A, and consider the principal ideal (x_1) of A generated by x_1. The ideal (x_1) cannot be nilpotent, and so

$$0 \neq (x_1)^3 \subseteq A(\mathbb{Z}x_1 + Ax_1 + x_1A + Ax_1A)A \subseteq Ax_1A.$$

Hence $(Ax_1A)^2 \neq 0$ because A is semiprime, and there must exist an element $y_1 \in A$ such that $x_1y_1x_1 \neq 0$. Define $x_2 = x_1y_1x_1$. This element x_2 is in (x_1). We repeat this process and obtain a sequence $x_1, \ldots, x_n, \ldots$ such that $0 \neq x_n \in (x_k)$ for every k with $1 \leq k \leq n$. Let S be the set of all ideals of A which do not meet $\{x_1, \ldots, x_n, \ldots\}$. Since 0 does not meet $\{x_1, \ldots, x_n, \ldots\}$, S is not empty. Then by Zorn's Lemma we pick a maximal element K of S. We plan to show that K is a prime ideal of A. If $L_1L_2 \subseteq K$ for ideals L_1, L_2 of A and if $L_1 + K$ and $L_2 + K$ both properly contain K, then some x_k is in $L_1 + K$ and some x_l is in $L_2 + K$ by the maximality of K. Let $n = \max(k, l)$. Then

$$x_n \in (x_k) \cap (x_l) \subseteq (L_1 + K) \cap (L_2 + K),$$

and therefore

$$x_{n+1} = x_ny_nx_n \in (L_1 + K)(L_2 + K) \subseteq L_1L_2 + K \subseteq K.$$

But this contradicts the maximality of K. Hence, if $L_1 L_2 \subseteq K$ then either $L_1 \subseteq K$ or $L_2 \subseteq K$, and so K is a prime ideal of A. Since $x_1 \notin K$, x_1 cannot be in the intersection Q of all the prime ideals of A. The choice of x_1 was arbitrary, so necessarily $Q = 0$. □

We need more information about prime rings.

PROPOSITION 3.8.2. *If $I \triangleleft A$ and A is a prime ring, then so is I.*

Proof: Let K and L be ideals of I with the property that $KL = 0$. Let $\overline{K}$ and $\overline{L}$ be the ideals of A generated by K and L, respectively. The Andrunakievich Lemma tells us that $\overline{K}^3 \cdot \overline{L}^3 \subseteq KL = 0$. Since A is prime, either $\overline{K}^3 = 0$ or $\overline{L}^3 = 0$. Then $\overline{K} = 0$ or $\overline{L} = 0$, again since A is prime. Thus $K = 0$ or $L = 0$ and I itself is prime. □

PROPOSITION 3.8.3. *If $I \triangleleft \cdot A$ and I is a prime ring, then so is A.*

Proof: Let K and L be ideals of A such that $KL = 0$. Suppose that $K \neq 0$. Then $I \cap K \neq 0$. Now

$$(I \cap K)(I \cap L) \subseteq KL = 0.$$

Since I is a prime ring, $I \cap L = 0$. Then $L = 0$ because I is essential in A. Therefore A is a prime ring. □

COROLLARY 3.8.4. *The class $\mathcal{P}$ of all prime rings is a weakly special class.*

Proof: The assertion follows immediately from Propositions 3.7.10, 3.8.2 and 3.8.3. □

Let us observe that because of Proposition 3.8.3, it is clear that Propositions 3.7.1 and 3.7.5 remain true if we write prime everywhere for semiprime.

Andrunakievich [1], [2] called a class ϱ of rings a *special class*, if ϱ is hereditary, consists of prime rings and satisfies condition (A) of Proposition 3.7.10. By Proposition 3.7.10, condition (A) can be replaced by condition (R) or by (λ) (being closed under essential extensions). This latter has been observed by E. H. Connell (see McKnight and Musser [1] Lemma 9). In view of Proposition 3.8.3 *in the definition of a special class ϱ condition (λ) can be substituted by the following requirement*:

(∗)   *if $0 \neq I \triangleleft A$, $I \in \varrho$ and A is a prime ring then $A \in \varrho$,*

because in a prime ring A every nonzero ideal I is obviously essential. Clearly, *every special class is a weakly special class.*

A radical γ will be called a *special radical*, if it is the upper radical of a special class. It is clear that *every special radical is supernilpotent and hence hereditary.* Furthermore, the results of Section 3.7 hold for special radicals, and in particular Corollary 3.7.4 gives us

COROLLARY 3.8.5. *Let ϱ be a special class of rings and $\gamma = \mathcal{U}\varrho$ its upper radical. If A is a γ-semisimple ring, then A is a subdirect sum of rings in ϱ.* □

One can give also a direct proof to Corollary 3.8.5 by proving that $\gamma = \mathcal{U}\varrho$ is hereditary and then that γ has the intersection property relative to the class ϱ. The proof runs as those of Proposition 3.7.2 and Theorem 3.7.3, but is much simpler.

Corollary 3.8.5 is the best general decomposition theorem for semisimple rings, this is what we expected and what we achieved. For the non-existence of better general decomposition results the reader is referred to Márki, Stewart and Wiegandt [1].

THEOREM 3.8.6. *Let γ be a hereditary radical. Then the class $\mathcal{S}\gamma \cap \mathcal{P}$ of prime γ-semisimple rings is a special class, and it is the largest special class contained in $\mathcal{S}\gamma$. If γ is a special radical, then $\gamma = \mathcal{U}(\mathcal{S}\gamma \cap \mathcal{P})$.*

Proof: The class $\mathcal{S}\gamma \cap \mathcal{P}$ consists, of course, of prime rings. It is hereditary (Proposition 3.8.2), and closed under essential extensions (Propositions 3.2.6 and 3.8.3). Therefore $\mathcal{S}\gamma \cap \mathcal{P}$ is a special class. The other assertions are clear. □

COROLLARY 3.8.7. *If ϱ is a hereditary class of prime rings, then its essential cover $\mathcal{E}\varrho$ is the smallest special class containing ϱ. In particular, if ϱ is a class of simple prime rings, then the smallest special class containing ϱ is the class of all subdirectly irreducible rings each having its heart in ϱ.*

Proof: The first part follows from Corollary 3.7.11 and Proposition 3.8.3. The second part follows from Proposition 3.6.6. □

We mention here that totally different special classes may yield the same special radical. Beidar [7] gave examples of special classes ϱ_1 and ϱ_2 such that $\varrho_1 \cap \varrho_2 = 0$ and $\mathcal{U}\varrho_1 = \mathcal{U}\varrho_2$.

The next Theorem 3.8.8 and Corollary 3.8.9 show that the intersection property with respect to a class ϱ of prime rings implies the hereditariness of the upper radical $\mathcal{U}\varrho$. This result will get an important application in Sections 3.13 and 4.9.

THEOREM 3.8.8 (Beidar [4]). *Let ϱ be a class of prime rings such that $\gamma = \mathcal{U}\varrho$ is a radical class. If the radical γ has the intersection property relative to ϱ, then $\mathcal{E}\varrho \subseteq \mathcal{SU}\varrho$.*

Proof: Suppose there is a ring $A \in \mathcal{E}\varrho$ with $\gamma(A) \neq 0$. Then there is a ring $B \in \varrho$ with $B \triangleleft \cdot A$ and thus $\gamma(A) \cap B \neq 0$.

Let $\mathcal{V}$ be the variety generated by ϱ, and let $\{A_\lambda \mid \lambda \in \Lambda\}$ be a *set* of rings from ϱ which generates $\mathcal{V}$. If a prime ring satisfies an identity, then every essential extension of this ring satisfies that identity. Hence $\mathcal{E}\varrho = \mathcal{V}$.

Let X be a set such that

$$|X| \geq \aleph_0, \ |X| \geq |A| \ \text{and} \ |X| \geq |A_\lambda| \ \text{for all} \ \lambda \in \Lambda,$$

and let F be the free ring in $\mathcal{V}$ generated by X. If $f = f(x_1, x_2, \ldots, x_n) \in F \setminus \{0\}$, $(x_1, x_2, \ldots, x_n \in X)$, then there is an A_λ and there are elements $a_1, a_2, \ldots, a_n \in A_\lambda$ for which $f(a_1, a_2, \ldots, a_n) \neq 0$ (since otherwise $f \equiv 0$ would be an identity of each A_λ, and hence of $\mathcal{V}$, and hence of F, that is, f would be 0 in F). We re-name this A_λ as H_f. As $|X| \geq |H_f|$, there is a surjective homomorphism $\varphi_f : F \to H_f$ such that

$$\varphi_f(x_1) = a_1, \ \varphi_f(x_2) = a_2, \ldots, \ \varphi_f(x_n) = a_n,$$

and hence $\varphi_f(f) = f(a_1, a_2, \ldots, a_n) \neq 0$. We therefore have

(1) $$\bigcap_{f \in F \setminus \{0\}} \ker \varphi_f = 0.$$

Now $A \in \mathcal{V}$ and $|X| \geq |A|$, so there is a surjective homomorphism $\varphi_0 : F \to A$. We re-name A as H_0, and then for *every* $f \in F$ (even for $f = 0$) we have a surjective homomorphism $\varphi_f : F \to H_f$, and hence a homomorphism

$$\psi : F \to \prod_{f \in F} H_f \ \text{ with } \ \psi(z) = (\ldots, (\varphi_f(z))_f, \ldots) \ \text{ for every } z \in F.$$

We identify $H_0(= A)$ with its copy in $\prod_{f \in F} H_f$, and let $\overline{F} = \psi(F) \cong F$. If $s \in \overline{F} \cap H_0$, then there exists $t \in F$ such that $s = \psi(t) = (\ldots, \varphi_f(t), \ldots)$ and $\varphi_f(t) = 0$ for all $f \neq 0$. But then by (1), $t = 0$, so $s = 0$. Thus $\overline{F} \cap A \ (= \overline{F} \cap H_0) = 0$.

Let $G = \overline{F} + B$. As B is an ideal of $\prod_{f \in F} H_f$ (being an ideal of the direct summand $H_0(= A)$), G is a subring of $\prod_{f \in F} H_f$. Clearly then $B \triangleleft G$ and, as $\overline{F} \cap B \subseteq \overline{F} \cap A = 0$, $G/B \cong \overline{F} \cong F$. From (1), as F is a subdirect sum of the H_f and these are in ϱ, $\gamma(F) = 0$. Hence $\gamma(G) \subseteq B$, so $\gamma(G) \subseteq \gamma(B)$. But $B \in \varrho$ so $\gamma(B) = 0$ and hence $\gamma(G) = 0$.

Now G must be a subdirect sum of rings in ϱ, so as $0 \neq \gamma(A) \cap B \subseteq G$, there is a ring $D \in \varrho$ and a surjective homomorphism $\vartheta : G \to D$ such that $\vartheta(\gamma(A) \cap B) \neq 0$.

If $y \in B$, then y is identified in $\prod_{f \in F} H_f$ with an element $(\ldots, b_f, \ldots)$ where $b_f = 0$ for all $f \neq 0$ and $b_0 = y$. Then for every $g \in F$ we have

$$y\psi(g) = (\ldots, b_f, \ldots)(\ldots, \varphi_f(g), \ldots) = (\ldots, b_f \varphi_f(g), \ldots).$$

But $b_f\varphi_f(g) = 0\varphi_f(g) = 0$ for $f \neq 0$, so $y\psi(g) = y(\ldots, z_f, \ldots)$, where $z_0 = \varphi(g)$ and $z_f = 0$ for $f \neq 0$. Our identification equates $(\ldots, z_f, \ldots)$ and $\varphi_0(g)$. Thus we have

(2) $\qquad y\psi(g) = y\varphi_0(g)$ for all $y \in B$, $g \in F$.

As $0 \neq \vartheta(\gamma(A) \cap B) \subseteq \vartheta(B)$ and ϑ is surjective, we have $0 \neq \vartheta(B) \triangleleft D$. But if $b \in B$, $g \in F$, then

$$\vartheta(b)(\vartheta\psi(g) - \vartheta\varphi_0(g)) = \vartheta(b(\psi(g) - \vartheta_0(g))) = 0$$

by (2). As D is a prime ring, this means that

(3) $\qquad \vartheta\psi(g) = \vartheta\varphi_0(g)$ for all $g \in F$.

In particular, if $\varphi_0(g) = 0$, then $\vartheta\psi(g) = 0$, that is, $\ker \varphi_0 \subseteq \ker \vartheta\psi$.

If $b \in B$ then $b \in A$ so $b = \varphi_0(f)$ for some $f \in F$ and then by (3) we have

$$\vartheta(b) = \vartheta\varphi_0(f) = \vartheta\psi(f) \in \vartheta(\overline{F}),$$

so that $\vartheta(B) \subseteq \vartheta(\overline{F})$ and thus

$$\vartheta(G) = \vartheta(\overline{F} + B) = \vartheta(\overline{F}) + \vartheta(B) = \vartheta(\overline{F}).$$

Noting that $A = \varphi_0(F)$ and $\vartheta\psi(F) = \vartheta(\overline{F}) = \vartheta(G) = D$, and using the fact about kernels just established, we conclude that there is a surjective homomorphism η making a commutative diagram

$$\begin{array}{ccc} & F & \\ \varphi_0 \downarrow & \searrow \vartheta\psi & \\ A & \xrightarrow{\eta} & D \end{array}.$$

Let $b = \varphi_0(f)$, $b \in B$, $f \in F$. Then by (3) we have

$$\vartheta(b) = \vartheta\varphi_0(f) = \vartheta\psi(f) = \eta\varphi_0(f) = \eta(b).$$

In particular, taking $b \in \gamma(A) \cap B$, we get

$$0 \neq \vartheta(\gamma(A) \cap B) = \eta(\gamma(A) \cap B) \triangleleft \eta(A) = D.$$

But then $0 \neq \eta(\gamma(A)) \subseteq \gamma(D)$, while D, being in ϱ, is γ-semisimple. From this contradiction we conclude that $\gamma(A) = 0$ for all $A \in \mathcal{E}\varrho$. $\square$

COROLLARY 3.8.9 (Beidar [4]). *A radical γ is special if and only if it is the upper radical $\gamma = \mathcal{U}\varrho$ defined by some class ϱ of prime rings and has the intersection property with respect to the class ϱ.*

Proof: Assume that the upper radical $\gamma = \mathcal{U}\varrho$ has the intersection property with respect to ϱ. Clearly, all accessible subrings in ϱ are in $\mathcal{SU}\varrho$. Let $\sigma = \mathcal{S}\gamma$ and π the class of all accessible subrings of rings in ϱ. Then $\varrho \subseteq \pi \subseteq \sigma$, so for every ring A we have

$$\mathcal{U}\varrho(A) = (A)\varrho \supseteq (A)\pi \supseteq (A)\sigma = \gamma(A).$$

Thus without loss of generality we may assume that ϱ is hereditary. Then by Corollary 3.8.7, $\mathcal{E}\varrho$ is a special class. By Theorem 3.8.8 we have $\varrho \subseteq \mathcal{E}\varrho \subseteq \mathcal{SU}\varrho$, so $\gamma = \mathcal{U}\varrho = \mathcal{UE}\varrho$. □

Semisimple classes of special radicals are characterized in

THEOREM 3.8.10 (Rjabuhin and Wiegandt [1]). *A class σ of rings is the semisimple class of a special radical if and only if*
  (i)   *σ is regular,*
  (ii) *σ is closed under subdirect sums,*
  (iii) *σ is closed under essential extensions,*
  (iv) *if $A \in \sigma$, then A is a subdirect sum*

$$A = \sum_{\text{subdirect}} (A_\alpha \mid A_\alpha \in \sigma \cap \mathcal{P})$$

of prime rings A_α, each in σ.

Proof: Corollaries 3.6.9 and 3.8.5 tell us that if σ is the semisimple class of a special radical, then all four conditions hold.

Conversely, suppose that the four conditions hold for σ. Then Theorem 3.2.7 tells us that σ is the semisimple class of the hereditary radical $\gamma = \mathcal{U}\sigma$. Hence $\mathcal{S}\gamma = \mathcal{SU}\sigma = \sigma$, and by Theorem 3.8.6 the class $\sigma \cap \mathcal{P}$ is a special class. The relation $\sigma \cap \mathcal{P} \subseteq \sigma$ implies $\mathcal{SU}(\sigma \cap \mathcal{P}) \subseteq \mathcal{SU}\sigma = \sigma$. Using condition (iv) we get $\sigma \subseteq \mathcal{SU}(\sigma \cap \mathcal{P})$. Thus $\sigma = \mathcal{SU}(\sigma \cap \mathcal{P})$ and so $\gamma = \mathcal{U}(\sigma \cap \mathcal{P})$ is a special radical. □

Theorem 3.7.9 helps us to construct the smallest semisimple class σ such that σ contains a given class ϱ of prime rings and $\gamma = \mathcal{U}\sigma$ is a special radical.

Up to now we have seen special radicals only as upper radicals of special classes. We now want to obtain an internal characterization of special radicals. To this end we define a ring B to be a *prime image* of a ring A, if B is a homomorphic image of A and B is a prime ring.

THEOREM 3.8.11 (Gardner and Wiegandt [1]). *A class γ of rings is a special radical if and only if*
  (i)   *γ is homomorphically closed,*
  (ii) *γ is hereditary,*
  (iii) *if every nonzero prime image of a ring A has a nonzero ideal in γ, then A is in γ.*

Proof: If γ is a special radical, then (i) and (ii) are immediate. To establish (iii), suppose that every nonzero prime image of A has a nonzero ideal in γ. If A is not in γ, then $A/\gamma(A)$ is a nonzero image of A. By Corollary 3.8.5 and Theorem 3.8.6, $A/\gamma(A)$ is a subdirect sum of rings $A_\alpha \in \mathcal{S}\gamma \cap \mathcal{P}$. Each A_α is a prime image of A, and is γ-semisimple. Thus none of them can have a nonzero ideal in γ. Then each A_α must be 0, and so $A/\gamma(A) = 0$, that is $A = \gamma(A) \in \gamma$, a contradiction. This proves (iii).

Conversely, suppose that γ is a class having the three conditions. From (i) the class γ satisfies (R1) of Theorem 2.1.5. From (iii) γ also satisfies (R2). Therefore γ is a radical class. From (ii) and Proposition 3.2.6 we know that the semisimple class $\mathcal{S}\gamma$ is closed under essential extensions. Then $\varrho = \mathcal{S}\gamma \cap \mathcal{P}$ is a special class by Theorem 3.8.6. Certainly $\gamma \subseteq \mathcal{U}\varrho$. To see that $\mathcal{U}\varrho \subseteq \gamma$, take any $A \in \mathcal{U}\varrho$. Then any nonzero prime image of A is not in ϱ and therefore not in $\mathcal{S}\gamma$. Thus every nonzero prime image of A does have a nonzero ideal in γ. By condition (iii), A itself must be in γ. Therefore $\gamma = \mathcal{U}\varrho$ and γ is a special radical. $\square$

Supernilpotent and special radicals have been constructed as upper radicals in Theorems 3.7.12 and 3.8.6. The smallest supernilpotent radical containing a given hereditary and homomorphically closed class δ is given simply as the lower radical $\mathcal{L}(\delta \cup \mathcal{Z})$ where $\mathcal{Z}$ is the class of all zero-rings. Moreover, by Corollary 3.4.9 we have $\mathcal{L}(\delta \cup \mathcal{Z}) = (\delta \cup \mathcal{Z})_2$. If we again start with a given class δ, it is not quite so easy to construct the smallest special radical containing δ. The solution is based on the characterization given in Theorem 3.8.11.

THEOREM 3.8.12 (Gardner and Wiegandt [1]). *Let δ be a hereditary and homomorphically closed class of rings. Then the smallest special radical containing δ is the class*

$$\mathcal{L}_{\mathrm{sp}}\delta = \left\{ A \;\middle|\; \begin{array}{l} \text{every nonzero prime image} \\ \text{of } A \text{ has a nonzero ideal in } \delta \end{array} \right\}$$

Proof: Clearly $\mathcal{L}_{\mathrm{sp}}\delta$ is homomorphically closed, and contains δ. To see that it is hereditary, let $I \triangleleft A \in \mathcal{L}_{\mathrm{sp}}\delta$, and let I/K be any nonzero prime image of I. We know that K must be an ideal of A by Proposition 3.6.4, and thus $I/K \triangleleft A/K$. Take J to be an ideal of A with $I \cap J = K$ and maximal for this. Then by Lemma 3.2.5 we have

$$I/K \cong (I+J)/J \triangleleft \cdot A/J \in \mathcal{L}_{\mathrm{sp}}\delta.$$

By Proposition 3.8.3 the ring A/J is a prime ring and is thus a nonzero prime image of A. Then there must be a nonzero ideal L/J of A/J with L/J in δ. Since $(I+J)/J \triangleleft \cdot A/J$, we have $(I+J)/J \cap L/J \neq 0$. Let us call this intersection M/J. Of course, M/J is an ideal of L/J, and since $L/J \in \delta$ and δ is hereditary, M/J must be in δ. Then M/J is a nonzero δ-ideal of $(I+J)/J \cong I/K$, and therefore I is in $\mathcal{L}_{\mathrm{sp}}\delta$. Thus $\mathcal{L}_{\mathrm{sp}}\delta$ is hereditary.

Next, we shall show that condition (iii) of Theorem 3.8.11 holds for $\mathcal{L}_{\mathrm{sp}}\delta$. Suppose that every nonzero prime image B of a ring A has a nonzero ideal I in $\mathcal{L}_{\mathrm{sp}}\delta$. Since B is a prime ring, so is I by Proposition 3.8.2. Then I must have a nonzero ideal K in δ by the definition of $\mathcal{L}_{\mathrm{sp}}\delta$. Let $\overline{K}$ be the ideal of B generated by K. Since B is prime, so is $\overline{K}$ by Proposition 3.8.2. Then $\overline{K}^3$ is nonzero, and $\overline{K}^3 \triangleleft K$ by the Andrunakievich Lemma. Since $K \in \delta$ and δ is hereditary, we have that $\overline{K}^3 \in \delta$. Therefore B has a nonzero ideal in δ. Thus A itself must be in $\mathcal{L}_{\mathrm{sp}}\delta$, and condition (iii) holds. Then by Theorem 3.8.11 $\mathcal{L}_{\mathrm{sp}}\delta$ is a special radical.

To see that this is the smallest special radical containing δ, let γ be any special radical containing δ. Take any ring $A \in \mathcal{L}_{\mathrm{sp}}\delta$. Then every nonzero prime image of A has a nonzero ideal in $\delta \subseteq \gamma$. Then by Theorem 3.8.11.(iii), A itself is in γ. Thus $\mathcal{L}_{\mathrm{sp}}\delta \subseteq \gamma$. □

So far all the supernilpotent radicals we met are special as seen from the following examples.

EXAMPLE 3.8.13. *The Baer radical β is the smallest special radical.* To see this we recall that the class $\mathcal{P}$ of all prime rings is the largest weakly special class consisting of prime rings (cf. Corollary 3.8.4), and so $\mathcal{P}$ is the largest special class, and $\mathcal{UP}$ is the smallest special radical. Clearly $\beta \subseteq \mathcal{UP}$ since β is the smallest supernilpotent radical (Example 3.6.1.(i)). On the other hand, take any ring $A \in \mathcal{UP}$. Let B be any nonzero homomorphic image of A. Then $B \in \mathcal{UP}$, and thus B has no nonzero prime ideals. By Theorem 3.8.1 B cannot be semiprime. Therefore B has a nonzero ideal I with $I^2 = 0$. Then $A \in \mathcal{LZ} = \beta$ follows, and so $\beta = \mathcal{UP}$.

EXAMPLE 3.8.14. (i). The class of all simple rings with unity element is a special class, and therefore *the Brown–McCoy radical* (Example 2.2.4) *is special.*

(ii) *The Thierrin radical* (Example 3.2.18) *is a special radical* because the class of all division rings is a special class.

(iii) *The antisimple radical* (Example 3.6.8.(i)) *is special,* because the class of all subdirectly irreducible rings with idempotent hearts is a special class.

(iv) *The Behrens radical* (Example 3.6.8.(ii)) *is special,* because the class of all subdirectly irreducible rings each with a nonzero idempotent in its heart, is a special class.

We have considered, for any ideal I of a ring A, the set

$$I^* = \{x \in A \mid xI = 0 = Ix\}$$

of *two-sided annihilator* of I in A. We can also consider the larger sets

$$I^r = \{x \in A \mid Ix = 0\}$$

of right annihilators of I in A and

$$I^l = \{x \in A \mid xI = 0\}$$

of left annihilators of I in A. (Later in this book I^l will be denoted also by $(0:I)_A$.) All three sets are ideals and

$$I^* = I^r \cap I^l.$$

PROPOSITION 3.8.15. *If I is a semiprime ring and $I \triangleleft A$, then $I^* = I^r = I^l$.*

Proof: We know that $II^r = 0$. Then

$$(I^r I)^2 = I^r I I^r I = 0,$$

and since $I^r I$ is an ideal of the semiprime ring I, we have $I^r I = 0$. Thus $I^r = I^*$. Similarly $I^l = I^*$. □

EXAMPLE 3.8.16. Let ϱ be the class of all domains. It is clear that all rings in ϱ are prime, and also that ϱ is hereditary (in fact, strongly hereditary). To show that ϱ is a special class, we need only show it satisfies condition (R) of Proposition 3.7.10. To this end, suppose that $I \triangleleft A$, $I \in \varrho$ and $I^* = 0$. Then $I^r = I^l = 0$ by Proposition 3.8.15. We must show that A is in ϱ. Let $a, b \in A$ be any two nonzero elements. Then $Ia \neq 0$ since $I^r = 0$, and $bI \neq 0$ since $I^l = 0$. Hence $IabI \neq 0$ since $Ia \subseteq I$, $bI \subseteq I$ and I has no zero divisors. Therefore $ab \neq 0$, and A has no zero divisors. Thus $A \in \varrho$ and condition (R) holds, whence ϱ is a special class. *The special radical $\mathcal{N}_g = \mathcal{U}\varrho$ is called the generalized nil radical* of Andrunakievich [2] and Thierrin [1]. A nil ring cannot be mapped onto a nonzero domain, so $\mathcal{N} \subseteq \mathcal{N}_g$. This justifies the terminology. *Every $\mathcal{N}_g$-semisimple ring is a subdirect sum of domains.*

Later we shall prove that Köthe's nil radical, Levitzki's locally nil radical and the Jacobson radical are all special radicals. However, there are supernilpotent radicals which are not special.

EXAMPLE 3.8.17 (Ryabukhin [1]). Let us consider the class

$$\mathcal{K} = \{\text{all Boolean rings } A \mid \text{every nonzero ideal of } A \text{ is infinite}\}.$$

First, we show that *the class $\mathcal{K}$ is not empty*. Let us consider the countably infinite dimensional vector space R over the two element field with basis $\{e_\lambda \mid \lambda \in \mathbb{Q}\}$, and define multiplication on R via $e_\lambda e_\mu = e_{\max(\lambda,\mu)}$. For any element $r = e_{\lambda_1} + \cdots + e_{\lambda_n} \in R$ we have

$$r^2 = e_{\lambda_1} + 3e_{\lambda_2} + \cdots + (2n-1)e_{\lambda_n} = e_{\lambda_1} + \cdots + e_{\lambda_n} = r,$$

and so R is a Boolean ring. Let $I \triangleleft R$ and $i = e_{\mu_1} + \cdots + e_{\mu_n}$ a nonzro element in I. If n is odd, then for any $\lambda \geq \mu_n$ we have

$$e_\lambda = ne_\lambda = ie_\lambda \in I,$$

and then I is infinite. If n is even, then for any λ with $\mu_{n-1} \leq \lambda < \mu_n$ we have
$$e_\lambda + e_{\mu_n} = (n-2)e_\lambda + e_\lambda + e_{\mu_n} = ie_\lambda \in I,$$
and again I is infinite. Hence $R \in \mathcal{K}$.

Next, we show that $\mathcal{K}$ is a weakly special class. It is obvious that $\mathcal{K}$ consists of semiprime rings and that $\mathcal{K}$ is hereditary ($K \triangleleft I \triangleleft A \in \mathcal{K}$ implies $K \triangleleft A$ because every ideal of A is idempotent). Let $I \triangleleft \cdot A$ and $I \in \mathcal{K}$. Every Boolean ring is a subdirect sum of copies of the two element field, so the Boolean rings form the semisimple class of the upper radical of the two element field. This is a special radical, so its semisimple class is closed under essential extensions (Proposition 3.2.6). Hence A is a Boolean ring. For any nonzero ieal K of A we have
$$0 \neq I \cap K \triangleleft I \in \mathcal{K},$$
whence $I \cap K$ is infinite. This implies $A \in \mathcal{K}$. Thus $\mathcal{K}$ is a weakly special class. If $\mathcal{UK}$ were a special radical, then by Theorem 3.8.6 $\mathcal{UK} = \mathcal{U}(\mathcal{SUK} \cap \mathcal{P})$. In view of the above subdirect sum representation of Boolean rings the only prime Boolean ring is the two element field and this is not in $\mathcal{K}$ but in $\mathcal{UK}$. Hence $\mathcal{SUK} \cap \mathcal{P}$ is empty, a contradiction. Thus $\mathcal{K}$ is not a special radical.

More examples of non-special supernilotent radicals were given by van Leeuwen and Jenkins [1], Ryabukhin [6], Beidar and Salavová [1], Gardner and Stewart [2] France-Jackson [1] and Beidar and Wiegandt [4]. Tumurbat [5] gave an example for a strongly hereditary supernilpotent radical which is not special.

The characteristic difference which makes a supernilpotent radical a special one is given in the following

THEOREM 3.8.18. *Let γ be a supernilpotent radical. Then γ is a special radical if and only if γ has the intersection property relative to the class $\varrho = \mathcal{S}\gamma \cap \mathcal{P}$.*

Proof: If γ is a special radical, then by Theorem 3.8.6, ϱ is a (weakly) special class, and so by Theorem 3.7.12 γ has the intersection property relative to ϱ.

Conversely, if γ has the intersection property relative to ϱ, then the semisimple class $\sigma = \mathcal{S}\gamma$ satisfies also condition (iv) of Theorem 3.8.10, and therefore γ is a special radical. □

Remark. If γ is any supernilpotent radical and γ is not special, then there exists a smallest special radical γ' containing γ. This special radical γ' can be obtained by Theorem 3.8.6 as
$$\gamma' = \mathcal{U}(\mathcal{S}\gamma \cap \mathcal{P}),$$

or by Theorem 3.8.12 as
$$\gamma' = \mathcal{L}_{\mathrm{sp}}\gamma.$$

In Example 3.8.17 the class $\mathcal{K}$ does not contain simple Boolean rings (that is, the two element field), and we concluded that $\mathcal{K}$ is not a special class. In general, however, *there are special classes which do not contain simple rings.* For instance, consider the ring $\mathbb{Q}[x]$ of rational polynomials and the class
$$\varrho = \{f(x)\mathbb{Q}[x] \mid f(x) \in \mathbb{Q}[x]\}.$$
The class ϱ does not contain simple rings. By Corollary 3.8.7 the essential cover $\mathcal{E}\varrho$ is the smallest special class containing ϱ, and also $\mathcal{E}\varrho$ does not contain simple rings.

PROPOSITION 3.8.19 (Leavitt and Watters [2]). *If ϱ is a special class which does not contain simple rings, then the semisimple class $\mathcal{SU}\varrho$ of the special radical $\gamma = \mathcal{U}\varrho$ does not contain subdirectly irreducible rings.*

Proof: If a ring $A \in \mathcal{SU}\varrho$ is subdirectly irreducible, then A must be prime. Hence so is its heart H, and by Proposition 1.2.8 H is a simple ring. Since $H \in \mathcal{SU}\varrho$, Corollary 3.8.5 implies $H \in \varrho$, a contradiction. □

Thus there are special radicals γ and rings A such that A is a subdirect sum of γ-semisimple prime but not subdirectly irreducible rings and in any subdirect decomposition of A into subdirectly irreducible rings none of the components is γ-semisimple.

A type of special radicals, called extraspecial radicals were introduced and studied by Gardner [17]. The theory of special radicals has been generalized to Ω-groups by Buys and Gerber [1] and Booth and Groenewald [3].

3.9. Supplementing and dual radicals

A radical $\gamma^\perp$ is called the *supplementing* (or *complementary*) *radical to the radical* γ, if $\gamma^\perp$ is the largest radical class among the radicals δ satisfying
$$\gamma(A) \cap \delta(A) = 0 \quad \text{for all rings } A.$$

The first question concerning the radical $\gamma^\perp$, supplementing to an arbitrary radical γ, is its existence. The next Theorem due to Cai [1], Chen [1] and Zhou [1] answers this question affirmatively (cf. also Wiegandt [8]). We shall need the *essential cover operator* $\mathcal{E}$ and the hereditary closure operator $\mathcal{H}$, both introduced in 3.7. Prior to the Theorem, however, we note that for a radical class γ it is clear that
$$\mathcal{E}\gamma = \{A \mid \gamma(A) \triangleleft \cdot A\},$$
so denoting the class of all subdirectly irreducible rings by s, we have obviously
$$s \cap \mathcal{E}\gamma = \{A \in s \mid \gamma(A) \neq 0\}.$$

THEOREM 3.9.1 (Cai [1], Chen [1] and Zhou [1]). *If γ is any radical class, then the supplementing radical $\gamma^\perp$ to γ is given by*

$$\gamma^\perp = \mathcal{UHE}\gamma = \mathcal{UH}(s \cap \mathcal{E}\gamma).$$

Moreover, $\gamma^\perp$ is contained in the semisimple class $\mathcal{S}\gamma$ and

$$\gamma^\perp \subseteq h\mathcal{S}\gamma = \{A \in \mathcal{S}\gamma \mid A \twoheadrightarrow B \text{ implies } B \in \mathcal{S}\gamma\},$$

that is, $\gamma^\perp$ is contained in the largest homomorphically closed subclass of $\mathcal{S}\gamma$.

Proof: Let $A \neq 0$ be any ring. By Lemma 3.2.5 there exists an ideal K of A such that $\gamma(A) \cap K = 0$ and

$$\gamma(A) \cong \gamma(A)/(\gamma(A) \cap K) \cong (\gamma(A) + K)/K \triangleleft \cdot A/K.$$

It follows that

$$(\gamma(A) + K)/K \subseteq \gamma(A/K) \triangleleft \cdot A/K,$$

that is, $A/K \in \mathcal{E}\gamma$. Moreover, by $\mathcal{E}\gamma \subseteq \mathcal{HE}\gamma \subseteq \mathcal{SUHE}\gamma$ we get $A/K \subset \mathcal{SUHE}\gamma$. Putting $\widehat{\gamma} = \mathcal{UHE}\gamma$, we have

$$\widehat{\gamma}(A)/(\widehat{\gamma}(A) \cap K) \cong (\widehat{\gamma}(A) + K)/K \triangleleft A/K \in \mathcal{S}\widehat{\gamma}.$$

Hence by $\widehat{\gamma}(A)/(\widehat{\gamma}(A) \cap K) \in \widehat{\gamma}$ and by the hereditariness of the semisimple class $\mathcal{S}\widehat{\gamma}$ we get that

$$(\widehat{\gamma}(A) + K)/K \in \widehat{\gamma} \cap \mathcal{S}\widehat{\gamma} = 0.$$

Hence $\widehat{\gamma}(A) \subseteq K$, and therefore

$$\gamma(A) \cap \widehat{\gamma}(A) \subseteq \gamma(A) \cap K = 0$$

holds where $A \neq 0$ was an arbitrary ring.

Next, let δ be a radical such that

$$\gamma(A) \cap \delta(A) = 0 \qquad \text{for all rings } A.$$

In particular, for $A \in \mathcal{E}\gamma$ we have

$$\gamma(A) \cap \delta(A) = 0 \quad \text{and} \quad \gamma(A) \triangleleft \cdot A$$

which yields $\delta(A) = 0$. Hence we have got that $\mathcal{E}\gamma \subseteq \mathcal{S}\delta$, and so also $\mathcal{HE}\gamma \subseteq \mathcal{S}\delta$ by the hereditariness of the semisimple class $\mathcal{S}\delta$. This yields

$$\delta = \mathcal{US}\delta \subseteq \mathcal{UHE}\gamma = \widehat{\gamma}.$$

Thus $\widehat{\gamma} = \mathcal{UHE}\gamma$ is the supplementing radical to γ.

For proving $\mathcal{UHE}\gamma = \mathcal{UH}(s \cap \mathcal{E}\gamma)$, we consider an arbitrary ring $A \in \mathcal{E}\gamma$. Now $\gamma(A) \triangleleft \cdot A$. By Birkhoff's Theorem 1.1.7, A can be represented as a subdirect sum of subdirectly irreducible rings, or equivalently, there is a set

$$\{K_\lambda \mid K_\lambda \triangleleft A, \lambda \in \Lambda\}$$

such that each A/K_λ is subdirectly irreducible and $\bigcap_{\lambda \in \Lambda} K_\lambda = 0$. Since A/K_λ is subdirectly irreducible, either $\gamma(A) \subseteq K_\lambda$ or $(\gamma(A) + K_\lambda)/K_\lambda \triangleleft \cdot A/K_\lambda$. The index set Λ decomposes into the union of two disjoint subsets Ξ and Θ such that $\gamma(A) \subseteq K_\lambda$ for all $\lambda \in \Xi$ and $(\gamma(A) + K_\lambda)/K_\lambda \triangleleft \cdot A/K_\lambda$ for all $\lambda \in \Theta$. Hence we have

$$\gamma(A) \cap \big(\bigcap_{\lambda \in \Theta} K_\lambda\big) \subseteq \big(\bigcap_{\lambda \in \Xi} K_\lambda\big) \cap \big(\bigcap_{\lambda \in \Theta} K_\lambda\big) = \bigcap_{\lambda \in \Lambda} K_\lambda = 0,$$

and so by $\gamma(A) \triangleleft \cdot A$ it follows $\bigcap_{\lambda \in \Theta} K_\lambda = 0$. Thus A is a subdirect sum of the subdirectly irreducible rings A/K_λ, $\lambda \in \Theta$. In view of

$$\gamma(A)/(\gamma(A) \cap K_\lambda) \cong (\gamma(A) + K_\lambda)/K_\lambda \triangleleft \cdot A/K_\lambda$$

and $\gamma(A)/(\gamma(A) \cap K_\lambda) \in \gamma$ for all $\lambda \in \Theta$ we get that

$$(\gamma(A) + K_\lambda)/K_\lambda \subseteq \gamma(A/K_\lambda) \triangleleft \cdot A/K_\lambda$$

which shows that $A/K_\lambda \in s \cap \mathcal{E}\gamma$ for all $\lambda \in \Theta$. Since the semisimple class $\mathcal{SUH}(s \cap \mathcal{E}\gamma)$ is closed under subdirect sums, we get $A \in \mathcal{SUH}(s \cap \mathcal{E}\gamma)$. Thus the inclusion $\mathcal{E}\gamma \subseteq \mathcal{SUH}(s \cap \mathcal{E}\gamma)$ has been established, which yields

$$\mathcal{SUHE}\gamma \subseteq \mathcal{SUHSUH}(s \cap \mathcal{E}\gamma) = \mathcal{SUH}(s \cap \mathcal{E}\gamma).$$

On the other hand, the inclusion $\mathcal{SUH}(s \cap \mathcal{E}\gamma) \subseteq \mathcal{SUHE}\gamma$ is trivial by $s \cap \mathcal{E}\gamma \subseteq \mathcal{E}\gamma$. Thus also $\mathcal{UHE}\gamma = \mathcal{UH}(s \cap \mathcal{E}\gamma)$ has been proved.

If $A \in \gamma^\perp$, then

$$\gamma(A) = \gamma(A) \cap \gamma^\perp(A) = 0,$$

yielding $\gamma^\perp \subseteq \mathcal{S}\gamma$. Since the radical class $\gamma^\perp$ is homomorphically closed, $\gamma^\perp$ is contained also in the largest homomorphically closed subclass $h\mathcal{S}\gamma$ of $\mathcal{S}\gamma$. □

EXAMPLE 3.9.2. Although $\gamma^\perp \subseteq h\mathcal{S}\gamma$ for every radical γ, it may happen that $\gamma^\perp \neq h\mathcal{S}\gamma$. We shall construct such a radical γ. Let $\mathcal{Z}^\infty$ denote the class of all zero-rings $Z(p^\infty)$ for all primes p. Further, let $\mathcal{I}$ stand for the class of all idempotent rings. *The supplementing radical to the lower radical* $\gamma = \mathcal{L}(\mathcal{Z}^\infty \cup \mathcal{I})$ *is* $\gamma^\perp = 0$. Assume that $Z(p) \in \gamma^\perp$ for some prime p. Since $\gamma^\perp$ is closed under extensions, also $Z(p^n) \in \gamma^\perp$ for $n = 1, 2, \ldots,$

and so by the inductive property of $\gamma^\perp$ it follows that $Z(p^\infty) \in \gamma^\perp$. Hence by $Z(p^\infty) \in \mathcal{L}(\mathcal{Z}^\infty \cup \mathcal{I})$ we have

$$0 = \gamma(Z(p^\infty)) \cap \gamma^\perp(Z(p^\infty)) = Z(p^\infty),$$

a contradiction. Hence $Z(p) \notin \gamma^\perp$ and $Z(p) \in \mathcal{S}\gamma^\perp$ for all primes p. Since the semisimple class $\mathcal{S}\gamma^\perp$ is closed under extensions, we get that $Z(p^n) \in \mathcal{S}\gamma^\perp$ for all primes and $n = 1, 2, \ldots$. By the construction of γ we have $Z(p^\infty) \in \gamma$, and so

$$0 = \gamma(Z(p^\infty)) \cap \gamma^\perp(Z(p^\infty)) = \gamma^\perp(Z(p^\infty)),$$

that is, $Z(p^\infty) \in \mathcal{S}\gamma^\perp$ for all primes p. The subdirectly irreducible zero-rings are exactly the rings $Z(p^n)$ for all primes and $n = 1, 2, \ldots, \infty$ and — as we have already seen — each of them is in $\mathcal{S}\gamma^\perp$. Since the semisimple class $\mathcal{S}\gamma^\perp$ is closed under subdirect sums and every zero-ring is a subdirect sum of subdirectly irreducible zero-rings, we have got that $\mathcal{S}\gamma^\perp$ contains all zero-rings. Hence $\gamma^\perp$ must consist of idempotent rings, whence

$$\gamma^\perp \subseteq \mathcal{I} \subseteq \mathcal{L}(\mathcal{Z}^\infty \cup \mathcal{I}) = \gamma$$

follows. Thus for every ring A we have $\gamma^\perp(A) \subseteq \gamma(A)$ and $0 = \gamma(A) \cap \gamma^\perp(A) = \gamma^\perp(A)$. Consequently $\gamma^\perp = 0$ must hold, although $Z(p) \in h\mathcal{S}\gamma$ and so $h\mathcal{S}\gamma \neq \{0\}$. Thus $\gamma^\perp \subset h\mathcal{S}\gamma$.

This Example indicates that a radical may be supplementing to several radicals.

We can tell much more on radicals supplementing to hereditary radicals. In the sequel we shall deal with the classical results of Andrunakievich [1].

We begin with an auxiliary statement.

PROPOSITION 3.9.3. *If σ is the semisimple class of a hereditary radical γ, then σ has the inductive property.*

Proof: Let $I_1 \subseteq \ldots \subseteq I_\lambda \subseteq \ldots$ be an ascending chain of ideals of a ring A such that $I_\lambda \in \sigma$ for each index λ. Put $I = \cup I_\lambda$, and suppose that $I \notin \sigma$, that is, $\gamma(I) \neq 0$. Then there exists an index λ such that $0 \neq \gamma(I) \cap I_\lambda$. By the hereditariness of γ and Proposition 3.2.3 we get $I_\lambda \cap \gamma(I) \subseteq \gamma(I_\lambda) = 0$, a contradiction. □

THEOREM 3.9.4. *If γ is a hereditary radical, then its supplementing radical $\gamma^\perp$ is the largest homomorphically closed subclass in $\mathcal{S}\gamma$, that is, $\gamma^\perp = h\mathcal{S}\gamma$.*

Proof: Put $\gamma' = h\mathcal{S}\gamma$. By definition the class γ' is homomorphically closed. For the inductive property, let $I_1 \subseteq \ldots \subseteq I_\lambda \subseteq \ldots$ be an ascending chain of ideals of a ring A such that $I_\lambda \in \gamma'$ for all indices λ. Take an arbitrary ideal K of $I = \cup I_\lambda$, and consider the chain

$$(I_1 + K)/K \subseteq \ldots \subseteq (I_\lambda + K)/K \subseteq \ldots$$

Radical Theory for Associative Rings

We have
$$(I_\lambda + K)/K \cong I_\lambda/(I_\lambda \cap K) \in \gamma' \subseteq \mathcal{S}\gamma,$$

because $I_\lambda \in \gamma'$. Since by Proposition 3.9.3 the semisimple class $\mathcal{S}\gamma$ has the inductive property, we have

$$I/K = (\cup(I_\lambda + K))/K \in \mathcal{S}\gamma.$$

Since K was an arbitrary ideal of I, it follows that $I \in \gamma'$.

Next we prove that γ' is closed under extensions, and so by Theorem 2.1.4 γ' is a radical class. Suppose that $I \in \gamma'$, $A/I \in \gamma'$, and take an arbitrary ideal K of A. We want to see that $A/K \in \mathcal{S}\gamma$. Since γ' is a homomorphically closed class, $A/I \in \gamma'$ implies that $A/(I+K) \in \gamma'$. Further, by $I \in \gamma'$ we have that

$$(I + K)/K \cong I/(I \cap K) \in \gamma' \subseteq \mathcal{S}\gamma.$$

Hence by

$$\frac{A/K}{(I+K)/K} \cong A/(I+K) \in \gamma' \subseteq \mathcal{S}\gamma$$

we conclude that $A/K \in \mathcal{S}\gamma$, since the semisimple class $\mathcal{S}\gamma$ is closed under extensions.

From the definition of γ' it is clear that γ' is the largest radical class in $\mathcal{S}\gamma$, so $\gamma^\perp \subseteq \gamma'$. Furthermore, by the hereditariness of γ and $\mathcal{S}\gamma$, for every ring A we have

$$\gamma(A) \cap \gamma'(A) \in \gamma \cap \gamma' \subseteq \gamma \cap \mathcal{S}\gamma = 0.$$

Hence $\gamma' \subseteq \gamma^\perp \subseteq \mathcal{S}\gamma$, and so $\gamma^\perp = \gamma' = h\mathcal{S}\gamma$. $\square$

Let us observe that for hereditary radicals γ, δ the condition

$$\gamma(A) \cap \delta(A) = 0 \quad \text{for all rings } A$$

is equivalent to $\gamma \cap \delta = 0$.

Next, we characterize supplementing radical classes as upper radicals of a certain class of subdirectly irreducible rings.

THEOREM 3.9.5. *Let γ be a hereditary radical class and $s(\gamma)$ be the class*

$$s(\gamma) = \{A \mid A \text{ is subdirectly irreducible with heart } H(A) \text{ in } \gamma\} = s \cap \mathcal{E}\gamma.$$

The supplementing radical $\gamma^\perp$ of γ is the upper radical $\mathcal{U}s(\gamma)$.

Proof: Although from the definition it is not obvious that $s(\gamma)$ is a regular class, one can easily prove it. As a matter of fact, we do not need to know it: the operator $\mathcal{U}$ can be applied to any class of rings, and the coincidence of $\gamma^\perp$ and $\mathcal{U}s(\gamma)$ justifies that $\gamma^\perp$ is the upper radical of $s(\gamma)$.

For proving that $\gamma^\perp \subseteq \mathcal{U}s(\gamma)$, let us consider any ring $A \in \gamma^\perp$ and a subdirectly irreducible homomorphic image B of A. Since $A \in \gamma^\perp$, we have $B \in \mathcal{S}\gamma$, and so also its heart $H(B)$ is in $\mathcal{S}\gamma$ and not in γ. Therefore $A \in \mathcal{U}s(\gamma)$ and $\gamma^\perp \subseteq \mathcal{U}s(\gamma)$ hold.

Next, suppose that $A \notin \gamma^\perp$. Since by Theorem 3.9.4 $\gamma^\perp$ is the largest homomorphically closed subclass of $\mathcal{S}\gamma$ and $A \notin \gamma^\perp$, A has a nonzero homomorphic image B (maybe $B = A$) which is not in $\mathcal{S}\gamma$. Let $B = \sum_{\text{subdirect}} B_\alpha$ be any subdirect decomposition into subdirectly irreducible rings B_α. Since $B \notin \mathcal{S}\gamma$ and the semisimple class $\mathcal{S}\gamma$ is closed under subdirect sums (Proposition 2.3.4), at least one subdirectly irreducible component B_α is not in $\mathcal{S}\gamma$, that is, $\gamma(B_\alpha) \neq 0$. Hence by the hereditariness of γ we get that $H(B_\alpha) \in \gamma$. Since B_α is a homomorphic image also of A, it follows that $A \notin \mathcal{U}s(\gamma)$. Thus also $\mathcal{U}s(\gamma) \subseteq \gamma^\perp$ has been established. □

A radical class γ is called *subidempotent*, if γ is hereditary and consists of idempotent rings. An example of a subidempotent radical is the class of all von Neumann regular rings (Example 3.2.1.(iv)) because by $a \in aAa$ it follows that $A \subseteq A^3 \subseteq A$ for every von Neumann regular ring A. Example 3.2.2 shows that the class $\mathcal{I}$ of all idempotent rings is a radical class but not a subidempotent one.

PROPOSITION 3.9.6. *The supplementing radical $\gamma^\perp$ to a supernilpotent radical γ is subidempotent.*

Proof: By Theorem 3.9.4, $\gamma^\perp$ is the largest homomorphically closed class in the semisimple class $\mathcal{S}\gamma$. Hence $A \in \gamma^\perp$ implies $A/A^2 \in \mathcal{S}\gamma$. Since γ is supernilpotent, also $A/A^2 \in \gamma$ holds, and therefore $A = A^2$ must hold. Thus $\gamma^\perp$ consists of idempotent rings.

Next, we prove that $\gamma^\perp$ is hereditary. For any $K \triangleleft I \triangleleft A \in \gamma^\perp$ and for the ideal $\overline{K}$ of A generated by K we have $A/\overline{K}^3 \in \mathcal{S}\gamma$. Hence by the hereditariness of $\mathcal{S}\gamma$ we have also $\overline{K}/\overline{K}^3 \in \mathcal{S}\gamma$. Since γ is supernilpotent, also $\overline{K}/\overline{K}^3 \in \gamma$ holds yielding $\overline{K} = \overline{K}^3$. Thus by the Andrunakievich Lemma we get that

$$K \subseteq \overline{K} = \overline{K}^3 \subseteq K, \quad \text{that is,} \quad K \triangleleft A.$$

Hence by $I/K \triangleleft A/K \in \mathcal{S}\gamma$ the hereditariness of $\mathcal{S}\gamma$ implies that $I/K \in \mathcal{S}\gamma$. Since K was arbitrary, we conclude that $I \in \gamma^\perp$. □

PROPOSITION 3.9.7. *The supplementing radical $\gamma^\perp$ to a subidempotent radical γ is supernilpotent.*

Proof: Since γ is subidempotent, every nilpotent ring must be in the semisimple class $\mathcal{S}\gamma$. The class of all nilpotent rings is homomorphically closed, so it is contained in the largest homomorphically closed subclass $h\mathcal{S}\gamma$ of $\mathcal{S}\gamma$ which is by Theorem 3.9.4 precisely the supplementing radical $\gamma^\perp$ of γ.

We still have to show that the radical $\gamma^\perp$ is hereditary. Let $K \triangleleft I \triangleleft A \in \gamma^\perp$, and $\overline{K}$ be the ideal of A generated by K. Then by the Andrunakievich Lemma the ring $\overline{K}/K$ is nilpotent, and hence in $\mathcal{S}\gamma$. Moreover, we have also

$$\frac{I/K}{\overline{K}/K} \cong I/\overline{K} \triangleleft A/\overline{K} \in \gamma^\perp \subseteq \mathcal{S}\gamma.$$

Since the semisimple class $\mathcal{S}\gamma$ is hereditary and closed under extensions, we get that $I/K \in \mathcal{S}\gamma$. This proves that $I \in \gamma^\perp$, and therefore $\gamma^\perp$ is hereditary. □

EXAMPLE 3.9.8. By Theorem 3.9.5 the supplementing radical $\beta^\perp$ to the Baer radical β is $\beta^\perp = \mathcal{U}s(\beta)$ where $s(\beta)$ is the class of all subdirectly irreducible rings whose hearts are zero-rings (that is, subdirectly irreducible rings which are not prime rings). Moreover, $\beta^\perp$ is a subidempotent radical by Proposition 3.9.6. Thus every ideal of any ring A in $\beta^\perp$ is idempotent. Such rings are referred to *hereditarily idempotent rings*. If B is any hereditarily idempotent ring, then it cannot be mapped homomorphically onto a nonzero subdirectly irreducible ring in $s(\beta)$. Thus *the radical class $\beta^\perp$ consists exactly of all hereditarily idempotent rings, and $\beta^\perp$ is obviously the largest subidempotent radical class*. In view of Proposition 3.9.7 the supplementing radical $\beta^{\perp\perp}$ to $\beta^\perp$ is supernilpotent and $\beta^{\perp\perp} = \mathcal{U}s(\beta^\perp)$ and $s(\beta^\perp)$ is clearly the class of all subdirectly irreducible rings with idempotent hearts. Hence *the supplementing radical $\beta^{\perp\perp}$ of the supplementing radical $\beta^\perp$ of β is the antisimple radical $\beta^{\perp\perp} = \beta_\varphi$ and β_φ is bigger than the class β* (cf. Proposition 4.11.9).

EXAMPLE 3.9.9. Let us consider the class

$$\tau = \{A \mid A^+ \text{ is a torsion group}\}.$$

The class τ is obviously homomorphically closed, has the inductive property, and is closed under extensions. Thus by Theorem 2.1.4 τ *is a radical*, called the *torsion radical*. Obviously, $\tau(A)$ is the maximal torsion ideal of the ring A. It is also clear that τ *is a strongly hereditary radical*. So by Theorem 3.9.5 its supplementing radical $\tau^\perp$ is $\mathcal{U}s(\tau)$. Let us consider the split-null extension $A = \mathbb{Q} * \mathbb{Q}$ of the field $\mathbb{Q}$ of rational numbers by itself (see the paragraph before Proposition 3.6.10). Then

$$(0, \mathbb{Q}) = \{(0, q) \mid q \in \mathbb{Q}\}$$

is an ideal in A and $(0, \mathbb{Q})$ is isomorphic to the zero-ring on $\mathbb{Q}^+$. Since for a given nonzero-element $(0, q) \in (0, \mathbb{Q})$ and an arbitrarily chosen element $(0, x) \in (0, \mathbb{Q})$ we have

$$(q^{-1}x, 0)(0, q) = (0, x),$$

the ideal $(0,\mathbb{Q})$ is a minimal one in A. Moreover, by $A/(0,\mathbb{Q}) \cong \mathbb{Q}$ we see that all the ideals of A are 0, $(0,\mathbb{Q})$ and A. Hence A is a subdirectly irreducible ring with heart $(0,\mathbb{Q})$. It is also clear that $A \in \mathcal{U}s(\tau) = \tau^\perp$. Taking the zero-ring on the infinite cyclic group, we have

$$Z(\infty) \triangleleft (0,\mathbb{Q}) \triangleleft A \in \tau^\perp.$$

Since $Z(\infty)$ has nonzero homomorphic images in τ, $Z(\infty)$ is not in $\tau^\perp$, and therefore *the supplementing radical $\tau^\perp$ is not hereditary*. Furthermore, the ring $\mathbb{Z}$ of integers is certainly in the semisimple class $\mathcal{S}\tau$ consisting of all torsionfree rings, but $\mathbb{Z} \notin \tau^\perp$ and $\tau^\perp(\mathbb{Z}) = 0$. Hence $\mathbb{Z} \in \mathcal{S}\tau^\perp$ and obviously $\mathbb{Z}$ is contained in the largest homomorphically closed subclass $h\mathcal{S}\tau^\perp$ of $\mathcal{S}\tau^\perp$. Thus $\tau \subset h\mathcal{S}\tau^\perp$ follows. It is not clear whether $h\mathcal{S}\tau^\perp$ is a radical class or not.

Gardner proved in [20] that for a hereditary radical γ such that $\beta \cap \tau \subseteq \gamma$, the supplementing radical $\gamma^\perp$ is hereditary if and only if $\gamma^\perp$ consists of hereditarily idempotent rings.

After these motivating examples we introduce the following notions.

A pair (γ, δ) of radicals is called a *pair of supplementing radicals*, if $\delta = \gamma^\perp$ is the supplementing radical to γ and $\gamma = \delta^\perp$ is the supplementing radical to δ. A radical γ is said to be a *dual radical*, if γ is the supplementing radical to its supplementing radical $\gamma^\perp$, that is $\gamma = \gamma^{\perp\perp}$. Thus (γ, δ) is a pair of supplementing radicals if and only if both γ and δ are dual radicals such that $\delta = \gamma^\perp$ and $\gamma = \delta^\perp$.

PROPOSITION 3.9.10. *If (γ, δ) is a pair of radicals such that γ is hereditary and $\delta = \gamma^\perp$ is the supplementing radical to γ, then $\delta^{\perp\perp} = \delta = h\mathcal{S}\delta^\perp$, and so δ is a dual radical.*

Proof: By definition we have $\gamma \subseteq \gamma^{\perp\perp} = \delta^\perp$ and $\delta \subseteq \delta^{\perp\perp}$. If $A \in \delta^{\perp\perp}$, then

$$\delta^\perp(A) = \delta^\perp(A) \cap A = \delta^\perp(A) \cap \delta^{\perp\perp}(A) = 0$$

which implies $\delta^{\perp\perp} \subseteq \mathcal{S}\delta^\perp$ as well as $\delta^{\perp\perp} \subseteq h\mathcal{S}\delta^\perp$ for $\delta^{\perp\perp}$ is homomorphically closed. Since γ is hereditary, Theorems 3.9.1 and 3.9.4 and the relation $\gamma \subseteq \gamma^{\perp\perp} = \delta^\perp$ yield

$$\delta^{\perp\perp} \subseteq h\mathcal{S}\delta^\perp = h\mathcal{S}\gamma^{\perp\perp} \subseteq h\mathcal{S}\gamma = \gamma^\perp = \delta \subseteq \delta^{\perp\perp}. \qquad \square$$

PROPOSITION 3.9.11. *If (γ, δ) is a pair of radicals such that γ is hereditary and $\gamma = \delta^\perp$ is the supplementing radical to δ, then $\delta = \gamma^\perp$ is the supplementing radical to γ and (γ, δ) is a pair of supplementing radicals.*

Proof: As in the proof of Proposition 3.9.10, we have $\delta \subseteq \delta^{\perp\perp}$ and also $\delta^{\perp\perp} \subseteq \mathcal{S}\delta^\perp = \mathcal{S}\gamma$. Hence by the hereditariness of γ, Proposition 3.9.10 is applicable, and so

$$\delta^{\perp\perp} \subseteq h\mathcal{S}\gamma = \gamma^\perp = \delta$$

follows. □

In Propositions 3.9.10 and 3.9.11 only the radical γ is supposed to be hereditary, and so δ may be not hereditary as we have seen in Example 3.9.9.

Now we return to hereditary radicals for which the theory of dual radicals develops more nicely. In view of Proposition 3.9.10, Propositions 3.9.6 and 3.9.7 tell us that *the supplementing radical to a supernilpotent (subidempotent radical) is always a subidempotent (supernilpotent, respectively) dual radical.*

PROPOSITION 3.9.12. *A supernilpotent dual radical is a special radical.*

Proof: If a supernilpotent radical γ is dual, then by Theorem 3.9.5 we have $\gamma = \gamma^{\perp\perp} = \mathcal{U}s(\gamma^{\perp})$. Since $\gamma^{\perp}$ is a subidempotent radical by Proposition 3.9.6, the heart $H(A)$ of each ring $A \in s(\gamma^{\perp})$ is a simple idempotent ring. Hence by Propositions 3.6.6 and 3.6.7 the class $s(\gamma^{\perp})$ is a special one. □

Summing up the previous results we arrive at Andrunakievich's full description of supernilpotent and subidempotent dual radicals ([1] Theorem 10).

THEOREM 3.9.13. *Let α be any class of simple idempotent rings, and define the classes $r(\alpha)$ and $t(\alpha)$ by*

$$r(\alpha) = \{every\ subdirectly\ irreducible\ ring\ A\ with\ heart\ in\ \alpha\}$$

and

$$t(\alpha) = \{every\ subdirectly\ irreducible\ ring\ A\ with\ heart\ not\ in\ \alpha\}.$$

Then $\mathcal{U}r(\alpha)$ and $\mathcal{U}t(\alpha)$ are mutually supplementing radicals, $\mathcal{U}r(\alpha)$ is a supernilpotent dual radical and $\mathcal{U}t(\alpha)$ is a subidempotent dual radical. Moreover, all supernilpotent and subidempotent dual radicals are obtained in this way.

Proof: $\gamma = \mathcal{U}r(\alpha)$ is obviously a supernilpotent radical, and by Theorem 3.9.5 $\gamma^{\perp} = \mathcal{U}s(\gamma)$ is a subidempotent dual radical. By Proposition 3.9.11 also γ is a dual radical. If $A \in s(\gamma)$, then $H(A) \in \gamma$ and so either $H(A)$ is a simple idempotent ring or a zero-ring, in both cases $H(A) \notin \alpha$. This shows that $s(\gamma) \subseteq t(\alpha)$. Conversely, if A is a subdirectly irreducible ring and $A \notin s(\gamma)$, then $H(A) \notin \gamma$. Hence $H(A) \in \alpha$, and therefore $A \notin t(\alpha)$. This proves that $t(\alpha) \subseteq s(\gamma)$. Thus $s(\gamma) = t(\alpha)$ and $\gamma^{\perp} = \mathcal{U}t(\alpha)$ has been established. The rest is obvious. □

EXAMPLE 3.9.14. *The Brown–McCoy, the Behrens, the Thierrin and the antisimple radicals are supernilpotent dual radicals. The antisimple radical is the smallest supernilpotent dual radical. A special radical need not* be a *dual radical:* the Baer radical is not a dual radical, because it does not contain the antisimple radical class (see Example 3.9.8).

Next, we give also another characterization of supernilpotent and subidempotent dual radicals.

THEOREM 3.9.15 (de la Rosa, Fong and Wiegandt [1]). *Let α be a class of simple idempotent rings, and define the classes $r(\alpha)$ and $t(\alpha)$ as in Theorem 3.9.13. The supernilpotent dual radical $\mathcal{U}r(\alpha)$ is the unique largest universal class ξ such that $\xi \cap \alpha = 0$. The subidempotent dual radical $\mathcal{U}t(\alpha)$ is the unique largest universal class η such that the simple rings of η are in α.*

Proof: $\mathcal{U}r(\alpha)$ and $\mathcal{U}t(\alpha)$ are certainly universal classes with the required properties. We have to prove that they are the largest such classes. Let B be a subdirectly irreducible homomorphic image of a ring A.

If $A \in \xi$, then $H(B) \notin \alpha$, and hence $B \notin r(\alpha)$. Thus $A \in \mathcal{U}r(\alpha)$ and also $\xi \subseteq \mathcal{U}r(\alpha)$ hold.

Suppose that $A \in \eta$. Then also $H(B) \in \eta$. If $H(B)$ is a zero-ring, then $H(B)$ has an ideal which has a homomorphic image C such that C is a simple zero-ring. Thus $C \notin \alpha$. By $H(B) \in \eta$ it follows then that $C \in \eta$, contradicting the assumption. Hence $H(B)$ is a simple idempotent ring and $H(B) \in \alpha$. This proves that $A \in \mathcal{U}t(\alpha)$ and also $\eta \subseteq \mathcal{U}t(\alpha)$. □

In particular, Theorem 3.9.15 characterizes the Brown–McCoy, the Thierrin, the Behrens and the antisimple radical classes as largest universal classes with specified properties imposed on simple rings.

More on supplementing radicals can be found also in the papers Birkenmeier [2], Groenewald and Olivier [1], Huang [1], Krachilov [1], [2], van Leeuwen [3], Ryabukhin and Kràchilov [1].

3.10. Subidempotent radicals

We start with two examples. In the first one we shall characterize the largest subidempotent radical which has been introduced in Example 3.9.8.

EXAMPLE 3.10.1. *The largest hereditary subclass*

$$\chi = \{\text{all hereditarily idempotent rings}\}$$

of the radical class $\mathcal{I}$ of all idempotent rings is the largest subidempotent radical. Moreover, $\chi = \mathcal{U}s(\beta)$ is the supplementing radical of the Baer radical β. Denoting the class of all simple idempotent rings by α_0, $\chi = \mathcal{U}t(\alpha_0)$ *is a dual subidempotent radical, the supplementing dual radical of χ is the antisimple radical β_φ, and the supplementing radical of β_φ is χ.* All these statements are reformulations of Example 3.9.8 or consequences of Theorem 3.9.13.

Following Blair [1], [2] we shall say that a ring A is *f-regular*, if every element $a \in A$ is contained in $(a)^2$ where (a) denotes the principal ideal generated by a. If A is a hereditarily idempotent ring, then by $a \in (a) = (a)^2$ the ring A is f-regular. Conversely, suppose that A is f-regular. Obviously A is also idempotent. Let B be any ideal of A, and for distinction, let us denote by $(b)_A$ and $(b)_B$ the principal ideal of A and B, respectively, generated by the element $b \in B$. Then by the f-regularity of A we have

$$b \in (b)_B \subseteq (b)_A = (b)_A^3 \subseteq B(b)_A B =$$
$$= B(\mathbb{Z}b + Ab + bA + AbA)B \subseteq BbB \subseteq (b)_B,$$

and so $b \in (b)_B = (b)_A = (b)_A^2 = (b)_B^2$. Hence every ideal B of an f-regular ring A is f-regular, and therefore also idempotent. Thus *the class χ of all hereditarily idempotent rings coincides with that of all f-regular rings.*

As expected, a subidempotent radical need not be a dual radical.

EXAMPLE 3.10.2. Let V be a countably infinite dimensional vector space over a division ring with a basis $x_1, \ldots, x_n, \ldots$. Further, let H denote the ring of all finite valued linear transformations of V. As we have seen in Example 1.2.11 (iii), H is a simple idempotent ring. Let us consider the linear transformation t defined by

$$tx_n = \begin{cases} x_{n+1} & \text{if } n \text{ is even} \\ 0 & \text{if } n \text{ is odd} \end{cases}$$

for $n = 1, 2, \ldots$. Clearly, t is *not* finite valued, and $t^2 = 0$. Let S denote the subring of linear transformations generated by H and the element t. It is easy to check that the only ideals of S are 0, H and S, and so S is a subdirectly irreducible prime ring with heart H. Moreover, $(S/H)^2 = 0$. Let us consider the class α consisting of 0 and H. The lower radical $\mathcal{L}\alpha$ is obviously a subidempotent radical (see Corollary 3.2.10). If $\mathcal{L}\alpha$ were a dual radical, then by Theorem 3.9.13 the only possible candidate would be $\mathcal{U}t(\alpha)$. Since $S \in \mathcal{U}t(\alpha)$, but $S \notin \mathcal{L}\alpha$, the subidempotent radical $\mathcal{L}\alpha$ is not a dual radical.

Now we focus our attention to characterizations of subidempotent radicals and of their semisimple classes.

THEOREM 3.10.3 (Veldsman [7]). *For a hereditary radical γ the following conditions are equivalent:*

(i) γ *is subidempotent,*

(ii) γ *satisfies condition* (F), *that is, if* $K \triangleleft I \triangleleft A$ *and* $I/K \in \gamma$, *then* $K \triangleleft A$,

(iii) *if* $K \triangleleft I \triangleleft A$ *and* $K \in \gamma$, *then* $K \triangleleft A$.

Proof: Since a subidempotent radical is regular (even hereditary), and consists of semiprime rings, the equivalence of (i) and (ii) follows immediately from Proposition 3.6.13.

(i) $\Longrightarrow$ (iii) Let $K \triangleleft I \triangleleft A$ and $K \in \gamma$. Denoting by $\overline{K}$ the ideal of A generated by K, we have $K/\overline{K}^3 \in \gamma$. Since K is an idempotent ring, it follows $K = \overline{K} \triangleleft A$.

(iii) $\Longrightarrow$ (i) If γ is not subidempotent, then it contains a ring C such that $0 \neq C/C^2 \in \gamma$. Putting $J = C/C^2$ we proceed as in the proof of Proposition 3.6.13. Let B be the Dorroh extension of J, and let A be the ring of all 2×2 matrices over B, I be the ring of all 2×2 matrices over J and
$$K = \begin{pmatrix} J & 0 \\ 0 & 0 \end{pmatrix}.$$
Clearly, $K \triangleleft I \triangleleft A$ and $K \cong J \in \gamma$. Nevertheless, K is not an ideal of A, because
$$\begin{pmatrix} j & 0 \\ 0 & 0 \end{pmatrix} \begin{pmatrix} 0 & 1 \\ 0 & 0 \end{pmatrix} = \begin{pmatrix} 0 & j \\ 0 & 0 \end{pmatrix} \notin K.$$
Hence (iii) is not satisfied. □

Combining Theorems 3.2.7 and 3.10.3 we get

COROLLARY 3.10.4 (Veldsman [7]). *For a class σ of rings the following three conditions are equivalent:*

(i) *σ is the semisimple class of a subidempotent radical,*

(ii) *σ is regular, closed under subdirect sums and essential extensions, and σ contains all nilpotent rings,*

(iii) *σ is regular, closed under subdirect sums and essential extensions, and if $K \triangleleft I \triangleleft A$ and $K = (K)\sigma$ then $K \triangleleft A$.* □

THEOREM 3.10.5 (Ryabukhin [7]). *A class σ of rings is the semisimple class of a subidempotent radical γ if and only if σ is regular, closed under subdirect sums and essential extensions, and the class*
$$t(\sigma) = \{\text{all subdirectly irreducible rings } A \text{ with heart } H(A) \notin \sigma\}$$
is a special class.

Proof: Suppose that σ is the semisimple class of a subidempotent radical γ. By Theorem 3.2.7 all what we have to prove is that the class $t(\sigma)$ is a special class. We have clearly
$$t(\sigma) = \{\text{all subdirectly irreducible rings } A \text{ with heart } H(A) \in \gamma\} = s(\gamma).$$
Since γ is subidempotent, $s(\gamma)$ consists of simple prime rings, and hence by Propositions 3.6.6 and 3.6.7 $t(\sigma) = s(\gamma)$ is a special class.

Conversely, let σ be a regular class of rings, closed under subdirect sums and essential extensions, and the class $t(\sigma)$ be a special class. By Theorem 3.2.7 σ is the semisimple class of a hereditary radical γ. Hence we have
$$t(\sigma) = \{\text{ all subdirectly irreducible rings } A \text{ with } H(A) \in \gamma\} = s(\gamma),$$

and by Theorem 3.9.5 it follows that the radical $\mathcal{U}s(\gamma)$ is a supplementing radical which is contained in $\mathcal{S}\gamma = \sigma$. Since $\mathcal{U}s(\gamma) = \mathcal{U}t(\sigma)$ is a special radical, $\sigma = \mathcal{S}\gamma$ contains all nilpotent rings. Hence γ is subidempotent. □

Further results on subidempotent radicals can be found in the papers of Beidar [2], Mendes [1], Sands [6], [13] and Veldsman [7].

3.11. Hypernilpotent and hypoidempotent radicals

A radical class which contains all zero-rings (or equivalently, all nilpotent rings), is called a *hypernilpotent radical*. We say that a radical class is *hypoidempotent*, if it consists of idempotent rings. Thus the hereditary hypernilpotent (hereditary hypoidempotent) radicals are the supernilpotent (subidempotent) ones.

PROPOSITION 3.11.1. *For a radical γ the following conditions are equivalent:*
  (i) γ *is hypernilpotent,*
  (ii) *if $I \triangleleft A$ and $I^n = 0$ then $I \subseteq \gamma(A)$,*
  (iii) *if $L \triangleleft_l A \in \mathcal{S}\gamma$ and $L^n = 0$ then $L = 0$.*

Proof: (i) $\implies$ (iii) Let $L \triangleleft_l A \in \mathcal{S}\gamma$ and $L^n = 0$. Then $L + LA \triangleleft A$, and by induction one readily shows that $(L + LA)^n = L^n + L^n A$. Hence by $L^n = 0$ the ideal $L + LA$ is nilpotent. Applying (i) it follows that $L + LA \in \gamma$, whence $L + LA \subseteq \gamma(A) = 0$. Thus also $L = 0$.

The implications (iii) $\implies$ (ii) $\implies$ (i) are straightforward. □

Let us recall that in Theorem 3.6.11 we have characterized semisimple classes of supernilpotent radicals as certain weakly homomorphically closed classes. By generalizing the notion of being weakly homomorphically closed, we shall characterize semisimple classes of hypernilpotent and hypoidempotent radicals.

THEOREM 3.11.2 (Veldsman [7]). *Let $\sigma = \mathcal{S}\gamma$ be the semisimple class of a radical γ. The radical γ is hypernilpotent or hypoidempotent if and only if σ satisfies the following condition:*

  (*) *if $I \triangleleft A \in \sigma$ and $A^2 = 0$, then also $A/I \in \sigma$.*

Proof: Obviously hypernilpotent and hypoidempotent radicals satisfy condition (*).

Conversely, suppose that σ satisfies condition (*). If σ does not contain a nonzero ring with zero-multiplication, then all such rings have to be in the class $\gamma = \mathcal{U}\sigma$, and so γ is hypernilpotent. Suppose, next, that σ contains a nonzero ring A with zero-multiplication. Then by the regularity of σ, the zero-ring $Z(n)$ over the cyclic group of order n belongs to σ for some $n = 2, \ldots, \infty$. As in the proof of Theorem 3.6.11 we get that also $Z(\infty)$ must be in σ. Every abelian group is a homomorphic image of a free abelian group, and clearly the same is true for rings with zero-multiplication. Hence every zero-ring is the homomorphic image of a free zero-ring, that is, of a direct sum of copies of $Z(\infty)$. Since $Z(\infty)$ is in σ, and semisimple classes are closed under subdirect sums, condition (∗) yields that every zero-ring must be in σ. Hence $\gamma = \mathcal{U}\sigma$ is hypoidempotent. □

McConnell and Stokes [1] proved that the class

$$\{A \mid \text{the semigroup } (A, \circ) \text{ has no non-trivial ideals}\}$$

is a nonhereditary hypernilpotent radical.

The next Theorem shows that life is much easier in the radical theory of algebras over a field.

THEOREM 3.11.3. *Let K be a field and $\mathbb{A}$ any universal class of K-algebras containing the one dimensional K-algebra K^0 with zero multiplication. Every radical γ in $\mathbb{A}$ is either hypernilpotent or hypoidempotent.*

Proof: If $K^0 \in \gamma$, then γ contains all algebras of $\mathbb{A}$ with zero-multiplication (a K-algebra is, in fact, a K-vector space, and hence a direct sum of copies of K^0). If $K^0 \notin \gamma$, then $K^0 \in \mathcal{S}\gamma$ (provided that $K^0 \in \mathbb{A}$), and so also every K-algebra with zero-multiplication is contained in $\mathcal{S}\gamma$. □

A careful analysis of the proofs of Theorems 3.11.2 and 3.11.3 tells us that there we did not use the ADS-Theorem, but only the Andrunakievich Lemma (or a suitable generalization of it) and the fact that $I^2 \triangleleft A$ (or more generally, $I^s \triangleleft A$ for some integer $s > 1$) whenever $I \triangleleft A$ (see, for instance, Anderson and Gardner [1], Ánh, Loi and Wiegandt [1], Buys and Gerber [1]).

3.12. Partition of simple rings, unequivocal rings

Given any radical γ, a simple ring S must declare itself, so to speak, to be either radical or semisimple, that is, either $\gamma(S) = S$ or $\gamma(S) = 0$. Thus every radical γ partitions the class of simple rings into two disjoint classes: the class δ of simple rings S with $\gamma(S) = S$ called the *lower class*, and the class μ of simple rings T with $\gamma(T) = 0$ called the *upper class*.

Conversely, let us consider a partition of the class of all simple rings into two disjoint subclasses δ and μ. Since both δ and μ are homomorphically closed and hereditary, we can consider the lower radical class $\mathcal{L}\delta$ and the upper radical class $\mathcal{U}\mu$ of the classes δ and μ, respectively.

PROPOSITION 3.12.1. *The lower radical class $\mathcal{L}\delta$ of any class δ of simple rings is a hereditary radical and $\mathcal{L}\delta = \delta_3$.*

Proof: Obvious by Corollary 3.2.10 and Theorem 3.4.13. □

Because of the peculiar properties of the zero-rings $Z(\infty)$ and $Z(p^\infty)$ we have

THEOREM 3.12.2. *For every partition (δ, μ) of the class of simple rings we have $\mathcal{L}\delta \subset \mathcal{U}\mu$.*

Proof: The containment $\mathcal{L}\delta \subseteq \mathcal{U}\mu$ is clear by definition.

Suppose that the class μ does not contain nonzero rings with zero-multiplication. We shall see that the zero-ring $Z(\infty)$ over the infinite cyclic group is in $\mathcal{U}\mu$ but not in $\mathcal{L}\delta$. Since every homomorphic image of $Z(\infty)$ is again a zero-ring, it cannot be mapped homomorphically onto any nonzero ring in μ. Hence $Z(\infty) \in \mathcal{U}\mu$. If $Z(\infty)$ were in $\mathcal{L}\delta$, then it would be a nonzero accessible subring in δ (cf. Theorem 3.3.2 and Corollary 3.3.4), in fact, an ideal in δ. But the ring $Z(\infty)$ does not contain nonzero simple rings. Thus $Z(\infty) \notin \mathcal{L}\delta$.

Next, assume that μ contains a simple zero-ring $Z(p)$ over the cyclic group of prime order p. The quasi-cyclic group $Z(p^\infty)$ cannot be mapped homomorphically onto a nonzero simple ring, so $Z(p^\infty) \in \mathcal{U}\mu$. If $Z(p^\infty) \in \mathcal{L}\delta$ would hold, then $Z(p^\infty)$ would have a nonzero simple subring in δ. But $Z(p^\infty)$ contains only one nonzero simple subring, namely $Z(p)$ which is in μ and not in δ. Hence $Z(p^\infty) \notin \mathcal{L}\delta$. □

The radicals $\mathcal{L}\delta$ and $\mathcal{U}\mu$ of a partition (δ, μ) of simple prime rings may differ even if further contraints are imposed, e.g. if both are special radicals which coincide on polynomial rings (see Tumurbat [3], Tumurbat and Wiegandt [2], [5]).

As is easily seen, beside the simple rings also the rings $Z(\infty)$ and $Z(p^\infty)$ must make up their minds and must go either to the radical or to the semisimple class with respect to any radical. A ring $A \neq 0$ is called *unequivocal*, if for every radical γ either $\gamma(A) = A$ or $\gamma(A) = 0$. This notion was introduced by Gardner [3] and investigated by Divinsky [4] and Puczyłowski [2].

PROPOSITION 3.12.3. *The following are equivalent:*

(1) *A is unequivocal,*

(2) *0 and A are the only ideals which may be the radicals of A for any radical,*

(3) *A is in the lower radical class $\mathcal{L}\{I\}$ for every nonzero ideal I of A,*

(4) *A is semisimple with respect to the upper radical determined by any nonzero homomorphic image B of A.*

Proof: The equivalence (1)⇔(2) is trivial.

(1)⇔(3) If A is unequivocal and $A \notin \mathcal{L}\{I\}$, then A is in the semisimple class of $\mathcal{L}\{I\}$. Since the semisimple class is hereditary, also I is in the semisimple class, contradicting $I \in \mathcal{L}\{I\}$.

Assume that the ring A satisfies (3). If A is not uniquivocal, then there exists a radical γ such that $0 \neq \gamma(A) \neq A$. Then by (3) we have $A \in \mathcal{L}\{\gamma(A)\} \subseteq \gamma$, and so $\gamma(A) = A$, again a contradiction.

(1)⇔(4) Let A be unequivocal. If A is radical with respect to any radical γ, then so is its homomorphic image B. Since B is in the semisimple class of the radical $\mathcal{U}\{B\}$, A cannot be in $\mathcal{U}\{B\}$. Consequently A is in the semisimple class of $\mathcal{U}\{B\}$.

Suppose (4), and suppose that there is a radical γ such that $0 \neq \gamma(A) \neq A$. Then $A/\gamma(A)$ is in the semisimple class of γ. Since $A/\gamma(A)$ is a homomorphic image of A, by (4) it follows that A is in the semisimple class of γ, contradicting $\gamma(A) \neq 0$. Thus A must be unequivocal. □

At this point we introduce some new radicals which are of motivating nature in this section and will be used in subsequent ones.

EXAMPLE 3.12.4. The class

$$\Delta = \{A \mid A^+ \text{ is a divisible group}\}$$

is easily seen to be a radical class, the *divisible radical*. For, the maximal divisible subgroup $\Delta(A)$ of a ring A is an ideal, the unique largest divisible ideal which satisfies conditions (b) and (c) of Definition 2.1.1; moreover, the class Δ is obviously homomorphically closed. *The divisible radical Δ is not hereditary*: cf. the proof for the divisible torsion radical τ_D in Example 3.2.2.

EXAMPLE 3.12.5. As in Example 3.9.9 we may see that the class

$$\tau_p = \{A \mid \forall a \in A,\ o(a) = p^n \text{ where } n \text{ is a natural number}\}$$

for any prime number p, is a strongly hereditary radical class, called the *p-torsion radical*. A ring $A \in \tau_p$ is said to be a *p-ring*.

An abelian group is called *reduced*, if its maximal divisible subgroup is zero. Having in mind that $\tau_p \cap \tau_q = 0$ for primes p, q, $(p \neq q)$, we have the following

PROPOSITION 3.12.6 (Divinsky [4]). *There are four kinds of unequivocal rings:*
 (1) *torsionfree and divisible,*
 (2) *torsionfree with reduced additive group,*
 (3) *divisible p-rings,*
 (4) *p-rings with reduced additive group.* □

Examples are
 (1) the ring $\mathbb{Q}$ of rationals,
 (2) the ring $Z(\infty)$ on the infinite cyclic group with zero-multiplication,
 (3) the zero-ring $Z(p^\infty)$ on the quasi-cyclic group $C(p^\infty)$,
 (4) the prime field $Z_p = \mathbb{Z}/(p)$ and the zero-ring $Z(p)$ of p elements.

Next, we shall see that unequivocal rings with zero-multiplication occur quite often.

LEMMA 3.12.7. *For any radical γ the class*

$$\gamma^0 = \{A \mid A^0 \in \gamma\}$$

is a radical class and

$$\gamma^0(A) = \sum(U^+ \subseteq A^+ \mid U^0 \in \gamma)$$

for all rings A.

Proof: The class γ^0 is obviously homomorphically closed. Set temporarily
$$\Sigma(A) = \sum(U^+ \subseteq A^+ \mid U^0 \in \gamma).$$
If $I \triangleleft A$ and $I \in \gamma^0$, then $I^0 \in \gamma$ so $I \subseteq \Sigma(A)$ and hence $\gamma^0(A) \subseteq \Sigma(A)$.

We are going to show that $\Sigma(A) \triangleleft A$. For any element $a \in A$ and subgroup $U^+ \subseteq A^+$ with $U^0 \in \gamma$ we consider the homomorphism

$$f_a : U^+ \longrightarrow aU^+$$

defined by $f_a(u) = au$ for all $u \in U^+$. The mapping f_a defines also a homomorphism from U^0 onto $(aU^+)^0$, and since $U^0 \in \gamma$ we have $(aU^+)^0 \in \gamma$. Hence $aU^+ \subseteq \Sigma(A)$ for all $a \in A$, which implies $a\Sigma(A) \subseteq \Sigma(A)$ for all $a \in A$. A similar reasoning yields that $\Sigma(A)a \subseteq \Sigma(A)$ for all $a \in A$, proving that $\Sigma(A) \triangleleft A$.

Since by definition we have

(∗) $$\bigl(\Sigma(A)\bigr)^0 = \Sigma(A^0) = \gamma(A^0) \in \gamma,$$

again by definition we conclude that $\Sigma(A) \in \gamma^0$. Thus also $\Sigma(A) \subseteq \gamma^0(A)$ holds, proving $\Sigma(A) = \gamma^0(A)$.

It is straightforward to see that the class γ^0 is also closed under extensions, whence by Proposition 2.1.2 the class γ_0 is a radical class. □

Let us observe that *Lemma 3.12.7 is valid also for not necessarily associative rings.*

THEOREM 3.12.8 (Puczyłowski [2]). *If A is an unequivocal ring, then so is the zero-ring A^0.*

Proof: If A is unequivocal, then either $\gamma^0(A) = 0$ or $\gamma^0(A) = A$ for any radical γ. Observing that the sum representations of $\gamma^0(A)$ in Lemma 3.12.7 and that of $\gamma(A^0)$ yield $(\gamma^0(A))^0 = \gamma(A^0)$, we have either $\gamma(A^0) = 0$ or $\gamma(A^0) = A$, proving that also A^0 is unequivocal. □

Proposition 3.12.6 tells us what kinds of groups may occur as additive groups of unequivocal rings. Feigelstock [3] proved that every group in (1), (3) and (4) of Proposition 3.12.6 may be the additive group of an unequivocal ring, as we shall see next.

PROPOSITION 3.12.9 (Feigelstock [3]). *A group G is the additive group of an unequivocal ring A if and only if for all subgroups J of G and* *for all nonzero homomorphic images H of G there exists a nonzero homomorphism $f\colon J \to H$.*

Proof: If some ring on G is unequivocal, then by Theorem 3.12.8, so is G^0. Then Proposition 3.12.3 (4) yields the existence of a nonzero homomorphism $f\colon J \to H$.

Conversely, if such an f exists for every subgroup J and homomorphic image H of G, then G^0 is in the semisimple class of the upped radical determined by H. Hence by Proposition 3.12.3 G^0 is unequivocal. □

LEMMA 3.12.10. *If G is a non-torsion group and H a non-reduced group, then there exists a nonzero homomorphism $f\colon G \to H$.*

Proof: Let $a \in G$ be an element with $o(a) = \infty$, and D a divisible subgroup of H. The subgroup $(a) \cong C(\infty)$ of G can be mapped homomorphically into D. Since D is an injective abelian group, the homomorphism $(a) \to D$ extends to a homomorphism $f\colon G \to D$. □

THEOREM 3.12.11 (Feigelstock [3]). *Every divisible torsionfree group or divisible p-group or reduced p-group is the additive group of an unequivocal ring.*

Proof: If G is divisible and torsionfree, then the result follows from Proposition 3.12.9 and Lemma 3.12.10.

Suppose that G is a divisible p-group, $0 \neq J \subseteq G$ and $G \twoheadrightarrow H \neq 0$. Since also H is divisible, for any nonzero element $a \in J$ there exists a nonzero homomorphism $g\colon (a) \to H$ which extends by the injectivity of H to a homomorphism $f\colon J \to H$. An application of Proposition 3.12.9 yields the assertion.

Let G be a reduced p-group, J and H as before. Since $pJ \neq J$, it follows that $J/pJ \neq 0$ is an elementary p-group. Since H possesses elements of order p, there exists a nonzero homomorphism $f\colon J \to H$. Hence by Proposition 3.12.9 G is the additive group of an unequivocal ring. □

In the case of reduced torsionfree groups only partial results are known, and we refer the reader to Feigelstock [3].

Following Puczyłowski [2] we say that I is a *P-ideal* in a ring A, if I is an ideal in every ring B with $A \triangleleft B$. A ring without trivial P-ideals will be called a *P-simple ring*. Notice that a quasi-semiprime ideal (cf. 3.6) is always a P-ideal, but a P-ideal need not be a quasi-semiprime ideal: let A be a ring with zero-multiplication, and γ be a radical such that $0 \neq \gamma(A) \neq A$. Then $\gamma(A)$ is a P-ideal in A but not a quasi-semiprime ideal of A. Thus, a P-simple ring has no non-trivial quasi-semiprime ideals but not conversely.

PROPOSITION 3.12.12. *If A is a P-simple ring, then A is unequivocal.*

Proof: Assume that A is not unequivocal. Then there exists a radical γ such that $0 \neq \gamma(A) \neq A$. Since by the ADS-Theorem $\gamma(A) \triangleleft B$ for every $A \triangleleft B$, $\gamma(A)$ is a P-ideal in A.

THEOREM 3.12.13 (Puczyłowski [2]). *A ring A is P-simple if and only if A is either simple or a zero-algebra over a field.*

Proof: Assume that the ring A is P-simple and that A is not simple. Then A has an ideal I such that $0 \neq I \neq A$. By definition AIA is a P-ideal of A, so the assumption yields

$$A \neq I \supseteq AIA = 0.$$

Hence for the middle annihilator M of A we have

$$0 \neq I \subseteq M = A$$

because M is a P-ideal of A. Thus $A^3 = 0$ follows implying $A^2 \neq A$. Since A^2 is a P-ideal in A, we get $A^2 = 0$. For any prime p the subring pA is a P-ideal in A. Hence either $pA = 0$ for some prime p or $pA = A$ for all primes p. In the first case A is a zero-algebra over the prime field Z_p. In the second case A^+ is a divisible and torsionfree group, so a direct sum of copies of the additive group $\mathbb{Q}^+$ of rational. Hence, in the second case, A is a zero algebra over the rationals.

Suppose, next, that A is a zero-algebra over a prime field F. Let I be a non-trivial ideal of A. If I is not a subalgebra in A, then there exists an element $k \in F$ such that $kI \not\subseteq I$, and the mapping $f \colon A \to A$ defined by $f(x) = kx$, $x \in A$, is obviously an endomorphism of the ring A. If I is a subalgebra, then I can be viewed also as a sub-vector space of A, and so by $I \neq A$ there exists a linear transformation f of A such that $f(I) \not\subseteq I$. Thus in both cases there exists a mapping $f \in \text{End}(A, A)$ such that $f(I) \not\subseteq I$. Let S be the subring of the endomorphism ring $\text{End}(A, A)$ generated by f. The set

$$B = \begin{pmatrix} S & A \\ 0 & 0 \end{pmatrix}$$

of matrices form a ring. Identifying the ring A with the ideal $\begin{pmatrix} 0 & A \\ 0 & 0 \end{pmatrix}$ of B and the ring S with the subring $\begin{pmatrix} S & 0 \\ 0 & 0 \end{pmatrix}$ we embed A and S into B such that $A \triangleleft B$, but $f(I) \not\subseteq I$. Hence $I \triangleleft A \triangleleft B$ and I is not an ideal in B. This means that I is not a P-ideal in A, and so A is a P-simple ring. □

Leavitt and van Leeuwen [1] investigated rings isomorphic with all proper factor rings. These rings are obviously unequivocal.

The full description of unequivocal rings is not known and it is unlikely that we will be able to classify them.

Looking at Proposition 3.12.3 (2) one may get the idea to study which ideals of a ring A may be radicals of A for some radical. Investigations in this direction were done by Gardner [3], Heinicke [3], N. R. McConnell [2], Propes [2] and Puczyłowski [16].

3.13. Minimally embeddable rings

Any class $\mathcal{M}$ of simple rings with unity element is special, so its the upper radical $\mathcal{UM}$ is hereditary and has the intersection property relative to $\mathcal{M}$. This suggests two questions to ask:

(i) *Does there exist a class $\mathcal{M}$ of simple prime rings not all with unity element such that the upper radical $\mathcal{UM}$ is hereditary* (van Leeuwen and Jenkins [2])?

(ii) *Does there exist a class $\mathcal{M}$ of simple prime rings not all with unity element such that every $\mathcal{UM}$-semisimple ring is a subdirect sum of $\mathcal{M}$-rings* (Leavitt [7])?

In his paper [8] Leavitt solved both problems. He answered the van Leeuwen–Jenkins problem (i) in the affirmative and his problem (ii) in the negative.

In this section first we shall address the van Leeuwen–Jenkins problem and examine the possibility of the existence of a class of simple idempotent rings defining a hereditary upper radical while containing rings without unity element. For the time being $\mathcal{M}$ is a class of simple idempotent rings and for notational convenience we call its upper radical class $\mathcal{R}$.

PROPOSITION 3.13.1. *If $\mathcal{R}$ is hereditary, $S \in \mathcal{M}$ and S satisfies the condition*

$$(*) \qquad S \triangleleft \cdot A \Rightarrow A/S \text{ is simple}$$

then $\mathcal{M}$ contains all such rings A/S.

Proof: If $S \triangleleft \cdot A$, then $\mathcal{R}(A) = 0$ as $\mathcal{R}$ is hereditary (Proposition 3.2.6) so A has an ideal $I \neq A$ with $A/I \in \mathcal{M}$. Now $S \cap I \neq 0$ so $S \subseteq I$. But A/S is simple so S is maximal, whence $I = S$ so that $A/S = A/I \in \mathcal{M}$. $\square$

COROLLARY 3.13.2. *If $\mathcal{R}$ is hereditary, $S \in \mathcal{M}$, S is an algebra over a field K and satisfies $(*)$ then $K \in \mathcal{M}$ or S has a unity element.*

Proof: Let $S * K$ be the ring on $S^+ \oplus K^+$ defined by

$$(s_1, k_1)(s_2, k_2) = (s_1 s_2 + k_1 s_2 + k_2 s_1, k_1 k_2)$$

(similarly to the Dorroh extension). Then

$$S \cong S_1 = (S, 0) \triangleleft S * K$$

and $(S*K)/S_1 \cong K$. If for some $(s,k) \in S*K$ we have $(s,k)S_1 = 0$, then for all $t \in S$, we have

$$(0,0) = (s,k)(t,0) = (st+kt,0)$$

so $st = -kt$. If $s \neq 0$, then as S is simple and $S^2 \neq 0$, $sS \neq 0$ so $k \neq 0$ and then $(-k^{-1}s)t = t$ for all $t \in S$, i.e. S has a left unity element whence, being simple, it has a unity element by Proposition 1.2.6. In the contrary case, $(0:S_1) = 0$ so $S_1 \triangleleft \cdot S*K$. But then Proposition 3.13.1 says that $K \in \mathcal{M}$. □

COROLLARY 3.13.3. *If $\mathcal{R}$ is hereditary, $S \in \mathcal{M}$, S has prime characteristic p and satisfies* (∗), *then $\mathbb{Z}_p \in \mathcal{M}$ or S has a unity element.* □

PROPOSITION 3.13.4 (van Leeuwen and Jenkins [2]). *A simple idempotent ring S without unity element which either*
  (i) *has characteristic 0 or*
  (ii) *has prime characteristic p and is an algebra over a field F which is not an algebraic extension of $\mathbb{Z}_p$,*
does not satisfy (∗).

Proof: If S has characteristic 0, then as in the proof of Corollary 3.13.2, $S*\mathbb{Q}$ is an essential extension of S. Hence identifying S with its copy in $S*\mathbb{Q}$ we get that the left annihilator ideal $(0:S)_{S*\mathbb{Q}}$ of S in $S*\mathbb{Q}$ is zero. Let R be a proper subring of $\mathbb{Q}$. Then $(0:S)_{S*R} = 0$ so $S \triangleleft \cdot (S*R)$ and $(S*R)/S$ is not simple. In the characteristic p case we can replace R by any nonzero subring of F which is not a field; for instance F contains a copy of $\mathbb{Z}_p[x]$. □

The simple rings with unity element, and just these, have no proper essential extensions (see Theorem 1.2.5) and every class of simple rings with unity element determines a hereditary upper radical. In these preliminary results we have relaxed the condition on the simple rings to being maximal ideals of all their proper essential extensions and observed that further restrictions are necessary if a hereditary upper radical is to be obtained. In particular if such a ring has no unity element it must have finite characteristic p and not admit an algebra structure over a non-algebraic extension of $\mathbb{Z}_p$. Restricting things a little further gives us a positive result.

THEOREM 3.13.5 (Leavitt [8]). *Let $\mathcal{M}$ be a class of simple idempotent rings, $\widetilde{\mathcal{M}}$ the subclass of rings in $\mathcal{M}$ without unity element. Then $\mathcal{M}$ defines a hereditary upper radical if $\widetilde{\mathcal{M}}$ satisfies the following conditions.*
  (i) *All rings in $\widetilde{\mathcal{M}}$ have finite characteristic.*
  (ii) *If $S \in \widetilde{\mathcal{M}}$, $S \triangleleft \cdot R$, $S \neq R$ and S has characteristic p, then $R/S \cong \mathbb{Z}_p$.*
  (iii) *If $S \in \widetilde{\mathcal{M}}$ and S has characteristic p, then $\mathbb{Z}_p \in \mathcal{M}$.*

Proof: As before, let $\mathcal{R}$ denote the upper radical class defined by $\mathcal{M}$. Suppose $I \triangleleft A \in \mathcal{R}$ but $I \notin \mathcal{R}$. Then I has an ideal K with $I/K \in \mathcal{M}$. Since

I/K is semiprime, K is an ideal of A and we now have $I/K \triangleleft A/K \in \mathcal{R}$. By the usual argument, A/K has a homomorphic image B which is an essential extension of I/K. We can't have $B = I/K$, since then $B \in \mathcal{R} \cap \mathcal{M} = \emptyset$ (so in particular I/K has no unity element). We also can't have $B/(I/K) \cong \mathbb{Z}_p$, where I/K has finite characteristic p, as this would put $\mathbb{Z}_p$ in $\mathcal{R} \cap \mathcal{M}$. We conclude that there is no such I/K, i.e. that I is in $\mathcal{R}$ if $I \triangleleft A \in \mathcal{R}$. □

We call attention to the fact that Theorem 3.13.5 refers to properties of the class $\mathcal{M}$ and not just to ring properties of its individual members. It remains unknown what conditions on a class of simple rings are necessary for its upper radical to be hereditary. We can however say precisely when a single simple ring defines a hereditary upper radical.

PROPOSITION 3.13.6. *Let S be a simple ring. The upper radical class $\mathcal{U}\{S\}$ defined by S is hereditary if and only if S has a unity element.*

Proof: If S has no unity element its Dorroh extension $S * \mathbb{Z}$ has a homomorphic image B which is an essential extension of S. Since every homomorphic image of B has a unity element, B is in $\mathcal{U}(S)$ which is therefore not hereditary, not having an essentially closed semisimple class. The converse is clear. □

Thus far we have been imposing lots of conditions on simple rings without considering problems of existence. Most of what follows is directed at establishing the existence of the kind of simple ring we need if Theorem 3.13.5 is not to be vacuous.

Let A be a ring with unity element, u an element of A which is cancellable (i.e. neither a left nor a right zero divisor).

We consider the ring $\operatorname{End}_{\mathbb{Z}}(uA)$ of abelian group endomorphisms of uA and its opposite ring $\operatorname{End}_{\mathbb{Z}}(uA)^{op}$. In the latter we have the same addition as in $\operatorname{End}_{\mathbb{Z}}(uA)$ but the multiplication $*$ is given by $f * g = gf$. An element (f', f'') of $\operatorname{End}_{\mathbb{Z}}(uA) \oplus \operatorname{End}_{\mathbb{Z}}(uA)^{op}$ will be called a *multiplier* of uA if

$$xf'(y) = f''(x)y \text{ for all } x, y \in uA.$$

If (f', f'') and (g', g'') are multipliers, then for all $x, y \in uA$ we have

$$x(f' - g')(y) = x(f'(y) - g'(y)) = xf'(y) - xg'(y)$$
$$= f''(x)y - g''(x)y = (f''(x) - g''(x))y = (f'' - g'')(x)y,$$

(so that $(f', f'') - (g', g'')$ is a multiplier) and

$$xf'g'(y) = f''(x)g'(y) = g''(f''(x))y = g''f''(x)y,$$

(so that $(f', f'')(g', g'') = (f'g', f'' * g'') = (f'g', g''f'')$ is a multiplier). Thus the multipliers form a subring, which we call $\operatorname{Mult}(uA)$ of $\operatorname{End}_{\mathbb{Z}}(uA) \oplus \operatorname{End}_{\mathbb{Z}}(uA)^{op}$.

Of course, all this is valid if an arbitrary ring and subring are substituted for A and uA. However, Mult(uA) has further properties which interest us.

In general, if B' is a subring of a ring B, we define the *idealizer* of B' in B to be the set

$$\{b \in B : bx \text{ and } xb \in B' \; \forall x \in B'\}.$$

This is a subring of B and is the largest subring which contains B' as an ideal.

For convenience, in what follows we will denote the idealizer of uA in A by T. As uA is a right ideal of A, $T = \{t \in A : tuA \subseteq uA\}$. But A has a unity element, so in fact $T = \{t \in A : tu \in uA\}$.

For a ring element a in general, we denote by L_a, R_a the *left and right multiplication functions*: $L_a(r) = ar$ and $R_a(r) = ra$ for all r. Returning to our particular case, if $t \in T$ then $L_t(x)$ and $R_t(x)$ are in uA for every $x \in uA$, so we can view L_t and R_t as members of $\text{End}_{\mathbb{Z}}(uA)$. But for $x, y \in uA$ we have, for all $t \in T$,

$$xL_t(y) = x(ty) = (xt)y = R_t(x)y,$$

so (L_t, R_t) is in Mult(uA). It turns out that every element of Mult(uA) has this form.

PROPOSITION 3.13.7. (Leavitt and van Leeuwen [1]) *Let A be a ring with unity element, u a cancellable element of A, and let T denote the idealizer of uA in A. Then the function*

$$\varphi : T \to \text{Mult}(uA) \, ; \; \varphi(t) = (L_t, R_t) \text{ for all } t$$

is a ring isomorphism.

Proof: Let t_1, t_2 be in T. Then for all $x \in uA$ we have

$$(L_{t_1} + L_{t_2})(x) = L_{t_1}(x) + L_{t_2}(x) = t_1 x + t_2 x$$
$$= (t_1 + t_2)x = L_{t_1+t_2}(x)$$

and so $L_{t_1+t_2} = L_{t_1} + L_{t_2}$. Similarly $R_{t_1+t_2} = R_{t_1} + R_{t_2}$. As well, we have

$$L_{t_1}L_{t_2}(x) = t_1 t_2(x) = L_{t_1 t_2}(x)$$

and

$$R_{t_2}R_{t_1}(x) = xt_1 t_2 = R_{t_1 t_2}(x),$$

so

$$L_{t_1 t_2} = L_{t_1}L_{t_2} \text{ and } R_{t_1 t_2} = R_{t_2}R_{t_1} = R_{t_1} * R_{t_2}$$

(where, as before, $*$ is the multiplication of $\text{End}_{\mathbb{Z}}(uA)^{op}$). From this it follows that φ is a homomorphism. If $\varphi(t) = 0$, i.e. $tx = 0 = xt$ for all $x \in uA$, then as u is cancellable, $t = 0$. Finally we show that φ is surjective. Let (g', g'') be in $\text{Mult}(uA)$. Then there exist $a, b \in A$ such that $g'(u) = ua$ and $g''(u) = ub$. Now

$$u(ua) = ug'(u) = g''(u)u = (ub)u = u(bu),$$

so by the cancellability of u, $ua = bu$. Thus $bu \in uA$ so $b \in T$. Now take any $r \in A$ and let $g'(ur) = us$. We have

$$u(us) = u\,g'(ur) = g''(u)ur = (ub)ur = u(b\,ur),$$

so again by the cancellability of u, $us = b\,ur$. Thus

$$g'(ur) = us = b\,ur,$$

whence by the arbitrariness of r, $g' = L_b$. Now let $g''(ur) = uq$. Then

$$u(qu) = (uq)u = g''(ur)u = ur\,g'(u) = ur(ua) = u(rua),$$

so

$$qu = rua = r(ua) = rg'(u) = r\,L_b(u) = (rb)u,$$

and hence $q = rb$. Finally,

$$g''(ur) = uq = u(rb) = (ur)b = R_b(ur)$$

so (as above) $g'' = R_b$. We therefore have $(g', g'') = (L_b, R_b)$. The proof is complete. $\square$

THEOREM 3.13.8. (Leavitt and van Leeuwen [1]) *Let A be a ring with unity element, u a cancellable element of A, T the idealizer of uA in A. If B is a ring such that for some $K \triangleleft B$ there is an isomorphism $g : K \to uA$, then there is a homomorphism $h : B \to T$ such that $h(z) = g(z)$ for all $z \in K$.*

Proof: Let z be in B, $x, y \in uA$. Then $g^{-1}(x), g^{-1}(y) \in K$ so $g^{-1}(x)z, zg^{-1}(y) \in K$. Thus there are uniquely determined $x', y' \in uA$ such that $g^{-1}(x)z = g^{-1}(x')$ and $zg^{-1}(y) = g^{-1}(y')$. We now have

$$g^{-1}(xy') = g^{-1}(x)g^{-1}(y') = g^{-1}(x) \cdot zg^{-1}(y) = g^{-1}(x)z \cdot g^{-1}(y),$$
$$= g^{-1}(x')g^{-1}(y) = g^{-1}(x'y),$$

so $xy' = x'y$. Now gL_zg^{-1} and gR_zg^{-1} take uA to uA and

$$x \cdot gL_zg^{-1}(y) = x \cdot g(zg^{-1}(y)) = x \cdot g(g^{-1}(y')) = xy'$$
$$= x''y = g(g^{-1}(x'))y = g(g^{-1}(x)z)y = gR_zg^{-1}(x)y.$$

Thus $(gL_zg^{-1}, gR_zg^{-1}) \in \text{Mult}(uA)$ (for every $z \in B$).

Define $f : B \to \text{Mult}(uA)$ by setting
$$f(z) = (gL_zg^{-1}, gR_zg^{-1}) \text{ for all } z \in B.$$

Then for $z, w \in B$ we have
$$\begin{aligned}
f(z+w) &= (gL_{z+w}g^{-1}, gR_{z+w}g^{-1}) \\
&= (g(L_z + L_w)g^{-1}, g(R_z + R_w)g^{-1}) \\
&= (gL_zg^{-1} + gL_wg^{-1}, gR_zg^{-1} + gR_wg^{-1}) \\
&= f(z) + f(w)
\end{aligned}$$

and

$$\begin{aligned}
f(zw) &= (gL_{zw}g^{-1}, gR_{zw}g^{-1}) \\
&= (gL_zL_wg^{-1}, gR_wR_zg^{-1}) \\
&= (gL_zg^{-1} \cdot gL_wg^{-1}, gR_wg^{-1} \cdot gR_zg^{-1}) \\
&= (gL_zg^{-1} \cdot gL_wg^{-1}, gR_zg^{-1} * gR_wg^{-1}) \\
&= (gL_zg^{-1}, gR_zg^{-1})(gL_wg^{-1}, gR_wg^{-1}) \\
&= f(z)f(w).
\end{aligned}$$

Thus f is a homomorphism.

If $z \in K$, then for all $x \in uA$ we have
$$gL_zg^{-1}(x) = g(zg^{-1}(x)) = g(z)gg^{-1}(x) = g(z)x = L_{g(z)}(x)$$

and

$$gR_zg^{-1}(x) = g(g^{-1}(x)z) = gg^{-1}(x)g(z) = xg(z) = R_{g(z)}(x).$$

Hence
$$f(z) = (gL_zg^{-1}, gR_zg^{-1}) = (L_{g(z)}, R_{g(z)}) = \varphi(g(z))$$

(where φ is the isomorphism of Proposition 3.13.7).

To complete the proof, we simple take $h = \varphi^{-1}f$. $\square$

To find an example of a family of simple rings satisfying the requirements of Theorem 3.13.5 we need only consider rings of prime characteristic. We shall now establish a connection between Theorems 3.13.5 and 3.13.8, but the latter will be applied only to rings with this restriction on characteristic.

PROPOSITION 3.13.9. *The following conditions are equivalent for a ring R of prime characteristic p.*

(i) *If $R \triangleleft W$ for some ring W, then there is an ideal J of W such that $J \cap R = 0$ and $R \oplus J = W$ or $W/(R \oplus J) \cong \mathbb{Z}_p$.*

(ii) If $R \triangleleft \cdot V$ then $V = R$ or $V/R \cong \mathbb{Z}_p$.

Proof: (i) $\Rightarrow$ (ii): If $R \triangleleft \cdot V$, the J of (i) must be 0.

(ii) $\Rightarrow$ (i): If $R \triangleleft W$, let J be an ideal of W which has zero intersection with R and is maximal for this. Then $R \cong R/(R \cap J) \cong (R+J)/J \triangleleft \cdot W/J$ so $(R+J)/J = W/J$ or $(W/J)/((R+J)/J) \cong \mathbb{Z}_p$, and thus $R \oplus J = R+J = W$ or $W/(R \oplus J) \cong (W/J)/((R+J)/J) \cong \mathbb{Z}_p$. $\square$

We shall call a ring satisfying the conditions of Proposition 3.13.9 *minimally embeddable*.

We return now to the situation described in Theorem 3.13.8, but now we take A to have prime characteristic p. The idealizer T of uA contains uA and 1_A so the smallest possibility for T is $uA + \mathbb{Z}_p 1_A$. (We include the degenerate case where $1_A \in uA$, i.e. $uA = A$.) It turns out that this smallest possibility can actually occur and it does so precisely when uA is minimally embeddable.

THEOREM 3.13.10. *Let A be a ring with unity element and with prime characteristic p containing a cancellable element u. Let T be the idealizer of uA in A. Then uA is minimally embeddable if and only if $T = uA + \mathbb{Z}_p 1_A$.*

Proof: Suppose $T = uA + \mathbb{Z}_p 1_A$, and let B be an essential extension of uA. By Theorem 3.13.8 there is a homomorphism $h : B \to T$ such that $h(ua) = ua$ for all $a \in A$. Now $\ker h \cap uA = 0$ so $\ker h = 0$, as $uA \triangleleft \cdot B$. Thus $B \cong h(B)$, while $uA \subseteq h(B) \subseteq T = uA + \mathbb{Z}_p 1_A$. As $\mathbb{Z}_p 1_A$ has no proper subrings, we conclude that $h(B) = uA$ or $uA + \mathbb{Z}_p 1_A$. But $B/uA \cong h(B)/uA$, so $B/uA \cong 0$ or $(uA + \mathbb{Z}_p 1_A)/uA$. Since $(uA + \mathbb{Z}_p 1_A)/uA \cong \mathbb{Z}_p$ or 0 (the latter only when $1_A \in uA$), we conclude that uA is minimally embeddable.

Now suppose uA is minimally embeddable. If $J \triangleleft T$ and $J \cap uA = 0$, then $Ju = Ju1_A \subseteq J \cap uA = 0$, so by the cancellability of u, $J = 0$. Thus $uA \triangleleft \cdot T$. If $uA \neq T$, then $T/uA \stackrel{\cong}{=} \mathbb{Z}_p$. Since then $uA \subsetneq uA + \mathbb{Z}_p 1_A \subseteq T$, we have

$$0 \neq (uA + \mathbb{Z}_p 1_A)/uA \subseteq T/uA \cong \mathbb{Z}_p,$$

so as $\mathbb{Z}_p$ has no proper subrings, $uA + \mathbb{Z}_p 1_A = T$. $\square$

It remains to be shown that the conditions of Theorem 3.13.10 with $1_A \notin uA$ can be met at all, and specifically that they can be met in instances where uA is simple. The next example will take care of all of this.

Let F_0 be a field, F the field of rational expressions over F_0 in commuting indeterminates y_n, $n \in \mathbb{Z}$. Let σ be the automorphism of F over F_0 given by

$$\sigma(y_n) = y_{n+1} \text{ for all } n.$$

We consider the *twisted Laurent polynomial ring* $W = F[x, x^{-1}; \sigma]$ in x, x^{-1} over F with respect to σ. This consists of polynomials in x and x^{-1} with coefficients in F with ordinary polynomial addition, but with the "twisted" polynomial multiplication given by

$$x\alpha = \sigma(\alpha)x \text{ for all } \alpha \in F.$$

Some more general multiplication rules follow from this. We have

$$x^2\alpha = x \cdot x\alpha = x \cdot \sigma(\alpha)x = x\sigma(\alpha) \cdot x = \sigma^2(\alpha)x \cdot x = \sigma^2(\alpha)x^2$$

and in general $x^n\alpha = \sigma^n(\alpha)x^n$ for $n = 1, 2, 3, \ldots$. Also

$$\alpha = x^{-1}x \cdot \alpha = x^{-1} \cdot x\alpha = x^{-1} \cdot \sigma(\alpha)x,$$

so

$$x^{-1}\alpha = x^{-1}\alpha \cdot xx^{-1} = x^{-1}\alpha x \cdot x^{-1} = x^{-1}\sigma\left(\sigma^{-1}(\alpha)\right)x \cdot x^{-1}$$
$$= x^{-1}x\sigma^{-1}(\alpha)x \cdot x^{-1} = \sigma^{-1}(\alpha)x^{-1},$$

and in general $x^{-n}\alpha = \sigma^{-n}(\alpha)x^{-n}$ for $n = 1, 2, 3, \ldots$. Since of course $x^0\alpha = \alpha = \sigma^0(\alpha)x^0$ we have

$$x^m\alpha = \sigma^m(\alpha)x^m \text{ for all } m \in \mathbb{Z}.$$

We wish to show that W is simple. First we consider the twisted polynomial ring $F[x, \sigma]$, the subring of W consisting of elements not involving negative powers of x. We know from Example 1.3.3 that in $F[x; \sigma]$ every nonzero ideal has the form $F[x; \sigma]x^k$.

Now let J be a nonzero ideal of $W = F[x, x^{-1}; \sigma])$. If $h \in J \setminus \{0\}$, then $hx^t \in F[x; \sigma]$ for some $t > 0$. It follows that $J \cap F[x; \sigma] \neq 0$. This is moreover a nonzero ideal of $F[x; \sigma]$. If it is generated by x^m (with m just possibly zero) then J contains $x^{-m}x^m = 1$, so $J = W$.

Thus W is a simple ring.

We are now ready for our example of minimal embeddability. We preserve all the notation from the above discussion, but specialize to the case $F_0 = \mathbb{Z}_p$.

THEOREM 3.13.11. (Leavitt and van Leeuwen [1]). *Let $F_0 = \mathbb{Z}_p$, so that the field F of rational expressions over F_0 in commuting indeterminates y_n, $n \in \mathbb{Z}$, the twisted polynomial ring $F[x; \sigma]$ and the twisted Laurent polynomial ring $W = F[x, x^{-1}; \sigma]$ have characteristic p. Let $S = (x-1)W$. Then S is a minimally embeddable simple ring without unity element.*

Proof: $F[x; \sigma]$ has no zero divisors by Theorem 1.3.1. Every element of W has the form $x^{-n}a = bx^{-n}$ for $a, b \in F[x; \sigma]$ and $n \geq 0$. (The relationship between a and b can be seen from the way σ is used to define the multiplication in $F[x, x^{-1}; \sigma]$). As x is a unit and $F[x; \sigma]$ has no zero divisors, W has no zero divisors, so in particular $x - 1$ is cancellable. As we indicated above, we let $S = (x-1)W$. If $0 \neq J \triangleleft S$, then since $SW = S$ and W (in particular S) has no zero divisors, we have

$$WJ \cdot SW = WJS \neq 0$$

and hence $WJ \cdot SW = W$, since W is simple. But then

$$S = SW = S(WJ \cdot SW) = SW \cdot J \cdot SW = SJS \subseteq J$$

so $J = S$ and S is simple. Now W has no idempotents other than 1_W and 0, while $1_W \notin S$. Thus S has no nonzero idempotents and in particular it has no unity element.

Since 1_W and x commute, we have

$$x^n - 1 = (x - 1)\left(x^{n-1} + x^{n-2} + \ldots + x + 1\right)$$

for all $n > 0$, so for $\alpha \in F$ we have

$$\begin{aligned}x^n \alpha &= (x^n - 1)\alpha + \alpha \\ &= (x - 1)\left(x^{n-1} + x^{n-2} + \ldots + x + 1\right)\alpha + \alpha.\end{aligned}$$

Similarly

$$\begin{aligned}x^{-n}\alpha &= \left(x^{-1} - 1\right)\left(x^{-n+1} + x^{-n+2} + \ldots + x^{-1} + 1\right)\alpha + \alpha \\ &= \left(x^{-1} - 1\right)(-x)\left(-x^{-1}\right)\left(x^{-n+1} + x^{-n+2} + \ldots + x^{-1} + 1\right)\alpha + \alpha \\ &= (x - 1)\left(-x^{-1}\right)\left(x^{-n+1} + x^{-n+2} + \ldots + x^{-1} + 1\right)\alpha + \alpha.\end{aligned}$$

Finally, $x^0 \alpha = (x - 1)0 + \alpha$. Since W consists of sums of terms of the form $x^m \alpha$ $(= \sigma^m(\alpha)x^m)$, $m \in \mathbb{Z}$, $\alpha \in F$, we conclude that every element of W has the form

$$(x - 1)a + \beta, \ a \in W, \ \beta \in F.$$

Let T be the idealizer of S in W. (This is consistent with our earlier notation.) If $t \in T$ then $t(x - 1) = (x - 1)v$ for some $v \in W$ and also $t = (x - 1)b + \gamma$ for some $b \in W$, $\gamma \in F$. Then $\gamma = t - (x - 1)b \in T + S = T$ so $\gamma(x - 1) = (x - 1)c$ for some $c \in W$. Clearly $c = 0$ if and only if $\gamma = 0$, while for $\gamma \neq 0$ we have

$$\partial(\gamma) + \partial(x - 1) = \partial(x - 1) + \partial(c)$$

so c is in F. Thus

$$\gamma x - \gamma = xc - c = \sigma(c)x - c$$

so equating coefficients

$$\sigma(c) = \gamma = c.$$

This means that c, as a fixed point of σ, is in $\mathbb{Z}_p$. Thus $t = (x - 1)b + c \in S + \mathbb{Z}_p 1_W$. We conclude that $T = S + \mathbb{Z}_p 1_W$ and so, by Theorem 3.13.10, that S is minimally embeddable. □

Thus by Theorem 3.13.5, $\{S, \mathbb{Z}_p\}$ defines a hereditary upper radical, though S has no unity element. □

The proof that W is simple comes from Jacobson [1] while the simplicity of S was proved by Robson [1]. The first example of a minimally embeddable ring was obtained by Leavitt [8].

COROLLARY 3.13.12 (Leavitt [8]). *There exists a class $\mathcal{M}$ of simple rings not all with unity element such that the upper radical $\mathcal{R} = \mathcal{UM}$ is hereditary.* □

Leavitt [8] used another, more involved ring construction for solving his problem (ii). Here we shall get a surprisingly easy solution of problem (ii) by applying Beidar's Corollary 3.8.9 and the following characterization of classes of simple rings with unity elment.

THEOREM 3.13.13 (Andrunakievich [1]). *For a class $\mathcal{M}$ of simple rings the following are equivalent:*

(1) $\mathcal{M}$ *consists of simple rings with unity element,*

(2) $\mathcal{M}$ *is a special class,*

(3) *the upper radical $\gamma = \mathcal{UM}$ is hereditary and its semisimple class is the subdirect closure of the class $\mathcal{M}$.*

Proof: The implications (1) $\Rightarrow$ (2) $\Rightarrow$ (3) are trivial consequences of Szendrei's Theorem 1.2.5 and Corollary 3.7.4, respectively.

(3) $\Rightarrow$ (1) Assume that $\mathcal{M}$ contains a simple ring A without unity element. Then by Theorem 1.2.5 there exists an overring B such that $A \triangleleft B$ and A is not a direct summand in B. Choosing an ideal K of B such that $A \cap K = 0$ and K is maximal for this, we have

$$A \cong (A+K)/K \triangleleft \cdot B/K.$$

We may confine ourselves to the case $0 \neq A \triangleleft \cdot B$ and $A \neq B$. If $\gamma(B) \neq 0$, then also $A \cap \gamma(B) \neq 0$. Since A is simple and γ is hereditary, we conclude that $A \in \gamma$, contradicting $A \in \mathcal{M}$. Thus $\gamma(B) = 0$. By the subdirect decomposition property there exist ideals $I_\lambda \triangleleft B$, $\lambda \in \Lambda$, such that each $B/I_\lambda \in \mathcal{M}$ and $\cap I_\lambda = 0$. Hence there exists an index $\mu \in \Lambda$ such that $A \not\subseteq I_\mu$, and so $A \cap I_\mu = 0$, contradicting $0 \neq A \triangleleft \cdot B$. Thus the class $\mathcal{M}$ must consists of simple rings with unity element. □

COROLLARY 3.13.14 (Leavitt [8]). *Let $\mathcal{M}$ be a class of simple idempotent rings. Then its upper radical $\mathcal{M}$ has the intersection property with respect to $\mathcal{M}$ if and only if every ring in $\mathcal{M}$ has a unity element.*

Proof: If $\mathcal{M}$ has the intersection property with respect to $\mathcal{M}$ then by Corollary 3.8.19, $\mathcal{UM}$ is special and thus hereditary. The result now follows from Theorem 3.13.13. □

3.14. Modules and radicals

Of the many known characterizations of the Jacobson radical, two of the most useful ones are based on one-sided concepts — modules (simple) and left ideals (modular maximal), see Theorems 4.4.12 and 4.4.9, respectively. It is possible to describe all radicals in terms of modules, and, at least the special radicals, by left ideals.

We begin our discussion with a general result relating modules over a ring and those over a factor ring. We denote by $(0:M)$ or $(0:M)_A$ the *annihilator*

$$(0:M)_A = \{a \in A \mid aM = 0\}$$

of an A-module M.

PROPOSITION 3.14.1. *Let A be a ring, $I \triangleleft A$.*

(i) *If M is an A/I-module, then under the scalar multiplication*

$$am = (a+I)m,$$

M becomes an A-module with $I \subseteq (0:M)_A$.

(ii) *If M is an A-module and $I \subseteq (0:M)_A$, then M is an A/I-module with respect to*

$$(a+I)m = am.$$

(iii) *Every submodule of the A/I-module M is a submodule of the A-module M, and in the case $I \subseteq (0:M)_A$ also the converse is true.*

(iv) $(0:M)_{A/I} = (0:M)_A/I$.

Proof: (i) is clear enough.

(ii) If $a + I = b + I$, then $a - b \in I \subseteq (0:M)_A$, so $am = bm$ for all $m \in M$ and scalar multiplication by elements of A is well defined.

(iii) is clear.

(iv) The following are equivalent: $a + I \in (0:M)_{A/I}$; $(a+I)m = 0$ for all $m \in M$; $am = 0$ for all $m \in M$; $a \in (0:M)_A$; $a + I \in (0:M)_A/I$. □

For each ring A, let Σ_A be a class of A-modules M with $AM \neq 0$, and let Σ be the union of all the Σ_A. Let

$$\ker(\Sigma_A) = \cap((0:M)_A \mid M \in \Sigma_A).$$

Consider now some conditions which Σ might satisfy.

(M1) If $M \in \Sigma_{A/I}$, then $M \in \Sigma_A$.
(M2) If $M \in \Sigma_A$ and $I \triangleleft A$, $I \subseteq (0:M)_A$, then $M \in \Sigma_{A/I}$.
(M3) If $\ker(\Sigma_A) = 0$, then $\Sigma_B \neq \emptyset$ for all nonzero ideals B of A.
(M4) If $\Sigma_B \neq \emptyset$ whenever $0 \neq B \triangleleft A$, then $\ker(\Sigma_A) = 0$.

An A-module M is *faithful* if $(0:M) = 0$.

Let Σ be a class of modules satisfying (M1), (M2), (M3). Let $F(\Sigma)$ denote the class of rings A for which Σ_A contains a faithful module.

Radical Theory for Associative Rings

PROPOSITION 3.14.2. $F(\Sigma)$ *is a regular class.*

Proof: If $A \in F(\Sigma)$, then clearly $\ker(\Sigma_A) = 0$, so if $0 \neq B \triangleleft A$, then $\Sigma_B \neq \emptyset$ by (M3). If $M \in \Sigma_B$, then by (M2), M is in $\Sigma_{B/(0:M)}$ and by Proposition 3.14.1 (iv) M is a faithful $B/(0:M)$-module. Thus $B/(0:M) \in F(\Sigma)$. $\square$

THEOREM 3.14.3 (Andrunakievich and Ryabukhin [3], Hentzel [1]). Let Σ be a class of modules satisfying (M1), (M2), (M3) and let

$$\gamma_\Sigma = \{A \mid \Sigma_A = \emptyset\}.$$

Then

(i) γ_Σ *is a radical class; in fact* $\gamma_\Sigma = \mathcal{U}F(\Sigma)$,

(ii) $\gamma_\Sigma(A) \subseteq \ker(\Sigma_A)$ *for every ring* A.

Proof: (i) If $A \in \gamma_\Sigma$ and $A/I \neq 0$, then by (M1) $\Sigma_{A/I} = \emptyset$, so certainly $\Sigma_{A/I}$ contains no faithful A/I-module, that is, $A/I \notin F(\Sigma)$. Hence $\gamma_\Sigma \subseteq \mathcal{U}F(\Sigma)$. If, on the other hand, $A \notin \gamma_\Sigma$, let N be in Σ_A. Then by (M2) and Proposition 3.14.1, N is in $\Sigma_{A/(0:N)}$ and is faithful, whence $A/(0:N) \in F(\Sigma)$ and $A \notin \mathcal{U}F(\Sigma)$.

(ii) For every ring A and for every $M \in \Sigma_A$, we have

$$\gamma_\Sigma(A)/(\gamma_\Sigma(A) \cap (0:M)) \cong (\gamma_\Sigma(A) + (0:M))/(0:M) \triangleleft A/(0:M) \in F(\Sigma).$$

Since $F(\Sigma)$ is regular, $(\gamma_\Sigma(A) + (0:M))/(0:M)$, if nonzero, has a nonzero homomorphic image in $F(\Sigma)$. But then so does $\gamma_\Sigma(A)$, a contradiction, as $\gamma_\Sigma(A) \in \mathcal{U}F(\Sigma)$. It follows that $\gamma_\Sigma(A) \subseteq (0:M)$ for all $M \in \Sigma_A$, that is, $\gamma_\Sigma(A) \subseteq \ker(\Sigma_A)$. $\square$

A better result can be achieved when Σ satisfies condition (M4) as well.

THEOREM 3.14.4 (Andrunakievich and Ryabukhin [3]). *Let* $\Sigma = \cup \Sigma_A$ *be a class of modules satisfying conditions* (M1)–(M4). *Then*

(i) A *is* γ_Σ-*semisimple if and only if* A *is a subdirect sum of rings in* $F(\Sigma)$.

(ii) $\gamma_\Sigma(A) = \ker(\Sigma_A)$ *for every ring* A.

Proof: (i) If $\gamma_\Sigma(A) = 0$, then A has no nonzero ideals in γ_Σ, that is $\Sigma_I \neq \emptyset$ whenever $0 \neq I \triangleleft A$. By (M4), $\ker(\Sigma_A) = 0$, so A is a subdirect sum of the rings $A/(0:M)$, $M \in \Sigma_A$, and these rings are in $F(\Sigma)$ by Proposition 3.14.1. Conversely, if $A \in F(\Sigma)$, then $\ker(\Sigma_A) = 0$, so by (M3) $\Sigma_I \neq \emptyset$ for every nonzero ideal I of A. Hence A is γ_Σ-semisimple. Since subdirect sums of γ_Σ-semisimple rings are γ_Σ-semisimple, we have the result.

(ii) Let I be an ideal of A such that $A/I \in F(\Sigma)$, and let M be a faithful module in $\Sigma_{A/I}$. Then by (M1), $M \in \Sigma_A$ and by Proposition 3.14.1

$$0 = (0:M)_{A/I} = (0:M)_A/I,$$

so $\ker(\Sigma_A) \subseteq (0:M)_A = I$. By (i), $A/\gamma_\Sigma(A)$ is a subdirect sum of rings $(A/\gamma_\Sigma(A))/(I_\lambda/\gamma_\Sigma(A))$ where for each λ,

$$A/I_\lambda \cong (A/\gamma_\Sigma(A))/(I_\lambda/\gamma_\Sigma(A)) \in F(\Sigma).$$

Hence (from the argument above)

$$\ker(\Sigma_A) \subseteq \cap I_\lambda \subseteq \gamma_\Sigma(A). \qquad \square$$

We now show that every radical arises as in the last two results from a class Σ of modules satisfying (M1)–(M4).

THEOREM 3.14.5 (Andrunakievich and Ryabukhin [3]). *Let γ be a radical class, and for every ring A, let*

$$\Sigma_A = \{M \mid \gamma(A/(0:M)) = 0\} \quad \text{and} \quad \Sigma = \cup \Sigma_A.$$

Then

(i) Σ *satisfies conditions* (M1)–(M4),
(ii) $\gamma = \gamma_\Sigma$.

Proof: (i) If $M \in \Sigma_{A/I}$, then $(A/I)M \neq 0$ and $(A/I)/(0:M)_{A/I}$ is γ-semisimple. By Proposition 3.14.1, $(0:M)_{A/I} = (0:M)_A/I$, so

$$A/(0:M)_A \cong (A/I)/(0:M)_{A/I}$$

is γ-semisimple. Thus, as clearly $AM \neq 0$, $M \in \Sigma_A$ and Σ satisfies (M1).

If $M \in \Sigma_A$ and $I \subseteq (0:M)$, then

$$(A/I)/(0:M)_{A/I} = (A/I)/((0:M)_A/I) \cong A/(0:M)_A$$

is γ-semisimple. Also $(A/I)M = AM \neq 0$, so $M \in \Sigma_A$. This gives us (M2).

Suppose $\ker(\Sigma_A) = 0$ and $0 \neq B \triangleleft A$. Then $B \not\subseteq \ker(\Sigma_A)$, so $BM \neq 0$ for some $M \in \Sigma_A$. Now

$$B/(0:M)_B = B/((0:M)_A \cap B) \cong (B + (0:M)_A)/(0:M)_A \triangleleft A/(0:M)_A.$$

Since $\gamma(A/(0:M)_A) = 0$, we have $\gamma(B/(0:M)_B) = 0$. Hence $M \in \Sigma_B$, $\Sigma_B \neq \emptyset$ and (M3) is satisfied.

Finally we consider (M4). Suppose $\Sigma_B \neq \emptyset$ for all nonzero $B \triangleleft A$. Then for each nonzero ideal B there exists an $M_B \in \Sigma_B$ and we have $\gamma(B/(0:M_B)_B) = 0$. Thus every nonzero ideal of A has a nonzero γ-semisimple homomorphic image. It follows that $\gamma(A) = 0$. Let A^1 be the Dorroh extension of A. This is, in an obvious way, an A-module, and $(0:A^1)_A = 0$. Hence $A/(0:A^1)_A \cong A$ is γ-semisimple, so $A^1 \in \Sigma_A$ and $\ker(\Sigma_A) = 0$.

(ii) If $A \in \gamma$ and M is an A-module such that $A/(0:M)_A$ is γ-semisimple, then of course $(0:M)_A = A$, that is, $AM = 0$. It follows that

$\Sigma_A = \emptyset$, that is, $A \in \gamma_\Sigma$. On the other hand, if $A \notin \gamma$, then there exists a nonzero A/I with $\gamma(A/I) = 0$. Since $(A/I)^1 \cong A^1/I$, it is thus in $\Sigma_{A/I}$. (M1), established in (i), implies that $(A/I)^1 \in \Sigma_A$, so that $\Sigma_A \neq \emptyset$ and $A \notin \gamma_\Sigma$. □

Of course the use of modules to describe radicals as in Theorem 3.14.5, where the class Σ of modules in as large as it can be, and is defined very directly in terms of the radical itself, is not all that illuminating. It would be much better to be able to use a class of modules which is smaller, or which has an interesting intrinsic characterization. There are cases where this can be done for well-known radicals, and we shall now consider two such cases in detail: the Jacobson and the Baer radicals.

A module M over a ring A is *irreducible* if M is simple and $AM \neq 0$. In this case of course $AM = M$.

PROPOSITION 3.14.6. *The class* $\Sigma = \cup \Sigma_A$ *of irreducible modules satisfies conditions* (M1)–(M4).

Proof: Let M be an irreducible A/I-module for a ring A and ideal I. By Proposition 3.14.1 (i), and (iii), M is a simple A-module. By Proposition 3.14.1 (iv)
$$(0:M)_A/I = (0:M)_{A/I},$$
so $(0:M) \neq A$. Hence M is an irreducible A-module and (M1) is satisfied. If, on the other hand, M is an irreducible A-module and $I \subseteq (0:M)_A$, then by Proposition 3.14.1 (ii) and (iii), M is a simple A/I-module. By Proposition 3.14.1 (iv)
$$(0:M)_{A/I} = (0:M)_A/I \neq A/I,$$
so M is irreducible over A/I. This gives us (M2).

Suppose a ring A satisfies the condition
$$\ker(\Sigma_A) = \cap((0:M)_A \mid M \text{ is an irreducible } A \text{ module}) = 0.$$
If $0 \neq B \triangleleft A$, then $BM \neq 0$ for some irreducible A-module M. Now M is a B-module in the obvious way. Let
$$N = \{m \in M \mid Bm = 0\}.$$
If $a \in A$, $m \in N$, then
$$B(am) = (Ba)m \subseteq Bm = 0.$$
Hence N is an A-submodule of M. Since M is simple and $BM \neq 0$, we have $N = 0$. Thus $Bm \neq 0$ for all nonzero $m \in M$. Since $ABm \subseteq Bm$ and M is a simple A-module, we therefore have $Bm = M$ for all nonzero $m \in M$.

Thus M is a simple B-module, and, since $BM \neq 0$, an irreducible one. Thus the class Σ of irreducible modules satisfies (M3).

Finally we consider (M4). Let A be a ring such that every nonzero ideal of A has an irreducible module. Suppose there is a nonzero ideal K such that $KM = 0$ for every irreducible A module M. Let L be an irreducible K module. Then $KL = L$. We make L into an A-module by defining

$$a \sum k_i l_i = \sum (ak_i) l_i, \qquad a \in A,\ k_i \in K,\ l_i \in L.$$

If $\sum k_i l_i = \sum k'_j l'_j$, $(k_i, k'_j \in K; l_i, l'_j \in L)$, then for every $k \in K$ we have

$$k(a \sum k_i l_i) = k \sum (ak_i) l_i = \sum (kak_i) l_i,$$

as L is a K-module and $k,\ ak_i \in K$. Further,

$$\sum (kak_i) l_i = (ka) \sum k_i l_i = ka \sum k'_j l'_j = \sum kak'_j l'_j = k \sum ak'_j l'_j = k(a \sum k'_j l'_j),$$

and so $k(a \sum k_i l_i - a \sum k'_j l'_j) = 0$ whence

$$a \sum k_i l_i = a \sum k'_j l'_j.$$

Thus scalar multiplication by elements of A is well-defined. It is now routine to verify that L is an A module. Let N be a nonzero A-submodule of L. Then N is a K-submodule, so $N = L$. Hence L is a simple A-module, and as $AL \supseteq KL = L$, we have $AL \neq 0$, so L is an irreducible A-module. Hence by the assumption on K we have $KL = 0$, a contradiction. Therefore

$$\ker(\Sigma_A) = \cap((0:M)_A \mid M \text{ is an irreducible } A \text{ module}) = 0.$$

This means that (M4) holds for the class of irreducible modules. The proof is therefore complete. □

We shall now show that when Σ is the class of irreducible modules, γ_Σ is the Jacobson radical class. Note that since all rings have simple modules, we can describe γ_Σ in this case as the class of all rings A such that $AM = 0$ for every simple A-module M.

PROPOSITION 3.14.7. *The following conditions are equivalent for an A-module M:*

(i) *M is irreducible,*

(ii) *$M = Am$ for all nonzero $m \in M$.*

Proof: (i) $\Rightarrow$ (ii) Let $N = \{m \in M \mid Am = 0\}$. Then N is a submodule and $N \neq M$ as $AM \neq 0$. Hence $N = 0$, so $Am \neq 0$ for all nonzero $m \in M$. This gives us (ii), as each Am is a submodule.

(ii) $\Rightarrow$ (i) This is clear. □

A left ideal L of a ring A is *modular* if there exists an element $e \in A$ such that $a - ae \in L$ for every $a \in A$. Note that if $L \neq A$ then $e \notin L$.

PROPOSITION 3.14.8. (i) *Let L be a modular maximal left ideal of A. Then A/L is an irreducible A-module.*

(ii) *If M is an irreducible A module, then $(0:m)$ is a modular maximal left ideal of A for every nonzero $m \in M$.*

Proof: (i) As L is maximal, A/L is certainly simple. Let $e \in A$ be such that $a - ae \in L$ for every $a \in A$. Then for each $a \in A \setminus L$ we have

$$a(e + L) = ae + L = a + L \neq 0,$$

so $A(A/L) \neq 0$. Hence A/L is irreducible.

(ii) By Proposition 3.14.7, $M = Am \cong A/(0:m)$ for each nonzero $m \in M$, so (as M is simple), $(0:m)$ is maximal. Since $m \in M = Am$, there exists an element $e \in A$ such that $m = em$. Then for all $a \in A$ we have $am = aem$, so $a - ae \in (0:m)$. Hence $(0:m)$ is modular. □

Let L be a modular left ideal of A, $a - ae \in L$ for all $a \in A$. If J is a left ideal of A and $L \subseteq J$, then $a - ae \in J$ for every $a \in A$, so in particular, J is modular. We say that a modular left ideal L of A is *maximal* if $L \neq A$ and if there is no modular left ideal $J \neq A$ with $L \subset J$. Thus by the above remark a maximal modular left ideal is a modular maximal left ideal and conversely. Hence we have

PROPOSITION 3.14.9. *The terms "modular maximal left ideal" and "maximal modular left ideal" are synonymous.* □

LEMMA 3.14.10. *Let L be a modular left ideal of A with $L \neq A$. Then there exists a maximal modular left ideal $\hat{L}$ of A for which $L \subseteq \hat{L}$.*

Proof: Let $a - ae \in L$ for all $a \in A$. By Zorn's Lemma there exists a left ideal $\hat{L}$ such that $L \subseteq \hat{L}$ and $\hat{L}$ is maximal with respect to $e \notin \hat{L}$. $\hat{L}$ is modular, for $L \subseteq \hat{L}$. Moreover, $\hat{L}$ is a maximal left ideal, because any left ideal J with $\hat{L} \subset J$ contains the element e, and so $a \in ae + L \subseteq J$ holds implying $J = A$. Hence by Proposition 3.14.9 $\hat{L}$ is what we seek. □

A consequence of Lemma 3.14.10 is crucial for our purposes.

COROLLARY 3.14.11. *The following conditions are equivalent for a ring A:*

(i) *A has no modular maximal left ideals,*

(ii) *A is the only modular left ideal of A.* □

We are now ready for our description of the radical defined by the class of irreducible modules. To be consistent, we call this nice theorem as

EXAMPLE 3.14.12. *Let Σ be the class of all irreducible modules. Then $\gamma_\Sigma = \mathcal{J}$, the Jacobson radical class.*

Proof: If $A \in \gamma_\Sigma$, then A has no irreducible modules, and hence by Proposition 3.14.8 (i) no modular maximal left ideals. On the other hand, if $A \notin \gamma_\Sigma$, that is, if A has an irreducible module, then by Proposition 3.14.8 (ii) it has a modular maximal left ideal. Thus, taking account of Corollary 3.14.11 we see that $A \in \gamma_\Sigma$ if and only if A is the only modular left ideal of A.

Now for every $a \in A$, the set

$$L_a = \{r - ra \mid r \in A\}$$

is a modular left ideal, as $r - ra \in L_a$ for all $r \in A$. If $A \in \gamma_\Sigma$, therefore, we have $L_a = A$ for all a, so for each $a \in A$ there exists an element r such that $a = r - ra$. Putting $b = -r$, we get $a = -b + ba$, that is

$$b \circ a = b + a - ba = 0.$$

Hence $\gamma_\Sigma \subseteq \mathcal{J}$.

Conversely, if $A \in \mathcal{J}$ and L is a modular left ideal of A with $a - ae \in L$ for each $a \in A$, then for some $f \in A$ we have

$$0 = f \circ e = f + e - fe$$

and so

$$e = -f + fe = (-f) - (-f)e \in L.$$

Thus $a \in ae + L = L$ follows for all $a \in A$, that is $L = A$. Hence $A \in \gamma_\Sigma$. □

As our second example, we shall present a characterization of the prime radical by modules.

An A-module M is *prime* if $AM \neq 0$ and for $m \in M$ and $J \triangleleft A$ we have

$$Jm = 0 \text{ implies } m = 0 \text{ or } JM = 0.$$

(see Andrunakievich [3] and Dauns [1]). Clearly, every irreducible module is prime.

PROPOSITION 3.14.13 (Andrunakievich [3]). *Let M be a prime A-module, $B \triangleleft A$. If $BM \neq 0$, then M is a prime B-module.*

Proof: Suppose $Jm = 0$, where $0 \neq m \in M$ and $J \triangleleft B$. Let $\bar{J}$ be the ideal of A generated by J. Then by the Andrunakievich Lemma we have $\bar{J}^3 \subseteq J$, so $\bar{J}^3 m = 0$. If $\bar{J}m = 0$, then, as M is a prime A-module and $m \neq 0$, $\bar{J}M = 0$. But then $JM \subseteq \bar{J}M = 0$. If $\bar{J}^2 m = 0 \neq \bar{J}m$, let $0 \neq m' \in \bar{J}m$. Then $\bar{J}m' = 0$ and by the previous argument $JM \subseteq \bar{J}M = 0$. The case when $\bar{J}^3 m = 0 \neq \bar{J}^2 m$ is handled similarly. □

PROPOSITION 3.14.14 (Andrunakievich [3]). *Let $B \triangleleft A$ and M be a prime B-module. Then BM is a prime A-module with respect to*

$$a \sum b_i m_i = \sum (ab_i) m_i \qquad a \in A,\ b_i \in B,\ m_i \in M,$$

and $(0:M)_B = B \cap (0:BM)_A$.

Proof: If $\sum b_i m_i = \sum c_j m'_j$, $(b_j, c_j \in B; m_i, m'_j \in M)$, then for all $a \in A$ we have
$$B(a \sum b_i m_i - a \sum c_j m'_j) = 0$$
as in the proof of Proposition 3.14.6 (verification of (M4)). The primeness of M now forces $a \sum b_i m_i = a \sum c_j m'_j$, so BM is an A-module.

If $J \triangleleft A$, $0 \neq x \in BM$ and $Jx = 0$, then $JB^2 x \subseteq Jx = 0$ and $JB^2 \triangleleft B$, so $JB^2 M = 0$. Since M is a prime B-module, there exists an element $z \in M$ for which $Bz \neq 0$. Let $0 \neq y \in Bz$. Then
$$JBy \subseteq JB^2 z \subseteq JB^2 M = 0$$
and $JB \triangleleft B$, so $(JB)M = 0$. But then
$$J(BM) = (JB)M = 0,$$
so BM is a prime A-module.

Let b be in $(0:M)_B$. Then
$$b(BM) \subseteq bM = 0,$$
so $b \in B \cap (0:BM)_A$. Conversely, if $b \in B \cap (0:BM)_A$, then $bBM = 0$, so $(b)BM = 0$, where (b) is the principal ideal of B generated by b. Since $BM \neq 0$, the primeness of M as a B-module requires that $(b)M = 0$, whence $b \in (0:M)_B$. □

PROPOSITION 3.14.15. *For each ring A, let Σ_A be the class of prime A modules. Then $\Sigma = \cup \Sigma_A$ satisfies conditions (M1)–(M4).*

Proof: Let M be in $\Sigma_{A/I}$. If $J \triangleleft A$, $m \in M$ and $Jm = 0$, then for every $j \in J$ we have
$$(j+I)m = jm = 0,$$
so $((J+I)/I)m = 0$. If $m \neq 0$, then $((J+I)/I)M = 0$, that is,
$$jM = (j+I)M = 0 \quad \forall j \in J,$$
that is, $JM = 0$. It follows that $M \in \Sigma_A$ and (M1) is satisfied.

Conversely, if $M \in \Sigma_A$, $I \triangleleft A$ and $I \subseteq (0:M)_A$, then if $(J/I)m = 0$, where $J/I \triangleleft A/I$ and $m \in M$, we have $Jm = 0$, so either $m = 0$ or $JM = 0$. In the latter case, $(J/I)M = 0$. This gives us (M2).

If $\ker(\Sigma_A) = 0$, let B be any nonzero ideal of A. Then $BM \neq 0$ for some $M \in \Sigma_A$. But then by Proposition 3.14.13, $M \in \Sigma_B$, so $\Sigma_B \neq \emptyset$ and (M3) is satisfied.

Finally we verify (M4). Suppose, on the contrary, that there exists a ring A with $\Sigma_B \neq \emptyset$ for all nonzero ideals B, and with $B_0 = \ker(\Sigma_A) \neq 0$.

Then in particular there is a prime B_0-module M, and by Proposition 3.14.14 $B_0 M$ is a prime A-module. But then

$$B_0 M = \ker{(\Sigma_A)} M = 0,$$

a contradiction. □

PROPOSITION 3.14.16. *If $I \triangleleft A$, then there is a prime A-module M with $(0 : M)_A = I$ if and only if I is a prime ideal.*

Proof: Let M be a prime A-module and let J, K be ideals of A with $JK \subseteq (0 : M)_A$. If $K \not\subseteq (0 : M)_A$, let $k \in K$, $m \in M$ be such that $km \neq 0$. Then

$$J(km) \subseteq JKm \subseteq (0:M)m = 0,$$

so $JM = 0$, that is, $J \subseteq (0 : M)_A$. Hence $(0 : M)_A$ is a prime ideal. Conversely, if I is a prime ideal, we consider A/I as an A-module. Since I is a prime ideal, we have $A \neq I$, so $A^2 \not\subseteq I$, that is, $A(A/I) \neq 0$. If $J(a + I) = 0$, $a \in A$, that is, if $Ja \subseteq I$, then $J(a) \subseteq I$, where (a) is the principal ideal of A generated by a. Hence $(a) \subseteq I$, in which case $a + I = I$, or $J \subseteq I$, in which case $J(A/I) \subseteq I(A/I) = 0$. We conclude that A/I is a prime A-module.

Clearly $I \subseteq (0 : A/I)_A$. But $(0 : A/I)_A \cdot (A/I) = 0$ implies $(0 : A/I)_A \cdot A \subseteq I$, so, I being a prime ideal, we have $(0 : A/I)_A \subseteq I$. Thus $I = (0 : A/I)_A$ is the annihilator of a prime module. □

From the proof we have also

COROLLARY 3.14.17. *If I is a prime ideal of A, then A/I is a prime A-module, and $(0 : A/I)_A = I$.* □

COROLLARY 3.14.18. *A ring A has a faithful prime module if and only if A is prime.* □

This completes the preparation for the next result which we call again an example.

EXAMPLE 3.14.19 (Adrunakievich [3]). *When Σ is the class of prime modules, $\gamma_\Sigma = \beta$ is the Baer radical class.*

Proof: A ring A is in γ_Σ if and only if A has no prime modules. By Proposition 3.14.16 this is equivalent to A having no prime ideals. This means by Example 3.8.13 that $A \in \beta$. □

As we have already noted, the module representation of a given radical is more effective the smaller and more independently characterized the class of modules. There is another way in which modules can be helpful — new radicals can be constructed from likely looking classes of modules. We present just one example.

An A-module M is *uniform* if $N \cap K \neq 0$ for all nonzero submodules N, K of M, that is, if M is an essential extension of all its nonzero submodules.

PROPOSITION 3.14.20 (Heinicke [1]). *For each ring A let Σ_A be the class of prime uniform A-modules and let $\Sigma = \cup \Sigma_A$. Then Σ satisfies conditions* (M1)–(M4).

Proof: In view of the fact that the class of prime modules satisfies (M1) and (M2), these properties for Σ are readily deducible from Proposition 3.14.1 (iii).

If $\ker(\Sigma_A) = 0$ and $0 \neq B \triangleleft A$, then there is a prime uniform A-module M with $BM \neq 0$. By Proposition 3.14.13, M is a prime B-module. Let N, K be nonzero B-submodules of M. Since M is a prime A-module, BN and BK are nonzero B-modules as $BM \neq 0$. Since BN and BK are contained in the A-module M, there is no problem with scalar multiplication by elements of A: BN and BK are thus nonzero submodules of M, so $BN \cap BK \neq 0$. Since N and K are B-modules, we then have

$$0 \neq BN \cap BK \subseteq N \cap K.$$

Thus M is a uniform B-module and we have (M3).

Turning to (M4), we consider a ring A such that $\Sigma_B \neq \emptyset$ for every nonzero ideal B of A. Suppose $B_0 = \ker(\Sigma_A) \neq 0$, and let M be a prime uniform B_0-module. By Proposition 3.14.14 $B_0 M$ is a prime A-module. Let N, K be nonzero A-submodules of $B_0 M$. Then N, K are also B_0-submodules of $B_0 M$, and hence of M. Since M is uniform as a B_0-module, we then have $N \cap K \neq 0$, so that $B_0 M$ is a uniform A-module and is therefore in Σ_A. But then $B_0^2 M = \ker(\Sigma_A) B_0 M = 0$. By arguments now familiar this can't happen, as M is a prime B_0-module. We conclude that in fact $\ker(\Sigma_A) = 0$ and (M4) is satisfied. □

EXAMPLE 3.14.21. We may call the radical γ_Σ defined by the class Σ of all prime uniform modules as Heinicke's *uniform radical*. Since irreducible modules are clearly prime uniform modules, from the definition of γ_Σ we see immediately that

$$\beta \subseteq \gamma_\Sigma \subseteq \mathcal{J}.$$

In [1] Heinicke proved that γ_Σ is a special radical and it is distinct from the Jacobson radical.

All the notable examples of module classes Σ which define radicals, including the three examples discussed above, actually define special radicals. By imposing a slightly different set of conditions on a class of modules, we can get a description of special radicals analogous to the description of radicals in general given by Theorems 3.14.4 and 3.14.5.

THEOREM 3.14.22 (Andrunakievich and Ryabukhin [1], [2]). *For each ring A, let Σ_A be a class of prime A-modules and let $\Sigma = \cup \Sigma_A$. Suppose that Σ satisfies* (M1), (M2) *and the following conditions:*
  (SM3) *if $M \in \Sigma_A$, $B \triangleleft A$ and $BM \neq 0$, then $M \in \Sigma_B$,*
  (SM4) *if $B \triangleleft A$ and $M \in \Sigma_B$, then $BM \in \Sigma_A$.*

Let $\gamma_\Sigma = \{A \mid \Sigma_A = \emptyset\}$, and let

$$F(\Sigma) = \{A \mid \Sigma_A \text{ contains a faithful } A\text{-module}\}.$$

Then

(i) $F(\Sigma)$ is a special class and γ_Σ is a special radical class, in fact $\gamma_\Sigma = \mathcal{U} F(\Sigma)$,

(ii) $\gamma_\Sigma(A) = 0$ if and only if A is a subdirect product of rings in $F(\Sigma)$,

(iii) $\gamma_\Sigma = \ker(\Sigma_A) = \cap((0:M)_A \mid M_A \in \Sigma_A)$ for all A.

Proof: We show firstly that (SM3) implies (M3) and (SM4) implies (M4).

If (SM3) is satisfied and $\ker(\Sigma_A) = 0$, let $0 \neq B \triangleleft A$. Then $B \not\subseteq \ker(\Sigma_A)$, so $BM \neq 0$ for some $M \in \Sigma_A$. By (SM3) $M \in \Sigma_B$, so $\Sigma_B \neq \emptyset$ and (M3) is thus satisfied. In the presence of (SM4), if $\Sigma_B \neq \emptyset$ for every nonzero $B \triangleleft A$, then for each such B we can choose $M_B \in \Sigma_B$ with BM_B therefore in Σ_A. But as M_B is a prime B-module, we have $B^2 M_B \neq 0$. Hence $B \not\subseteq \ker(\Sigma_A)$, so $\ker(\Sigma_A) = 0$ and (M4) holds.

Next we show that $F(\Sigma)$ is a special class. By Proposition 3.14.16 $F(\Sigma)$ consists of prime rings. If $0 \neq B \triangleleft A \in F(\Sigma)$, let M be in Σ_A with $(0:M)_A = 0$. By (SM3), M is in Σ_B. Since $(0:M)_B \subseteq (0:M)_A = 0$, B is in $F(\Sigma)$. Now if $R \in F(\Sigma)$ and $R \triangleleft \cdot S$, let N be in Σ_R with $(0:N)_R = 0$. By (SM4), RN is in Σ_S. Let $I = (0:RN)_S$. Then $(IR)N = I(RN) = 0$, so $IR \subseteq (0:N)_R = 0$ whence $(I \cap R)^2 \subseteq IR = 0$. But $R \triangleleft \cdot S$, so $I = 0$ and hence S is in $F(\Sigma)$. Thus $F(\Sigma)$ is a special class.

Since Σ (M1)–(M4), everything follows from Theorems 3.14.3, 3.14.4 and 3.14.5. □

As in the case of Theorem 3.14.5, we have a converse.

THEOREM 3.14.23 (Andrunakievich and Ryabukhin [1], [2]). *Let γ be a special radical class, and for each ring A let*

$$\Sigma_A = \{M \mid M \text{ is a prime } A\text{-module and } \gamma(A/(0:M)_A) = 0\}$$

and $\Sigma = \cup \Sigma_A$. Then

(i) Σ *satisfies* (M1), (M2), (SM3) *and* (SM4),

(ii) $\gamma = \gamma_\Sigma$.

Proof: (i) For each A, let

$$\Gamma_A = \{M \mid AM \neq 0 \text{ and } \gamma(A/(0:M)) = 0\}.$$

Then by Theorem 3.14.5, Γ_A satisfies (M1) and (M2). By Proposition 3.14.15 the prime modules also satisfy (M1) and (M2), so Σ does also.

Let M be in Σ_A, $B \triangleleft A$ and $BM \neq 0$. Then by Proposition 3.14.13, M is a prime B-module. Since

$$B/(0:M)_B = B/(B \cap (0:M)_A) \cong (B + (0:M)_A)/(0:M)_A \triangleleft A/(0:M)_A,$$

we see that $\gamma(B/(0:M)_B) = 0$, so that $M \in \Sigma_B$. Thus we have (SM3).

Finally we consider (SM4). Let $B \triangleleft A$ and $M \in \Sigma_B$. By Proposition 3.14.14, BM is a prime A-module, so by Proposition 3.14.16, $A/(0:BM)_A$ is a prime ring. Further, again by Proposition 3.14.14, $(0:M)_B = B \cap (0:BM)_A$, therefore

$$B/(0:M)_B = B/(B \cap (0:BM)_A)$$
$$\cong (B + (0:BM)_A)/(0:BM)_A \triangleleft A/(0:BM)_A,$$

and by $M \in \Sigma_B$ also $\gamma(B/(0:M)_B) = 0$. Hence

$$(B + (0:BM)_A)/(0:BM)_A \in \mathcal{S}\gamma.$$

Since $A/(0:BM)_A$ is a prime ring and in view of Theorem 3.8.6 $\mathcal{S}\gamma \cap \mathcal{P}$ is a special class, condition (∗) before Corollary 3.8.5 yields $A/(0:BM)_A \in \mathcal{S}\gamma$, that is $\gamma(A/(0:BM)_A) = 0$. Thus $BM \in \Sigma_A$ and (SM4) is satisfied.

(ii) By Theorem 3.14.5 (ii) we have $\gamma = \gamma_\Gamma$. Since $\Sigma \subseteq \Gamma$, we therefore have $\gamma = \gamma_\Gamma \subseteq \gamma_\Sigma$. If, on the other hand, $A \notin \gamma_\Gamma = \gamma$, then let M be in Γ_A. Then $AM \neq 0$ and $\gamma(A/(0:M)_A) = 0$. Since γ is a special radical, in view of Corollary 3.8.5, A has a prime ideal I such that $(0:M)_A \subseteq I$ and $\gamma(A/I) = 0$. By Proposition 3.14.17, A/I is a prime A-module and $(0:A/I)_A = I$. Since now

$$\gamma(A/(0:A/I)_A) \cong \gamma(A/I) = 0,$$

we see that $A/I \in \Sigma_A$. Thus $A \notin \gamma_\Sigma$. We conclude that $\gamma_\Sigma = \gamma$. $\square$

Next, applying Theorem 3.14.22 we shall show that the Jacobson radical $\mathcal{J}$ is a special radical.

EXAMPLE 3.14.24. *The Jacobson radical $\mathcal{J}$ is a special radical and $\mathcal{J} = \mathcal{U}F(\Sigma)$ where Σ is the class of all irreducible modules and*

$$F(\Sigma) = \{A \mid \text{there exists a faithful irreducible } A\text{-module}\}.$$

In view of Example 3.14.12 and Theorem 3.14.22 it suffices to show that the class Σ of all irreducible modules satisfies conditions (M1), (M2), (SM3) and (SM4). The validity of (M1) and (M2) follows readily from Theorem 3.14.5 and Example 3.14.12, (or from a direct verification).

For proving (SM3), suppose that $M \in \Sigma_A$, $B \triangleleft A$ and $BM \neq 0$. Let $bm \in BM$ be any nonzero element. Since $M \in \Sigma_A$, it follows that $M = Abm$, and so

$$0 \neq BM = B(Abm) \subseteq Bbm$$

holds. Moreover, Bbm is an A-submodule of the irreducible and hence simple A-module M, whence $BM \neq 0$ implies $BM = Bbm = M$. Thus M is irreducible as a B-module, that is $M \in \Sigma_B$ which means that (SM3) is satisfied.

The demonstration of (M4) in Proposition 3.14.6 actually establishes that if $K \triangleleft A$ and L is an irreducible K-module, then $KL = L$ and L is an irreducible A-module. This proves (SM4).

In 4.7 we shall apply Theorem 3.14.22 to show that the non-trivial compressible modules define a special radical (cf. Theorem 4.7.13).

It will have been observed that in the proofs of Propositions 3.14.7 and 3.14.15 (M4) was somewhat harder to verify than (M1), (M2) and (M3), while a certain amount (Proposition 3.14.3) could be proved without (M4). Life would be easier, therefore, were it possible to dispense with (M4), that is, if (M4) were a consequence of the other three conditions. Such is not the case, however.

PROPOSITION 3.14.25 (Hentzel [1]). *Let ρ be a regular class of rings. For each $R \in \rho$ let M_R be a fixed faithful R-module, and for a ring $A \notin \rho$, let*

$$\Sigma_A = \{M_R \mid R \text{ is a homomorphic image of } A\},$$

(where we understand that the A-module structure of M_R is that induced by a homomorphism from A onto R). Then $\Sigma = \cup \Sigma_A$ satisfies (M1), (M2) and (M3).

Proof: If $M \in \Sigma_{A/I}$, then $M = M_R$ for some R which is a homomorphic image of A/I and hence of A. Thus $M \in \Sigma_A$ and we have (M1). On the other hand, if $M \in \Sigma_A$ and $I \subseteq (0 : M)_A$, we can assume $M = M_R$ where $R = A/J$ for some ideal J. Now M_R is a faithful R-module, so by Proposition 3.14.1

$$(0 : M)_A / J = (0 : M)_{A/J} = (0 : M_R)_R = 0.$$

Hence $I \subseteq (0 : M)_A = J$, so $R = A/J$ is a homomorphic image of A/I and $M = M_R \in \Sigma_{A/I}$. This gives us (M2).

It follows from the argument just used that for every ring A we have

$$\ker(\Sigma_A) = \cap (J \mid J \triangleleft A \text{ and } A/J \in \rho).$$

Hence if $\ker(\Sigma_A) = 0$, A is in the semisimple class generated by ρ, (see Proposition 3.2.3), so if $0 \neq B \triangleleft A$, then B has a nonzero homomorphic image $C \in \rho$ and then $M_C \in \Sigma_B$, so $\Sigma_B \neq \emptyset$. Thus (M3) holds as well. □

EXAMPLE 3.14.26 (Hentzel [1]). Let ρ be a regular class of rings having a single nonzero member, the zeroring $Z(p)$ on the cyclic group of prime order p and let Σ be constructed as in Proposition 3.14.25. The zeroring $Z(p^2)$ has just two nonzero ideals: $Z(p^2)$ and $Z(p)$. But while $\Sigma_{Z(p^2)}$ and $\Sigma_{Z(p)}$ are non-empty, we have

$$\ker(\Sigma_{Z(p^2)}) = \cap \left(J \triangleleft Z(p^2) \mid 0 \neq Z(p^2)/J \in \rho \right) = Z(p).$$

Thus (M4) *is not satisfied*.

G. G. Emin [1] has characterized those classes Σ of modules for which the corresponding radicals γ_Σ are strict (see section 3.17).

3.15. Radicals defined by properties of ring elements

In many cases radicals may be defined in terms of properties of ring elements. We have introduced, for instance, the nil radical $\mathcal{N}$, the Jacobson radical $\mathcal{J}$ and the von Neumann radical ν via properties of ring elements. It is our purpose in this section to investigate the question when will a property of elements define a radical or a hereditary, supernilpotent and special radical. Applying the results, we shall introduce de la Rosa's λ-radical, and prove that the nil radical is a special one. We shall follow Veldsman [5] and Wiegandt [5].

Let $\mathbb{P}$ be an abstract property that an element of a ring may possess. $\neg\mathbb{P}$ will denote "not $\mathbb{P}$", the logical negation of $\mathbb{P}$. We assume, for convenience, that the element 0 of any ring has both $\mathbb{P}$ and $\neg\mathbb{P}$. An element a of a ring A is a $\mathbb{P}$-*element*, if a has property $\mathbb{P}$. A ring A (subring or ideal of A) is a $\mathbb{P}$-ring ($\mathbb{P}$-subring or $\mathbb{P}$-ideal of A), if each of its elements is a $\mathbb{P}$-element of A. Caution: if $a \in B$, B a subring of A and a is a $\mathbb{P}$-element of A, then a need not be a $\mathbb{P}$-element of the ring B, and conversely. For instance, a zero divisor of A need not be a zero divisor of B, and the unity element of B is not necessarily a unity element of A.

Let $I \triangleleft A$. The following conditions will be frequently used:

(a) if a is a $\mathbb{P}$-element of A, then the coset $a + I$ is also a $\mathbb{P}$-element of A/I,

(b) if $a \in I$ is a $\mathbb{P}$-element of A, then a is also a $\mathbb{P}$-element of I,

(c) if a is a $\mathbb{P}$-element of I, then a is also a $\mathbb{P}$-element of A,

(d) if I is a $\mathbb{P}$-ideal in A and the coset $a + I$ is a $\mathbb{P}$-element of A/I, then a is a $\mathbb{P}$-element of A,

(s) if a is a $\neg\mathbb{P}$ element of A, then there exists a prime ideal I of A such that $a \notin I$ and A/I has no nonzero $\mathbb{P}$-ideals,

(t) if a is a $\neg\mathbb{P}$ element of A, then there exists a semiprime ideal I of A such that $a \notin A$ and A/I has no nonzero $\mathbb{P}$-ideals,

THEOREM 3.15.1 (Veldsman [5] and Wiegandt [5]). *Let $\mathbb{P}$ be a property which satisfies conditions* (a), (c) *and* (d). *Then*

$$\gamma_\mathbb{P} = \{A \mid A \text{ is a } \mathbb{P}\text{-ring}\}$$

is a radical class. If $\mathbb{P}$ also satisfies condition (b), *then the radical class $\gamma_\mathbb{P}$ is hereditary. Conversely, if γ is a radical class, then there is a property $\mathbb{P}$ which satisfies conditions* (a), (c) *and* (d) *such that $\gamma = \gamma_\mathbb{P}$. If γ is also hereditary, then $\mathbb{P}$ also satisfies condition* (b).

Proof: By condition (a) the class $\gamma_\mathbb{P}$ is homomorphically closed. Condition (c) implies readily that $\gamma_\mathbb{P}$ has the inductive property. Moreover, it is straightforward to see that by condition (d) the class $\gamma_{\mathbb{P}}$ is closed under extensions. Thus by Theorem 2.1.4 $\gamma_{\mathbb{P}}$ is a radical class. Condition (b) implies clearly that radical class $\gamma_{\mathbb{P}}$ is hereditary.

Conversely, let γ be a radical class. We define a property $\mathbb{P}$ as follows: a is a $\mathbb{P}$-element of A if and only if $a \in \gamma(A)$. Clearly, $\gamma = \gamma_{\mathbb{P}}$. Because $f(\gamma(A)) \subseteq \gamma(f(A))$ for every homomorphism $f : A \to f(A)$, $\mathbb{P}$ satisfies condition (a). Condition (c) follows from $\gamma(I) \subseteq \gamma(A)$ for any ideal I of A, (here also the ADS-Theorem 3.1.2 is used). To show that $\mathbb{P}$ also satisfies condition (d), let I be a $\mathbb{P}$-ideal in A, that is, $I \subseteq \gamma(A)$, and let $a + I$ be a $\mathbb{P}$-element of A/I. By (a) the coset $(a + I) + \gamma(A)/I$ is a $\mathbb{P}$-element in

$$(A/I)/(\gamma(A)/I) \cong A/\gamma(A).$$

Because $\gamma(A/\gamma(A)) = 0$, we have $a + I \in \gamma(A)/I$, and hence $a \in \gamma(A)$ follows. If γ is hereditary, then $I \cap \gamma(A) \subseteq \gamma(I)$ for every $I \triangleleft A$, hence $\mathbb{P}$ has also (b). □

Starting with a property $\mathbb{P}_1$ which satisfies conditions (a), (c), (d), $\gamma = \gamma_{\mathbb{P}_1}$ is a radical class. Using 3.15.1, there is a property $\mathbb{P}_2$ (viz. $a \in \gamma(A)$) which satisfies conditions (a), (c) and (d) such that $\gamma = \gamma_{\mathbb{P}_2}$. These properties $\mathbb{P}_1$ and $\mathbb{P}_2$ may be different. Obviously, if a is a $\mathbb{P}_2$-element of A, then a is also a $\mathbb{P}_1$-element of A. The converse is not true. For example, let $\mathbb{P}_1$ mean that the element is nilpotent. Then $\gamma_{\mathbb{P}_1}$ is the nil radical. Property $\mathbb{P}_2$ is then $a \in \mathcal{N}(A)$ that is the principal ideal (a) is nil. If A is a 2×2 matrix ring over a division ring, then $a = \begin{pmatrix} 0 & 1 \\ 0 & 0 \end{pmatrix}$ is nilpotent, but $(a) = A$ is not a nil ring. Thus a $\mathbb{P}_1$-element need not be a $\mathbb{P}_2$-element.

The radicals $\mathcal{N}$, $\mathcal{J}$, ν, τ and Δ have been introduced by properties of elements as given in Theorem 3.15.1, (see Examples 2.1.6, 2.1.7, 2.1.9, 3.9.9 and 3.12.4). In Example 3.10.1 we have seen that also the largest subidempotent radical χ can be defined by means of elements (viz. f-regularity). As an application of Theorem 3.15.1, we shall introduce de la Rosa's λ-radical.

EXAMPLE 3.15.2. An element a of a ring A is said to be a λ-*element*, if $a \in AaA$, (caution: AaA consists of sums $\sum x_i a y_i$; $x_i, y_i \in A$). The ring A is called a $\lambda - ring$, if every element of A is a λ-element. By definition, for instance, every ring with unity element is a λ-ring. *The class*

$$\lambda = \{A \mid A \text{ is a } \lambda\text{-ring}\}$$

is a radical class, called de la Rosa's λ-radical ([1], [2]). To prove that λ is a radical class, by Theorem 3.15.1 we have to show that the property λ fulfils (a), (c) and (d). Conditions (a) and (c) are obviously satisfied. To show (d), for any coset $a + I \in A/I$ we have

$$a + I \in (A/I)(a + I)(A/I),$$

that is, $a \in AaA + I$, that is, $a - \sum x_i a y_i \in I$ with suitable elements $x_i, y_i \in A$. Since I is a λ-ideal of A, there exist elements $u_j, v_j \in A$ such that
$$a - \sum_i x_i a y_i = \sum_j u_j \left(a - \sum_i x_i a y_i \right) v_j \in AaA.$$
Hence it follows that $a \in AaA$, proving that also a is a λ-element of A. Thus also condition (d) is fulfilled.

THEOREM 3.15.3 (Veldsman [5]). *If a property $\mathbb{P}$ satisfies conditions (a), (b), (c) and (s), then $\gamma_\mathbb{P}$ is a special radical. If $\mathbb{P}$ satisfies (a), (b), (c) and (t), then $\gamma_\mathbb{P}$ is a supernilpotent radical. Conversely, if γ is a special (supernilpotent) radical, then there exists a property $\mathbb{P}$ which satisfies (a), (b), (c) and (s) (and (t), respectively) such that $\gamma = \gamma_\mathbb{P}$.*

Proof: First, assume that $\mathbb{P}$ fulfils conditions (a), (b), (c) and (s). It follows from (a) and (b) that $\gamma_\mathbb{P}$ is homomorphically closed and hereditary. If $A \notin \gamma_\mathbb{P}$, then there is a nonzero $\neg\mathbb{P}$-element a of A. By (s) there is a prime ideal I in A such that $a \notin I$ and A/I has no nonzero $\mathbb{P}$-ideals. Hence A/I is a nonzero prime factor ring of A which has no nonzero $\gamma_\mathbb{P}$-ideals by condition (c), and condition (iii) of Theorem 3.8.11 is satisfied. Thus Theorem 3.8.11 implies that $\gamma_\mathbb{P}$ is a special radical.

Next, assume that $\mathbb{P}$ satisfies conditions (a), (b), (c) and (t). Clearly $\gamma_\mathbb{P}$ is homomorphically closed and hereditary. For proving that $\gamma_\mathbb{P}$ is a radical class, we show that $\gamma_\mathbb{P}$ satisfies condition (R2) in Theorem 2.1.5. Suppose that every nonzero homomorphic image of a ring A has a nonzero $\mathbb{P}$-ideal, but $A \notin \gamma_\mathbb{P}$. Then A is not a $\mathbb{P}$-ring, so A has a nonzero $\neg\mathbb{P}$-element a. By (t) there exists a semiprime ideal I of A such that $a \notin I$ and A/I has no nonzero $\mathbb{P}$-ideals, contradicting our assumption. Thus $\gamma_\mathbb{P}$, being homomorphically closed, satisfies conditions (R1) and (R2), and therefore by Theorem 2.1.5, $\gamma_\mathbb{P}$ is a radical class. We still have to show that $\gamma_\mathbb{P}$ is hypernilpotent. By condition (t) for every nonzero $\neg\mathbb{P}$-element $a \in A$ there exists a semiprime ideal I_a such that $a \notin I_a$ and A/I_a has no nonzero $\mathbb{P}$-ideals. Hence
$$J = \cap (I_a \mid a \text{ is a nonzero } \mathbb{P}\text{-element of } A)$$
is a semiprime $\mathbb{P}$-ideal of A. Since $\gamma_\mathbb{P}(A)$ is a $\mathbb{P}$-ring, by (c) we conclude that $\gamma_\mathbb{P}(A)$ is a $\mathbb{P}$-ideal of A, and therefore by (a) we get that $\gamma_\mathbb{P}(A) \subseteq I_a$ for each nonzero $\neg\mathbb{P}$-element $a \in A$. Thus $\gamma_\mathbb{P}(A) \subseteq J$. Since J is a $\mathbb{P}$-ideal of A, by (b) J is a $\mathbb{P}$-ring, whence $J \subseteq \gamma_\mathbb{P}(A)$. Thus $\gamma_\mathbb{P}(A) = J$, and $\gamma_\mathbb{P}(A)$ is a semiprime ideal of A. This shows that $\beta \subseteq \gamma_\mathbb{P}$, whence $\gamma_\mathbb{P}$ is hypernilpotent.

If γ is a special radical, then the property $\mathbb{P} : a \in \gamma(A)$, satisfies condition (a) and in view of $\gamma(I) = \gamma(A) \cap I$ for $I \triangleleft A$, also conditions (b) and (c). Using that
$$\gamma(A) = \cap (I \triangleleft A \mid I/A \text{ is a prime ring without nonzero ideals in } \gamma),$$
also condition (s) follows.

Analogous reasoning can be applied for a supernilpotent radical. □

EXAMPLE 3.15.4. *Köthe's nil radidal $\mathcal{N}$ is a special radical.* Let $\mathbb{P}$ be the property that the element $a \in A$ is nilpotent. Conditions (a), (b) and (c) are trivially fulfilled. We show that also condition (s) is satisfied. Suppose that a is a nonzero $\neg\mathbb{P}$-element of A. Then the set $M = \{a, a^2, \ldots, a^n, \ldots\}$ does not contain 0. By Zorn's Lemma we may choose an ideal I of A maximal with respect to the property $M \cap I = \emptyset$. We show that I is a prime ideal of A. Let J and K be ideals of A both properly containing I. Then J as well as K contain elements a^j and a^k in M, and so

$$a^j a^k = a^{j+k} \in M \setminus I$$

implies that $JK \not\subseteq I$, proving that I is a prime ideal of A. Let B/I be a nonzero $\mathbb{P}$-ideal of A/I. Then for each $b \in B$ there is a positive integer n such that $b^n \in I$. Because $B/I \neq 0$, by the choice of I there is an element $a^k \in M \cap B$, and hence a power of a^k is in $I \cap M$, a contradiction. Thus A/I does not contain nonzero $\mathbb{P}$-ideals, and so condition (s) is satisfied. Hence Theorem 3.15.3 yields that $\mathcal{N}$ is a special radical.

Andrunakievich [2] attributed the result that $\mathcal{N}$ is a special radical to Hsieh Pang Chieh [1]. It was Wang Xion-Hau [1] who first proved this result. We shall reprove it in Theorem 4.1.10. The proof given above is due to Veldsman [5].

In Example 3.14.12 we have seen that the Jacobson radical $\mathcal{J}$ is special. Veldsman [5] proved this statement by using Proposition 3.15.3. Later we shall prove that $\mathcal{J}$ is the upper radical of the special class of all primitive rings (Theorem 4.4.14).

There are several other ways for defining radicals by means of elements. If $\mathbb{P}$ is the property that the element is idempotent, then

$$\gamma_\mathbb{P} = \{A \mid A \text{ is a } \mathbb{P}\text{-ring}\} = \{A \mid A \text{ is a Boolean ring}\}$$

(see Andrunakievich [2]) and, as we shall see in Proposition 4.11.2,

$$\mathcal{B} = \left\{ A \mid \begin{array}{l} \text{every nonzero homomorphic image of } A \\ \text{has no nonzero } \mathbb{P}\text{-elements} \end{array} \right\}$$

is the Behrens radical (cf. Examples 3.6.8 (ii) and 3.8.14 (iv)). De la Rosa and Wiegandt [2] characterized supernilpotent dual radicals in terms of properties of element, and in particular, they got various characterizations of the Behrens radical. Aburawash [1] characterized semisimple classes of hereditary, supernilpotent, special and supernilpotent dual radicals via properties of ring elements.

Recently Drazin and Roberts [1] introduced and investigated polynomial and multiplicative radicals which can be viewed also as radicals defined by means of elements (see Tumurbat and Wiegandt [4]).

3.16. One-sided hereditary radicals and stable radicals

In Examples 3.2.12 and 3.2.13 we have seen that the Baer, Köthe and Jacobson radicals are left (and right) hereditary, and clearly so is the Levitzki radical. Also we know from Corollary 3.2.11 that the lower radical of a left hereditary class of rings is again left hereditary. The torsion radical and the p-torsion radicals, as strongly hereditary radicals, are left hereditary (cf. Examples 3.9.9 and 3.12.5).

EXAMPLE 3.16.1. (i) *The Brown–McCoy radical $\mathcal{G}$ is not left or right hereditary.* Let us consider the ring H of all finite valued linear transformations of a countably infinite dimensional vector space V over a division ring D. From Example 1.2.11 we know that H is a simple ring without unity element. Fixing a basis $\{v_1, \ldots, v_k, \ldots\}$ for V, we consider the subsets

$$L = \{t \in H \mid tV \subseteq Dv_k\}$$

and
$$F = \{t \in L \mid tv_k = 0\}.$$

It is straightforward to check that

$$F \triangleleft L \triangleleft_r H \quad \text{and} \quad L/F \cong D.$$

Since $D \in \mathcal{SG}$, it follows that $L \notin \mathcal{G}$, although $H \in \mathcal{G}$.

(ii) The matrix ring $M_2(D)$ over a division ring D is von Neumann regular, as we know from Example 2.1.9. However, the left ideal

$$L = \begin{pmatrix} D & 0 \\ D & 0 \end{pmatrix}$$

of $M_2(D)$ is not von Neumann regular:

$$\begin{pmatrix} 0 & 0 \\ b & 0 \end{pmatrix} L \begin{pmatrix} 0 & 0 \\ b & 0 \end{pmatrix} = 0$$

for any nonzero element $G \in D$. Thus *the von Neumann regular radical ν is not left hereditary.*

A proof similar to that of Proposition 3.2.3 gives us

PROPOSITION 3.16.2. *A radical γ is left hereditary if and only if $L \cap \gamma(A) \subseteq \gamma(L)$ for every left ideal L of every ring A.* □

In view of Corollary 3.2.4 it is natural to ask: under what condition imposed on γ, will the inclusion become an equality in Proposition 3.16.2? This happens precisely when $\gamma(L) \subseteq \gamma(A)$ for all $L \triangleleft_l A$.

PROPOSITION 3.16.3. *For a radical γ the following conditions are equivalent:*

(i) $\gamma(L) \subseteq \gamma(A)$ holds for all $L \triangleleft_l A$,

(ii) $I \triangleleft L \triangleleft_l A$ and $I \in \gamma$ imply $\overline{I} \subseteq \gamma(A)$ where $\overline{I}$ stands for the ideal of A generated by I,

(iii) the semisimple class $\sigma = \mathcal{S}\gamma$ is left hereditary.

Proof: (i) $\Longrightarrow$ (ii) By (i) we have $I \subseteq \gamma(L) \subseteq \gamma(A)$ which implies $\overline{I} \subseteq \gamma(A)$.

(ii) $\Longrightarrow$ (iii) Applying (ii) to $I = \gamma(L)$ and assuming that $A \in \sigma$ we get $\gamma(L) \subseteq \overline{\gamma(L)} \subseteq \gamma(A) = 0$, and so $L \in \sigma$.

(iii) $\Longrightarrow$ (i) For any $L \triangleleft_l A$ one has

$$L/(L \cap \gamma(A)) \cong (L + \gamma(A))/\gamma(A) \triangleleft_l A/\gamma(A) \in \sigma.$$

Hence (iii) yields $L/(L \cap \gamma(A)) \in \sigma$, whence $\gamma(L) \subseteq \gamma(A)$. □

A radical γ which satisfies any of the equivalent conditions of Proposition 3.16.3, is called a *left stable radical*. Puczyłowski [3] has shown that a class γ of rings is a left stable radical if and only if γ is homomorphically closed and if every nonzero homomorphic image of a ring A has a nonzero subring in γ which is accessible by left ideals, then $A \in \gamma$.

Propositions 3.16.2 and 3.16.3 yield immediately

COROLLARY 3.16.4. *A radical γ is left hereditary and left stable if and only if $\gamma(L) = L \cap \gamma(A)$ holds for every left ideal L of every ring A.*

EXAMPLE 3.16.5. (i) *A radical γ is a left stable radical whenever the semisimple class $\sigma = \mathcal{S}\gamma$ consists of commutative rings*, for instance, if γ is the upper radical of a special class $\mathcal{M}$ of commutative rings.

(ii) The semisimple class $\mathcal{SN}_g$ of the generalized nil radical $\mathcal{N}_g$ is strongly hereditary, so $\mathcal{N}_g$ *is a left and right stable radical*. Let us consider any subring S of a ring $A \in \mathcal{SN}_g$. Since by Example 3.8.16 $\mathcal{N}_g$ is the upper radical of the special class of all domains (i.e. rings without zero divisors), A is a subdirect sum of domains, and so is its subring S as one can readily verify. Hence also $S \in \mathcal{SN}_g$ holds.

Puczyłowski [3] has proved that $\mathcal{N}_g$ is the smallest left stable supernilpotent radical.

(iii) Having in mind that division rings have no proper one-sided ideals, an analogous reasoning shows that *the Thierrin radical $\mathcal{F}$* (cf. Examples 3.2.18 and 3.8.14.(ii)) *is left and right stable*.

(iv) Let γ be a hypernilpotent radical. If the matrix ring $M_n(D)$ over a division D ring is in the semisimple class $\sigma = \mathcal{S}\gamma$, then σ is not left hereditary for

$$0 \neq D^0 \cong \begin{pmatrix} 0 & \cdots & 0 \\ \vdots & & \\ 0 & \cdots & 0 \\ D & 0 \cdots & 0 \end{pmatrix} \triangleleft \begin{pmatrix} D & 0 & \cdots & 0 \\ \vdots & & & \\ D & 0 & \cdots & 0 \end{pmatrix} \triangleleft_l M_n(D) \in \sigma$$

and $D^0 \in \gamma$. In particular, *the Baer, Levitzki, Köthe, Jacobson, Behrens, Brown–McCoy and antisimple radicals are not left stable.*

Our next aim is to describe the left hereditary and left stable radicals. Recall from Lemma 3.12.7 that

$$\gamma^0 = \{A \mid A^0 \in \gamma\}$$

is a radical class for every radical γ.

LEMMA 3.16.6 (Jaegermann [1] and Krempa [3]). *Let γ be a hereditary radical satisfying the following condition*

$(l-s)$ $L \in \gamma$ *implies* $L \subseteq \gamma(A)$ *for all* $L \triangleleft_l A$.

Then $A \in \gamma$ implies $A^0 \in \gamma$, that is, $\gamma \subseteq \gamma^0$.

Proof: The left ideal $\begin{pmatrix} A & 0 \\ 0 & 0 \end{pmatrix}$ of the ring $\begin{pmatrix} A & A \\ 0 & 0 \end{pmatrix}$ is isomorphic to A, and so $\begin{pmatrix} A & 0 \\ 0 & 0 \end{pmatrix} \in \gamma$. Since γ satisfies condition $(l-s)$, we have $\begin{pmatrix} A & 0 \\ 0 & 0 \end{pmatrix} \subseteq \gamma \begin{pmatrix} A & A \\ 0 & 0 \end{pmatrix}$. Hence

$$\begin{pmatrix} A & A^2 \\ 0 & 0 \end{pmatrix} = \begin{pmatrix} A & 0 \\ 0 & 0 \end{pmatrix} \begin{pmatrix} A & A \\ 0 & 0 \end{pmatrix} + \begin{pmatrix} A & 0 \\ 0 & 0 \end{pmatrix}$$

$$\subseteq \gamma \begin{pmatrix} A & A \\ 0 & 0 \end{pmatrix} \begin{pmatrix} A & A \\ 0 & 0 \end{pmatrix} + \gamma \begin{pmatrix} A & A \\ 0 & 0 \end{pmatrix} \subseteq \gamma \begin{pmatrix} A & A \\ 0 & 0 \end{pmatrix}.$$

Let us define the mapping

$$f: A \longrightarrow \begin{pmatrix} A & A \\ 0 & 0 \end{pmatrix} \Big/ \gamma \begin{pmatrix} A & A \\ 0 & 0 \end{pmatrix}$$

by $f(x) = \begin{pmatrix} 0 & x \\ 0 & 0 \end{pmatrix} + \gamma \begin{pmatrix} A & A \\ 0 & 0 \end{pmatrix}$ for all $x \in A$. Since

$$\begin{pmatrix} z & x \\ 0 & 0 \end{pmatrix} + \gamma \begin{pmatrix} A & A \\ 0 & 0 \end{pmatrix} = \begin{pmatrix} 0 & x \\ 0 & 0 \end{pmatrix} + \gamma \begin{pmatrix} A & A \\ 0 & 0 \end{pmatrix}, \qquad z \in A,$$

the mapping f is surjective and obviously an additive homomorphism. Furthermore,

$$f(xy) = \begin{pmatrix} 0 & xy \\ 0 & 0 \end{pmatrix} + \gamma \begin{pmatrix} A & A \\ 0 & 0 \end{pmatrix} \in \left(\begin{pmatrix} A & A^2 \\ 0 & 0 \end{pmatrix} \right.$$

$$\left. + \gamma \begin{pmatrix} A & A \\ 0 & 0 \end{pmatrix} \right) \Big/ \gamma \begin{pmatrix} A & A \\ 0 & 0 \end{pmatrix} = 0$$

and

$$f(x)f(y) =$$
$$= \left(\begin{pmatrix} 0 & x \\ 0 & 0 \end{pmatrix} \begin{pmatrix} 0 & y \\ 0 & 0 \end{pmatrix} + \gamma \begin{pmatrix} A & A \\ 0 & 0 \end{pmatrix} \right) \Big/ \gamma \begin{pmatrix} A & A \\ 0 & 0 \end{pmatrix} = 0$$

shows that f preserves also multiplication. Since $A \in \gamma$, we conclude that

$$\begin{pmatrix} A & A \\ 0 & 0 \end{pmatrix} \Big/ \gamma \begin{pmatrix} A & A \\ 0 & 0 \end{pmatrix} \in \gamma \cap \mathcal{S}\gamma$$

which implies

$$\begin{pmatrix} A & A \\ 0 & 0 \end{pmatrix} = \gamma \begin{pmatrix} A & A \\ 0 & 0 \end{pmatrix} \in \gamma.$$

Hence by the hereditariness of γ and by $\begin{pmatrix} 0 & A \\ 0 & 0 \end{pmatrix} \triangleleft \begin{pmatrix} A & A \\ 0 & 0 \end{pmatrix}$ we get that

$$A^0 \cong \begin{pmatrix} 0 & A \\ 0 & 0 \end{pmatrix} \in \gamma. \qquad \square$$

Let us observe that *every left stable radical γ trivially satisfies condition $(l-s)$* in Lemma 3.16.6.

THEOREM 3.16.7 (Gardner [8]). *Let γ be a left hereditary and left stable radical. Then $A \in \gamma$ if and only if $A^0 \in \gamma$, that is, $\gamma = \gamma^0$.*

Proof: The implication $A \in \gamma \Rightarrow A^0 \in \gamma$ follows from Lemma 3.16.6.

Let us suppose that $A^0 \in \gamma$. Then also $\begin{pmatrix} 0 & 0 \\ A & 0 \end{pmatrix} \in \gamma$. Further, we have

$$\begin{pmatrix} 0 & 0 \\ A & 0 \end{pmatrix} \triangleleft \begin{pmatrix} A & 0 \\ A & 0 \end{pmatrix} \triangleleft_l \begin{pmatrix} A^1 & A^1 \\ A^1 & A^1 \end{pmatrix}$$

where A^1 denotes the Dorroh extension of A. The ideal of $\begin{pmatrix} A^1 & A^1 \\ A^1 & A^1 \end{pmatrix}$ generated by $\begin{pmatrix} 0 & 0 \\ A & 0 \end{pmatrix}$ is clearly $\begin{pmatrix} A & A \\ A & A \end{pmatrix}$, and so by the left stability of γ Proposition 3.16.3 yields

$$\begin{pmatrix} A & A \\ A & A \end{pmatrix} \subseteq \gamma \begin{pmatrix} A^1 & A^1 \\ A^1 & A^1 \end{pmatrix} \in \gamma.$$

Since γ is left hereditary, from

$$\begin{pmatrix} A & 0 \\ A & 0 \end{pmatrix} \triangleleft_l \begin{pmatrix} A & A \\ A & A \end{pmatrix} \triangleleft \gamma \begin{pmatrix} A^1 & A^1 \\ A^1 & A^1 \end{pmatrix}$$

we conclude that $\begin{pmatrix} A & 0 \\ A & 0 \end{pmatrix} \in \gamma$. The mapping

$$g: \begin{pmatrix} A & 0 \\ A & 0 \end{pmatrix} \longrightarrow A$$

defined by $g\begin{pmatrix} x & 0 \\ y & 0 \end{pmatrix} = x$ is obviously a homomorphism onto A. Hence $\begin{pmatrix} A & 0 \\ A & 0 \end{pmatrix} \in \gamma$ implies $A \in \gamma$, proving the implication $A^0 \in \gamma \implies A \in \gamma$. $\square$

Theorem 3.16.7 indicates that supernilpotent and subidempotent radicals cannot be simultaneously left hereditary and left stable. Also the converse of Theorem 3.16.7 is true:

THEOREM 3.16.8 (Gardner [8]). *Let γ be a hereditary radical such that $A \in \gamma$ if and only if $A^0 \in \gamma$, that is, $\gamma = \gamma^0$. Then γ is left hereditary and left stable.*

Proof: If $L \triangleleft_l A \in \gamma$, then by the assumption $L^0 \triangleleft A^0 \in \gamma$. Hence the hereditariness of γ yields $L^0 \in \gamma$ and again by the assumption $L \in \gamma$. Thus γ is left hereditary.

For γ left stable, we prove that $\mathcal{S}\gamma$ is left hereditary. Suppose that this is not true. Then there exist rings A, L, I such that

$$0 \neq I \triangleleft L \triangleleft_l A \in \mathcal{S}\gamma$$

and $I \in \gamma$. By the assumption also $I^0 \in \gamma$, and so

$$0 \neq I \subseteq \gamma^0(A) \triangleleft A$$

in view of Lemma 3.12.7. Since $\gamma^0(A) \in \gamma^0$, by hypothesis we have $\gamma^0(A) \in \gamma$. Thus we get

$$0 \neq \gamma^0(A) \subseteq \gamma(A) = 0.$$

This contradiction proves that $\mathcal{S}\gamma$ is left hereditary. $\square$

Theorems 3.16.7 and 3.16.8 characterize the left hereditary and left stable radicals by symmetric terms. Thus by Corollary 3.16.4 we have also the following

COROLLARY 3.16.9. *For a radical γ the following conditions are equivalent:*

(i) γ *is left hereditary and left stable,*
(ii) γ *is right hereditary and right stable,*
(iii) γ *is hereditary and $\gamma = \gamma^0$,*
(iv) $\gamma(L) = L \cap \gamma(A)$ *for every $L \triangleleft_l A$,*
(v) $\gamma(R) = R \cap \gamma(A)$ *for every $R \triangleleft_r A$.* $\square$

Left stable radicals can be constucted as upper radicals of left hereditary classes.

PROPOSITION 3.16.10. *If ϱ is a left hereditary class of rings, then the upper radical $\gamma = \mathcal{U}\varrho$ is a left stable radical.*

Proof: Let us consider the class $\sigma = \mathcal{SU}\varrho$ as constucted in Theorem 3.1.11. A routine easy proof shows that σ is left hereditary and so by Proposition 3.16.3 γ is left stable. $\square$

EXAMPLE 3.16.11. (i) *The torsion radical τ* of Example 3.9.9 as well as *the p-torsion radicals τ_p* of Example 3.12.5 *are obviously left hereditary and left stable radicals.*

(ii) *The generalized nil radical $\mathcal{N}_g$ is a left (and right) stable special radical* (cf. Examples 3.8.16 and 3.16.5 (ii)) which is by Theorem 3.16.7 certainly *not left (or right) hereditary.*

All the radicals introduced so far are left and right symmetric. In the rest of this section we are going to introduce a radical which is not left and right symmetric.

The notion of strongly prime rings was introduced independently by Handelman and Lawrence [1] and Rubin [1]. We say that a ring A is *(left) strongly prime*, if for every nonzero element $a \in A$ there exists a finite subset $F \subseteq A$ such that $xFa = 0$ implies $x = 0$, that is $(0 : Fa)_A = 0$. The subset F is called a *(left) insulator* of the element $a \in A$. *Right strongly prime rings* are defined correspondingly. In view of Lemma 1.2.9 *a strongly prime ring is always a prime ring*, and for finite rings these notions coincide.

LEMMA 3.16.12 (Parmenter, Stewart and Wiegandt [1]). *A ring A is strongly prime if and only if every nonzero ideal I of A contains a finite subset G such that $(0 : G)_A = 0$.*

Proof: Suppose that A is strongly prime and $a \neq 0$ is any element of $I \triangleleft A$ with an insulator $F \subseteq A$ for the element a. Then the required condition is fulfilled with the set $G = Fa \subseteq I$.

Conversely, assume that the ring A satisfies the condition required in the Lemma and $a \in A$ is a nonzero element. Then the ideal $(a)_A$ contains a finite subset G such that $(0 : G)_A = 0$, and so certainly $ya \neq 0$ for some $y \in A$. Now the ideal $(ya)_A$ contains a finite subset H such that $(0 : H)_A = 0$, and we may assume that H is of the form

$$H = \{ya, x_1ya, \ldots, x_kya\}$$

for some $x_1, \ldots, x_k \in A$. Hence the set

$$F = \{y, x_1y, \ldots, x_ky\}$$

is an insulator for the element $a \in A$. $\square$

Notice that by Lemma 3.16.12 *every simple ring with unity element is strongly prime.* (Since A is simple, the only ideal of A generated by any nonzero element is A itself. Hence for the one-ements set $G = \{e\}$ consisting of the unity element we have $(0 : G)_A = 0$.)

PROPOSITION 3.16.13. *The class of all strongly prime rings is hereditary.*

Proof: Let A be a strongly prime ring, $I \triangleleft A$ and $a \in I$ a nonzero element. If $Ia = 0$, then for any finite subset G of A we have $0 \neq I \subseteq (0 : Ga)_A$, and so A is not strongly prime, a contradiction. Hence there exists an element $y \in I$ such that $ya \neq 0$. By Lemma 3.16.12 there exists a finite subset H of the ideal $(ya)_A$ such that $(0 : H)_A = 0$. As in the proof of Lemma 3.16.12 we may assume that H is of the form

$$H = \{ya, x_1 ya, \ldots, x_k ya\}$$

for some $x_1, \ldots, x_k \in A$. Hence the subset

$$F = \{y, x_1 y, \ldots, x_k y\}$$

of I is an insulator for $a \in I$. Thus I is strongly prime.

EXAMPLE 3.16.14. *The upper radical s_ℓ of the class of all (left) strongly prime rings is called the Groenewald–Heyman strongly prime radical* (cf. Groenewald and Heyman [1]). Similarly, we may define the *right strongly prime radical s_r*.

(i) *The left strongly prime radical s_ℓ does not coincide with the right strongly prime radical s_r.* Let D be a division ring and A the ring of infinite matrices over D such that the matrices of A contain only finitely many nonzero columns. The ring A is simple, as one can easily see. Let $(e)_1 \in A$ denote the matrix having the unity element e everywhere in the first column and only zeros in the other columns. For the set $G = (e)_1$ we have $(0 : G)_A = 0$, and therefore by Lemma 3.16.12 the ring A is left strongly prime. Thus $s_\ell(A) = 0$. However, for every finite subset H of A there exists a column index n such that the n-th column of each matrix in H consists of zero entries only. Hence for the matrix $(e)_{nn}$ having the unity element $e \in D$ at the (n, n) position and zeros everywhere else, we have $H(e)_{nn} = 0$ and

$$(e)_{nn} \in \{x \in A \mid Hx = 0\}.$$

Hence applying Lemma 3.16.12 to the ideal A, we get that A is not right strongly prime. Since A is simple, necessarily $s_r(A) = A$.

(ii) The following negative result will be used in Section 3.18. *The left strongly prime radical s_ℓ is not left hereditary and the right strongly prime radical s_r is not right hereditary.* We shall consider a ring which

was constructed by Handelman and Lawrence [1] and used in Parmenter, Passman and Stewart [1] and Veldsman [4]. Let us consider the two element field Z_2, the free noncommutative Z_2-algebra $Z_2\langle x_1, x_2, \ldots \rangle$ the ideal I of $Z_2\langle x_1, x_2, \ldots \rangle$ generated by the monomials of the form $x_i x_j x_k$ where $i < j < k$, and the factor algebra $A = Z_2\langle x_1, x_2, \ldots \rangle / I$. For any coset $f(x_{i_1}, \ldots, x_{i_n}) + I \in A$ and finite subset $F \subseteq A$ there is a large number j such that
$$(f(x_{i_1}, \ldots, x_{i_k}) + I) F(x_j x_{j+1} + I) \subseteq I.$$
Consequently A is not right strongly prime. One easily sees that also no nonzero homomorphic image of A is right strongly prime, and therefore $A \in s_r$.

Let us consider the right ideal $R = [x_1)$ of A. The elements of R are cosets of the form $x_1 f(x_{i_1}, \ldots, x_{i_n}) + I$. If $x_1 f(x_{i_1}, \ldots, x_{i_n}) \notin I$, then the relation
$$(x_1 f(x_{i_1}, \ldots, x_{i_n})) x_1 (x_1 g(x_{j_1}, \ldots, x_{j_k})) \in I$$
holds only if the last factor is in I. This shows that $F = \{x_1\}$ is a right insulator for every nonzero element of R, and so R is a right strongly prime ring. Thus the radical s_r is not right hereditary.

Negative answer to the question as whether left hereditariness implies right hereditariness was given by Beidar [2], Parmenter, Passman and Stewart [1], Puczyłowski [10] and Watters [7]. Puczyłowski and Zand [3] introduced and studied left (right) subhereditary radicals, see also Beidar, Ke and Puczyłowski [1], [2], Mendes and Tumurbat [1], Tumubat and Zand [2]. Semisimple classes of left hereditary radicals were described by Sands and Tumurbat [1].

In the subsequent Section 3.17 we shall see that left stability of a radical does not imply right stability.

3.17. Strong radicals and strict radicals

A radical γ which satisfies condition

$(l - s)$ $\qquad L \in \gamma$ implies $L \subseteq \gamma(A)$ for all $L \triangleleft_l A$

of Lemma 3.16.3 is said to be a *left strong radical*. *Right strong radicals* are defined correspondingly. As mentioned after Lemma 3.16.6, left stable radicals are left strong.

PROPOSITION 3.17.1 (Amitsur [3]). *A radical γ is left strong if and only if*
$$0 \neq L \triangleleft_l A \in \mathcal{S}\gamma \text{ implies } L \notin \gamma.$$

Proof: Straightforward. □

EXAMPLES 3.17.2 (i) *The Baer radical β is left (and right) strong.* We are going to prove that the condition of Proposition 3.17.1 is satisfied.

Suppose the contrary: there exists an $L \triangleleft_l A \in \mathcal{S}\beta$ such that $0 \neq L \in \beta$. Then there exists a nonzero ideal of L which is a zero-ring. Zorn's Lemma is applicable to the set of ideals of L which are zero-rings, and so we can choose an ideal K of L which is maximal with respect to the property $K^2 = 0$. Now $LK + LKA \triangleleft A$ and

$$(LK + LKA)^2 = (LK)^2 + (LK)(LK)A + LK(ALK) \\ + LK(ALK)A \subseteq (LK)_A^2 = 0.$$

But A is semiprime, so $LK + LKA = 0$ and thus $LK = 0$. Since A is a semiprime ring and $L \neq 0$, we have

$$(L + LA)^2 \subseteq L^2 + L^2 A$$

which implies $L^2 \neq 0$. Hence by $LK = 0$ it follows that $K \neq L$, and so $0 \neq L/K \in \beta$. Hence there exists a nonzero ideal J/K of L/K such that $J^2 \subseteq K$. Then

$$(LJ)^2 \subseteq LJ^2 \subseteq LK = 0,$$

and so $LJ + LJA \triangleleft A$ and we have

$$(LJ + LJA)^2 \subseteq (LJ)^2 + (LJ)^2 A = 0.$$

Since A is semiprime, we have $LJ + LJA = 0$ and also $LJ = 0$. Therefore $J^2 = 0$ and so by the maximality of K it follows $K = J$, contradicting $J/K \neq 0$. This proves that $L \notin \beta$.

(ii) *The Levitzki radical $\mathcal{L}$ is left (and right) strong.* Let $L \triangleleft_l A$ and $0 \neq L \in \mathcal{L}$, and consider any finitely generated subring $S = \langle s_1, \ldots, s_k \rangle$ of the ideal $L + LA$ of A. Then each s_i, $(i = 1, \ldots, k)$, has the form

$$s_i = l_i + \sum_{j=1}^{n_j} l_{ij} a_{ij}, \qquad (l_i, l_{ij} \in L;\ a_{ij} \in A).$$

Take the finitely generated subrings

$$T = \langle l_i, l_{ij}, a_{ij}l_{rt} \mid i, r = 1, \ldots, k;\ j, t = 1, \ldots, n_j \rangle$$

and

$$B = \langle l_{ij} a_{ij} \mid i = 1, \ldots, k;\ j = 1, \ldots, n_j \rangle$$

of L. T is nilpotent because $T \subseteq L \in \mathcal{L}$. Clearly $BT \subseteq T$ and

$$S^2 \subseteq T^2 + T^2 B.$$

Hence by induction we get

$$S^{2N} \subseteq T^{2N} + T^{2N}B$$

for every $N \geq 1$. Since T is nilpotent, so also is S. Thus we have

$$L \subseteq L + LA \subseteq \mathcal{L}(A),$$

which proves the assertion.

KÖTHE'S PROBLEM 3.17.3. *Is Köthe's nil radical left strong?* This problem was raised in 1930 (Köthe [1]), and seems to be one of the hardest problems in ring theory. Köthe's Problem has many equivalent formulations (cf. for instance Rowen [1], [2], Krempa [1] and Puczyłowski [13]). An affirmative answer for Köthe's Problem has been verified for many different classes of rings. Recently Sakhajev [1] announced the negative solution of Köthe's Problem.

EXAMPLE 3.17.4. *The Jacobson radical $\mathcal{J}$ is left and right strong.* We prove this assertion in four steps.

(i) *The sum of finitely many Jacobson radical left ideals is a Jacobson radical left ideal.* It suffices to show that if L and K are Jacobson radical left ideals of a ring A, then so is $L + K$. Let $a + b \in L + K$, $a \in L$, $b \in K$, be arbitrary elements and let $x \in L$ be an element such that $x \circ a = 0$. To the element $b - xb \in K$ let us choose an element $y \in K$ such that $y \circ (b - xb) = 0$. Now we have

$$(y \circ x) \circ (a + b) = (y + x - yx) + a + b - (y + x - yx)(a + b) =$$
$$= x + a - xa - y(x + a - xa) + y + (b - xb) - y(b - xb) =$$
$$= x \circ a - y(x \circ a) + y \circ (b - xb) = 0.$$

Since the choice of $a + b \in L + K$ was arbitrary, $L + K$ is a Jacobson radical ring.

(ii) *If L is a Jacobson radical left ideal of a ring A, then for any element $a \in L$ the left ideal La is a Jacobson radical ring.* Take an element $ba \in La$, $b \in L$. Since $ab \in L$, there exists an element $u \in L$ such that $u \circ (ab) = 0$. We claim that $v \circ (ba) = 0$ with the element $v = bua - ba$. Indeed,

$$v \circ (ba) = (bua - ba) + ba - (bua - ba)ba =$$
$$= b(u + ab - uab)a = b(u \circ ab)a = 0.$$

Thus La is a Jacobson radical ring.

(iii) *If L is a Jacobson radical left ideal of a ring A, then so is the ideal $L + LA$ of A generated by L.* Any element $d = b + \sum_{\text{finite}} c_i a_i$, $(b, c_i \in L$; $a_i \in A)$, is in the finite sum $L + \sum_{\text{finite}} La_i$ of left ideals each of which is a Jacobson radical ring in view of (ii). Hence (i) yields the assertion.

(iv) *If L is a Jacobson radical left ideal of a ring A, then $L \subseteq \mathcal{J}(A)$.*
In fact, by (iii) we have

$$L \subseteq L + LA \subseteq \mathcal{J}(A).$$

THEOREM 3.17.5 (Divinsky, Krempa and Suliński [1]). *The following four conditions are equivalent:*
   (i) γ *is a left strong radical.*
   (ii) γ *is a radical whose semisimple class $\sigma = \mathcal{S}\gamma$ is left regular (that is, if $0 \neq L \triangleleft_l A \in \sigma$, then L can be mapped homomorphically onto a nonzero ring in σ).*
   (iii) γ *is a homomorphically closed class and if every nonzero homomorphic image of a ring A has a nonzero left ideal in γ then $A \in \gamma$.*
   (iv) γ *is a radical and if $L \triangleleft_l A$, $L \in \gamma$ then the ideal $L + LA$ of A is in γ.*

Proof: (i) $\Longrightarrow$ (ii) Obvious by Proposition 3.17.1.

(ii) $\Longrightarrow$ (iii) Let A be a ring such that every nonzero homomorphic image of A has a nonzero left ideal in γ but $A \notin \gamma$. Then $0 \neq A/\gamma(A) \in \sigma$ and by the assumption $A/\gamma(A)$ has a nonzero left ideal $L/\gamma(A)$ in γ. Applying (ii), $L/\gamma(A)$ has a nonzero homomorphic image in σ, a contradiction.

(iii) $\Longrightarrow$ (iv) Now conditions (R1) and (R2) of Theorem 2.1.5 are trivially fulfilled, so γ is a radical class. We have to prove only that $L + LA \in \gamma$. For that purpose, suppose that $L \triangleleft_l A$ and $0 \neq L \in \gamma$. We have to show that $L + LA \in \gamma$. Let us consider a nonzero homomorphic image $(L + LA)/K$ of $L + LA$. If $L \not\subseteq K$, then

$$0 \neq L/(L \cap K) \cong (L + K)/K \triangleleft_l (L + LA)/K$$

and by $L \in \gamma$ also $(L + K)/K \in \gamma$. If $L \subseteq K$, then $LA \not\subseteq K$ and so there exists an element $a \in A$ such that $La \not\subseteq K$. Consider the mapping

$$f \colon L \longrightarrow (La + K)/K$$

defined by $f(l) = la + K$ for all $l \in L$. f is obviously an additive homomorphism. Since by $L \subseteq K$ we have

$$f(l_1 l_2) = l_1 l_2 a + K \in (K(LA) + K)/K = 0$$

and

$$f(l_1)f(l_2) = l_1 a l_2 a + K \in (L(AL)A + K)/K \subseteq (K(LA) + K)/K = 0,$$

the mapping f is also a multiplicative homomorphism. Hence by $L \in \gamma$ also the left ideal $(La + K)/K$ of $(L + LA)/K$ is in γ. Thus by (iii) we conclude that $L + LA \in \gamma$.

(iv) $\Longrightarrow$ (i) Trivial. □

We can easily generate left strong radicals as upper radicals. A left-regular class ϱ is also regular, so $\gamma = \mathcal{U}\varrho$ is a radical class.

PROPOSITION 3.17.6 (Divinsky, Krempa and Suliński [1]). *If ϱ is a left regular class of rings, then the upper radical $\gamma = \mathcal{U}\varrho$ is left strong.*

Proof: Take a ring A not in γ. Then A has a nonzero homomorphic image B in ϱ, and so B has no nonzero left ideal in γ since ϱ is left regular. This proves that if every nonzero homomorphic image of A has a nonzero left ideal in γ, then A is itself in γ. Thus by Theorem 3.17.5 (i) and (iii) the radical γ is left strong. □

From Lemma 3.16.3 we know that all left stable radicals are left strong. The converse of this statement is false.

EXAMPLE 3.17.7. *The Baer, the Levitzki and the Jacobson radicals are left hereditary and left strong* (cf. Examples 3.2.12, 3.2.13 (ii), 3.17.2 and 3.17.4), *but in view of Example 3.16.5(iv) or Theorem 3.16.7 they are not left stable.*

Not all the good radicals are left strong as seen from

EXAMPLES 3.17.8 (i) (Divinsky, Krempa and Suliński [1]). *The Brown–McCoy radical $\mathcal{G}$ is neither left nor right strong.* As seen in Example 1.3.2, the Weyl algebra W_1 is a simple ring with unity element and every nonzero left ideal $L = W_1 f(x)$ of W_1 is also a simple ring but without nonzero idempotents. Thus W_1 is a Brown–McCoy semisimple ring whose nonzero left ideals L ($\neq W_1$) are Brown–McCoy radical rings.

(ii) Reasoning as in (i), one gets that *The Behrens radical is not left (or right) strong.*

For more general results we refer to Tumurbat [1], Tumurbat and Wiegandt [1].

EXAMPLE 3.17.9 (Sands [4]). Let F be a prime field. The class

$$\varrho = \left\{ \begin{pmatrix} F & 0 \\ F & 0 \end{pmatrix}, \begin{pmatrix} 0 & 0 \\ F & 0 \end{pmatrix}, 0 \right\}$$

of matrix rings is clearly left hereditary, so the radical $\gamma = \mathcal{U}\varrho$ is left stable by Proposition 3.16.10. Since

$$F \cong \begin{pmatrix} F & 0 \\ 0 & 0 \end{pmatrix} \triangleleft_r \begin{pmatrix} F & 0 \\ F & 0 \end{pmatrix} \in \varrho \subseteq \mathcal{S}\mathcal{U}\gamma$$

and $F \in \gamma$, the radical γ is not right strong. Thus γ *is an example of a radical which is left stable and left strong but not right strong and not right stable.*

EXAMPLE 3.17.10 (Puczyłowski [10]). *The strongly prime radical s_ℓ is left strong but neither left nor right stable.*

First, we prove that *if $0 \neq L \triangleleft_l A$ and A is a strongly prime ring, then so is the factor ring $L/\beta(L)$.*

Let L^r denote the right annihilator of the ring L. We show that $\beta(L) = L^r$ for every prime ring A. The inclusion $L^r \subseteq \beta(L)$ is clear. Conversely, $L\beta(L) \triangleleft \beta(L)$, so $L\beta(L) \in \beta$. Moreover, $L\beta(L) \triangleleft_l A$ and β is a left strong radical by Example 3.17.2. Hence, taking into account that A is a (semi)prime ring, we get that $L\beta(L) \subseteq \beta(A) = 0$ and $\beta(L) \subseteq L^r$. Thus $\beta(L) = L^r$.

Now, suppose that $0 \neq I/\beta(L) \triangleleft L/\beta(L)$. We claim that $LIA \neq 0$. If $LIA = 0$, then LI is in the left annihilator of the prime ring A, and so $LI = 0$. Hence $I \neq 0$ is in the right annihilator L^r of L. Clearly, $L^r \triangleleft_r A$. Since $L^r = \beta(L)$, as we have already seen, and β is right strong by Example 3.17.2, we conclude that

$$0 \neq I \subseteq L^r \subseteq \beta(A) = 0,$$

a contradiction. Thus $LIA \neq 0$, as claimed.

Applying Lemma 3.16.12 to the strongly prime ring A and nonzero ideal LIA of A, we see that there exists a finite subset $F = \{f_1, \ldots, f_n\} \subseteq LIA$ such that the left annihilator $(0 : F)_A$ is 0. Let

$$f_k = \sum_j l_{kj} i_{kj} a_{kj},$$

where $l_{kj} \in L, i_{kj} \in I, a_{kj} \in A$ and $k = 1, 2, \ldots, n$ and

$$G = \{l_{kj} i_{kj} \mid k, l \text{ runs over all relevant subscripts}\}.$$

Obviously $G \subseteq I$. If $bG \subseteq \beta(L)$ for some $b \in L \setminus \beta(L)$, then

$$LbG \subseteq L\beta(L) = LL^r = 0.$$

In consequence, $LbF = 0$. Now $Lb \neq 0$ as $b \notin \beta(L) = L^r$ so, as $(0 : F)_A = 0$ we have a contradiction. Hence the left annihilator $(0 : (G + \beta(L))/\beta(L))_{L/\beta(L)} = 0$. Thus, again by Lemma 3.16.12, the ring $L/\beta(L)$ is strongly prime.

In order to prove that s_ℓ is left strong, by Proposition 3.17.1 we have to show that if $0 \neq L \triangleleft_l A \in Ss_\ell$ then $L \not\subseteq s_\ell$. Since s_ℓ is the upper radical of strongly prime rings, every s_ℓ-semisimple ring $A \neq 0$ has a factor ring $A/I \neq 0$ which is strongly prime and $L/I \triangleleft_l A/I$. Thus we can confine ourselves to the case when A is a strongly prime ring and $0 \neq L \triangleleft_l A$.

Now, using the statement proved above, either $L/\beta(L) \neq 0$ is a strongly prime ring, or $L = \beta(L)$. In the first case $L \not\subseteq s_\ell$. Since β is a left strong radical by Example 3.17.2, in the second case we have

$$L = \beta(L) \subseteq \beta(A) = 0,$$

a contradiction. Thus $L \not\subseteq s_\ell$.

To prove that s_ℓ is neither left nor right stable, in view of Example 3.16.5 (iv) it suffices to show that *the matrix ring $M_n(D)$ over a division ring D is not left strongly prime*. But this is true, since simple rings with unity element are left strongly prime (as noted after Lemma 3.16.2).

A radical γ is said to be *strict*, if for every ring A, $\gamma(A)$ contains every γ-subring of A. A *verbatim* repetition of the proof of Proposition 2.3.11 shows that γ *is a strict radical if and only if the semisimple class $S\gamma$ is strongly hereditary*. A strict radical is, of course, left and right strong. Strict radicals can be obtained as upper radicals of strongly hereditary classes, as we shall see it in the following.

THEOREM 3.17.11 (Stewart [2]). *If ϱ is a strongly hereditary class of rings, then its upper radical $\gamma = \mathcal{U}\varrho$ is strict, or equivalently, the smallest semisimple class $\sigma = \mathcal{S}\mathcal{U}\varrho$ containing the class ϱ is strongly hereditary. Every strict radical is the upper radical of a strongly hereditary class.*

Proof: First we show that the class $\gamma = \mathcal{U}\varrho$ coincides with the class

$$\hat{\gamma} = \left\{ A \;\middle|\; \begin{array}{l} \text{every nonzero homomorphic image of } A \text{ has} \\ \text{a nonzero subring which is not in } \varrho \end{array} \right\}.$$

The containment $\gamma \subseteq \hat{\gamma}$ is trivial. Let $A \notin \gamma$. Then A has a nonzero homomorphic image B in ϱ. Since ϱ is strongly hereditary, every nonzero subring of B is in ϱ, whence $A \notin \hat{\gamma}$.

To prove the Theorem, we consider a ring A and a subring S of A such that $S \in \gamma$. If we prove that also the ideal $\overline{S}$ of A generated by S is in γ, we are done. For that purpose it suffices to find a nonzero homomorphism of S into any nonzero factor ring $\overline{S}/K$, for γ is homomorphically closed and $\gamma = \hat{\gamma}$.

If $S \not\subseteq K$, then

$$S \longrightarrow S/(S \cap K) \cong (S + K)/K \subseteq \overline{S}/K$$

provides a desired homomorphism. Next, we look at the case $S \subseteq K$. Then we also have $\overline{S} \subseteq \overline{K}$, so $\overline{K} = \overline{S}$. Thus the Andrunakievich Lemma implies $\overline{S}^3 \subseteq K$. Again, we have two cases: either $\overline{S}^2 \not\subseteq K$ or $\overline{S}^2 \subseteq K$.

If $\overline{S}^2 \not\subseteq K$, then we have

$$(AS + SA + ASA)\overline{S} \not\subseteq K$$

and there are elements $a, b \in A$ and $c \in \overline{S}$ such that at least one of the mappings

$$f \colon S \longrightarrow (aSc + K)/K, \qquad f(x) = axc + K, \qquad \forall x \in S$$

$$g\colon S \longrightarrow (Sac+K)/K, \qquad g(x) = xac + K, \qquad \forall x \in S$$
$$h\colon S \longrightarrow (aSbc+K)/K, \qquad h(x) = axbc + K, \qquad \forall x \in S$$

is a homomorphism with nonzero image, as one can readily check.

If $\overline{S}^2 \subseteq K$ then we can proceed as above (but we do not need an element $c \in \overline{\overline{S}}$). $\qquad\square$

EXAMPLE 3.17.12 (i). By definition, *the torsion radical τ, the p-torsion radical τ_p as well as the divisible radical Δ are strict radicals.*

(ii) *The generalized nil radical $\mathcal{N}_g$, as the upper radical of all domains, is strict* (cf. Example 3.8.16).

A lot of work has been done on the dependence and independence among radicals involving one-sided ideals (one-sided hereditariness, stability and strength) first of all by Beidar [2], [8], Puczyłowski [10], [11], [13], and Sands [3], [4], [8], [9], [11]; see also Beidar, Fong and Wang [1], Beidar and Salavová [2], Puczyłowski and Zand [2]. For lower radical constructions involving one-sided ideals we refer the reader to Divinsky, Krempa and Suliński [1], Osłowski and Puczyłowski [1], Puczyłowski [5].

3.18. Normal radicals

Morita equivalence and Morita contexts [1], [2] play a key role in the theory of module categories. Morita contexts have an impact also on radical theory. More precisely, a wide class of radicals, including the Baer, Levitzki, Jacobson, and antisimple radicals and also the A-radicals (see Section 3.19) fulfil a condition imposed on Morita contexts.

For the sake of convenience, in this section we change the convention, and rings will be denoted by R and S, the Dorroh extension of R by R^1.

Let R and S be rings, $V = {}_RV_S$ and $W = {}_SW_R$ an R-S-bimodule and an S-R-bimodule, respectively. The quadruple (R, V, W, S) is called a *Morita context*, if the set $\begin{pmatrix} R & V \\ W & S \end{pmatrix}$ of matrices forms a ring under matrix addition and multiplication. This definition will make sense, if we assume the existence of mappings

$$V \times W \to R \quad \text{and} \quad W \times V \to S$$

denoted simply by vw and wv for all $v \in V$ and $w \in W$ such that for every $v, v_1, v_2 \in V$; $w, w_1, w_2 \in W$, $r \in R$, $s \in S$ the following identities are satisfied:

$$(v_1 + v_2)w = v_1w + v_2w \qquad v(w_1 + w_2) = vw_1 + vw_2$$
$$r(vw) = (rv)w \qquad (vw)r = v(wr)$$
$$(vs)w = v(sw) \qquad (v_1w)v_2 = v_1(wv_2)$$

and their duals

$$(w_1 + w_2)v = w_1v + w_2v \qquad w(v_1 + v_2) = wv_1 + wv_2$$
$$s(wv) = (sw)v \qquad (wv)s = w(vs)$$
$$(wr)v = w(rv) \qquad (w_1v)w_2 = w_1(vw_2).$$

Following Jaegermann [2], a radical γ is said to be *normal*, if

$$V\gamma(S)W \subseteq \gamma(R)$$

for every Morita context (R, V, W, S). Since this condition is demanded for every Morita context and along with (R, V, W, S) also (S, W, V, R) is a Morita context, normality of the radical γ can be defined also by requiring

$$W\gamma(R)V \subseteq \gamma(S)$$

for all Morita contexts (R, V, W, S). The Morita context (S, W, V, R) is called the *dual* to (R, V, W, S).

PROPOSITION 3.18.1 (Sands [2]). *If γ is a normal radical, then γ is left and right strong.*

Proof: Let $L \triangleleft_l R \in \mathcal{S}\gamma$ and $L \in \gamma$. Consider the Morita context (R, L, R^1, L) with the naturally defined multiplication. Since γ is normal, we have

$$L\gamma(L)R^1 \subseteq \gamma(R) = 0$$

which, since $L \in \gamma$, implies that $L^2 \subseteq L^2R^1 = 0$. Hence

$$L \triangleleft L + LR \triangleleft R \in \mathcal{S}\gamma,$$

so $L \in \gamma \cap \mathcal{S}\gamma = 0$. Thus γ is left strong in view of Proposition 3.17.1.

For right strongness the proof is analogous. □

PROPOSITION 3.18.2. *If γ is a hypernilpotent normal radical, then γ is left and right hereditary.*

Proof: Let $L \triangleleft_l R \in \gamma$, and let us consider the Morita context (L, R, L, R). Then we have $R\gamma(R)L \subseteq \gamma(L)$. Since $R \in \gamma$, it now follows from Proposition 3.18.1 that

$$L^3 \subseteq R^2L \subseteq \gamma(L).$$

Hence

$$(L/\gamma(L))^3 = (L^3 + \gamma(L))/\gamma(L) = 0,$$

and therefore Proposition 3.11.1 (iii) yields that

$$L/\gamma(L) \subseteq \gamma(L/\gamma(L)) = 0,$$

that is, $L = \gamma(L) \in \gamma$. Thus γ is left hereditary.

The right hereditariness of γ can be proved similarly. □

Dropping hypernilpotency, we can prove somewhat less. A radical γ is called *principally left* hereditary, if $R \in \gamma$ implies $Ra \in \gamma$ for every element $a \in R$. *Principally right hereditary radicals* may be defined similarly. In Example 3.17.4 (ii) we have seen that the Jacobson radical is principally left hereditary.

PROPOSITION 3.18.3 (Sands [2]). *Every normal radical γ is principally left hereditary and principally right hereditary.*

Proof: Let $R \in \gamma$ and $a \in R$. Consider the Morita context (Ra, R^1, Ra, R) with the natural definition of multiplication. Since γ is normal, we have
$$R^1 \gamma(R) Ra \subseteq \gamma(Ra),$$
that is, $R^2 a \subseteq \gamma(Ra)$. Consider the mapping
$$f \colon R \to Ra/\gamma(Ra)$$
given by $f(x) = xa + \gamma(Ra)$ for all $x \in R$. It is easily verified, using $R^2 a \subseteq \gamma(Ra)$, that f is a surjective ring homomorphism. Hence as $R \in \gamma$ it follows that $Ra/\gamma(Ra) \in \gamma \cap S\gamma = 0$, that is $Ra = \gamma(Ra) \in \gamma$.

Principally right hereditariness can be proved similarly. □

We shall see that left strongness and principally left hereditariness characterize the normal radicals. Prior to this, we prove the auxiliary

LEMMA 3.18.4. *Let γ be a left strong and principally left hereditary radical.*
(i) *If $L \triangleleft_l R$, $L \in \gamma$ and $a \in R$, then $La \in \gamma$.*
(ii) *If $R \in \gamma$, then also $R^0 \in \gamma$, whence $\gamma \subseteq \gamma^0$.*

Proof: (i) Since γ is left strong and $L \in \gamma$, we have $L \subseteq \gamma(R)$. Therefore, for each $l \in L$ we have $la \subseteq \gamma(R)$. Since γ is principally left hereditary, it follows that $\gamma(R)la \in \gamma$. Further, $\gamma(R)la \triangleleft_l \gamma(R)La$. Taking into consideration that γ is left strong, we get that $\gamma(R)la \subseteq \gamma(\gamma(R)La)$. But then
$$\gamma(R)La = \sum(\gamma(R)la \mid l \in L) \subseteq \gamma(\gamma(R)La),$$
so $\gamma(R)La = \gamma(\gamma(R)La)) \in \gamma$. Consider the mapping $f \colon L \to La/\gamma(R)La$ given by $f(l) = la + \gamma(R)La$ for all $l \in L$. As one easily verifies, f is a surjective ring homomorphism. Since $L \in \gamma$, it follows that $La/\gamma(R)La \in \gamma$. Hence by $\gamma(R)La \triangleleft La$, $\gamma(R)La \in \gamma$ and $La/\gamma(R)La \in \gamma$ we conclude that $La \in \gamma$.

(ii) Since $\begin{pmatrix} R & 0 \\ 0 & 0 \end{pmatrix} \triangleleft_l \begin{pmatrix} R & R^1 \\ 0 & 0 \end{pmatrix}$ and
$$R^0 \cong \begin{pmatrix} 0 & R \\ 0 & 0 \end{pmatrix} = \begin{pmatrix} R & 0 \\ 0 & 0 \end{pmatrix} \begin{pmatrix} 0 & 1 \\ 0 & 0 \end{pmatrix},$$

by (i) it follows that $R^0 \in \gamma$. □

THEOREM 3.18.5 (Sands [2]). *A radical γ is normal if and only if γ is left strong and principally left hereditary (or right strong and principally right hereditary).*

Proof: Propositions 3.18.1 and 3.18.3 yield that normal radicals are left strong and principally left hereditary. Assume that the radical γ is left strong and principally left hereditary. Let (R, V, W, S) be a Morita context. By Lemma 3.18.4 (ii), $(\gamma(S))^0$ is in γ. For every $v \in V$, $\begin{pmatrix} 0 & v\gamma(S) \\ 0 & 0 \end{pmatrix}$ is a homomorphic image of $(\gamma(S))^0$, and so is in the γ-radical of $\begin{pmatrix} 0 & V\gamma(S) \\ 0 & 0 \end{pmatrix}$. Hence $\begin{pmatrix} 0 & V\gamma(S) \\ 0 & 0 \end{pmatrix} \in \gamma$. $\begin{pmatrix} 0 & V\gamma(S) \\ 0 & \gamma(S) \end{pmatrix}$ is the extension of $\begin{pmatrix} 0 & 0 \\ 0 & \gamma(S) \end{pmatrix} \cong \gamma(S)$ by $\begin{pmatrix} 0 & V\gamma(S) \\ 0 & 0 \end{pmatrix}$, so $\begin{pmatrix} 0 & V\gamma(S) \\ 0 & \gamma(S) \end{pmatrix} \in \gamma$ follows. Since

$$\begin{pmatrix} 0 & V\gamma(S) \\ 0 & \gamma(S) \end{pmatrix} \triangleleft_l \begin{pmatrix} R & V \\ W & S \end{pmatrix}$$

and for every $w \in W$

$$\begin{pmatrix} 0 & V\gamma(S) \\ 0 & \gamma(S) \end{pmatrix} \begin{pmatrix} 0 & 0 \\ w & 0 \end{pmatrix} = \begin{pmatrix} V\gamma(S)w & 0 \\ \gamma(S)w & 0 \end{pmatrix},$$

from Lemma 3.18.4 (i) it follows that $\begin{pmatrix} V\gamma(S)w & 0 \\ \gamma(S)w & 0 \end{pmatrix}$ is in γ, and therefore so is its homomorphic image $V\gamma(S)w$. Since $V\gamma(S)w \triangleleft_l V\gamma(S)W$ and γ is left strong, we conclude that

$$V\gamma(S)W = \sum(V\gamma(S)w \mid w \in W) \in \gamma.$$

Moreover, $V\gamma(S)W \triangleleft R$, and therefore

$$V\gamma(S)W \subseteq \gamma(R).$$

Thus γ is a normal radical.

A similar reasoning proves the right version of the Theorem. □

If it would not have been clear from the definition that *the notion of normal radicals is left and right symmetric* as revealed by Theorem 3.18.5.

Let us have a short break, and give some examples.

EXAMPLE 3.18.6 (i) As we have seen in Example 3.17.7, *the Baer, Levitzki and Jacobson radicals are* left strong and left hereditary. Hence by Theorem 3.18.5 they are *normal radicals*.

(ii) Since Köthe's nil radical is left hereditary, *Köthe's problem 3.17.3 is equivalent to the question as whether the nil radical is normal.*

(iii) In Example 3.16.14 we have seen that *the strongly prime radical s_ℓ is not left hereditary*. Hence in view of Theorem 3.18.5 s_ℓ *is not a normal radical.*

(iv) We have seen in Example 3.17.8 that neither the Brown–McCoy nor the Behrens radical is left strong, so by Theorem 3.18.5 *the Brown–McCoy and Behrens radicals are not normal.*

EXAMPLE 3.18.7. In view of Examples 3.16.9 and 3.17.12 *the torsion radical and the p-torsion radicals* are left hereditary and left strong, so by Theorem 3.18.5 they *are normal radicals.*

Now, we continue characterizing normal radicals, but first we prove two Lemmas.

LEMMA 3.18.8 (Jaegermann [3]). *If (R, V, W, S) is a Morita context and γ is any radical, then*

$$\gamma \begin{pmatrix} R & V \\ W & S \end{pmatrix} = \begin{pmatrix} A & B \\ C & D \end{pmatrix}$$

with appropriate ideals A, D of R, S and bisubmodules B, C of V, W, respectively.

Proof: It is clear that

$$\gamma \begin{pmatrix} R & V \\ W & S \end{pmatrix} \triangleleft \begin{pmatrix} R & V \\ W & S \end{pmatrix} \triangleleft \begin{pmatrix} R^1 & V \\ W & S^1 \end{pmatrix},$$

so by the ADS-Theorem 3.1.2 also

$$\gamma \begin{pmatrix} R & V \\ W & S \end{pmatrix} \triangleleft \begin{pmatrix} R^1 & V \\ W & S^1 \end{pmatrix}.$$

Now, multiplying $\gamma \begin{pmatrix} R & V \\ W & S \end{pmatrix}$ by $\begin{pmatrix} 1 & 0 \\ 0 & 0 \end{pmatrix}$ and $\begin{pmatrix} 0 & 0 \\ 0 & 1 \end{pmatrix}$ from the left and right in all four possible ways one gets that $\gamma \begin{pmatrix} R & V \\ W & S \end{pmatrix}$ has the form as stated in the Lemma. □

Remember the convention that $\mathbb{Z}$ and $Z(\infty)$ denote the ring of integers and that with zero-multiplication on the infinite cyclic group. The latter is unequivocal.

LEMMA 3.18.9. *Let γ be a normal radical and $R \triangleleft \overline{R}$.*
(i) *If $Z(\infty) \in \gamma$ then $\gamma(R) = \gamma(\overline{R}) \cap R$.*
(ii) *If $Z(\infty) \in \mathcal{S}\gamma$ and $\overline{R}/R \cong \mathbb{Z}$, then $\gamma(R) = \gamma(\overline{R})$.*

Proof: (i) In this case every zero-ring — as a homomorphic image of a direct sum of copies of $Z(\infty)$ — must be in γ, whence γ is hypernilpotent. Hence by Proposition 3.18.2 γ is (left) hereditary, and so by Corollary 3.2.4 we have that $\gamma(R) = \gamma(\overline{R}) \cap R$.

(ii) Consider the radical $\gamma(\mathbb{Z}) = (n)$ of $\mathbb{Z}$. Then by Lemma 3.18.4 (ii) we have $(n)^0 \in \gamma$. If $n \neq 0$, then $(n)^0 \cong Z(\infty) \in \mathcal{S}\gamma$, a contradiction. Therefore $n = 0$ and $\mathbb{Z} \in \mathcal{S}\gamma$. By $\overline{R}/R \cong \mathbb{Z} \in \mathcal{S}\gamma$ we conclude that $\gamma(\overline{R}) \subseteq R$ whence also $\gamma(\overline{R}) \subseteq \gamma(R)$. Moreover, the ADS-Theorem 3.1.2 implies that $\gamma(R) \triangleleft \overline{R}$ yielding $\gamma(R) \subseteq \gamma(\overline{R})$. Thus $\gamma(R) = \gamma(\overline{R})$. $\square$

THEOREM 3.18.10 (Jaegermann [3]). *For a radical γ the following statements are equivalent:*
  (i)   *γ is a normal radical,*
  (ii) *$\gamma(R) = \gamma(R^1) \cap R$ for every ring R and $\gamma(eRe) = e\gamma(R)e$ for every idempotent $e \in R$,*
  (iii) *for every Morita context (R, V, W, S) we have*

$$\gamma\begin{pmatrix} R & V \\ W & S \end{pmatrix} = \begin{pmatrix} \gamma(R) & B \\ C & \gamma(S) \end{pmatrix}$$

where B and C are bisubmodules of V and W, respectively.

Proof: (i) $\Longrightarrow$ (ii) Either $Z(\infty) \in \gamma$ or $Z(\infty) \in \mathcal{S}\gamma$, Lemma 3.18.9 infers $\gamma(R) = \gamma(R^1) \cap R$.

Let us consider the Morita context (R, Re, eR, eRe). Since γ is normal, we have that
$$Re\gamma(eRe)eR \subseteq \gamma(R)$$
and
$$eR\gamma(R)Re \subseteq \gamma(eRe).$$
Since e is a unity element in eRe,
$$\gamma(eRe) = e^3\gamma(eRe)e^3 \subseteq eRe\gamma(eRe)eRe \subseteq e\gamma(R)e$$
and
$$e\gamma(R)e = e^2\gamma(R)e^2 \subseteq eR\gamma(R)Re \subseteq \gamma(eRe)$$
which yield $\gamma(eRe) = e\gamma(R)e$.

(ii) $\Longrightarrow$ (iii) We have obviously the relations
$$\begin{pmatrix} R & V \\ W & S \end{pmatrix} \triangleleft \begin{pmatrix} R^1 & V \\ W & S \end{pmatrix} \triangleleft \begin{pmatrix} R^1 & V \\ W & S^1 \end{pmatrix}.$$

Since both the factor rings are isomorphic to $\mathbb{Z}$, Lemma 3.18.8 and a twofold application of Lemma 3.18.9 gives us

$$\begin{pmatrix} A & B \\ C & D \end{pmatrix} = \gamma\begin{pmatrix} R & V \\ W & S \end{pmatrix} = \gamma\begin{pmatrix} R^1 & V \\ W & S^1 \end{pmatrix} \cap \begin{pmatrix} R & V \\ W & S \end{pmatrix}$$

where A, B, C, D are as in Lemma 3.18.8.

Hence, applying the equalities of (ii) for the idempotent $\begin{pmatrix} 1 & 0 \\ 0 & 0 \end{pmatrix} \in \begin{pmatrix} R^1 & V \\ W & S^1 \end{pmatrix}$ and for $\gamma \begin{pmatrix} R & V \\ W & S \end{pmatrix}$, we get

$$\begin{pmatrix} A & 0 \\ 0 & 0 \end{pmatrix} = \begin{pmatrix} 1 & 0 \\ 0 & 0 \end{pmatrix} \gamma \left(\begin{pmatrix} R & V \\ W & S \end{pmatrix} \right) \begin{pmatrix} 1 & 0 \\ 0 & 0 \end{pmatrix} =$$
$$= \begin{pmatrix} 1 & 0 \\ 0 & 0 \end{pmatrix} \left(\gamma \left(\begin{pmatrix} R^1 & V \\ W & S^1 \end{pmatrix} \right) \cap \begin{pmatrix} R & V \\ W & S \end{pmatrix} \right) \begin{pmatrix} 1 & 0 \\ 0 & 0 \end{pmatrix} \subseteq$$
$$\subseteq \begin{pmatrix} 1 & 0 \\ 0 & 0 \end{pmatrix} \gamma \begin{pmatrix} R^1 & V \\ W & S^1 \end{pmatrix} \begin{pmatrix} 1 & 0 \\ 0 & 0 \end{pmatrix} \cap \begin{pmatrix} R & 0 \\ 0 & 0 \end{pmatrix} =$$
$$= \gamma \begin{pmatrix} R^1 & 0 \\ 0 & 0 \end{pmatrix} \cap \begin{pmatrix} R & 0 \\ 0 & 0 \end{pmatrix} = \begin{pmatrix} \gamma(R^1) \cap R & 0 \\ 0 & 0 \end{pmatrix} = \begin{pmatrix} \gamma(R) & 0 \\ 0 & 0 \end{pmatrix}.$$

Thus $A = \gamma(R)$. A similar reasoning gives $D = \gamma(S)$.

(iii) $\implies$ (i) We have

$$\begin{pmatrix} 0 & V \\ 0 & 0 \end{pmatrix} \gamma \begin{pmatrix} R & V \\ W & S \end{pmatrix} \begin{pmatrix} 0 & 0 \\ W & 0 \end{pmatrix} \subseteq \gamma \begin{pmatrix} R & V \\ W & S \end{pmatrix} = \begin{pmatrix} \gamma(R) & B \\ C & \gamma(S) \end{pmatrix},$$

and so

$$\begin{pmatrix} 0 & V \\ 0 & 0 \end{pmatrix} \begin{pmatrix} \gamma(R) & B \\ C & \gamma(S) \end{pmatrix} \begin{pmatrix} 0 & 0 \\ W & 0 \end{pmatrix} = \begin{pmatrix} V\gamma(S)W & 0 \\ 0 & 0 \end{pmatrix} \subseteq \begin{pmatrix} \gamma(R) & B \\ C & \gamma(S) \end{pmatrix}.$$

Hence $V\gamma(S)W \subseteq \gamma(R)$ follows proving that γ is normal. □

The following Proposition expresses a kind of weak homomorphic closedness of semisimple classes of normal radicals (cf. Theorem 3.16.11). We shall make use of this Proposition only in Section 3.19.

PROPOSITION 3.18.11 (Jaegermann and Sands [1]). *Let γ be a normal radical. If $R \in S\gamma$, then also $R/(0:R) \in S\gamma$ where $(0:R) = \{x \in R \mid xR = 0\}$.*

Proof: Consider the Morita context $(R, R^1/(0:R), R, R/(0:R))$. Since γ is normal, we have

$$R^1/(0:R) \cdot \gamma(R/(0:R)) \cdot R \subseteq \gamma(R) = 0.$$

Hence for every coset $x + (0:R) \in \gamma(R/(0:R))$ we have

$$R^1/(0:R) \cdot (x + (0:R)) \cdot R = 0$$

which implies $xR \subseteq (0:R)$. So $x \in (0:R)$ and $\gamma(R/(0:R)) = 0$. Thus $R/(0:R) \in S\gamma$. □

In the next bulk of this section we shall focus our attention to hypernilpotent normal radicals.

Sands [1] calls a radical γ an *N-radical*, if γ is hypernilpotent, left hereditary and left strong.

THEOREM 3.18.12. *A radical γ is an N-radical if and only if γ is a supernilpotent normal radical.*

Proof: By Theorem 3.18.5 every N-radical is normal. Conversely, Propositions 3.18.1 and 3.18.2 imlpy that every hypernilpotent normal radical is left strong and left hereditary. □

COROLLARY 3.18.13. *Any N-radical is left and right symmetric, in particular, right hereditary and right strong.*

Proof: Obvious by Theorem 3.18.12. □

Let us mention that the Brown–McCoy radical $\mathcal{G}$ is not left hereditary, though supernilpotent (cf. Example 3.16.1). Hence by Corollary 3.18.13, $\mathcal{G}$ *is not an N-radical*. In view of Example 3.18.6 (iv) the Behrens radical is not an N-radical. Beidar [7] constructed a supernilpotent radical which is principally left hereditary but not left hereditary and so not an N-radical (cf. Propositions 3.18.2 and 3.18.3).

The N-radical version of Theorem 3.18.10 is the following

THEOREM 3.18.14 (Jaegermann [3]). *A radical γ is an N-radical if and only if for every Morita context (R, V, W, S)*

$$\gamma \begin{pmatrix} R & V \\ W & S \end{pmatrix} = \begin{pmatrix} \gamma(R) & B \\ C & \gamma(S) \end{pmatrix}$$

where either

$$B = \{v \in V \mid Wv \subseteq \gamma(S)\}, \ or \ B = \{v \in V \mid vW \subseteq \gamma(R)\};$$

or

$$C = \{w \in W \mid wV \subseteq \gamma(S)\}, \ or \ C = \{w \in W \mid Vw \subseteq \gamma(R)\}.$$

Moreover, if any of these equalities holds, then all of them hold.

Proof: Assume that γ is an N-radical. Then by Theorem 3.18.10 for every Morita context (R, V, W, S)

$$\gamma \begin{pmatrix} R & V \\ W & S \end{pmatrix} = \begin{pmatrix} \gamma(R) & B \\ C & \gamma(S) \end{pmatrix} \triangleleft \begin{pmatrix} R & V \\ W & S \end{pmatrix}.$$

Hence from

$$\begin{pmatrix} R & V \\ W & S \end{pmatrix} \begin{pmatrix} \gamma(R) & B \\ C & \gamma(S) \end{pmatrix} = \begin{pmatrix} * & RB + V\gamma(S) \\ * & WB + S\gamma(S) \end{pmatrix}$$

it follows that $WB \subseteq \gamma(S)$, that is,
$$B \subseteq \{v \in V \mid Wv \subseteq \gamma(S)\}.$$

Put $B_0 = \{v \in \mid Wv \subseteq \gamma(S)$. Then also, $V\gamma(S) \subseteq B \subseteq B_0$. Since γ is hypernilpotent, from
$$(B_0)^0 \cong \begin{pmatrix} 0 & B_0 \\ 0 & 0 \end{pmatrix} \triangleleft \begin{pmatrix} 0 & B_0 \\ 0 & \gamma(S) \end{pmatrix}$$

and
$$\begin{pmatrix} 0 & B_0 \\ 0 & \gamma(S) \end{pmatrix} \Big/ \begin{pmatrix} 0 & B_0 \\ 0 & \gamma(S) \end{pmatrix} \cong \gamma(S) \in \gamma$$

the extension property of γ yields $\begin{pmatrix} 0 & B_0 \\ 0 & \gamma(S) \end{pmatrix} \in \gamma$. By $V\gamma(S) \subseteq B \subseteq B_0$ we have
$$\begin{pmatrix} 0 & B_0 \\ 0 & \gamma(S) \end{pmatrix} \triangleleft_l \begin{pmatrix} R & V \\ W & S \end{pmatrix},$$

so the left strongness of γ yields
$$\begin{pmatrix} 0 & B_0 \\ 0 & \gamma(S) \end{pmatrix} \subseteq \gamma \begin{pmatrix} R & V \\ W & S \end{pmatrix} = \begin{pmatrix} \gamma(R) & B \\ C & \gamma(S) \end{pmatrix}.$$

Thus also $B_0 \subseteq B$. A similar argument (using also the left and right symmetry of N-radicals) proves the further equalities.

Conversely, assume that for every Morita context (R, V, W, S) the radical γ satisfies
$$\gamma \begin{pmatrix} R & V \\ W & S \end{pmatrix} = \begin{pmatrix} \gamma(R) & B \\ C & \gamma(S) \end{pmatrix}$$

and one of the four equalities, say
$$B = \{v \in V \mid Wv \subseteq \gamma(S)\},$$

is fulfilled. In particular, for the Morita context $(0, T, 0, 0)$, where T is any ring with zero-multiplication, we have
$$\gamma \begin{pmatrix} 0 & T \\ 0 & 0 \end{pmatrix} = \begin{pmatrix} 0 & B \\ 0 & 0 \end{pmatrix},$$
$$B = \{v \in T \mid 0v = 0\} = T.$$

Hence
$$T \cong \begin{pmatrix} 0 & T \\ 0 & 0 \end{pmatrix} = \gamma \begin{pmatrix} 0 & T \\ 0 & 0 \end{pmatrix} \in \gamma$$

follows, whence γ is a hypernilpotent radical. By Theorem 3.18.10, γ is a normal radical, so by Theorem 3.18.12 γ is an N-radical. □

Our next aim is to determine normal radicals, in particular N-radicals as upper radicals. For that purpose we need some preparations.

Let (R, V, W, S) be a Morita context and $P \triangleleft R$ and $Q \triangleleft S$. Define

$$\triangle(P) = \{s \in S \mid VsW \subseteq P\}$$
$$\nabla(Q) = \{r \in R \mid WrV \subseteq Q\}.$$

Clearly $\triangle(P) \triangleleft S$, $\nabla(Q) \triangleleft R$ and the assignments

$$P \longrightarrow \triangle(P) \quad \text{and} \quad Q \longrightarrow \nabla(Q)$$

preserve inclusion and intersection. Moreover, for

$$T = \{v \in V \mid vW \subseteq P\} \quad \text{and} \quad U = \{w \in W \mid Vw \subseteq P\}$$

we have

$$\begin{pmatrix} P & T \\ U & \triangle(P) \end{pmatrix} \triangleleft \begin{pmatrix} R & V \\ W & S \end{pmatrix}.$$

This leads to the Morita context $(R/P, V/T, W/U, S/\triangle(P))$ which is called the *factor context* of (R, V, W, S).

A Morita context (R, V, W, S) is said to be *S-faithful*, if $S \neq 0$ and $VsW = 0$, $s \in S$, implies $s = 0$.

PROPOSITION 3.18.15 (Nicholson and Watters [1]). *The following are equivalent for a class ϱ of rings:*

(a) *if (R, V, W, S) is a Morita context, $P \triangleleft R$ and $R/P \in \varrho$, then $S/\triangle(P) \in \varrho$,*

(b) *if (R, V, W, S) is a Morita context and $R \in \varrho$, then $S/\triangle(0) \in \varrho$,*

(c) *if (R, V, W, S) is an S-faithful Morita context and $R \in \varrho$, then $S \in \varrho$.*

Proof: Clearly (a) $\Longrightarrow$ (b) $\Longrightarrow$ (c).

(c) $\Longrightarrow$ (a) Let (R, V, W, S) be a Morita context, $P \triangleleft R$ and $R/P \in \varrho$. Then the factor context $(R/P, V/T, W/U, S/\triangle(P))$ satisfies the condition that $(V/T)(s + \triangle(P))(W/U) = 0$ implies $s \in \triangle(P)$. Hence the contex $(R/P, V/T, W/U, S/\triangle(P))$ is $S/\triangle(P)$-faithful, provided that $S \neq \triangle(P)$. So by (c) it follows that $S/\triangle(P) \in \varrho$. □

A class ϱ of rings is called a *normal class*, if for every Morita context (R, V, W, S) the equivalent conditions of Proposition 3.18.15 are fulfilled. A normal class of semiprime rings is called an *N-class*.

THEOREM 3.18.16. *A radical γ is normal if and only if its semisimple class $S\gamma$ is a normal class. In particular, γ is an N-radical if and only if $S\gamma$ is an N-class.*

Proof: Let γ be a normal radical, and (R, V, W, S) an S-faithful Morita context such that $R \in S\gamma$. Since γ is normal,

$$V\gamma(S)W \subseteq \gamma(R) = 0.$$

Taking into account that the context is S-faithful, we conclude that $\gamma(S) = 0$, that is, $S \in \mathcal{S}\gamma$. Thus by Proposition 3.18.15, $\mathcal{S}\gamma$ is a normal class.

Assume that $\mathcal{S}\gamma$ is a normal class, and let (R, V, W, S) be any Morita context. Let us consider the factor context $(R/\gamma(R), V/T, W/U, S/\triangle(\gamma(R)))$. Since $\mathcal{S}\gamma$ is a normal class, by Proposition 3.18.15, $R/\gamma(R) \in \mathcal{S}\gamma$ implies $S/\triangle(\gamma(R)) \in \mathcal{S}\gamma$. Hence $\gamma(S) \subseteq \triangle(\gamma(R))$ follows yielding $V\gamma(S)W \subseteq \gamma(R)$. Thus γ is a normal radical. □

LEMMA 3.18.17 (Sands [1]). *Let (R, V, W, S) be a Morita context. Then*

(i) *for every semiprime ideal P of R, $\triangle(P)$ is a semiprime ideal of S,*

(ii) *if $WV = S$ then for every semiprime ideal Q of S, $\nabla(Q)$ is a semiprime ideal of R such that $\triangle\nabla(Q) = Q$,*

(iii) *if $VW = R$ and $WV = S$ then $P \mapsto \triangle(P)$ establishes a bijection between the semiprime ideals of R and S. In particular, $\triangle(\beta(R)) = \beta(S)$.*

Proof: (i) Consider the factor context $(R/P, V/T, W/U, S/\triangle(P))$. Since the Baer radical is normal (Example 3.18.6), we have

$$(V/T)\beta(S/\triangle(P))(W/U) \subseteq \beta(R/P) = 0.$$

Hence, setting $D/A(P) = \beta(S/\triangle(P))$, it follows that $VDW \subseteq P$. Therefore $D \subseteq \triangle(P)$, that is, $\beta(S/\triangle(P)) = 0$. Thus $\triangle(P)$ is a semiprime ideal of S.

(ii) Consider the factor context $(S/Q, W/U, V/T, R/\nabla(Q))$. As in (i), we conclude that $\nabla(Q)$ is a semiprime ideal of R. Further,

$$\triangle\nabla(Q) = \{s \in S \mid VsW \subseteq \nabla(Q)\} = \{s \in S \mid WVsWV \subseteq Q\} =$$
$$= \{s \in S \mid SsS \subseteq Q\} = Q,$$

since Q is a semiprime ideal of S (cf. Lemma 1.2.10).

(iii) If also $VW = R$ then, as in (ii), we get that $\nabla\triangle(P) = P$ for every semiprime ideal P of R. Thus the correspondence $P \mapsto \triangle(P)$ is one-to-one. Since this correspondence preserves intersection, we have

$$\beta(S) = \cap(Q \triangleleft S \mid S/Q \in \mathcal{S}\beta) = \cap(\triangle(P) \mid P \triangleleft R, R/P \in \mathcal{S}\beta) =$$
$$= \triangle(\cap(P \triangleleft R \mid R/P \in \mathcal{S}\beta)) = \triangle(\beta(R)). \qquad □$$

THEOREM 3.18.18 (Puczyłowski [12]). *For a class ϱ of rings the following two conditions are equivalent.*

1) *ϱ is an N-class,*

2) *ϱ is a weakly special class and for a Morita context (R, V, W, S) such that $VW = R$ and $WV = S$, $R/\beta(R) \in \varrho$ if and only if $S/\beta(S) \in \varrho$.*

Proof: 1) $\Longrightarrow$ 2) Let us consider the Morita context (R, I, R^1, I) where $0 \neq I \triangleleft R \in \varrho$ and the multiplication is the ring multiplication. Then

$$\triangle(I) = \{i \in I \mid IiR^1 = 0\},$$

and so $\triangle(I) \subseteq \triangle(I)R^1$ is an ideal of I such that $(\triangle(I))^2 = 0$. By $I \triangleleft R \in \varrho$, I is a semiprime ring, so $\triangle(I) = 0$. Since $R \in \varrho$ and ϱ is an N-class, Proposition 3.18.15 yields that $I \cong I/\triangle(I) \in \varrho$. Thus the class ϱ is hereditary.

Let I be an essential ideal in R and $I \in \varrho$. For the Morita context (I, R^1, I, R) one gets that

$$\triangle(0) = \{r \in R \mid R^1 r I = 0\}.$$

Hence

$$(\triangle(0) \cap I)^2 \subseteq \triangle(0)I \subseteq R^1 \triangle(0)I = 0.$$

Since I is a semiprime ring, necessarily $\triangle(0) \cap I = 0$. Hence $I \triangleleft \cdot A$ implies that $\triangle(0) = 0$, and so by Proposition 3.18.15 we conclude that $R \in \varrho$. Thus ϱ is a weakly special class.

Let (R, V, W, S) be a Morita context such that $VW = R$ and $WV = S$. Then by Lemma 3.18.17 we have that $\triangle(\beta(R)) = \beta(S)$, and if $S = \beta(S)$ then $R = \beta(R)$. Since ϱ is an N-class, $R/\beta(R) \in \varrho$ implies

$$S/\beta(S) = S/\triangle(\beta(S)) \in \varrho.$$

Similarly, considering the dual context, one gets that if ϱ is an N-class and $S/\beta(S) \in \varrho$ then $R/\beta(R) \in \varrho$.

2) $\Longrightarrow$ 1) Let (R, V, W, S) be an S-faithful Morita context and $R \in \varrho$. Since R is a semiprime ring, 0 is a semiprime ideal of R. Hence by Lemma 3.18.17, $\triangle(0)$ is a semiprime ideal of S. Moreover, since (R, V, W, S) is S-faithful, $\triangle(0) = 0$. Thus S is a semiprime ring. Again by the S-faithfulness, $WV \neq 0$ and also $(WV)^2 \neq 0$ for $WV \triangleleft S \in S\beta$. Hence also $VW \neq 0$ and from the hereditariness of ϱ and $VW \triangleleft R \in \varrho$ we conclude that $VW \in \varrho$. Considering the Morita context (VW, V, W, WV), by 2) the relation $VW \in \varrho$ implies $WV \in \varrho$. Let

$$L = \{s \in S \mid s(WV) = 0\}$$

denote the left annihilator of WV in S, and take any $s \in L$. Then

$$(VsW)^2 = V(sWV)sW = 0.$$

Since $VsW \triangleleft S$ and S is semiprime, it follows that $VsW = 0$. But (R, V, W, S) is S-faithful, so $s = 0$. Hence $L = 0$ and also the two-sided annihilator of WV in S is 0. Thus Proposition 3.7.10 implies that $S \in \varrho$ and so by Proposition 3.18.15 ϱ is a normal class. □

COROLLARY 3.18.19. *A class ϱ of rings is a normal class of prime rings (that is, an N-class of prime rings) if and only if ϱ is a special class satisfying condition 2) in Theorem 3.18.18.* □

Radical Theory for Associative Rings

PROPOSITION 3.18.20 (Amitsur [6]). *The class $\mathcal{P}$ of all prime rings and the class $\mathcal{S}\beta$ of all semiprime rings are normal classes.*

Proof: Let (R, V, W, S) be an S-faithful Morita context and $R \in \mathcal{P}$. Since the context is S-faithful, $VW \neq 0$. From $VW \triangleleft R \in \mathcal{P}$ it follows that $VW \in \mathcal{P}$. Hence $(VW)^2 \neq 0$ which implies $WV \neq 0$.

Consider the Morita context (VW, V, W, WV). Lemma 3.18.18 (iii) is applicable yielding $WV \in \mathcal{P}$. Let $s \in S$ be an element which annihilates WV from the left. Then $(VsW)^2 = V(sWV)sW = 0$. Since (R,V,W,S) is S-faithful, we have that $s = 0$. Thus the (left) annihilator of WV in S is 0. Since $\mathcal{P}$ is a special class (Example 3.8.13), by Proposition 3.7.10 we get that $S \in \mathcal{P}$. Thus by Proposition 3.18.15 $\mathcal{P}$ is a normal class.

The same proof applies to the class $\mathcal{S}\beta$ of all semiprime rings. □

THEOREM 3.18.21. *If ϱ is a normal class of (semi)prime rings, then the upper radical $\mathcal{U}\varrho$ is a normal and (supernilpotent) special radical, (respectively). If γ is a normal and special radical, then the class $\mathcal{S}\gamma \cap \mathcal{P}$ of γ-semisimple prime rings is a normal and special class. In particular, a radical is normal and special if and only if it is the upper radical determined by a normal class of prime rings.*

Proof: Let ϱ be a normal class of prime rings. Then by Corollary 3.18.19 ϱ is a special class. Hence by Corollary 3.8.5 the semisimple class $\mathcal{SU}\varrho$ consists of subdirect sums of rings from ϱ. Since the conditions of Proposition 3.18.15 are inherited by subdirect sums and ϱ is a normal class, so also $\mathcal{SU}\varrho$ is a normal class. Hence by Theorem 3.18.16 $\mathcal{U}\varrho$ is a normal as well as a special radical.

In the semiprime case the proof is similar.

Let γ be a normal and special radical. By Theorem 3.18.16 $\mathcal{S}\gamma$ is an N-class, so $\mathcal{S}\gamma$ is a normal class. By Proposition 3.18.20 also $\mathcal{P}$ is a normal class. Since the conditions of Proposition 3.18.15 are preserved by intersection, also $\mathcal{S}\gamma \cap \mathcal{P}$ is a normal class. Moreover, by Theorem 3.8.6 the class $\mathcal{S}\gamma \cap \mathcal{P}$ is a special class. □

EXAMPLE 3.18.22 (i) (Nicholson and Watters [1]). *The antisimple radical β_φ is a normal radical.* Let (R,V,W,S) be an S-faithful Morita context such that R is a subdirectly irreducible prime ring with heart H and $S \neq 0$. Since $WHV \triangleleft S$, R is prime and (R,V,W,S) is S-faithful, we have

$$V(WHV)W = (VW)H(VW) \neq 0.$$

If I is any nonzero ideal in S, then by the S-faithfulness we get

$$0 \neq VIW \triangleleft R$$

and so the subdirect irreducibility of R yields that $H \subseteq VIW$. Hence

$$0 \neq WHV \subseteq WVIWV \subseteq I$$

which shows that S is subdirectly irreducible and also prime by Proposition 3.18.20. Hence by Proposition 3.18.15 *the class of all subdirectly irreducible prime rings is a normal class*, and so by Theorem 3.18.21 its upper radical, namely *the antisimple radical β_φ is a normal radical*.

(ii) By Example 3.18.6 (iv) the Behrens radical $\mathcal{B}$ is not normal. Hence by Theroem 3.18.21 *the class of all subdirectly irreducible rings having a nonzero idempotent in their heart, is not a normal class*.

In the proof of Theorem 3.18.24 we shall make use of

LEMMA 3.18.23 (Nicholson and Watters [1] and Hongan [1]). *Let (R, V, W, S) be an S-faithful Morita context such that*

1) R *is a (semi)prime ring,*
2) $Vw = 0$, $w \in W$ *implies* $w = 0$,
3) $vW = 0$, $v \in V$ *implies* $v = 0$.

Then $\begin{pmatrix} R & V \\ W & S \end{pmatrix}$ *is a (semi)prime ring.*

Proof: To see that $\begin{pmatrix} R & V \\ W & S \end{pmatrix}$ is a prime ring if R is prime and 2), 3) are satisfied, suppose that

$$\begin{pmatrix} r_1 & v_1 \\ w_1 & s_1 \end{pmatrix} \begin{pmatrix} R & V \\ W & S \end{pmatrix} \begin{pmatrix} r_2 & v_2 \\ w_2 & s_2 \end{pmatrix} = 0.$$

In view of Lemma 1.2.9 we have to show that

$$\begin{pmatrix} r_1 & v_1 \\ w_1 & s_1 \end{pmatrix} \neq 0 \quad \text{implies} \quad \begin{pmatrix} r_2 & v_2 \\ w_2 & s_2 \end{pmatrix} = 0.$$

Computing the above product of the three matrices, we have

$$\begin{pmatrix} r_1Rr_2+v_1Wr_2+r_1Vw_2+v_1Sw_2 & r_1Rv_2+v_1Wv_2+r_1Vs_2+v_1Ss_2 \\ w_1Rr_2+s_1Wr_2+w_1Vw_2+s_1Sw_2 & w_1Rv_2+s_1Wv_2+w_1Vs_2+s_1Ss_2 \end{pmatrix} = 0.$$

Assume that $r_1 \neq 0$. $R \in \mathcal{P}$ and $r_1Rr_2 = 0$ yield that $r_2 = 0$. Also $r_1Vs_2 = 0$, and so $r_1(Vs_2W) = 0$. We have $0 \neq (r_1)_R \triangleleft R$, $Vs_2W \triangleleft R$ and $(\dot{r})_R(Vs_2W) = 0$. Hence by $R \in \mathcal{P}$ it follows that $Vs_2W = 0$. So the S-faithfulness of (R, V, W, S) implies $s_2 = 0$. From $(r_1)_R(Vw_2) = 0$ a similar reasoning gives that $Vw_2 = 0$. Hence by 2) we get that $w_2 = 0$. Since $r_1Rv_2 = 0$ and so $r_1(Rv_2W) = 0$, we have also $(Rv_2W) \triangleleft R$ and $(r_1)_R(Rv_2W) = 0$. Hence the primeness of R yields that $Rv_2W = 0$. Using again the primeness of R we get that $v_2W = 0$, whence by 3) $v_2 = 0$ follows.

Arguing similarly in the cases $v_1 \neq 0$, $w_1 \neq 0$ and $s_1 \neq 0$, we can always conclude that $\begin{pmatrix} r_2 & v_2 \\ w_2 & s_2 \end{pmatrix} = 0$. Thus $\begin{pmatrix} R & V \\ W & S \end{pmatrix}$ is a prime ring.

For semiprime rings the proof is similar but simpler. □

We shall denote by $\operatorname{ann}_R X$ the two-sided annihilator of the subset X in a ring R.

THEOREM 3.18.24 (Nicholson and Watters [1] and Hongan [1]). *A class ϱ of (semi)prime rings is normal if and only if ϱ satisfies the following conditions:*

(N1) $L \triangleleft_l K \triangleleft_r R$, $R \in \varrho$ and L (semi)prime imply $L \in \varrho$.

(N2) $L \triangleleft_l K \triangleleft_r R$, $(\operatorname{ann}_K L = \operatorname{ann}_R K = 0)$, R (semi)prime and $L \in \varrho$ imply $R \in \varrho$.

Proof: First, we shall prove the Theorem for prime rings. Let ϱ be a normal class of prime rings, $0 \neq L \triangleleft_l K \triangleleft_r R \in \varrho$ and L prime. One readily sees that (R, RL, K, L) is a Morita context. We claim that this context is L-faithful. For that purpose, assume that $RLlK = 0$ for some $l \in L$. Since R is a prime ring, it follows that $LlK = 0$. Therefore

$$(Ll)^2 = (LlL)l \subseteq (LlK)l = 0.$$

Since L is a prime ring and $Ll \triangleleft L$ (because $LlL = 0$), we have that $Ll = 0$. Again, the primeness of L yields $l = 0$. Thus (R, RL, K, L) is L-faithful. Since ϱ is a normal class, Proposition 3.18.15 implies that $L \in \varrho$. Thus condition (N1) is fulfilled.

Next, consider the dual context (L, K, RL, R) and assume that $0 \neq L \in \varrho$, R prime. We show that this dual is R-faithful. Let $r \in R$ and $KrRL = 0$. Then since $L \neq 0$, Lemma 1.2.9 implies that $Kr = 0$. Moreover,

$$0 \neq L \subseteq K + RK \triangleleft R \in \mathcal{P}$$

and r annihilates $K + RK$ from the right. This implies $r = 0$, whence (L, K, RL, R) is R-faithful. Taking into consideration that ϱ is a normal class, Proposition 3.18.15 yields that $R \in \varrho$. Thus also Condition (N2) is fulfilled.

Conversely, suppose that the class ϱ of prime rings satisfies conditions (N1) and (N2). Let (R, V, W, S) be an S-faithful Morita context with $R \in \varrho$. In the notation preceding Proposition 3.18.15 (with $P = 0$) we have the factor context $(R, V/T, W/U, S)$ for which all the assumptions of Lemma 3.18.23 are fulfilled. Hence $\begin{pmatrix} R & V/T \\ W/U & S \end{pmatrix}$ is a prime ring. Now, $R \in \varrho$ and

$$R \cong \begin{pmatrix} R & 0 \\ 0 & 0 \end{pmatrix} \triangleleft_l \begin{pmatrix} R & V/T \\ 0 & 0 \end{pmatrix} \triangleleft_r \begin{pmatrix} R & V/T \\ W/U & S \end{pmatrix},$$

so condition (N2) implies that $\begin{pmatrix} R & V/T \\ W/U & S \end{pmatrix} \in \varrho$. Since

$$S \cong \begin{pmatrix} 0 & 0 \\ 0 & S \end{pmatrix} \triangleleft_l \begin{pmatrix} 0 & 0 \\ W/T & S \end{pmatrix} \triangleleft_r \begin{pmatrix} R & V/T \\ W/T & S \end{pmatrix}$$

and since S is a prime ring by Propositions 3.18.15 and 3.18.20, condition (N1) yields that $S \in \varrho$. Thus ϱ is a normal class in view of Proposition 3.18.15.

In the semiprime case we argue similarly. Here we demonstrate only the validity of condition (N2) where the annihilator conditions are involved.

Let $L \triangleleft_l K \triangleleft_r R$, $0 \neq L \in \varrho$, R semiprime and $\operatorname{ann}_K L = \operatorname{ann}_R K = 0$. We claim that the Morita context (L, K, RK, R) is R-faithful, that is, $KrRL = 0$ and $r \in R$ imply $r = 0$. KrR is clearly in the left annihilator of L in K. Since R is semiprime and

$$(LKr)R(LKr) = L(KrRL)Kr = 0,$$

Lemma 1.2.10 implies that $LKr = 0$. Hence Kr as well as KrR are in the right annihilator of L in K. Thus

$$KrR \subseteq \operatorname{ann}_K L = 0.$$

Furthermore, $(Kr)R(Kr) = (KrR)Kr = 0$ and R is semiprime. Hence again by Lemma 1.2.10 we get that $Kr = 0$. Since $RrK \triangleleft R$ and $(RrK)^2 \subseteq RrKrK = 0$ and R is semiprime, we conclude that $RrK = 0$. Also $rKRrK = 0$, so Lemma 1.2.10 and the semiprimeness of R give us that $rK = 0$. Thus r annihilates K from both sides, whence $r = 0$, as claimed. An application of Proposition 3.18.15 (c) yields that $R \in \varrho$. □

Lajos and Szász [1] called a subring B of a ring R a *biideal*, denoted by $B \triangleleft_b R$, if $BRB \subseteq B$. If $B \triangleleft_b R$ then $B \triangleleft_l (B + BR) \triangleleft_r R$, and if $L \triangleleft_l K \triangleleft_r R$ then $LRL \subseteq KL \subseteq L$ whence L is a biideal of R. Hence conditions (N1) and (N2) can be given also as

(N1) $B \triangleleft_b R \in \varrho$ and B prime imply $B \in \varrho$,

(N2) $B \triangleleft_b R$, R prime and $B \in \varrho$ imply $R \in \varrho$,

whenever ϱ is a class of prime rings.

Not every supernilpotent dual radical is normal, for instance, *the Brown–McCoy radical is a dual special radical which is not normal* (cf. Example 3.9.14 and the remark following Corollary 3.18.13). Thus by Theorem 3.18.21 *the class of all simple rings with unity element is not a normal class.*

The question arises naturally as whether every supernilpotent normal radical is special. Jaegermann and Sands [1] constructed an example of an N-radical which is not special. Another example was given by Beidar and Wiegandt [4].

In the characterizations of normal radicals the attributes left and right can be used arbitrarily as shown in the papers of Jaegermann and Sands [1], Beidar [8] and Tumurbat [4].

For more results on normal radicals the reader is referred to the already cited papers and to Beidar and Salavová [2] and Sokol'skiĭ [1], Tumurbat and Wiegandt [1], Tumurbat and Zand [4]. A more general setting is treated in Gardner and Kelarev [1].

3.19. A-radicals

The concept of an A-radical was introduced by Gardner [2] as a radical which depends only on the additive structure of the ring.

DEFINITION 3.19.1. A radical γ is called an *A-radical*, if for any ring $A \in \gamma$ and any additive homomorphism $f: A \to B$ such that $f(A)$ is a subring of B also $f(A) \in \gamma$.

Equivalent conditions are given in the following

PROPOSITION 3.19.2. *For a radical γ the following conditions are equivalent:*
  (i)   *γ is an A-radical,*
  (ii) *$A \in \gamma$ and $A^+ \cong B^+$ imply $B \in \gamma$,*
  (iii) *$A \in \gamma$ if and only if $A^0 \in \gamma$,*
  (iv) *$\gamma = \gamma^0$.*

Proof: The implications (i) $\implies$ (ii) $\implies$ (iii) are straightforward.

(iii) $\implies$ (i) If $f: A \to B$ is an additive homomorphism and $f(A)$ is a subring of B and $A \in \gamma$, then $A^0 \in \gamma$ and f induces a ring homomorphism from A^0 to B^0 with image $f(A)^0 \in \gamma$, so that $f(A) \in \gamma$.

(iii) $\implies$ (iv) Take any ring $A \in \gamma$. Then by (iii) also $A^0 \in \gamma$ holds implying $A \in \gamma^0$. Thus $\gamma \subseteq \gamma^0$. Consider a ring $A \in \gamma^0$. We have $A^0 \in \gamma$ and so by (iii) also $A \in \gamma$. Hence $\gamma^0 \subseteq \gamma$.

(iv) $\implies$ (iii) $A \in \gamma = \gamma^0$ if and only if $A^0 \in \gamma$. $\square$

We can get easily more characteristic properties.

PROPOSITION 3.19.3. *Each of the following conditions is equivalent to any of the conditions of Proposition 3.19.2.*
  (v) *$(\gamma(A))^+ = (\gamma(A^0))^+$ for every ring A,*
  (vi) *$A \in \mathcal{S}\gamma$ if and only if $A^0 \in \mathcal{S}\gamma$,*
  (vii) *$A \in \mathcal{S}\gamma$ and $A^+ \cong B^+$ imply $B \in \mathcal{S}\gamma$.*
Moreover, γ^0 is an A-radical for every radical γ.

Proof: (iv) $\implies$ (v) In view of Lemma 3.12.7 we have
$$(\gamma^0(A))^+ = (\gamma^0(A^0))^+.$$
Hence by (iv) condition (v) follows.

The implications (v) $\implies$ (vi) $\implies$ (vii) are trivial.

(vii) $\implies$ (iii) Suppose that (iii) is not satisfied. Then there exists a ring A such that either $A \in \gamma$ and $A^0 \notin \gamma$, or $A \notin \gamma$ and $A^0 \in \gamma$.

Case $A \in \gamma$: We have $(A/\gamma^0(A))^+ \cong (A^0/\gamma^0(A^0))^+$, and by the definition of γ^0 in Lemma 3.12.7 also $\gamma(A^0) = \gamma^0(A^0)$. Thus
$$A^0/\gamma^0(A^0) = A^0/\gamma(A^0) \in \mathcal{S}\gamma.$$

Hence by (vii) we get that $A/\gamma^0(A) \in \mathcal{S}\gamma$ which implies that $A = \gamma(A) \subseteq \gamma^0(A)$. Thus $\gamma^0(A) = A$, that is, $A \in \gamma^0$, that is, $A \in \gamma$, a contradiction.

Case $A^0 \in \gamma$: We have $(A/\gamma(A))^+ \cong (A^0/(\gamma(A))^0)^+$ and $A/\gamma(A) \in \mathcal{S}\gamma$. Hence by (vii) it follows that $A^0/(\gamma(A))^0 \in \mathcal{S}\gamma$ yielding $A^0 = \gamma(A^0) \subseteq (\gamma(A))^0$. Thus $\gamma(A) = A$ follows, a contradiction.

Since $A \in \gamma^0 \Leftrightarrow A^0 \in \gamma \Leftrightarrow A^0 \in \gamma^0 \Leftrightarrow A \in \gamma^{00}$, also the last statement is true. □

EXAMPLE 3.19.4. *The torsion radical τ (Example 3.9.9), the p-torsion radicals τ_p (Example 3.12.5) are hereditary A-radicals. The divisible radical Δ (Example 3.12.4) and the divisible torsion radical τ_D (Example 3.2.2) are non-hereditary A-radicals.*

In the previous sections we have got acquainted with a fair amount of knowledge of A-radicals. Let us recall Corollary 3.16.9.

COROLLARY 3.19.5. *For a radical γ the following conditions are equivalent:*
  (i) *γ is a hereditary A-radical,*
  (ii) *γ is left hereditary and left stable,*
  (iii) *$\gamma(L) = L \cap \gamma(A)$ for every $L \triangleleft_l A$.* □

(Hereditary) A-radicals can be characterized also via a property which can be regarded as a one-sided version of the ADS-Theorem 3.1.2. We may call a radical γ a *left-ADS-radical*, if $\gamma(L) \triangleleft_l A$ for all left ideals L of any ring A. Right-ADS-radicals are defined correspondingly.

PROPOSITION 3.19.6 (Sands [9]). *Let γ be a radical and σ its semisimple class. Then the following conditions are equivalent:*
  (i) *γ is a left-ADS-radical ,*
  (ii) *$A^0 \in \gamma$ implies $A \in \gamma$, that is, $\gamma^0 \subseteq \gamma$,*
  (iii) *$A \in \sigma$ implies $A^0 \in \sigma$.*

Proof: (i) $\Longrightarrow$ (iii) Assume that $A \in \sigma$ and $A^0 \notin \sigma$ Then by Lemma 3.12.7 we have
$$0 \neq \gamma(A^0) = \gamma^0(A) \triangleleft A.$$
Hence $0 \neq \gamma^0(A) \in \sigma$ and by $\gamma^0(A) \in \gamma^0$ also $(\gamma^0(A))^0 \in \gamma$. Thus, without loss of generality we may assume that $A \in \sigma$ and $A^0 \in \gamma$. Let us consider the matrix rings
$$B = \begin{pmatrix} A & A \\ A & A \end{pmatrix}, \quad L = \begin{pmatrix} A & 0 \\ A & 0 \end{pmatrix}, \quad K = \begin{pmatrix} 0 & 0 \\ A & 0 \end{pmatrix}.$$
Then $K \triangleleft L \triangleleft_l B$, and so by (i) we have $\gamma(L) \triangleleft_l B$. Further $L/K \cong A \in \sigma$ holds implying that $\gamma(L) \subseteq K$. Since $K \cong A^0 \in \gamma$, it follows that $K \subseteq \gamma(L)$. Thus $K = \gamma(L) \triangleleft_l B$, yielding $BK \subseteq K$, that is, $\begin{pmatrix} A^2 & 0 \\ A^2 & 0 \end{pmatrix} \subseteq K$. Hence $A^2 = 0$ follows, and so $A^0 = A \in \sigma$, a contradiction.

(iii) $\Longrightarrow$ (ii) Assume that $A^0 \in \gamma$. Then $A/\gamma(A) \in \sigma$, and so (iii) yields $A^0/(\gamma(A))^0 \in \sigma$. But $A^0/(\gamma(A))^0 \in \gamma$, whence $A^0 = (\gamma(A))^0$ and also $A = \gamma(A) \in \gamma$.

(ii) $\Longrightarrow$ (i) Suppose that $\gamma^0 \subseteq \gamma$. Let $L \triangleleft_l A$, and set $\gamma(L) = K$. Since $K/K^2 \in \gamma$ and K/K^2 is a zero-ring, by definition $K/K^2 \in \gamma^0$. For any element $a \in A$, we define the mapping

$$f \colon K/K^2 \to (L/K)^0$$

by $f(x + K^2) = ax + K$. Since $AK^2 \subseteq ALK \subseteq LK \subseteq K$, the mapping f is well defined, and f is obviously a homomorphism. Hence by $K/K^2 \in \gamma^0$ we have that

$$f(K/K^2) = ((aK + K)/K)^0 \in \gamma^0$$

and summing over all $a \in A$ it follows that

$$((AK + K)/K)^0 \triangleleft (L/K)^0 \quad \text{and} \quad ((AK + K)/K)^0 \in \gamma^0.$$

Hence by definition

$$((AK + K)/K)^0 = (((AK + K)/K)^0)^0 \in \gamma.$$

Therefore by $\gamma^0 \subseteq \gamma$ it follows that $(AK + K)/K \in \gamma$. But

$$(AK + K)/K \triangleleft L/K \in \mathcal{S}\gamma,$$

so $AK \subseteq K$, that is, $\gamma(L) = K \triangleleft_l A$. $\square$

Since the condition $\gamma^0 \subseteq \gamma$ in Proposition 3.19.6 is left and right symmetric, we have also

COROLLARY 3.19.7. *A radical is a left-ADS-radical if and only if it is a right-ADS-radical.* $\square$

COROLLARY 3.19.8. *Every A-radical and every hypoidempotent radical is a left-ADS-radical. A radical γ is an A-radical if and only if γ is a left-ADS-radical and $\gamma \subseteq \gamma^0$.*

Proof: By Propositions 3.19.3 and 3.19.6 A-radicals are left-ADS-radicals. For any hypoidempotent radical γ and any $L \triangleleft_l A$ we have

$$A\gamma(L) = A(\gamma(L))^2 \subseteq AL\gamma(L) \subseteq L\gamma(L) \subseteq \gamma(L),$$

whence $\gamma(L) \triangleleft_l A$. The last statement is obvious by Proposition 3.19.6. $\square$

We get also the following characterization of hereditary A-radicals.

COROLLARY 3.19.9. *A hereditary radical γ is an A-radical if and only if γ is a left strong and left-ADS-radical.*

Proof: If γ is a hereditary A-radical then by Corollary 3.19.5, γ is a left-ADS-radical which is left stable, and so left strong. Conversely, if γ is a hereditary left strong and left-ADS-radical then Lemma 3.16.6 implies $\gamma \subseteq \gamma^0$, and Proposition 3.19.3 yields $\gamma^0 \subseteq \gamma$. Thus by Proposition 3.19.2, γ is an A-radical. □

COROLLARY 3.19.10. *If a subidempotent radical γ is left strong (in particular, normal), then $\gamma = 0$. If a left strong supernilpotent radical γ is a left-ADS-radical, then $\gamma = \{$all rings$\}$.*

Proof: Trivial by Corollaries 3.19.8 and 3.19.9. □

A connection between A-radicals and strict radicals is given in the following which may be compared with Corollary 3.19.5 (i) and (ii).

PROPOSITION 3.19.11 (Gardner [2]). *Every A-radical γ is strict and hence left and right stable (and also left and right strong).*

Proof: Let U be a subring of a ring A such that $U \in \gamma$. Since γ is an A-radical, also $U^0 \in \gamma$ holds. Hence by Lemma 3.12.7 and by $\gamma = \gamma^0$ we conclude that
$$U \subseteq \gamma^0(A) = \gamma(A).$$ □

The converse of Proposition 3.19.11 fails to be true even for hereditary radicals: *the generalized nil radical $\mathcal{N}_g$, as we have seen in Example 3.16.15, is a strict and special radical* which is not an A-radical.

PROPOSITION 3.19.12. *Every A-radical γ is principally left hereditary and principally right hereditary.*

Proof: The mapping $f\colon A \to Aa$ given by $f(x) = xa$, $x \in A$, preserves addition. Hence by Definition 3.19.1, $A \in \gamma$ entails $Aa \in \gamma$, that is, γ is principally left hereditary. □

Comparing Theorem 3.18.5 with the following Theorem 3.19.13, we see the sharp contrast between left strongness and left stability of a radical.

THEOREM 3.19.13 (Jaegermann and Sands [1]). *A radical γ is an A-radical if and only if γ is left stable and principally left hereditary.*

Proof: Assume, first, that γ is principally left hereditary and left stable. Since left stable implies left strong, by Lemma 3.18.4 (ii) it follows that $R \in \gamma$ implies $R^0 \in \gamma$. Hence, to prove that γ is an A-radical, we have to show that $R^0 \in \gamma$ implies $R \in \gamma$. Let $R^0 \in \gamma$ and put $S = R/\gamma(R)$. By Theorem 3.18.5, γ is a normal radical, so Theorem 3.18.10 is applicable to the Morita context (S, S, S, S) yielding

$$\gamma \begin{pmatrix} S & S \\ S & S \end{pmatrix} = \begin{pmatrix} \gamma(S) & B \\ C & \gamma(S) \end{pmatrix} = \begin{pmatrix} 0 & B \\ C & 0 \end{pmatrix},$$

for suitable B, C. Hence by $\begin{pmatrix} S & S \\ S & S \end{pmatrix} \triangleleft \begin{pmatrix} S^1 & S^1 \\ S^1 & S^1 \end{pmatrix}$ the ADS-Theorem 3.1.2 gives us

$$\begin{pmatrix} 0 & B \\ C & 0 \end{pmatrix} = \gamma \begin{pmatrix} S & S \\ S & S \end{pmatrix} \triangleleft \begin{pmatrix} S^1 & S^1 \\ S^1 & S^1 \end{pmatrix},$$

and so

$$\begin{pmatrix} B & 0 \\ 0 & 0 \end{pmatrix} = \begin{pmatrix} 0 & B \\ 0 & 0 \end{pmatrix} \begin{pmatrix} 0 & 0 \\ 1 & 0 \end{pmatrix} \subseteq \gamma \begin{pmatrix} S & S \\ S & S \end{pmatrix} = \begin{pmatrix} 0 & B \\ C & 0 \end{pmatrix}$$

follows, implying $B = 0$. Similarly, also $C = 0$. Hence $\begin{pmatrix} S & S \\ S & S \end{pmatrix} \in \mathcal{S}\gamma$. Since γ is left stable, from

$$S^0 \cong \begin{pmatrix} 0 & 0 \\ S & 0 \end{pmatrix} \triangleleft \begin{pmatrix} S & 0 \\ S & 0 \end{pmatrix} \triangleleft_l \begin{pmatrix} S & S \\ S & S \end{pmatrix} \in \mathcal{S}\gamma$$

we conclude that $S^0 \in \mathcal{S}\gamma$. Also

$$S^0 \cong R^0/(\gamma(R))^0 \in \gamma$$

holds, whence $S = 0$ follows. Thus $R = \gamma(R) \in \gamma$, and so γ is an A-radical. The converse follows from Propositions 3.19.11 and 3.19.12. □

COROLLARY 3.19.14. *Every A-radical is a normal radical. For a radical γ the following are equivalent:*
  (i) *γ is an A-radical,*
  (ii) *γ is a normal and left stable radical,*
  (iii) *γ is a normal and left-ADS-radical.*

Proof: The first assertion and the equivalence of (i) and (ii) follows immediately from Theorems 3.18.5 and 3.19.13. By Lemma 3.18.4 (ii) and Theorem 3.18.5 we have that $\gamma \subseteq \gamma^0$ for every normal radical γ, so Proposition 3.19.6 and Corollary 3.19.8 jointly imply the equivalence of (i) and (iii). □

Note that the generalized nil radical $\mathcal{N}_g$ is left stable (Example 3.16.5 (ii)) but not an A-radical. Hence by Theorem 3.19.13, $\mathcal{N}_g$ is not principally left hereditary, and so not an N-radical.

Our next aim is to show that every normal radical can be represented as an intersection of an N-radical and an A-radical. For this purpose we need a Proposition which provides a hidden property of Baer's prime radical β.

PROPOSITION 3.19.15. *Let γ be any radical. If $A \in \gamma^0 \cap \mathcal{S}\gamma$, then*

$$\beta(A) = (0 : A)_A = A^*$$

where $(0 : A)_A$ is the left annihilator and A^* is the two-sided annihilator of A.

Proof: It is obvious that $(0 : A)_A \subseteq \beta(A)$.

Let $b \in \beta(A)$ be an arbitrary element and $B = (0 : bA)_{bA}$. Let us consider an ideal X of bA such that $X^2 \subseteq B \subseteq X$. For every element $x \in X$ we have that $AxbA \lhd A$ and

$$(AxbA)^2 = AxbAAxbA \subseteq AX^2bA = 0.$$

Since
$$AxbA = \sum(axbA \mid a \in A)$$

and each $axbA$ is a homomorphic image of A^0 which is in γ since $A \in \gamma^0$, we have that $AxbA \in \gamma$. Hence by $A \in \mathcal{S}\gamma$ we conclude that $AxbA = 0$. Thus $xbA \lhd A$ and $(xbA)^2 = 0$. Hence, as above, $A \in \gamma^0$ implies $A^0 \in \gamma$ as well as $xbA \in \gamma$. This and $xbA \lhd A \in \mathcal{S}\gamma$ yield that $xbA = 0$. Thus $x \in B$, so that $X = B$. Hence bA/B has no nonzero ideal with trivial multiplication, and so in view of Example 2.2.2 and Corollary 3.4.8 the ring bA/B must be β-semisimple. Since $bA \subseteq \beta(A)$ and $bA \lhd_r \beta(A)$, by Example 3.2.12 we see that $bA \in \beta$. Thus as $bA/B \in \mathcal{S}\beta$ we conclude that $bA = B$, whence $(bA)^2 = B(bA) = 0$. Since $A \in \gamma^0$ implies $A^0 \in \gamma$ and since bA is a homomorphic image of A^0, we get that $bA \in \gamma$. Thus $(bA)^2 = 0$ and

$$AbA = \sum(abA \mid a \in A)$$

yield that $AbA \in \gamma$. Since also $AbA \lhd A \in \mathcal{S}\gamma$, it follows that $AbA = 0$. Thus $bA \lhd A \in \mathcal{S}\gamma$ holds yielding $bA \in \mathcal{S}\gamma$. Hence $bA = 0$, whence $\beta(A) \subseteq (0 : A)$.

Analogous reasoning proves that $\beta(A)$ is also the right annihilator ideal of A and so also $\beta(A) = A^*$ has been proved. □

THEOREM 3.19.16 (Jaegermann and Sands [1]). *Every normal radical γ is the intersection of an A-radical and an N-radical, more precisely, $\gamma = \gamma^0 \cap \mathcal{L}(\gamma \cup \beta)$. Moreover, if $\gamma = \alpha \cap \mathcal{L}(\gamma \cup \beta)$, then $\alpha = \gamma^0$.*

Proof: γ^0 is clearly an A-radical by Proposition 3.19.3. We show that $\mathcal{L}(\gamma \cup \beta)$ is an N-radical. By Theorem 3.19.16 it suffices to show that the semisimple class $\mathcal{SL}(\gamma \cup \beta)$ is a normal class. Theorem 3.4.7 tells us that $\mathcal{L}(\gamma \cup \beta) = \mathcal{U}(\mathcal{S}\gamma \cap \mathcal{S}\beta)$, and so $\mathcal{SL}(\gamma \cup \beta) = \mathcal{S}\gamma \cap \mathcal{S}\beta$. Since both $\mathcal{S}\gamma$ and $\mathcal{S}\beta$ are normal classes by Theorem 3.18.16, an application of Proposition 3.18.15 yields that also $\mathcal{S}\gamma \cap \mathcal{S}\beta$ is a normal class. Thus, again by Theorem 3.18.16 $\mathcal{L}(\gamma \cup \beta)$ is an N-radical.

Since γ is a normal radical, Theorem 3.18.5 and Lemma 3.18.4 (ii) yield that $\gamma \subseteq \gamma^0$, whence $\gamma \subseteq \gamma^0 \cap \mathcal{L}(\gamma \cup \beta)$.

Take a ring $A \in \gamma^0 \cap \mathcal{L}(\gamma \cup \beta)$ and let $B = A/\gamma(A)$. Then $B \in \gamma^0 \cap \mathcal{S}\gamma$, and so Proposition 3.19.15 is applicable, yielding that $\beta(B) = (0 : B)$. By Proposition 3.18.11 we get that $B/\beta(B) \in \mathcal{S}\gamma \cap \mathcal{S}\beta$. Hence

$B/\beta(B) \in \mathcal{S}(\mathcal{L}(\gamma \cup \beta))$. But $B/\beta(B)$, being a homomorphic image of A, is in $\mathcal{L}(\gamma \cup \beta)$. Therefore $B = \beta(B) = (0 : B)$, that is, $B = B^0$. Since $B = B^0 = A^0/\gamma(A)^0 \in \gamma$, it follows that $A/\gamma(A) = B \in \gamma$, that is, $A = \gamma(A) \in \gamma$. This shows that $\gamma^0 \cap \mathcal{L}(\gamma \cup \beta) \subseteq \gamma$, and so the required equality is proved.

Now, let $\gamma = \alpha \cap \delta$ where α is an A-radical and δ is an N-radical. Since $\beta \subseteq \delta$, then δ^0 is the class of all rings. Therefore

$$\gamma^0 = (\alpha \cap \delta)^0 = \alpha^0 \cap \delta^0 = \alpha^0 = \alpha$$

which proves the last statement. □

A non-trivial A-radical is, of course, not comparable with hypernilpotent or hypoidempotent radicals. In the sequel we shall relate an A-radical γ to subradicals of the Baer-radical β as well as to the overradical $\mathcal{U}(\mathcal{S}\gamma \cap \mathcal{Z})$ of the idempotent radical $\mathcal{I} = \mathcal{UZ}$ (where $\mathcal{Z}$ denotes the class of rings with zero-multiplication). For this end we need some preparations. We start with a Lemma which is not surprising but remained hidden so far.

LEMMA 3.19.17 (Gardner [7]). *A nilpotent ring A belongs to a radical γ if and only if $A^0 \in \gamma$.*

Proof: Let A be a nilpotent ring in γ, and suppose $A^0 \notin \gamma$. Then

$$(A/\gamma^0(A))^0 \cong A^0/\gamma(A^0) \in \mathcal{S}\gamma \quad \text{and} \quad A/\gamma^0(A) \in \gamma.$$

Thus we may assume without loss of generality that $A \in \gamma$ and $A^0 \in \mathcal{S}\gamma$.

If $A^2 = 0$ there is nothing to prove, so let $A^{m+1} = 0 \neq A^m$ for some $m > 1$. As $A^m \neq 0$, there exists an element $b \in A^{m-1}$ such that $bA \neq 0$. If $a_1, a_2 \in A$ then $ba_1a_2 \in A^{m-1}A^2 = 0$ while $ba_1ba_2 \in A^{m-1}AA^{m-1}A \subseteq A^{m+1} = 0$. Thus the function $f : A \to bA$ given by $f(a) = ba$ is a surjective ring homomorphism, so as A is in γ, so is bA. But $(bA)^2 = 0$, so $bA \cong (bA)^0 \triangleleft A^0 \in \mathcal{S}\gamma$, a contradiction. Thus $A^0 \in \gamma$.

Next, suppose that $A^0 \in \gamma$, and let A be a nilpotent ring. We proceed by induction on the nilpotence degree n of A. For $n = 2$ the statement is trivial. So, assume that $n > 2$ and that $B^0 \in \gamma$ implies $B \in \gamma$ whenever $B^{n-1} = 0$. The mapping

$$f_a : A^0 \longrightarrow (Aa)^0$$

defined by $f_a(x) = xa$ is clearly a homomorphism, so $(Aa)^0 \in \gamma$ by $A^0 \in \gamma$. We have, further,

$$(A^2)^0 = \sum ((Aa)^0 \mid a \in A).$$

For any nonzero factor ring $(A^2)^0/S^0$ there exists an element $a \in A$ such that $(Aa)^0 \not\subseteq S^0$. Hence

$$0 \neq ((Aa)^0 + S^0)/S^0 \cong (Aa)^0/((Aa)^0 \cap S^0) \in \gamma.$$

Thus every nonzero homomorphic image of $(A^2)^0$ has a nonzero ideal in γ, and therefore $(A^2)^0 \in \gamma$. Since $(A^2)^{n-1} = 0$, the hypothesis implies that $A^2 \in \gamma$. Thus from
$$A/A^2 \cong A^0/(A^2)^0 \in \gamma$$
we conclude that $A \in \gamma$. □

PROPOSITION 3.19.18. *If γ is an A-radical, then $\gamma \cap \beta$ is a principally left hereditary radical.*

Proof: Obvious by Proposition 3.19.12 and Example 3.2.12. □

PROPOSITION 3.19.19. *Let γ be a principally left hereditary radical and γ_N the class of all nilpotent rings contained in γ. If $\gamma \subseteq \beta$ then $\gamma = \mathcal{L}\gamma_N$.*

Proof: By Theorem 3.3.2 $\mathcal{L}\gamma_N$ consists of all rings A such that every nonzero homomorphic image of A has a nonzero accessible subring in γ_N. Since γ is homomorphically closed, every nonzero ring A in γ has a nonzero accessible subring in γ_N. Clearly only a non-nilpotent ring A needs to be considered. Such a ring A does not coincide with its right annihilator A^r, so $A/A^r \in \gamma \subseteq \beta$, and A/A^r, has a nonzero nilpotent ideal I/A^r. Let $a \in I$ be an element such that $a \notin A^r$. Then
$$Aa/(Aa \cap A^r) \cong (Aa + A^r)/A^r \subseteq I/A^r$$
and so Aa is nilpotent. Since also A^r is nilpotent, it follows that both I and Aa are nilpotent. If the degree of nilpotence of I is n, then
$$Aa \triangleleft Aa + I^{n-1} \triangleleft \cdots \triangleleft Aa + I^2 \triangleleft I \triangleleft A.$$
Thus Aa is a nonzero accessible subring of A. Since γ is principally left hereditary, $A \in \gamma$ imples $Aa \in \gamma$. Moreover, Aa is nilpotent, whence $Aa \in \gamma_N$. This proves that $\gamma \subseteq \mathcal{L}\gamma_N$. The containment $\mathcal{L}\gamma_N \subseteq \gamma$ is trivially true. □

THEOREM 3.19.20. *If γ is an A-radical, then $\mathcal{L}(\gamma \cap \mathcal{Z}) = \gamma \cap \beta$.*

Proof: Lemma 3.19.17 is applicable, yielding $\gamma_N \subseteq \mathcal{L}(\gamma \cap \mathcal{Z})$. Hence $\mathcal{L}\gamma_N = \mathcal{L}(\gamma \cap \mathcal{Z})$. Moreover, by Propositions 3.19.18 and 3.19.19 we have that $\gamma \cap \beta = \mathcal{L}\gamma_N$. □

Next, we shall see that corresponding to each A-radical there is an interval of radicals such that each A-radical belongs to exactly one such interval.

THEOREM 3.19.21 (Sands [9]). *For every radical γ we have*
$$\gamma \cap \beta \subseteq \gamma \subseteq \mathcal{U}(\mathcal{S}\gamma \cap \mathcal{Z}).$$
Let γ be an A-radical. If $\delta^0 = \gamma$ for a radical δ, then $\gamma \cap \beta \subseteq \delta \subseteq \mathcal{U}(\mathcal{S}\gamma \cap \mathcal{Z})$ and conversely.

Radical Theory for Associative Rings

Proof: The first statement can be readily verified. If $\delta^0 = \gamma$, then $\delta \cap \mathcal{Z} = \delta^0 \cap \mathcal{Z} = \gamma \cap \mathcal{Z}$ holds implying by Theorem 3.19.20 that

$$\gamma \cap \beta = \mathcal{L}(\gamma \cap \mathcal{Z}) = \mathcal{L}(\delta \cap \mathcal{Z}) \subseteq \delta.$$

Also $\mathcal{S}\delta \cap \mathcal{Z} = \mathcal{S}\delta^0 \cap \mathcal{Z} = \mathcal{S}\gamma \cap \mathcal{Z}$ and consequently

$$\delta \subseteq \mathcal{U}(\mathcal{S}\delta \cap \mathcal{Z}) = \mathcal{U}(\mathcal{S}\gamma \cap \mathcal{Z}).$$

Assume that $\gamma \cap \beta \subseteq \delta \subseteq \mathcal{U}(\mathcal{S}\gamma \cap \mathcal{Z})$. Then we have

$$\gamma \cap \mathcal{Z} \subseteq \delta \cap \mathcal{Z} \subseteq \mathcal{U}(\mathcal{S}\gamma \cap \mathcal{Z}) \cap \mathcal{Z}.$$

If $A \in \mathcal{U}(\mathcal{S}\gamma \cap \mathcal{Z}) \cap \mathcal{Z}$ then A has no nonzero homomorphic image in $\mathcal{S}\gamma$, whence $A \in \mathcal{U}\mathcal{S}\gamma = \gamma$. Thus

$$\mathcal{U}(\mathcal{S}\gamma \cap \mathcal{Z}) \cap \mathcal{Z} \subseteq \gamma \cap \mathcal{Z},$$

whence

$$\gamma \cap \mathcal{Z} = \delta \cap \mathcal{Z} = \mathcal{U}(\mathcal{S}\gamma \cap \mathcal{Z}) \cap \mathcal{Z}.$$

Hence we conclude that $\gamma = \gamma^0 = \delta^0$. $\square$

Observe that the upper radical $\mathcal{U}(\mathcal{S}\gamma \cap \mathcal{Z})$ contains always the radical class $\mathcal{I} = \mathcal{U}\mathcal{Z}$ of all idempotent rings. Thus the interval $[\gamma \cap \beta, \mathcal{U}(\mathcal{S}\gamma \cap \mathcal{Z})]$ is a considerable one; meanwhile $[0, \beta]$ is a tiny interval.

3.20. Radical semisimple classes

As we have already seen, hereditary radicals are very common. So it is natural to ask about the dual notion, about homomorphically closed semisimple classes. Remember that in Theorem 3.9.4 the supplementing radical $\overline{\gamma}$ of a hereditary radical γ has been given as the largest homomorphically closed subclass in the semisimple class $\mathcal{S}\gamma$, but the possible case $\overline{\gamma} = \mathcal{S}\gamma$ has not been treated. In fact, in the case of coincidence, the supplementing radical class $\overline{\gamma}$ turns out to be simultaneously a radical and a semisimple class.

Semisimple classes of hereditary radicals have the inductive property (Proposition 3.9.3). Which hereditary radical classes have the coinductive property? In view of Theorem 3.1.8 such radical classes are also semisimple classes.

We may ask also for *radical semisimple classes*, that is, for classes which are both radical and semisimple classes. In this section we shall determine explicitly the radical semisimple classes. Prior to this, we shall give the interrelation between homomorphically closed semisimple classes and subdirectly closed radical classes. The key to this is Kogalovskiĭ's Theorem [1] which has been proved for universal algebras.

LEMMA 3.20.1. *If a homomorphically closed class ϱ of rings is closed under subdirect sums, then ϱ is strongly hereditary.*

Proof: Let B be a subring of a ring $A \in \varrho$, and consider the direct product $C = \prod_{i=1}^{\infty} A_i$ of copies A_i of A by the isomorphisms $f_i \colon A_i \to A$, $i = 1, 2, \ldots$. Take the diagonal subsets

$$D_A = \{(a_1, \ldots, a_i, \ldots) \in C \mid f_i(a_i) = a \in A;\ i = 1, 2, \ldots\}$$

and

$$D_B = \{(b_1, \ldots, b_i, \ldots) \in C \mid f_i(b_i) = b \in B.\ i = 1, 2, \ldots\}$$

Clearly the subrings D_A and D_B are isomorphic to A and B, respectively. Denoting by C_0 the direct sum $C_0 = \bigoplus_{i=1}^{\infty} A_i$, we consider the subring $D_B + C_0$ of C. Since

$$C_0 \subseteq D_B + C_0 \subseteq C,$$

the subring $D_B + C$ is obviously a subdirect sum of the rings A_i, $(i = 1, 2, \ldots)$. Hence by the assumption $D_B + C_0 \in \varrho$. Furthermore, $D_B \cap C_0 = 0$ implies that

$$B \cong D_B = D_B/(D_B \cap C_0) \cong (D_B + C_0)/C_0.$$

Since $D_B + C_0 \in \varrho$ and ϱ is homomorphically closed, we conclude that $B \in \varrho$. □

THEOREM 3.20.2 (Kogalovskiĭ [1]). *A subclass ϱ of rings is a variety if and only if ϱ is closed under homomorphic images and subdirect sums.*

Proof: By definition, a variety is a class of rings which is closed under homomorphic images, subrings and subdirect sums. Hence the assertion is an immediate consequence of Lemma 3.20.1. □

THEOREM 3.20.3 (Armendariz [1], Gardner and Stewart [1], Wiegandt [3]). *For a class ϱ of rings the following conditions are equivalent:*
  (i) ϱ *is a variety which is closed under extensions,*
  (ii) ϱ *is a radical class which is closed under subdirect sums,*
  (iii) ϱ *is a radical semisimple class,*
  (iv) ϱ *is a homomorphically closed semisimple class.*

Proof: Only the implication (i) $\Longrightarrow$ (ii) requires proof, the implications (ii) $\Longrightarrow$ (iii) $\Longrightarrow$ (iv) $\Longrightarrow$ (i) are trivial consequences of Theorems 2.1.4, 3.1.6 and 3.20.2.

(i) $\Longrightarrow$ (ii) In view of Theorem 2.1.4 we have to prove that ϱ has the inductive property. By Birkhoff's Theorem 1.1.6 on varieties, ϱ is a class of rings satisfying certain identities, and hence ϱ has the inductive property.

For the sake of completeness, however, we shall prove that ϱ has the inductive property without making use of Birkhoff's Theorem 1.1.6 (cf. Cohn [1]).

Let $I_1 \subseteq \ldots \subseteq I_\lambda \subseteq \ldots$ be an ascending chain of ideals (or just subrings) of a ring A such that $\cup I_\lambda = A$ and $I_\lambda \in \varrho$ for every index λ. Further, let $C = \prod_\lambda J_\lambda$ be the direct product of rings J_λ isomorphic to I_λ by isomorphisms $f_\lambda \colon J_\lambda \to I_\lambda$ for all λ. Take the set

$$B = \left\{ b = (b_1, \ldots, b_\lambda, \ldots) \in C \,\middle|\, \begin{array}{l} \text{there is an index } \mu \text{ depending on } b \\ \text{such that } f_\mu(b_\mu) = f_\nu(b_\nu) \text{ for } \nu \geq \mu \end{array} \right\}.$$

Then B is a subring of C such that each projection $\pi_\lambda \colon C \to J_\lambda$ maps B onto J_λ, and so B is a subdirect sum of the rings J_λ each in ϱ. Since the variety ϱ is closed under subdirect sums, $B \in \varrho$ follows.

Define the mapping $f \colon B \to A$ by $f(b) = f_\mu(b_\mu)$ where μ is the index as required in the definition of B. Since every element $a \in A = \cup I_\lambda$ belongs to an I_λ, there exists an element $b \in B$ with a component b_μ satisfying $f(b) = f_\mu(b_\mu) = a$. Thus f maps B onto A. One readily sees that f is, in fact, an isomorphism. Hence $B \in \varrho$ implies $A \in \varrho$. □

As mentioned before Theorem 3.17.11, a radical is strict if and only if its semisimple class is strongly hereditary. Thus by Lemma 3.20.1 *the upper radical of a homomorphically closed semisimple class, that is, of a radical semisimple class is a strict radical.*

The next Proposition shows that a non-trivial radical semisimple class must be a very peculiar subidempotent radical.

PROPOSITION 3.20.4 (Stewart [1]). *Let ϱ be a radical semisimple class. If the class ϱ contains a ring A which possesses a nonzero nilpotent element, then ϱ is the class of all associative rings.*

Proof: By the assumption we may choose an element $a \in A$ such that $a \neq 0 = a^2$. The subring $\langle a \rangle$ of A is a zero-ring isomorphic to $Z(n)$ for some $n = 2, 3, \ldots, \infty$. Moreover, $\langle a \rangle \in \varrho$ for by Lemma 3.20.1 the class ϱ is stongly hereditary. We claim that $Z(\infty) \in \varrho$. If the additive order of a is infinite, we are done. Let the additive order of a be a finite number n. Then $Z(n) \in \varrho$. Using that ϱ is homomorphically closed and also closed under extensions, we get that $Z(p^k) \in \varrho$ for any prime divisor p of n and every $k = 1, 2, \ldots$. Since ϱ is closed under subdirect sums and the zero-ring $Z(\infty)$ is a subdirect sum of the rings $Z(p^k)$, $k = 1, 2, \ldots$, we conclude that $Z(\infty) \in \varrho$.

Taking into account that ϱ is a radical class, every direct sum of copies of $Z(\infty)$, that is, every zero-ring on free abelian groups is in ϱ. Hence ϱ contains all zero-rings, because ϱ is homomorphically closed. Using repeatedly the extension closed property of ϱ, it follows that every nilpotent ring is in ϱ.

Let F be any free associative ring. The factor ring F/F^n is obviously nilpotent for every $n = 1, 2, 3, \ldots$. A moment's reflection shows that

$\bigcap_{n=1}^{\infty} F^n = 0$, and so F is a subdirect sum of the nilpotent rings F/F^n, each in ϱ. Hence $F \in \varrho$ follows. Thus ϱ contains all free associative rings, whence ϱ is the variety of all associative rings. □

A ring without nonzero nilpotent elements is called a *reduced ring*.

THEOREM 3.20.5 (Abian [1], Andrunakievich and Ryabukhin [4], Bell [1], Stewart [1]). *A ring A is reduced if and only if A is a subdirect sum of domains, that is, A is semisimple with respect to the generalized nil radical $\mathcal{N}_g$.*

Proof: We shall follow Bell's proof. We start with an observation. Let A be a reduced ring. For any two elements $x, y \in A$, $xy = 0$ implies $(yx)^2 = 0$ and so by the hypothesis on A also $yx = 0$ holds. Thus for any $z \in A$ we have $zyx = 0$ and $xzy = 0$. Hence

(∗) if a finite product of elements of A is zero, the insertion of additional factors in any position leaves a product of zero.

Now consider any nonzero element $a \in A$. The multiplicative subsemigroup

$$\{a, a^2, a^3, \dots\}$$

does not contain 0. A straightforward application of Zorn's Lemma shows that this multiplicative subsemigroup is contained in one, say M_a, which is maximal with respect to the exclusion of 0. Define

$$I_a = \{x \in A \mid cx = 0 \text{ for at least one } c \in M_a\}.$$

In showing that I_a is an ideal, closure under substraction is not quite trivial. Let $x, y \in I_a$. Then there exist elements $c, d \in M_a$ with $cx = 0 = dy$. Hence by (∗), $cdx = 0 = cdy$, so that $cd(x - y) = cdx - cdy = 0$.

If $x \notin M_a$, then the multiplicative subgroup generated by M and x must contain 0. Since A has no nonzero nilpotent elements, some finite product containing x as at least one factor and having at least one factor from m, must be 0. Repeated application of (∗) established the existence of an element $m \in M_a$ such that mx is nilpotent, and hence 0. This shows that the set theoretic complement of I_a is M_a, and therefore the factor ring A/I_a has no zero divisors.

Observe that $a \notin I_a$. Hence $\bigcap(I_a \mid a \in A) = 0$, and so A is a subdirect sum of domains A/I_a.

Since a subdirect sum is a subring of the direct product, a subdirect sum of domains cannot contain a nonzero nilpotent element.

Reminding the reader to Example 3.8.16, the proof is complete. □

THEOREM 3.20.6 (Stewart [1]). *A class ϱ of associative rings which is neither $\{0\}$ nor the class of all rings is a radical semisimple class if and* only if ϱ is the variety generated by a strongly hereditary finite set of finite fields. Moreover, ϱ is the semisimple class of the upper radical of a strongly hereditary finite set $\mathcal{F}$ of finite fields and $\mathcal{F}$ is a special class.

Proof: In the first part we shall follow Gardner [18]. Let ϱ be a radical semisimple class. By Theorem 3.20.3 ϱ is a variety closed under extensions. By the assumption on ϱ and Proposition 3.20.4 a ring $A \in \varrho$ cannot possess nonzero nilpotent elements, so in view of Theorem 3.20.5 A is a subdirect sum of domains. Also the subdirect components, as homomorphic images of $A \in \varrho$, are in ϱ. Obviously, domains, which are in ϱ, generate ϱ as a variety.

We shall show that a ring $B \neq 0$ in ϱ which has no zero-divisors, is a division ring. For any nonzero element $a \in B$ we have $\langle a \rangle / \langle a \rangle^2 \in \varrho$ because $B \in \varrho$ and ϱ is a variety. Since rings in ϱ have no nonzero nilpotent elements, it follows that $\langle a \rangle = \langle a \rangle^2$. In particular, $a \in \langle a \rangle^2$, so there is an equation $a = \sum_{i=2}^{k} n_i a^i$, $n_i \in \mathbb{Z}$. Let $e = \sum_{i=2}^{k} n_i a^{i-1}$. Clearly $e \in \langle a \rangle$ and $ae = a = ea$. Thus e is a unity element for the subring $\langle a \rangle$. For any nonzero element $b \in B$ we have $ba = b(ea) = (be)a$ and $ab = (ae)b = a(eb)$. So as there are no zero-divisors in B, $be = b = eb$ and e is a unity element in B. Since $e \in \langle a \rangle = \langle a \rangle^2 = \langle a \rangle a$, there is an element $a' \in \langle a \rangle \subset B$ such that $e = a'a$. But a is an arbitrary nonzero element of B, so B is a division ring.

We show that B is a finite field. Since

$$\langle 2e \rangle = \langle 2e \rangle^2 = \langle 4e \rangle,$$

we can write

$$2e = \sum_{i=1}^{m} l_i (4e)^i = \left(\sum_{i=1}^{m} 4^i l_i \right) e$$

where $l_i \in \mathbb{Z}$, so

$$\left(2 - \sum_{i=1}^{m} 4^i l_i \right) e = 0.$$

This means that B has finite characteristic. Since $e \in \langle a \rangle$ for all nonzero elements $a \in B$, B is algebraic over its prime subfield. Hence by Jacobson's Commutativity Theorem 1.1.10, B is a field.

Suppose B is infinite, and consider the polynomial ring $B[x]$. For each $b \in B$ we define the evaluation mapping $f_b \colon B[x] \to B$ by

$$f_b \left(\sum_{i=0}^{n} a_i x^i \right) = \sum_{i=0}^{n} a_i b^i.$$

f_b maps $B[x]$ homomorphically onto B and B is infinite, so

$$\bigcap (\ker f_b \mid b \in B) = 0.$$

Thus $B[x]$ is a subdirect sum of copies of $B \in \varrho$, so also $B[x] \in \varrho$. But $B[x]$ has no zero-divisors, so as above $B[x]$ is a field, a contradiction. Thus B is a finite field.

Suppose ϱ contains fields of infinitely many (finite) characteristics. Let $F_i \in \varrho$ be fields of characteristic p_i such that $p_i \neq p_j$ for $i,j = 1,2,\ldots$. Then the direct product of these fields F_i contain the element $e = (e_1, e_2, \ldots)$ where each component e_i is the unity element of F_i. Obviously

$$\mathbb{Z} \cong \langle e \rangle \in \varrho,$$

and so the nilpotent ring $2\mathbb{Z}/(4)$ is in ϱ, contradicting Proposition 3.20.4. Thus only finitely many characteristics are involved.

Assume that for some prime p there are infinitely many non-isomorphic fields of characteristic p in ϱ. Let $K_1, K_2, \ldots$ be such fields with $|K_1| < |K_2| < \cdots$. Then we can find elements $u_1 \in F_1$, $u_2 \in F_2$, ... with minimum polynomials over the prime field Z_p of unbounded degrees. But then the element $u = (u_1, u_2, \ldots)$ and the unity element $e = (e_1, e_2, \ldots)$ of the direct product $\prod F_i$ generate the subring $\langle u, e \rangle$ which is isomorphic to the polynomial ring $Z_p[x]$. Since $F_i \in \varrho$, so $\prod F_i \in \varrho$ and also $Z_p[x] \cong \langle u, e \rangle \in \varrho$. Hence taking into consideration that $Z_p[x]$ has no zero-divisors, we conclude as above that $Z_p[x]$ is a field, a contradiction. Thus there are only finitely many non-isomorphic fields of characteristic p in ϱ.

Conversely, let ϱ be a variety generated by a finite set $\mathcal{F}$ of finite fields. Then ϱ consists of commutative rings.

Each field $F_i \in \mathcal{F}$ satisfies an identity $x^{n_i} = x$ with $n_i = |F_i|$, $i = 1, \ldots, k$. Then the nonzero elements $x \in F_i$ satisfy $x^{n_i - 1} = 1$. So if m is the lowest common multiple of $n_1 - 1, \ldots, n_k - 1$, each F_i satisfies the identity $x^{m+1} = x$. This is passed on to subrings, homomorphic images and direct products, so every ring in ϱ satisfies $x^{m+1} = x$. Hence any ring $A \in \varrho$ has no nonzero nilpotent element. The ring A is a subdirect sum of subdirectly irreducible rings by Birkhoff's Theorem, and being ϱ a variety, each subdirectly irreducible component is in ϱ.

Let B be an arbitrary subdirectly irreducible ring in ϱ. Since B is commutative, and the heart H of B is a simple ring (B satisfies $x^{m+1} = x$!), so H is a field with unity element e. Then $H = B$ as $H \triangleleft \cdot B$, so B is a field.

The subdirect closure $\overline{\mu}$ of the class μ of all subdirectly irreducible rings of ϱ is exactly ϱ. Since μ consists of fields, μ is a special class and $\varrho = \overline{\mu}$ is a semisimple class (cf. Corollary 3.8.5). Hence ϱ is closed under extensions and so by Theorem 3.20.3 ϱ is a radical semisimple class.

The last statement is now obvious. $\square$

Now we can easily determine the identities which define the extension closed subvarieties of the variety of associative rings.

COROLLARY 3.20.7 (Gardner and Stewart [1]). *A class ϱ of rings is a radical semisimple class with $\{0\} \neq \varrho \neq \{\text{all rings}\}$ if and only if ϱ is a subvariety in the variety of associative rings given by a finite nonempty set $P = \{p_1, \ldots, p_k\}$ of primes and a finite nonempty set $N = \{N_1, \ldots, N_k\}$ of positive integers and determined by the identities*

(1) $$\prod(p \mid p \in P)x = 0$$

and

(2) $$x^{m+1} = x$$

where m is the lowest common multiple of the positive integers $p_i^{n_i} - 1$, $1 \leq n_i \leq N_i$ and $i = 1, \ldots, k$.

Proof: In virtue of Theorem 3.20.6 the radical semisimple class determines the sets P and N such that ϱ satisfies the identities (1) and (2).

Conversely, suppose that the ρ is determined by the identities (1) and (2). By Jacobson's Commutativity Theorem 1.1.10, ϱ consists of commutative rings. As we have seen in the second part of the proof of Theorem 3.20.6, all subdirectly irreducible rings of ϱ are fields (mind that by (2) rings in ϱ do not contain nonzero nilpotent elements). By (2) these fields are finite of bounded order, and by (1) they have only finitely many prime characteristics. Since every ring of ϱ is a subdirect sum of subdirectly irreducible rings in ϱ, ϱ is generated by a finite set of finite fields. Hence Theorem 3.20.6 implies that ϱ is a radical semisimple class. □

The next Theorem will tell us that the radical classes and varieties closed under essential extensions are precisely the radical semisimple classes.

THEOREM 3.20.8 (Loi [2]). *For a subclass ϱ of associative rings the following conditions are equivalent:*

(i) ϱ *is a radical class closed under essential extensions,*

(ii) ϱ *is closed under homomorphic images, direct sums and essential extensions,*

(iii) ϱ *is a variety closed under essential extensions,*

(iv) ϱ *is variety closed under extensions.*

Proof: (i) $\Longrightarrow$ (ii) Trivial.

(ii) $\Longrightarrow$ (iii) We adopt the proof of Gardner [18]. We first show that ϱ is strongly hereditary. Let B be a subring of a ring $A \in \varrho$. Let D be the Dorroh extension of A and K an ideal of D with $A \cap K = 0$ and maximal for this property. Then as we know from Lemma 3.2.5

$$A = A/(A \cap K) \cong (A+K)/K \triangleleft \cdot D/K$$

and D/K is, of course, a ring with unity element. Thus without loss of generality we may assume that

$$B \subseteq A \triangleleft \cdot C$$

where C is a ring with unity element.

Let S denote the direct sum $S = \bigoplus_{i=1}^{\infty} C_i$ of copies C_i of ring C. As in the proof of Corollary 3.13.2, we define a ring $R = S * B$ on $S^+ \oplus B^+$ by multiplication

$$(s_1, b_1)(s_2, b_2) = (s_1 s_2 + b_1 s_2 + s_1 b_2, b_1 b_2)$$

for all $s_1, s_2 \in S$ and $b_1, b_2 \in B$. Here the elements of B act componentwise; this makes sense as $B \subseteq A \triangleleft C$. It is routine to check that R is a ring,

$$S \cong (S, 0) \triangleleft R \quad \text{and} \quad R/(S, 0) \cong B$$

where, as usual, $(S, 0) = \{(s, 0) \in R \mid s \in S\}$. We claim that $(S, 0)$ is an essential ideal in R. Let $I \triangleleft R$ with $(S, 0) \cap I = 0$, and (s, b) an arbitrary element of I. Then for all $t \in S$ we have

$$(ts + tb, 0) = (t, 0)(s, b) = 0 = (s, b)(t, 0) = (st + bt, 0).$$

Let j be such that s has zero j-th component and let t be chosen to have all components zero except (possibly) the j-th. Then $ts = 0 = st$, and as above $ts + tb = 0 = st + bt$. Thus $tb = 0 = bt$. Since B acts on S componentwise, and t is, in effect, an arbitrary element of C_j, this means that b annihilates C_j. But $C_j \cong C$ has a unity element, so $b = 0$. Hence

$$(s, b) = (s, 0) \in I \cap (S, 0) = 0,$$

so $s = 0$ and $(s, b) = 0$. As (s, b) is anything in I, we see that $I = 0$, so that $(S, 0) \triangleleft \cdot R$.

Since ϱ is closed under essential extensions, $A \in \varrho$ implies $C \in \varrho$. Taking into account that ϱ is closed under direct sums, we conclude that $(S, 0) \cong S \in \varrho$ and also $R \in \varrho$. The class ϱ is closed under homomorphic images, so $B \cong R/(C, 0) \in \varrho$ follows. Thus the class ϱ is strongly hereditary.

All we need now is to prove that ϱ is closed under direct products. For this purpose, let us consider a direct product $A = \prod(A_\lambda \mid \lambda \in \Lambda)$ of rings $A_\lambda \in \varrho$, and let us embed each A_λ into a ring C_λ with unity such that $A_\lambda \triangleleft \cdot C_\lambda$. We have seen that this is possible. Using the properties of ϱ we get that the direct sum $S = \bigoplus(C_\lambda \mid \lambda \in \Lambda)$ is in ϱ. We show that S is an essential ideal in the direct product $C = \prod(C_\lambda \mid \lambda \in \Lambda)$. Let $J \triangleleft C$ such that $S \cap J = 0$. Take an arbitrary element $(\ldots, x_\lambda, \ldots) \in J$ and the element $(0, \ldots, 0, e_\mu, 0, \ldots)$ in S where e_μ is the unity element of C_μ. Then

$$(0, \ldots, 0, x_\mu, 0, \ldots) = (\ldots, x_\lambda, \ldots)(0, \ldots, 0, e_\mu, 0, \ldots) \in J \cap S = 0,$$

whence $x_\mu = 0$. Doing so for all μ we see that $J = 0$. Thus $S \triangleleft \cdot C$. This and $S \in \varrho$ imply $C \in \varrho$. Since A is a subring of C and, as we have already seen, ϱ is strongly hereditary, we conclude that $A \in \varrho$.

(iii) $\implies$ (iv) Now the class ϱ is hereditary and closed under subdirect sums and essential extensions. Hence the proof of Theorem 3.2.7 is applicable yielding (iv).

(iv) $\implies$ (i) By Theorem 3.20.3 ϱ is a radical semisimple class, and so in view of Theorem 3.20.6 ϱ is the semisimple class of a special radical. Since special radicals are hereditary, by Proposition 3.2.6 the class ϱ is closed under essential extensions. □

EXAMPLE 3.20.9. *The Jacobson radical $\mathcal{J}$ is not strongly hereditary* (what we have already seen in Example 3.2.13 (ii)). Since any direct product of Jacobson radical rings is again a Jacobson radical ring by its definition in Example 2.1.7, the class $\mathcal{J}$ is closed under direct products. If $\mathcal{J}$ were strongly hereditary, $\mathcal{J}$ would be a variety closed under extensions, and so by Theorem 3.20.8 $\mathcal{J}$ would be a radical semisimple class, and so a subidempotent radical. Nevertheless, the Jacobson radical rings $A(,+,\cdot,\circ)$ considered as rings with one more operation $\circ$ form a variety because they are defined by identities.

EXAMPLE 3.20.10. Let $\mathcal{F}$ be a strongly hereditary finite set of finite fields. By Theorem 3.20.6 the semisimple class $\mathcal{SUF}$ is also a radical class, and so homomorphically closed, that is, $h\mathcal{SUF} = \mathcal{SUF}$. Further, $\mathcal{F}$ is a special class, so the radical $\mathcal{UF}$ is hereditary. Hence by Theorem 3.9.4 *the supplementing radical to the special radical $\mathcal{UF}$ is the radical semisimple class $\mathcal{SUF}$*.

The reader might observe that studying closure properties led to useful results. Closure properties characterized, among others, radical classes (Theorem 2.1.4), semisimple classes (Theorem 3.1.6), semisimple classes of hereditary and supernilpotent radicals (Proposition 3.2.8 and Theorem 3.6.11), supplementing radicals (Theorem 3.9.1) and A-radicals (Proposition 3.19.2 and Corollary 3.19.5).

In this section we investigated semisimple classes closed under homomorphic images, radical classes closed under subdirect sums or essential extensions and varieties closed under extensions: all these classes are the radical semisimple classes. Further questions concerning closure properties may yield important information on classes of rings. Birkenmeier [1] showed that the essential cover of a supernilpotent radical class is nearly a semisimple class: it is hereditary, closed under extensions, finite subdirect sums and arbitrary direct sums and direct products. He raised the question: which radical classes have semisimple essential covers? These radicals were fully described by Beidar, Fong, Ke and Shum [1], and in an alternative way by Birkenmeier and Wiegandt [1]. Radicals having semisimple essential covers are either supernilpotent or subidempotent; radical semisimple classes are trivial examples of such radicals (cf. Theorem 3.20.8), but none of the so far introduced supernilpotent radicals have semisimple essential covers. See also Beidar, Fong and Ke [1], Birkenmeier [2] and Wu [1].

Also Majumdar [1] studied radical and semisimple classes.

For further results on radical semisimple classes and extension closed subvarieties (not in the variety of associative rings) the reader is referred to papers of Gardner [6], [10], [13].

Chapter IV
Concrete Radicals and Structure Theorems

In the previous chapters we introduced quite a few concrete radicals as examples arising from general radical theoretical considerations. Thus the reader has become familiar with many concrete radicals and has got a fair amount of knowledge of them. In this chapter we shall discuss concrete radicals *per se*, and therefore we shall change our point of view: we shall promote the examples of the previous chapters to theorems.

The main purpose of this chapter will be to examine concrete radicals, to prove structure theorems for semisimple rings, to study the interrelations between the concrete radicals as well as the behaviour of radicals of related rings, in particular of matrix rings and of polynomial rings. Finally we shall position the concrete radicals treated in this book.

4.1. The principal nil radicals

In this section we shall recapitulate facts on the Baer, Levitzki and nil radicals, and add more statements concerning these radicals.

The content of Examples 2.2.2, 3.2.12, 3.6.1 (i), 3.7.13, 3.8.13, 3.14.19, 3.16.5 (iv), 3.17.2 (i), 3.17.7 and 3.18.6 (i) can be summarized as follows.

DEFINITION AND THEOREM 4.1.1. *The Baer (or prime) radical β is the lower radical of all nilpotent rings as well as of all zero-rings, and hence every Baer radical ring is a nil ring. The Baer radical β is the smallest supernilpotent radical and the smallest special radical. The semisimple class $S\beta$ of β is the class of all semiprime rings. β is the upper radical $\mathcal{UP}$ of all prime rings. For any ring A*

$$\beta(A) = \cap(P \triangleleft A \mid A/P \in \mathcal{P}),$$

that is, $\beta(A)$ is the intersection of all prime ideals of A. In particular, $\beta(A) = 0$ if and only if A is a subdirect sum of prime rings. The Baer radical class β consists of all those rings which have no prime modules. β is a strongly hereditary radical and also a left (and right) strong radical, and hence a normal radical. β is, however, not left (or right) stable. □

As we have seen in Theorem 3.3.5, Watters' construction provides the lower radical of a given class for every ring. This applies, in particular, to the Baer radical which is the lower radical of all nilpotent rings (see Example 2.2.2). To construct the Baer radical $\beta(A)$ of a ring A, there is also a somewhat simpler construction.

Let A be any ring, and let δ denote the class of all nilpotent rings. Then we define an ideal $\mathcal{V}(A)$ of A as follows:

$$\mathcal{V}_1(A) = \sum(I \triangleleft A \mid I \in \delta),$$
$$\mathcal{V}_\lambda(A) = \sum(I \triangleleft A \mid I/\mathcal{V}_{\lambda-1}(A) \in \delta)$$

for non-limit ordinals, and

$$\mathcal{V}_\lambda(A) = \bigcup_{\mu < \lambda} \mathcal{V}_\mu(A)$$

for limit ordinals. Since A is a set, there exists an ordinal ζ depending on A such that

$$\mathcal{V}(A) = \mathcal{V}_\zeta(A) = \mathcal{V}_{\zeta+1}(A).$$

The construction of $\mathcal{V}(A)$ is analogous to Watters' construction $\mathcal{W}(A)$ in Section 3.3, but it uses ideals instead of accessible subrings.

THEOREM 4.1.2 (Baer [1]). $\beta(A) = \mathcal{V}(A)$ for every ring A.

Proof: $\mathcal{V}_1(A) \subseteq \beta(A)$ is clear in view of $\beta = \mathcal{L}\delta$. Suppose that $\mathcal{V}_\mu(A) \in \beta$ for all $\mu < \lambda$. If $\lambda - 1$ exists, then $\mathcal{V}_{\lambda-1}(A) \in \beta$ and by definition

$$\mathcal{V}_\lambda(A)/\mathcal{V}_{\lambda-1}(A) = \mathcal{V}_1(A/\mathcal{V}_{\lambda-1}(A)) \in \beta.$$

So, as β is closed under extensions, $\mathcal{V}_\lambda$ is in β. If λ is a limit ordinal, then $\mathcal{V}_\lambda(A) \in \beta$ by the inductive property of β (cf. Theorem 2.1.4). Thus $\mathcal{V}_\lambda(A) \in \beta$ in any case. By transfinite induction then $\mathcal{V}(A) \subseteq \beta(A)$.

Suppose that $\mathcal{V}(A) \neq \beta(A)$. Putting $B = A/\mathcal{V}(A)$ and $C = \beta(A)/\mathcal{V}(A)$ we have $0 \neq C \triangleleft B$. Since by Corollary 3.4.8 we have $\beta = \mathcal{L}\delta = \delta_2$, the relation $0 \neq C \in \delta_2$ yields the existence of a nonzero ideal I of C in δ. Let $\bar{I}$ denote the ideal of B generated by I. By the Andrunakievich Lemma $\bar{I}$ is nilpotent, so B contains a nonzero nilpotent ideal in δ. This is, however, impossible because $B = A/\mathcal{V}(A) = A/\mathcal{V}_\zeta(A)$ has no nonzero δ-ideals. Thus $\mathcal{V}(A) = \beta(A)$. □

One of the first radicals we have met in this book was the Levitzki radical (Example 2.1.8) introduced by Levitzki [1]. Recollecting the statements of Examples 3.2.1 (iii), 3.2.13 (i), 3.6.1 (ii), 3.16.5 (iv), 3.17.2 (ii), 3.17.7 and 3.18.6 (i), we have

DEFINITION AND THEOREM 4.1.3. *The Levitzki radical class $\mathcal{L}$ is the class of all locally nilpotent rings. The Levitzki radical $\mathcal{L}$ is supernilpo-*

tent, strongly hereditary, left (and right) strong, hence a normal radical. The Levitzki radical is not left (or right) stable. □

THEOREM 4.1.4 (Babich [1]). *The Levitzki radical $\mathcal{L}$ is a special radical.*

Proof: We shall base our proof on that given by Andrunakievich and Ryabukhin [2]. Let $\mathcal{C}$ be the class

$$\mathcal{C} = \left\{ A \;\middle|\; \begin{array}{l} \text{there exists a non-nilpotent finitely generated subring } S \\ \text{of } A \text{ such that for every nonzero ideal } I \text{ of } A,\, S^m \subseteq I \text{ for} \\ \text{some positive integer } m \end{array} \right\}.$$

We shall prove that $\mathcal{C}$ is a special class and $\mathcal{L} = \mathcal{UC}$.

Let A be an arbitrary ring in $\mathcal{C}$. If I and J are nonzero ideals of A, then for a non-nilpotent finitely generated subring S, $S^m \subseteq I$ and $S^n \subseteq J$ for suitable m, n, so that $S^{m+n} \subseteq IJ$. Since S is non-nilpotent, we have $IJ \neq 0$, whence A is a prime ring, and $\mathcal{C}$ consists of prime rings.

Let $0 \neq I \triangleleft A \in \mathcal{C}$. If S is a finitely generated non-nilpotent subring such that $S^m \subseteq I$ then also S^m is finitely generated and not nilpotent. Moreover, if $0 \neq K \triangleleft I$ and $\overline{K}$ is the ideal of A generated by K, then $\overline{K}^3 \subseteq K$ by the Andrunakievich Lemma, and $0 \neq \overline{K}^3 \triangleleft A$ because A is a prime ring. Hence $S^k \subseteq \overline{K}^3$ for some k, so that

$$(S^m)^k = (S^k)^m \subseteq \overline{K}^3 \subseteq K.$$

This proves that I is in $\mathcal{C}$, whence $\mathcal{C}$ is a hereditary class.

If $A \in \mathcal{C}$ is an essential ideal of a ring B and $0 \neq L \triangleleft B$, then $0 \neq L \cap A$, so $S^\ell \subseteq L \cap A \subseteq L$ with a suitable non-nilpotent finitely generated subring S and integer ℓ. Hence B is in $\mathcal{C}$, and so the class $\mathcal{C}$ is closed under essential extensions. Thus $\mathcal{C}$ is a special class.

For any ring $A \in \mathcal{C}$, we must have $\mathcal{L}(A) = 0$ as otherwise $\mathcal{L}(A)$ contains a non-nilpotent finitely generated subring S^m, a power of an appropriate subring S. Thus the special class $\mathcal{C}$ consists of $\mathcal{L}$-semisimple rings, so $\mathcal{L} \subseteq \mathcal{UC}$.

We complete the proof by showing that if a ring A is not in $\mathcal{L}$, then $A \notin \mathcal{UC}$. Let a be in $A \setminus \mathcal{L}(A)$. Then the ideal (a) of A generated by a is not locally nilpotent, so it contains a non-nilpotent finitely generated subring T. Let us consider the set

$$\{I \triangleleft A \mid I \text{ contains no power of } T\}.$$

This set is not empty; it contains 0 as T is not nilpotent. Let $I_1 \subseteq \cdots \subseteq I_\lambda \subseteq \ldots$ be an ascending chain of ideals from this set. If $T^k \subseteq \cup I_\lambda$, then as T^k is finitely generated, $T^k \subseteq I_\lambda$ for some λ, a contradiction. Thus by Zorn's Lemma we can choose a maximal element M of this set. Now M contains no power of T, so $(T+M)/M$ is not nilpotent; also it is finitely generated as a homomorphic image of T. If $0 \neq L/M \triangleleft A/M$, then $M \subset L \triangleleft A$, so L

contains a power of T and therefore L/M contains a power of $(T+M)/M$. Hence $A/M \in \mathcal{C}$. But then $A \notin \mathcal{UC}$. □

Passman [1], [2] defined the N^*-*radical* of a ring A with unity element by

$$N^*(A) = \{a \in A \mid aS \text{ is nilpotent for all finitely generated subrings } S \text{ of } A\}$$

which is an ideal of A. It is known that $N^*(A)$ lies between the Levitzki radical and Jacobson radical of A. The N^*-radical turned out to be very useful in the theory of group rings. The shortcoming (of the terminology) is that $N^*(A)$ is not a radical assignment in the sense of Kurosh and Amitsur: although $N^*(A/N^*(A)) = 0$ for every group ring A over a field of finite characteristic, this is not true in general for arbitrary rings A. Next, we shall show the relation of the N^*-radical to the Levitzki radical, and in this way we shall get (not left and right symmetric) characterizations of the Levitzki radical and semisimple class.

A subset F of a ring A is said to be *nilpotent*, if there exists an integer $n \geq 1$ such that each product $f_1 \cdots f_n$ of n factors is 0 whenever $f_1, ..., f_n \in F$.

THEOREM 4.1.5 (Beidar and Wiegandt [5]). *The following three conditions are equivalent for a ring A:*

(i) *A is a Levitzki radical ring,*
(ii) *$N^*(A) = A$,*
(iii) *for all $a \in A$ and finite subsets $F \subseteq A$, aF is nilpotent,*
(iv) *every finite subset $G \subseteq A$ is nilpotent.*

Proof: (i) $\Rightarrow$ (ii) If $a \in A \in \mathcal{L}$ and S is a subring of A generated by the elements $s_1, \ldots, s_n$, then the subring aS is contained in the subring generated by $a, s_1, \ldots, s_n$, and this subring is nilpotent because $A \in \mathcal{L}$. Hence $A = N^*(A)$.

The implications (ii) $\Rightarrow$ (iii) and (i) $\Rightarrow$ (iv) $\Rightarrow$ (iii) are obvious.

(iii) $\Rightarrow$ (i) Let us suppose that the ring A satisfies condition (iii). Further, let $a \in A$ and $F = \{f_1, \ldots, f_n\} \subseteq A$ be arbitrarily chosen. By the assumption the set aF is nilpotent, and so the subring generated by aF is also nilpotent. Hence the right ideal aA of A is locally nilpotent. Since $\mathcal{L}$ is a right strong radical (see Example 3.17.2 (ii)), it follows that $aA \subseteq \mathcal{L}(A)$. The element $a \in A$ has been chosen arbitrarily, so

$$A^2 = \sum(aA \mid a \in A) \subseteq \mathcal{L}(A).$$

Taking into account that $\mathcal{L}$ is a supernilpotent radical (see Example 3.6.1 (ii)), we have that $A^2 \in \mathcal{L}$ and $A/A^2 \in \mathcal{L}$. Thus also $A \in \mathcal{L}$ holds as radical classes are closed under extensions. □

From Theorem 4.1.5 it follows trivially

COROLLARY 4.1.6. *The Levitzki radical class can be given as*

$$\mathcal{L} = \{A \mid N^*(A) = A\};$$

further

$$\mathcal{L}(A) = \sum(I \triangleleft A \mid N^*(I) = I) =$$
$$= \sum(K \triangleleft A \mid \text{for all } k \in K \text{ and all finite subsets}$$
$$F \subseteq K, kF \text{ is nilpotent})\qquad\square$$

Let us consider the following two classes of rings:

$$\mathcal{M} = \left\{ A \;\middle|\; \begin{array}{l}\text{every nonzero ideal } I \text{ of } A \text{ contains}\\ \text{an element } a \text{ and a finite subset } F\\ \text{such that } aF \text{ is not nilpotent}\end{array}\right\}$$

and

$$\mathcal{M}^* = \left\{ A \;\middle|\; \begin{array}{l}\text{every nonzero ideal } I \text{ of } A \text{ contains an element } a \text{ and a finitely generated subring } S\\ \text{such that } aS \text{ is not nilpotent}\end{array}\right\}.$$

THEOREM 4.1.7. $\mathcal{M} = \mathcal{M}^* = \mathcal{SL}$ *is the Levitzki semisimple class.*

Proof: The equality $\mathcal{M} = \mathcal{M}^*$ is obvious by Theorem 4.1.5 (ii) and (iv).

If $A \notin \mathcal{SL}$, then $0 \neq I = \mathcal{L}(A) \in \mathcal{L}$ holds, and then by Theorem 4.1.5 $I = N^*(I)$. Hence A has an ideal $I \neq 0$ such that for every $a \in I$ and finitely generated subring S, aS is nilpotent. Thus $A \notin \mathcal{M}^*$, and so $\mathcal{M}^* \subseteq \mathcal{SL}$.

Assume that $a \notin \mathcal{M}$. Then there exists an ideal $I \neq 0$ of A such that for all $a \in I$ and finite subsets $F \subseteq I$, aF is nilpotent. Hence by Theorem 4.1.5 (i) and (iii) it follows that $I \in \mathcal{L}$ and so $A \notin \mathcal{SL}$. Thus also $\mathcal{SL} \subseteq \mathcal{M}$ has been proved. $\square$

An immediate consequence of Theorems 4.1.4 and 4.1.7 is

COROLLARY 4.1.8. $\mathcal{M} \cap \mathcal{P} = \mathcal{M}^* \cap \mathcal{P}$ *is the special class of all Levitzki semisimple prime rings where $\mathcal{P}$ denotes the class of all prime rings.* $\square$

One may say by right that radical theory started when Köthe [1] introduced the concept of the nil radical. Examples 2.1.6, 3.2.1, 3.2.13 (i), 3.6.10 (ii), 3.14.4, 3.15.4, 3.16.5 (iv) and 3.18.6 (ii) with Köthe's Problem 3.17.3 tell us the following.

DEFINITION AND THEOREM 4.1.9. *Köthe's nil radical is defined by the radical class $\mathcal{N}$ of all nil rings. The nil radical is a strongly hereditary special radical. Köthe's Problem is to determine whether the nil radical $\mathcal{N}$ is left (or right) strong, and this is equivalent to showing whether or not $\mathcal{N}$ is a normal radical.* $\square$

In Example 3.15.4 we have seen that the nil radical is a special one. This can be proved also by similar argument to that used in the proof of Theorem 4.1.4. We can therefore be a bit briefer.

PROPOSITION 4.1.10 (Wang Xion Hau [1], Hsieh Pang Chieh [1]). *Köthe's nil radical $\mathcal{N}$ is a special radical.*

Proof: Let $\mathcal{C}$ be the class of rings given by

$$\mathcal{C} = \left\{ A \;\middle|\; \begin{array}{l} \text{there exists a non-nilpotent element } a \in \\ A \text{ such that for every nonzero ideal } I \triangleleft A, \\ a^m \in I \text{ for some integer } m \end{array} \right\}.$$

If $A \in \mathcal{C}$ and I, J are nonzero ideals of A, let $a^m \in I$ and $a^n \in J$. Then $a^{m+n} \in IJ$, so $IJ \neq 0$. Hence A is a prime ring. Also $\mathcal{N}(A) = 0$, as otherwise $\mathcal{N}(A)$ contains some a^t and then a^t is nilpotent, so a is. As in the proof of Theorem 4.1.4 one shows that $\mathcal{C}$ is a special class and $\mathcal{N} \subseteq \mathcal{UC}$.

If $\mathcal{N}(A) \neq A$, and if $r \in A \backslash \mathcal{N}(R)$, then again in imitation of the previous proof, we find an ideal M maximal with respect to containing no power of r. It then transpires that $A/M \in \mathcal{C}$, so that $A \notin \mathcal{UC}$. □

4.2. Separation of the nil radicals

From Corollary 3.4.8 we know that every nonzero homomorphic image of a ring in the Baer radical class β has a nonzero nilpotent, hence locally nilpotent ideal. Thus we have $\beta \subseteq \mathcal{L}$, the Levitzki radical. Further, if $A \in \mathcal{L}$, then every element $a \in A$ generates a nilpotent subring, so A consists of nilpotent elements and $A \in \mathcal{N}$, the nil radical, that is, $\mathcal{L} \subseteq \mathcal{N}$.

In this section we shall prove nothing more than $\beta \neq \mathcal{L} \neq \mathcal{N}$. It is comparatively easy to show that a locally nilpotent ring need not be in β. The first example of such a ring seems to have been given by Baer [1]. Here we shall present a simpler example, due to Zel'manov [1].

THEOREM 4.2.1. *There exists a ring A which is locally nilpotent as well as prime, so that $\beta(A) = 0$ while $\mathcal{L}(A) = A$. In particular $\beta \neq \mathcal{L}$.*

Proof: Let F be the free ring on two generators, x and y, S the free semigroup on x and y, so that we view S as a subset of F. For any $w \in S$, let $\partial_x(w)$ and $\partial(w)$ denote the degree of w in x and the degree of w, respectively, and let

$$f(w) = \frac{\partial(w)}{\partial_x(w)^2 + 1}.$$

Finally, let I be the ideal of F generated by the set

$$\left\{ w \in S \;\middle|\; f(w) < \frac{1}{2} \right\}.$$

Our first step is to show that F/I is a prime ring, that is, I is a prime ideal of F. To this end, suppose there exist elements $a, b \in F \backslash I$ such that $aFb \subseteq I$. Then as $IF \subseteq I$ and $FI \subseteq I$, we can assume that $a = \sum_i n_i a_i$, $b = \sum_j m_j b_j$ where n_i and m_j are integers and the a_i and b_j are in $S \backslash I$.

Thus neither any a_i nor any b_j has a subword c with $f(c) < \dfrac{1}{2}$. Let

$$N = \left(\max_{i,j}(\partial_x(a_i) + \partial_x(b_j))\right)^2 + 1.$$

Consider the subword e of $a_i y^N b_j$ for some i,j. If e is a subword of a_i or b_j, then, as noted above, $f(e) \geq \dfrac{1}{2}$. If $e = e'y^k$, where e' is a subword of a_i and $1 \leq k \leq N$, then

$$f(e) = f(e'y^k) = \frac{\partial(e'y^k)}{\partial_x(e'y^k)^2 + 1} = \frac{\partial(e') + k}{\partial_x(e')^2 + 1} > \frac{\partial(e')}{\partial_x(e')^2 + 1} = f(e') \geq \frac{1}{2}.$$

In the same way, $f(e) \geq \dfrac{1}{2}$ if $e = y^k e''$ and e'' is a subword of b_j. Finally, if $e = \hat{a} y^N \hat{b}$, where $\hat{a}, \hat{b}$ are subwords of a_i, b_j respectively, then

$$f(e) = f(\hat{a} y^N \hat{b}) = \frac{\partial(\hat{a} y^N \hat{b})}{\partial_x(\hat{a} y^N \hat{b})^2 + 1} =$$

$$= \frac{\partial(\hat{a}) + \partial(\hat{b}) + N}{(\partial_x(\hat{a}) + \partial_x(\hat{b}))^2 + 1} \geq$$

$$\geq \frac{\partial(\hat{a}) + \partial(\hat{b}) + (\partial_x(a_i) + \partial_x(b_j))^2 + 1}{(\partial_x(\hat{a}) + \partial_x(\hat{b}))^2 + 1} \geq$$

$$\geq \frac{\partial(\hat{a}) + \partial(\hat{b}) + (\partial_x(a_i) + \partial_x(b_j))^2 + 1}{(\partial_x(a_i) + \partial_x(b_j))^2 + 1} \geq$$

$$\geq 1 > \frac{1}{2}.$$

Thus $a_i y^N b_j$ has no subword e with $f(e) < \dfrac{1}{2}$, so $a_i y^N b_j \notin I$. This being so for all i,j, we conclude, as F is additively free on S, that $a y^N b \notin I$. We therefore have a contradiction, so F/I is indeed a prime ring.

Now let J be the ideal of F generated by x. If $w \in S$ and $f(w) < \dfrac{1}{2}$, then $\partial_x(w) \neq 0$. Thus $I \subseteq J$. However $f(x) = \dfrac{1}{2}$, so $I \neq J$. Thus J/I, as an ideal of the prime ring F/I, is a prime ring by Proposition 3.8.2.

We shall now prove that J/I is locally nilpotent. Let $W = \{w_1, \ldots, w_n\}$ be a finite set of words in J. Let $M = 2\max\{\partial(w_1), \ldots, \partial(w_n)\}$, and consider a product $w_{i_1} w_{i_2} \cdots w_{i_M}$, where $\{w_{i_1}, w_{i_2}, \ldots, w_{i_M}\} \subseteq W$ and some

w_{i_j} may coincide. Since each w_{i_j} is in J and so $\partial_x(w_{i_j}) \geq 1$, we have

$$f(w_{i_1} \cdots w_{i_M}) = \frac{\sum_{j=1}^{M} \partial(w_{i_j})}{(\sum_{j=1}^{M} \partial_x(w_{i_j}))^2 + 1} \leq \frac{M \max\{\partial(w_1), \ldots, \partial(w_n)\}}{M^2 + 1} <$$

$$< \frac{1}{M} \max\{\partial(w_1), \ldots, \partial(w_n)\} = \frac{1}{2}.$$

Thus $w_{i_1} \cdots w_{i_M} \in I$. Now if we take any finite subset E of J, the argument just given shows that there is an integer k such that I contains every product of k monomials occuring in elements of E. This means that E generates a subring of I which is nilpotent modulo I.

Thus we conclude that the ring $A = J/I$ is locally nilpotent as well as prime. □

The existence of rings which are nil but not locally nilpotent, was proved by Golod and Shafarevich [1], Golod [1]. Our demonstration of existence of such rings will be based largely on the account by Fischer and Struik [1] of aspects of the work in the above cited papers of Golod and Shafarevich.

Let K be a finite field. We denote by $K\langle x, y\rangle$ the polynomial algebra over K in non-commuting indeterminates x and y. Since $K\langle x, y\rangle$ is countable, we can write the set R of polynomials with zero constant term as $\{u_1, u_2, \ldots\}$. We shall continue to use the symbol ∂ to denote the total degree of a polynomial. Let

$$u_1^{10} = s_{11} + s_{12} + \cdots + s_{1k_1}$$

where the s_{1j} are homogeneous polynomials and

$$10 \leq \partial(s_{11}) < \partial(s_{12}) < \cdots < \partial(s_{1k_1}).$$

Putting $N_1 = \partial(s_{1k_1})$, we have

$$u_2^{N_1+1} = s_{21} + s_{22} + \cdots + s_{2k_2},$$

where the s_{2j} are homogeneous and

$$N_1 + 1 \leq \partial(s_{21}) < \partial(s_{22}) < \cdots < \partial(s_{2k_2}).$$

Since $\partial(s_{1k_1}) = N_1 < N_1 + 1 \leq \partial(s_{21})$, all the s_{ij} introduced so far have different degrees. This process can be continued: let $N_2 = \partial(s_{2k_2})$ and let

$$u_3^{N_2+1} = s_{31} + s_{32} + \cdots + s_{3k_3},$$

where s_{3j} are homogeneous and

$$N_2 + 1 \leq \partial(s_{31}) < \partial(s_{32}) < \cdots < \partial(s_{3k_3}),$$

and so on. We get a countable set $\{s_{11}, \ldots, s_{ij}, \ldots\}$ of polynomials such that

(i) $\partial(s_{ij}) \geq 10$ for all i, j,

and

(ii) $\partial(s_{ij}) = \partial(s_{kl})$ implies $i = k$ and $j = l$.

Let I denote the ideal of the algebra R generated by all the s_{ij}. Then $u_1^{10} \in I$, $u_2^{N+1} \in I$, $u_3^{N_2+1} \in I$, and so on, so R/I is nil. We wish to have R/I not locally nilpotent. Although it is by no means obvious at this stage, starting with u_1^{10}, rather then some lower powers, will ensure this. (Of course, R/I is still nil if any power of u_1 is used.)

Now R/I is generated by the finite set $\{x + I, y + I\}$, so, if locally nilpotent, it must be nilpotent. Suppose it is nilpotent: let $(R/I)^m = 0$. Then as a K-vector space, R/I is generated by the words in $x + I$ and $y + I$ of length $< m$. The number of such words is $2 + 2^2 + \cdots + 2^{m-1}$ or less. In any case, it is finite. This leads to the following

CONCLUSION 4.2.2. *If R/I is locally nilpotent, then it is, as a K-vector space, finite dimensional.*

Thus we need to show that R/I has infinite dimension. This is equivalent to showing that $K\langle x, y\rangle/I$ is infinite dimensional.

Let $K\langle x, y\rangle = T_0 \oplus T_1 \oplus \ldots$ (vector space decomposition) where T_n is the set of polynomials which are homogeneous of (total) degree n. (It is clear that every polynomial is uniquely a sum of homogeneous ones.) We re-name the s_{ij} as $f_1, f_2, \ldots$. Then the ideal I consists of all sums of polynomials uf_iv with $u, v \in K\langle x, y\rangle$. Since the u and v can be written as sums of homogeneous polynomials, we lose no generality by assuming they *are* homogeneous.

Let $w = w_1 + w_2 + \cdots + w_r$, where w_j are homogeneous of different degrees. If $w \in I$, and if some uf_iv occurring in the representation of w makes a contribution to w_1, then as f_i is homogeneous and thus uf_iv is too, all of uf_iv must go to w_1. Since on the other hand, every monomial in w_1 must come from some uf_iv (as w is assumed to be in I) it follows that w_1 is in I and, of course, similarly $w_2, \ldots, w_r \in I$. Let $A_n = (T_n + I)/I$. Since clearly $K\langle x, y\rangle/I$ is the vector space sum of $A_0, A_1, \ldots$, the above arguments shows that

$$K\langle x, y\rangle/I = A_0 \oplus A_1 \oplus \ldots \text{ (vector space decomposition)}.$$

LEMMA 4.2.3. *Let $b_n = \dim(A_n)$, $n = 0, 1, 2, \ldots$. Then $b_n > 2b_{n-1} - \sum_{i=10}^{n} b_{n-i}$ for $n = 2, 3, 4, \ldots$.* (Here the sum is to be interpreted as 0 for $n < 10$.)

Proof: Let $I_n = T_n \cap I$, and let $T_n = I_n \oplus S_n$ (vector space decomposition). Then

$$\dim(S_n) = \dim(T_n/I_n) =$$
$$= \dim(T_n/(T_n \cap I)) = \dim((T_m + I)/I) = \dim(A_n) = b_n.$$

Since I is generated by polynomials of degree ≥ 10, we have $I_2 = 0$ and $I_1 = 0$. Now

$$b_2 = \dim(T_2) = 4 = 2\dim(T_1) = 2b_1 = 2b_1 + \sum_{i=10}^{1} b_{i-1}.$$

Thus our inequality is valid for $n = 2$. Assume, for induction, that

$$b_{n-1} \geq 2b_{n-2} - \sum_{i=10}^{n-1} b_{n-1-i}.$$

Let $\{s_1, s_2, \ldots, s_{b_{n-1}}\}$ be a basis for S_{n-1}, $\{g_1, \ldots, g_m\}$ a basis for I_{n-1} so that

$$b_{n-1} + m = \dim(T_{n-1}) = 2^{n-1}.$$

Then the set

$$\{s_1 x, s_1 y, \ldots, s_{b_{n-1}} x, s_{b_{n-1}} y, g_1 x, g_1 y, \ldots, g_m x, g_m y\}$$

is a basis for T_n: it contains 2^n elements, and the vanishing of a linear combination of its elements amounts to an equality $c_1 x = c_2 y$, where c_1 and c_2 are linear combinations of things from the basis $\{s_1, \ldots, s_{b_{n-1}}, g_1, \ldots, g_m\}$ of T_{n-1}. In such an equality we must have $c_1 = c_2 = 0$ and then all coefficients of c_1 and c_2 are zero. The same then goes for the coefficients of the original linear combination.

Let V be a basis, consisting of homogeneous polynomials, of $S_0 + \cdots + S_{n-1}$. As $S_0 \subseteq T_0$, $S_1 \subseteq T_1$, $\ldots$, there do exist such bases. Let L be the space spanned by

(*) $$\{t f_i \mid t \in V; \quad i = 1, 2, \ldots\} \cap T_n,$$

and let J be spanned by

$$\{g_1 x, g_1 y, \ldots, g_m x, g_m y\}.$$

This set is a basis for J, as the $g_i x$, $g_i y$ form part of a basis for T_n. Then $L \subseteq I_n$, as the f_i are in I, and $J \subseteq I_n$, as $g_1, \ldots, g_m \in I_{n-1}$. We shall show that, in fact, $I_n = L + J$.

Let u, v be monomials, $u f_i v \in I_n$. We consider two cases.

Case I: $\partial(v) > 0$. Here we can assume without loss of generality that $v = v' x$ for some monomial v'. Then $u f_i v = (u f_i v') x$, where $\partial(u f_i v') = n - 1$ and so $u f_i v' \in I_{n-1}$. But then $u f_i v \in I_{n-1} x \subseteq J$.

Case II: $\partial(v) = 0$. Here we can assume $u f_i v = u f_i$. Let $\partial(u) = k$. Then $u \in T_k$, so $u = v' + w'$, where $v' \in S_k$ and $w' \in I_k$. We have

$uf_iv = uf_i = v'f_i + w'f_i$. Let $f_i = f'x + f''y$. Then $w'f_i = w'f'x + w'f''y$, where $w'f', w'f'' \in I$ and $\partial(w'f') = n - 1 = \partial(w'f'')$. Thus $w'f_i \in I_{n-1}x + I_{n-1}y \subseteq J$. Also $v'f_i \in S_k f_i \cap T_n \subseteq L$, so $uf_iv = uf_i = v'f_i + w'f_i \in L + J$.

We have shown that every uf_iv, where u and v are monomials, is in $L + J$ as asserted. We therefore have

$(**)$ $\qquad \dim(I_n) \leq \dim(L) + \dim(J).$

For every tf_i in the specified spanning set $(*)$ for L, we have

$$n = \partial(tf_i) = \partial(t) + \partial(f_i) \geq \partial(t) + 10.$$

Since different f_i have different degrees, we have

$$L \subseteq S_{n_1}f_{i_1} + \cdots + S_{n_c}f_{i_c}$$

for some $\{n_1, \ldots, n_c\} \subseteq \{n - 10, n - 11, \ldots, 0\}$ and $i_1, \ldots, i_c$ for which

$$n_1 + \partial(f_{i_1}) = n_2 + \partial(f_{i_2}) = \cdots = n_c + \partial(f_{i_c}) = n.$$

Thus
$(***)$
$\dim(L) \leq \dim(S_{n_1}f_{i_1} + \cdots + S_{n_c}f_{i_c}) \leq$
$\leq \dim(S_{n_1}f_{i_1}) + \cdots + \dim(S_{n_c}f_{i_c}) = \dim(S_{n_1}) + \cdots + \dim(S_{n_c}) \leq$
$\leq \dim(S_{n-10}) + \dim(S_{n-11}) + \cdots + \dim(S_1) + \dim(S_0) =$
$= \dim(A_{n-10}) + \dim(A_{n-11}) + \cdots + \dim(A_1) + \dim(A_0) =$
$= b_{n-10} + b_{n-11} + \cdots + b_1 + b_0.$

Since
$$\{g_1x, g_1y, \ldots, g_mx, g_my\}$$
is a basis for J and
$$\{s_1x, s_1y, \ldots, s_{b_{n-1}}x, s_{b_{n-1}}y, g_1x, g_1y, \ldots, g_mx, g_my\}$$
is a basis for T_n, we have
$$\dim(T_n) = 2b_{n-1} + \dim(J),$$
whence from $(**)$ and $(***)$ we get

$\dim(I_n) \leq \dim(L) + \dim(J) \leq b_{n-10} + b_{n-11} + \cdots + b_1 + b_0 + (\dim(T_n) - 2b_{n-1}),$

and, finally

$b_n = \dim(A_n) = \dim(T_n/I_n) = \dim(T_n) - \dim(I_n) \geq 2b_{n-1} - \sum_{i=10}^{n} b_{n-i}.$ $\qquad \square$

Remember: what we want to show is that $\dim(K\langle x, y \rangle / I)$ is infinite, because then R/I is infinite dimensional and hence not locally nilpotent.

This amounts to showing, in the notation just introduced, that there are infinitely many nonzero b_i. We establish this by proving the following equivalent statement:

$\sum_{i=1}^{\infty} b_i z^i$ is not a polynomial in the ring $\mathbb{R}[[z]]$ of real formal power series.

Let

$$\left(\sum_{i=0}^{\infty} b_i z^i\right)\left(1 - 2z + \sum_{i=10}^{\infty} z^i\right) = D(z) = \sum_{i=0}^{\infty} d_i z^i.$$

Now $d_0 = b_0$, $d_1 = b_1 - 2b_0$, $d_2 = b_2 - 2b_1$, ..., $d_9 = b_9 - 2b_8$, and for $n \geq 10$

$$d_n = b_n - 2b_{n-1} + b_{n-10} + b_{n-11} + \cdots + b_0.$$

By Lemma 4.2.3 all of these coefficients are non-negative. Let

$$C(z) = \left(1 - 2z + \sum_{i=10}^{\infty} z^i\right)^{-1} = \sum_{i=0}^{\infty} c_i z^i.$$

Then $\sum_{i=0}^{\infty} b_i z^i = D(z)C(z)$, so $\sum_{i=0}^{\infty} b_i z^i$ is a non-polynomial provided that all coefficients of $C(z)$ are strictly positive. This we shall now prove, but the initial step in the proof is the demonstration of strict positivity of coefficients of a related series. Let

$$r_i = \frac{1}{16}\left(\frac{3}{2}\right)^{i-2} = \frac{3^{i-2}}{2^{i+2}}, \quad i = 2, 3, 4, \ldots.$$

We shall examine $(1 - 2z + \sum_{i=2}^{\infty} r_i z^i)^{-1}$. The point is that this latter series is easier to calculate with (the r_i being powers of a fixed element) and each r_i is $\geq$ the corresponding coefficient in the series we want to find out about. It turns out that the positivity of the coefficients in $(1 - 2z + \sum_{i=10}^{\infty} z^i)^{-1}$ therefore can be deduced from the positivity of the coefficients of $(1 - 2z + \sum_{i=2}^{\infty} r_i z^i)^{-1}$, which we now establish.

We have

$$1 - 2z + \sum_{i=2}^{\infty} r_i z^i =$$

$$= 1 - 2z + \frac{1}{16}z^2 \sum_{i=0}^{\infty} \left(\frac{3}{2}\right)^i z^i = 1 - 2z + \frac{1}{16}z^2 \left(1 - \frac{3}{2}z\right)^{-1} =$$

Concrete Radicals and Structure Theorems 195

$$= \frac{1 - \frac{3}{2}z - 2z + 3z^2 + \frac{1}{16}z^2}{1 - \frac{3}{2}z} = \frac{1 - \frac{7}{2}z + \frac{49}{16}z^2}{1 - \frac{3}{2}z} = \frac{\left(1 - \frac{7}{4}z\right)^2}{1 - \frac{3}{2}z}.$$

Thus

$$\left(1 - 2z + \sum_{i=2}^{\infty} r_i z^i\right)^{-1} = \frac{1 - \frac{3}{2}z}{\left(1 - \frac{7}{4}z\right)^2}.$$

But $\left(1 - \frac{7}{4}z\right)^{-2} = \sum_{i=1}^{\infty} i \left(\frac{7}{4}z\right)^{i-1}$, so

$$\left(1 - 2z + \sum_{i=2}^{\infty} r_i z^i\right)^{-1} =$$

$$= \left(1 - \frac{3}{2}z\right) \sum_{i=1}^{\infty} i \left(\frac{7}{4}z\right)^{i-1} =$$

$$= \left(1 - \frac{3}{2}z\right) \left(1 + \sum_{i=1}^{\infty} (i+1) \left(\frac{7}{4}z\right)^i\right) =$$

$$= 1 + \sum_{i=1}^{\infty} (i+1) \left(\frac{7}{4}z\right)^i - \sum_{i=1}^{\infty} (i+1) \frac{3}{2} \left(\frac{7}{4}z\right)^i z - \frac{3}{2}z =$$

$$= 1 + 2 \cdot \frac{7}{4}z + \sum_{i=2}^{\infty} (i+1) \left(\frac{7}{4}z\right)^i - \sum_{i=1}^{\infty} (i+1) \frac{3}{2} \left(\frac{7}{4}z\right)^i z - \frac{3}{2}z =$$

$$= 1 + 2 \cdot z + \sum_{i=1}^{\infty} \left((i+2) \left(\frac{7}{4}z\right)^{i+1} - (i+1) \frac{3}{2} \left(\frac{7}{4}z\right)^i z \right) =$$

$$= 1 + 2 \cdot z + \sum_{i=1}^{\infty} \left(i + 2 - (i+1) \frac{3}{2} \cdot \frac{4}{7} \right) \left(\frac{7}{4}z\right)^{i+1} =$$

$$= 1 + 2 \cdot z + \sum_{i=1}^{\infty} \left(i + 2 - \frac{6}{7}(i+1) \right) \left(\frac{7}{4}z\right)^{i+1}.$$

Let $H = 1 - 2z + \sum_{i=10}^{\infty} z^i$, $G = 1 - 2z + \sum_{i=2}^{\infty} r_i z^i$,

$$U = r_2 z^2 + \cdots + r_9 z^9 + \sum_{i=10}^{\infty} (r_i - 1) z^i.$$

Then $U + H = G$. As shown above, G^{-1} has strictly positive coefficients, and we have to prove that H^{-1} does also. Now

$$H^{-1} = (G - U)^{-1} = (G(1 - UG^{-1}))^{-1} = G^{-1}(1 - UG^{-1})^{-1}.$$

For $i \geq 10$ we have

$$r_i = \frac{3^{i-2}}{2^{i+2}} \geq \frac{3^8}{2^{12}} = \frac{6561}{4096} > 1,$$

so U has positive coefficients, and thus UG^{-1} does also. Let $UG^{-1} = \sum_{i=2}^{\infty} h_i z^i$ and

$$(1 - UG^{-1})^{-1} = \sum_{i=0}^{\infty} e_i z^i.$$

Then

$$\left(1 - \sum_{i=2}^{\infty} h_i z^i\right) \sum_{i=0}^{\infty} e_i z^i = 1,$$

so $e_0 = 1 > 0$. Assume for induction that $e_0, e_1, \ldots, e_{n-1} > 0$. Then

$$0 = 1 \cdot e_n - \sum_{i=2}^{\infty} h_i e_{n-i},$$

so $e_n = \sum_{i=2}^{n} h_i e_{n-i} > 0$.

Summarizing, then, G^{-1} and $(1 - UG^{-1})^{-1}$ have strictly positive coefficients and so therefore does $H^{-1} = C$, whence $\sum_{i=0}^{\infty} b_i z^i$ is not a polynomial and $\dim(K\langle x, y\rangle/I)$ is infinite. Thus R/I is infinite dimensional and so is not locally nilpotent, even though it is nil.

THEOREM 4.2.4. *The finitely generated ring $A = R/I$ is in $\mathcal{N}$ but is not in $\mathcal{L}$, whence $\mathcal{L} \neq \mathcal{N}$.* □

4.3. Coincidence of the nil radicals

In this section we shall prove the coincidence of the principal nil radicals β, $\mathcal{L}$ and $\mathcal{N}$ on noetherian rings and on commutative rings.

A ring A satisfies the *ascending chain condition* (briefly a.c.c.) *on left ideals*, if every strictly ascending chain of left ideals terminates in a finite number of steps. Rings with a.c.c. on left ideals are called also *(left) noetherian rings*. The result that the three radicals coincide on noetherian rings is due to Levitzki [2]; we follow the proof of Utumi [1].

LEMMA 4.3.1. *Let A be a noetherian ring. If A has a nonzero nil one-sided ideal, it has a nonzero nilpotent ideal.*

Proof: First, suppose A has a nil right ideal $R \neq 0$. Let
$$(0:c) = \{x \in A \mid xc = 0\}$$
be the left annihilator of the element $c \in A$. Then $(0:c)$ is clearly a left ideal of A. Since A is noetherian, the set
$$\{(0:r) \mid 0 \neq r \in R\}$$
has a maximal element, $(0:a)$. If $aA = 0$, then the ideal generated by a is nilpotent. In the contrary case, let $r \in R$ be such that $ar \neq 0$. Then $ar \in R$, so for some integer $n > 1$ we have $(ar)^n = 0 \neq (ar)^{n-1}$. Since $(ar)^{n-1} \in R$ and $(0:a) \subseteq (0:(ar)^{n-1})$, the maximality of $(0:a)$ requires that $(0:(ar)^{n-1}) = (0:a)$. Thus
$$ar \in (0:(ar)^{n-1}) = (0:a),$$
so $ara = 0$. This is so whatever $ar \neq 0$. Since $asa = 0$ for $as = 0$, we see that $aAa = 0$. Let
$$(a) = \mathbb{Z}a + aA + Aa + aAa$$
be the ideal of A generated by a. Then one readily sees that $(a)^3 = 0$.

If A has a nil left ideal $L \neq 0$, then Ab is nil for every $b \in L$. Therefore, as
$$(bu)^{k+1} = b(ub)^k u \quad \text{for all integers } k > 0 \text{ and } u \in A,$$
the right ideal bA is nil. If $bA = 0$, then the ideal (b) is nilpotent. If not, then bA is a nonzero nil right ideal, and therefore, as already proved, A has a nonzero nilpotent ideal. □

THEOREM 4.3.2 (Levitzki [2]). *If A is a noetherian ring, then $\mathcal{N}(A)$ is nilpotent.*

Proof: As A is noetherian, the set of nilpotent ideals of A has a maximal element N. Since the sum of nilpotent ideals is nilpotent, N contains every nilpotent ideal. If $\mathcal{N}(A/N) \neq 0$, then, as also A/N is noetherian, by Lemma 4.3.1 A/N has a nonzero nilpotent ideal I/N, where I is a suitable ideal of A. But then $I^k \subseteq N$ for some integer $k > 1$, and so I is a nilpotent ideal of A. Hence $I = N$, and $I/N = 0$, a contradiction. Thus we conclude that $\mathcal{N}(A) \subseteq N$. The inclusion $N \subseteq \mathcal{N}(A)$ is clear. □

Recall from §4.1 that $\mathcal{V}_1(A) = \sum (I \triangleleft A \mid I \text{ is nilpotent})$. It was Brauer [1] who first proved the nilpotency of $\mathcal{V}_1(A)$ under certain chain condition.

COROLLARY 4.3.3. *If A is a noetherian ring, then*
$$\mathcal{N}(A) = \mathcal{L}(A) = \beta(A) = \mathcal{V}_1(A).$$
□

There is an analogous result for rings with descending chain condition (d.c.c.) on left ideals, but in this case the Jacobson radical is also nilpotent.

This will be proved in Lemma 4.5.8. In Section 4.10 we shall discuss the coincidence of radicals with the Jacobson radical on the class $\mathcal{A}$ of all rings with d.c.c. on left ideals. Exact lower and upper bounds of radicals for the coincidence with the Jacobson radical on $\mathcal{A}$ will be given in Theorem 4.10.8.

The principal nil radicals β, $\mathcal{L}$ and $\mathcal{N}$ coincide also on commutative rings. We shall prove, in fact, a stronger statement.

THEOREM 4.3.4. *Let A be a commutative ring, N the set of all nilpotent elements of A. Then*

$$\mathcal{N}_g(A) = \mathcal{N}(A) = \mathcal{L}(A) = \beta(A) = \mathcal{V}_1(A) = N.$$

Proof: If $a^n = 0 = b^m$, then $(a \pm b)^{n+m} = 0$ and $(ra)^n = 0$ for all $r \in A$. Hence $N \triangleleft A$. Since N is clearly nil, we have $\mathcal{N}(A) \subseteq N$. But clearly if $a^n = 0$ then $(\mathbb{Z}a + Aa)^n = 0$, so

$$N = \sum(\mathbb{Z}a + Aa \mid a \text{ is nilpotent}) \subseteq \mathcal{V}_1(A).$$

Next, we shall prove that $\beta(A) = \mathcal{N}_g(A)$ for every commutative ring A. The containment $\beta(A) = \mathcal{N}(A) \subseteq \mathcal{N}_g(A)$ is obvious. Assume that there exists a commutative ring A such that $\beta(A) \neq \mathcal{N}_g(A)$. Then $\mathcal{N}_g(A)/\beta(A)$ is a nonzero semiprime ring which has a nonzero homomorphic image B such that B is a prime ring in $\mathcal{N}_g$. Let $ab = 0$ for some elements $a, b \in B$. Since B is commutative, the product of the principal ideals (a) and (b) is 0. But B is a prime ring, so $a = 0$ or $b = 0$ follows. This means that B has no zero divisors, contradicting $0 \neq B \in \mathcal{N}_g$. □

The radicals β, $\mathcal{L}$ and $\mathcal{N}$ coincide not only for commutative rings which are rings satisfying the identity $xy - yx = 0$, but also for PI-rings.

We shall call a ring A a *PI-ring* if A satisfies a *proper* polynomial identity, that is, an identity of the form $\sum n_i f_i = 0$ where the f_i are monomials of different types and there is no nonzero element $a \in A$ for which $n_i a = 0$ for all i. Equivalently, A satisfies a homogeneous multilinear polynomial identity with all coefficients ± 1 (see e.g. Procesi [1] pp 43–44).

PROPOSITION 4.3.5 (Amitsur [1]). *Let A be a PI-ring. Then there exists an integer ℓ such that $B^\ell \subseteq \mathcal{V}_1(A)$ for every nilpotent subring B of A.*

Proof: Let A satisfy a homogeneous multilinear identity of degree d with all coefficients ± 1. We shall prove that $B^{[\frac{d}{2}]} \subseteq \mathcal{V}_1(A)$, where $[\frac{d}{2}]$ is the greatest integer $\leq \frac{d}{2}$.

Now $B^m = 0$ for some integer m, so $AB^m A = 0$. Let n be the smallest integer such that $AB^n A$ is nilpotent. For $i = 1, 2, \ldots, n$, let

$$R_{2i-1} = B^{n-i} A B^{i-1}, \qquad R_{2i} = B^{n-i} A B^i$$

(where $AB^0 = A$, etc.). If $j > k$, then

$$R_{2j}R_{2k} = B^{n-j}AB^jB^{n-k}AB^k \subseteq B^{n-j}AB^kB^{n-k}AB^k \subseteq AB^nA,$$
$$R_{2j}R_{2k-1} = B^{n-j}AB^jB^{n-k}AB^{k-1} \subseteq B^{n-j}AB^kB^{n-k}AB^{k-1} \subseteq AB^nA,$$
$$R_{2j-1}R_{2k} = B^{n-j}AB^{j-1}B^{n-k}AB^k \subseteq B^{n-j}AB^kB^{n-k}AB^k \subseteq AB^nA.$$

It follows that

(∗) $\qquad\qquad R_r R_s \subseteq AB^n A \quad \text{whenever } r > s.$

Also, if $h < 2n$, then

$$R_1R_2R_3\ldots R_h = B^{n-1}AB^{n-1}ABB^{n-2}ABB^{n-2}AB^2\cdots =$$
$$= \begin{cases} B^{n-1}AB^{n-1}AB^{n-1}A\ldots AB^i & \text{if } h = 2i \text{ is even} \\ B^{n-1}AB^{n-1}AB^{n-1}A\ldots AB^{i-1} & \text{if } h = 2i-1 \text{ is odd} \end{cases} =$$
$$= (B^{n-1}A)^h B^{[\frac{h}{2}]}.$$

Suppose $n > [\frac{d}{2}]$. Then $2n > d$, so taking $h = d$ above, we get

(∗∗) $\qquad\qquad R_1 R_2 \ldots R_d = (B^{n-1}A)^d B^{[\frac{d}{2}]}.$

Let $x_1 x_2 \ldots x_d + g(x_1, x_2, \ldots, x_d) = 0$ be a homogeneous multilinear identity for A (it is understood that g is a sum of monomials of the form $\pm x_{\sigma(1)} x_{\sigma(2)} \ldots x_{\sigma(d)}$ where the σ are non-trivial permutations of $\{1, 2, \ldots, d\}$). From (∗), we see that

$$g(R_1, R_2, \ldots, R_d) \subseteq AB^n A,$$

whence $R_1 R_2 \ldots R_d \subseteq AB^n A$. If then follows from (∗∗) that

$$(B^{n-1}A)^d B^{[\frac{d}{2}]} \subseteq AB^n A.$$

Now we are assuming $n > [\frac{d}{2}]$, so $[\frac{d}{2}] \leq n-1$ and thus

$$(B^{n-1}A)^d B^{n-1} \subseteq AB^n A.$$

Therefore

$$(AB^{n-1}A)^{d+1} \subseteq A(B^{n-1}A)^d B^{n-1}A \subseteq A(AB^n)A^2 \subseteq AB^n A.$$

Since $AB^n A$ is nilpotent, there exists an integer s such that $((AB^{n-1}A)^{d+1})^s = 0$. This means that $AB^{n-1}A$ is nilpotent, contrary to the assumed minimality of n. We conclude, therefore, that $n \leq [\frac{d}{2}]$, so that $AB^{[\frac{d}{2}]}A \subseteq AB^n A$ and thus $AB^{[\frac{d}{2}]}A$ is nilpotent. Now the ideal I generated by $B^{[\frac{d}{2}]}$ is

$$B^{[\frac{d}{2}]} + AB^{[\frac{d}{2}]} + B^{[\frac{d}{2}]}A + AB^{[\frac{d}{2}]}A,$$

so $I^3 \subseteq AB^{[\frac{d}{2}]}A$ and therefore I is nilpotent. This means that $B^{[\frac{d}{2}]} \subseteq I \subseteq \mathcal{V}(1)$. □

THEOREM 4.3.6 (Levitzki [3], Kaplansky [1]). *If A is a PI-ring, then $\mathcal{L}(A) = \beta(A)$.*

Proof: Let l be as in Proposition 4.3.5. Then $S^l \subseteq \mathcal{V}_1(A)$ for every finitely generated subring of $\mathcal{L}(A)$. If $a_1, a_2, \ldots, a_l \in \mathcal{L}(A)$ and S is the subring they generate, then $a_1 a_2 \ldots a_l \in S^l \subseteq \mathcal{V}_1(A)$. Thus $\mathcal{L}(A)^l \subseteq \mathcal{V}_1(A)$, so $\mathcal{L}(A) \subseteq \mathcal{V}_2(A) \subseteq \beta(A)$. □

COROLLARY 4.3.7. *For every PI-ring A we have $\beta(A) = \mathcal{V}_2(A)$.* □

This is the best that can be done: there are PI-rings A for which $\beta(A) \neq \mathcal{V}_1(A)$.

Kaplansky [1] proved that every finitely generated nil PI-algebra over a field is nilpotent. This result has been generalized recently by D. M. Riley [1].

EXAMPLE 4.3.8 (Amitsur [1]). Let R be a commutative ring with unity element such that $\beta(R)$ is not nilpotent. (For example, R could be Dorroh extension of the Zassenhaus ring, Example 3.4.14). Let

$$A = \begin{pmatrix} \beta(R) & R \\ \beta(R) & \beta(R) \end{pmatrix}, \qquad B = \begin{pmatrix} \beta(R) & \beta(R) \\ \beta(R) & \beta(R) \end{pmatrix}.$$

Then as $\beta(R)$ is locally nilpotent, so is B, since every finite set of matrices over $\beta(R)$ can be viewed as matrices over the (nilpotent) ring generated by their entries. But B is also a PI-ring — by the Amitsur–Levitzki Theorem 1.1.8 B satisfies the standard identity of degree 4 — so $B \in \beta$ by Theorem 4.3.6. Now define a function f from A to the zero-ring $(R/\beta(R))^0$ on $R/\beta(R)$ by setting

$$f\left(\begin{pmatrix} a & b \\ c & d \end{pmatrix}\right) = b + \beta(R).$$

Clearly f is surjective and respects addition. Since also

$$f\left(\begin{pmatrix} a & b \\ c & d \end{pmatrix} \begin{pmatrix} a' & b' \\ c' & d' \end{pmatrix}\right) =$$
$$= ab' + bd' + \beta(R) = 0 = f\left(\begin{pmatrix} a & b \\ c & d \end{pmatrix}\right) f\left(\begin{pmatrix} a' & b' \\ c' & d' \end{pmatrix}\right),$$

f is a ring homomorphism with kernel B. Since B and $(R/\beta(R))^0 \cong A/B$ are in β, so is A. As with B, also A is a PI-ring. But the ideal I of A generated by $\begin{pmatrix} 0 & 1 \\ 0 & 0 \end{pmatrix}$ contains all elements $\begin{pmatrix} 0 & 0 \\ a & 0 \end{pmatrix} \begin{pmatrix} 0 & 1 \\ 0 & 0 \end{pmatrix} = \begin{pmatrix} 0 & 0 \\ 0 & a \end{pmatrix}$, with $a \in \beta(R)$ and thus an isomorphic copy of $\beta(R)$. Thus I is not nilpotent. But if I where contained in $\mathcal{V}_1(A)$ there would be nilpotent ideals J_λ such that $\begin{pmatrix} 0 & 1 \\ 0 & 0 \end{pmatrix} \in I \subseteq \sum J_\lambda$ and therefore there would be finitely many nilpotent

Concrete Radicals and Structure Theorems

ideals $J_{\lambda_1}, \ldots, J_{\lambda_k}$ with $\begin{pmatrix} 0 & 1 \\ 0 & 0 \end{pmatrix} \in J_{\lambda_1} + \cdots + J_{\lambda_k}$ and thus $I \subseteq J_{\lambda_1} + \cdots + J_{\lambda_k}$. This would make I nilpotent. *We conclude, therefore that $\beta(A) \neq \mathcal{V}_1(A)$.*

We turn now to an examination of the nil radical of a PI-ring.

THEOREM 4.3.9 (Levitzki [3], Kaplansky [1]). *Let A be a PI-ring. Then $\mathcal{N}(A) = \mathcal{L}(A)$.*

Proof: Since $\mathcal{L}(A) = \mathcal{L}(\mathcal{N}(A))$ and $\mathcal{N}(A)$ is a PI-ring, we lose no generality by considering the case $A = \mathcal{N}(A)$. Furthermore $A/\mathcal{L}(A)$ is a nil PI-ring and (as $\mathcal{L}(A) \subseteq \mathcal{N}(A)$), $\mathcal{N}(A/\mathcal{L}(A)) = \mathcal{N}(A)/\mathcal{L}(A)$. Thus we may further restrict attention to the case when $\mathcal{L}(A) = 0$. The assertion of our Theorem is then equivalent to the statement that if A is a nil PI-ring with $\mathcal{L}(A) = 0$, then $A = 0$, that is, *if A is a nonzero nil PI-ring, then $\mathcal{L}(A) \neq 0$.* This we shall now prove.

Let $a \in A$ with $a^2 = 0 \neq a$. If $Aa = 0$, then a is in the right annihilator of A which is a nilpotent ideal. We may therefore suppose that $Aa \neq 0$. Let

$$(\Diamond) \qquad x_1 f(x_2, \ldots, x_n) + g(x_1, \ldots, x_n) = 0$$

be a homogeneous multilinear identity for A with all coefficients ± 1 and where $g(x_1, \ldots, x_n)$ is the term of monomials with x_1 not occurring first. (Note that by a suitable labelling of the variables we can arrange things so that f is not zero.)

For all $r_2, \ldots, r_n \in A$ we have $g(a, r_2 a, \ldots, r_n a) = 0$, as the monomials of g don't start with x_1 and thus $g(a, r_2 a, \ldots, r_n a)$ is a sum of terms of the form $\ldots r_i a a \cdots = 0$, (as $a^2 = 0$). But also, by $(\Diamond)$

$$af(r_2 a, \ldots, r_n a) + g(a, r_2 a, \ldots, r_n a) = 0.$$

Hence $af(r_2 a, \ldots, r_n a) = 0$. This means that in the ring Aa, $f(r_2 a, \ldots, r_n a)$ is a part of the right annihilator. Calling the latter I, we can conclude that

$$f(x_2, \ldots, x_n) = 0 \quad \text{is an identity of degree } n-1 \text{ for the nil } PI\text{-ring } Aa/I.$$

Now either $Aa/I = 0$ in which case Aa is nilpotent, or $Aa/I \neq 0$ and by induction on the degrees of identities we can assume Aa/I has a nonzero locally nilpotent ideal L/I. In this latter case, since I is nilpotent, L is a locally nilpotent ideal of A. □

Recall that we have confined our attention to rings A satisfying a *proper* polynomial identity. It is perhaps natural to ask what happens in the presence of an identity which is merely non-trivial, that is, one such that the coefficients don't annihilate the whole ring.

EXAMPLE 4.3.10. Let R be a ring of characteristic 2 which is nil but not locally nilpotent, and let A be the Dorroh extension of R. Then

$\mathcal{N}(A/R) \cong \mathcal{N}(\mathbb{Z}) = 0$, so $R = \mathcal{N}(A)$, but for $(r,n),(q,m) \in A$ we have

$$2(r,n)(q,m) - 2(q,m)(r,n) = 2(rq+nq+mr, nm) - 2(qr+mr+nq, mn) =$$
$$= (0, 2nm) - (0, 2mn) = 0,$$

so A satisfies the identity $2xy - 2yx$ while $2A \neq 0$. Thus $\mathcal{N}(A) \neq \mathcal{L}(A)$ is possible when A satisfies a non-trivial identity. Similarly, the Levitzki and Baer radical can differ.

In the theory of PI-rings and also of group rings one of the major issues is to decide whether the Jacobson radical of a PI-ring or a group ring is nil or even nilpotent.

4.4. The Jacobson radical

The Jacobson radical has already been introduced in Example 2.1.7 via the circle operation $a \circ b = a+b-ab$, and quite a few important properties have been derived from the general theory by using only its definition. We have seen also a characterization of the Jacobson radical in terms of modules (cf. Example 3.14.12). Before compiling these results, we add a bit more to the elementary theory. An element $a \in A$ of a ring A is said to be *left quasi-regular*, if there exists an element $x \in A$ such that $x \circ a = 0$. In this case the element x is called a *left quasi-inverse* of the element a. A *right quasi-regular element* and a *right quasi-inverse* are defined correspondingly. A ring A is called *quasi-regular*, if each of its elements is left quasi-regular. Obviously, A is quasi-regular if and only if $(A, \circ)$ is a group.

PROPOSITION 4.4.1. *$(A, \circ)$ is a monoid for every ring A and 0 is the unity element of the circle operation $\circ$, that is,*

$$(a \circ b) \circ c = a \circ (b \circ c) \quad \forall a,b,c \in A,$$
$$0 \circ a = a \circ 0 = a \quad \forall a \in A.$$

The proof is straightforward. □

PROPOSITION 4.4.2. *An idempotent element e of a ring is left quasi-regular if and only if $e = 0$.*

Proof: If $e = e^2$ and x is a left-quasi-inverse of e, then $x + e - xe = 0$, and therefore $xe + e^2 - xe^2 = 0$. Thus $e = e^2 = 0$. □

COROLLARY 4.4.3. *A Jacobson radical ring does not contain nonzero idempotents. Every simple ring with a nonzero idempotent, in particular, every matrix ring over a division ring, is always Jacobson semisimple.* □

If a, b are elements of a ring with unity element e, then

$$(b-e)(a-e) = ba - b - a + e = e - (b \circ a).$$

Thus $(b-e)(a-e) = 0$ if and only if $b \circ a = 0$, whence it follows that a is quasi-regular if and only if $a - e$ is a unit. In particular, in a division ring all elements but the unity element e, are quasi-regular (and e is not).

PROPOSITION 4.4.4. *Every nilpotent element of a ring is left quasi-regular.*

Proof: If $a^n = 0$ with some exponent $n \geq 2$, then put $x = -a - a^2 - \ldots - a^{n-1}$. Now $x \circ a = 0$, and therefore x is a left quasi-inverse of a. □

Summarizing the results of Examples 2.1.7, 3.2.1 (ii), 3.2.13 (ii), 3.6.1 (iii), 3.14.12, 3.14.24, 3.16.5 (iv), 3.17.4, 3.18.6 (i), 3.20.9 and of Proposition 4.4.4 we arrive at

DEFINITION and THEOREM 4.4.5. *The Jacobson radical class $\mathcal{J}$ is the class of all quasi-regular rings. The Jacobson radical $\mathcal{J}(A)$ of a ring A is the sum of all quasi-regular ideals of A, and also the sum of all quasi-regular left (right) ideals of A. $\mathcal{J}$ is a left (right) hereditary and left (right) strong radical. Moreover, $\mathcal{J}$ is a special and a normal radical, but not a left stable and not a strongly hereditary radical. $\mathcal{J}$ is the class of all rings having no irreducible modules. The class $\mathcal{J}$ contains properly Köthe's nil radical class $\mathcal{N}$.* □

In order to obtain more information on the Jacobson radical and also on the structure of Jacobson semisimple rings, we need to introduce further notions.

DEFINITION 4.4.6. A ring $A \neq 0$ is called *left primitive* (for short only *primitive*), if A contains a maximal left ideal L such that $xA \subseteq L$ implies $x = 0$. An ideal I of a ring A is said to be a *left primitive* (shortly, *primitive*) ideal of A, if the factor ring A/I is a primitive ring.

Reformulating Definition 4.4.6, we may say that *the ring A is primitive if and only if A possesses a maximal left ideal L such that*

$$(L:A)_A = \{x \in A \mid xA \subseteq L\} = 0,$$

if and only if the annihilator $(0:A/L)_A$ of the simple A-module A/L is 0.

PROPOSITION 4.4.7. *A ring A is primitive if and only if there exists a faithful simple A-module.*

Remark. Memorizing the definition of primitivity, the easiest way is to think of Proposition 4.4.7.

Proof: If A is a primitive ring, then A/L is obviously a faithful simple A-module where L is a maximal left ideal of A as required in the definition.

Let M be a faithful simple A-module. For any fixed element $m \in M$, $m \neq 0$, the mapping

$$f: a \mapsto am \qquad \forall a \in A$$

maps A, as an A-module, homomorphically onto M. Since $A/\ker f \cong M$ and M is simple, $L = \ker f$ is a maximal left ideal of A. Moreover, since M is faithful, we have $(0 : A/L) = 0$. Thus A is primitive. □

It is clear from the definition of primitivity that if the maximal left ideal L happens to be a two-sided ideal, then necessarily $L = 0$, and the primitive ring A is a division ring. In particular, *a commutative ring is primitive if and only if it is a field.* Thus primitive rings are non-commutative generalizations of fields.

EXAMPLE 4.4.8. Let V be any vector space over a division ring. *The ring T of all linear transformations of V is a primitive ring.* We show that V is a faithful simple T-module. This is fairly obvious. Since for any nonzero vector $v \in V$ we have $Tv = V$, V is a simple T-module. In addition, V is faithful because $0 \in T$ is the only linear transformation which sends all vectors of V into $0 \in V$.

As we have seen in Example 1.2.11, rings of linear transformations of infinite dimensional vector spaces are not simple. Thus *primitive rings need not be simple.*

The next result is related to Example 3.14.12.

THEOREM 4.4.9. *The Jacobson radical $\mathcal{J}(A)$ of a ring A is the intersection M of all modular maximal left ideals of A, and also the intersection P of all primitive ideals of A.*

Proof: We shall proceed by showing the inclusions

$$\mathcal{J}(A) \supseteq M \supseteq P \supseteq \mathcal{J}(A).$$

i) Let us suppose that $M \not\subseteq \mathcal{J}(A)$. Then since M is a left ideal, there exists an element $a \in M$ which is not left quasi-regular, that is, $a \neq xa - x$ for all $x \in A$, that is, a is not contained in the left ideal

$$L = \{xa - x \mid x \in A\}$$

of A. By Zorn's Lemma there exists a left ideal L_0 of A containing L and being maximal with respect to the exclusion of the element a. Since $xa - x \in L \subseteq L_0$, the left ideal L_0 is modular. Let K be any left ideal of A with $L_0 \subset K \subseteq A$. By the choice of L_0 we have $a \in K$, and so for every element $x \in A$ it holds

$$x = -(xa - x) + xa \in L_0 + K \subseteq K.$$

Hence $K = A$, and L_0 is a modular maximal left ideal of A such that $a \notin L_0 \supseteq M$ which contradicts $a \in M$. Thus $M \subseteq \mathcal{J}(A)$.

ii) We shall prove that every modular maximal left ideal of a ring A contains a primitive ideal which yields clearly $P \subseteq M$. Let L be a modular maximal ideal of the ring A, and consider the set

$$I = (L : A)_A = \{x \in A \mid xA \subseteq L\}.$$

One readily sees that I is a two-sided ideal of A. Moreover, we claim that $I \subseteq L$. The modularity of L assures the existence of an element $e \in A$ such that $xe - x \in L$ for every $x \in I$. In particular, if $x \in I$, then $xe \in L$ and

$$x = xe - (xe - x) \in L + L = L.$$

Since by definition $xA \subseteq L$ implies $x \in I$, the factor module A/L is an irreducible A/I-module (cf. Propositions 3.14.1 and 3.14.8), so the factor ring A/I is primitive, that is, I is a primitive ideal of A which is contained in the modular maximal left ideal L.

iii) Let us assume that $\mathcal{J}(A) \not\subseteq P$. Then there exists a primitive ideal P_0 of A such that $\mathcal{J}(A) \not\subseteq P_0$. Since P_0 is primitive, there exists a maximal left ideal $L \supseteq P_0$ of A such that $xA \subseteq L$ implies $x \in P_0$. Since $\mathcal{J}(A) \not\subseteq P_0$, it follows $\mathcal{J}(A)A \not\subseteq L$, and the maximality of L implies $\mathcal{J}(A)A + L = A$. Furthermore, by $P_0 \neq A$ we have $AA \not\subseteq L$ and also

$$A(\mathcal{J}(A)A + L) = AA \not\subseteq L.$$

Hence by $A(\mathcal{J}(A)A + L) \subseteq A\mathcal{J}(A)A + AL$ we conclude that

$$A\mathcal{J}(A)A \not\subseteq L.$$

Thus there are elements $a \in \mathcal{J}(A)$ and $b \in A$ such that $Aab \not\subseteq L$, and by the maximality of L we have $Aab + L = A$. Hence there exists an element $x \in A$ such that $b \in xab + L$, that is, $b - xab \in L$. Taking into account that $a \in \mathcal{J}(A)$, also $xa \in \mathcal{J}(A)$, and hence xa has a left quasi-inverse y. Now, by $b - xab \in L$ we get

$$b = b - (y \circ (xa))b = (b - xab) - y(b - xab) \in L,$$

and consequently $Aab \subseteq L$, contradicting $Aab \not\subseteq L$. This proves $\mathcal{J}(A) \subseteq P$. $\square$

Right primitivity of a ring and of an ideal as well as *modular right ideals* can be defined analogously to their left counterparts. Although, as it will be shown in § 4.6, *a right primitive ring need not be left primitive*, the symmetric definition of the Jacobson radical given by the circle operation $\circ$ and Theorem 4.4.9 give immediately

COROLLARY 4.4.10. *The Jacobson radical of a ring A is the intersection of all modular maximal right ideals of A, as well as the intersection of all right primitive ideals of A.* $\square$

The Jacobson radical of a ring has many further characterizations. Kertész gave a cyclic proof for ten characterizations (see the books of Kertész [1] and Szász [6]).

In accordance with the notion of semiprimeness (cf. Theorem 3.8.1) a ring A is said to be *semiprimitive*, if the intersection of all primitive ideals of A is 0. In view of Theorem 4.4.9 *semiprimitivity is just another name for Jacobson semisimplicity.*

COROLLARY 4.4.11. *The Jacobson radical $\mathcal{J}$ is the upper radical $\mathcal{J} = \mathcal{UQ}$ of the class $\mathcal{Q}$ of all primitive rings. A ring A is semiprimitive if and only if A is a subdirect sum of primitive rings.*

The assertions are clear in view of Theorem 4.4.9. □

An external characterization of the Jacobson radical in terms of irreducible modules has been given in Example 3.14.12 (and Theorem 3.14.4 (ii)). A reformulation of these statements is the following

THEOREM 4.4.12. *The Jacobson radical $\mathcal{J}(A)$ of a ring A is given as the intersection*

$$\mathcal{J}(A) = \cap (a \in A \mid aM = 0 \text{ for every irreducible } A\text{-module } M).$$

Moreover, A is a Jacobson radical ring if and only if A there are no irreducible A-modules. □

PROPOSITION 4.4.13. *Every primitive ring is a prime ring.*

Proof: Let A be a primitive ring and M a faithful irreducible A-module. Suppose that $I \neq 0$ and $K \neq 0$ are ideals of A such that $IK = 0$. Then $KM \neq 0$, so $KM = M$ and then $IM \neq 0$ so $M = IM = I(KM) = 0$, a contradiction. □

We already know that the Jacobson radical is special (see Example 3.14.24). We can deduce this also from Theorem 3.8.11: if a ring A does not satisfy condition (iii) of Theorem 3.8.11, then $A/\mathcal{J}(A) \neq 0$ which is a subdirect sum of primitive and hence prime rings (Corollary 4.4.11 and Proposition 4.4.13). Thus $A/\mathcal{J}(A)$ as well as A has a primitive prime image which is, in view of Theorem 4.4.9, $\mathcal{J}$-semisimple. We can give explicitly a special class which generates $\mathcal{J}$ as an upper radical (beside the obvious choice $\mathcal{P} \cap \mathcal{SJ}$). In the light of the earlier results we may have a strong feeling that the class $\mathcal{Q}$ of all primitive rings is a special class. Indeed, one can prove this directly. Nevertheless, results of Section 3.14 provide an easier way. By Example 3.14.24 we have $\mathcal{J} = \mathcal{U}F(\Sigma)$ where

$$F(\Sigma) = \{A \mid \text{there exists a faithful irreducible } A\text{-module}\}.$$

Now by Proposition 4.4.7

$$F(\Sigma) = \{A \mid A \text{ is primitive}\} = \mathcal{Q}.$$

Thus we have got the following.

THEOREM 4.4.14. *The class $\mathcal{Q}$ of all primitive rings is a special class.* □

There are, however, prime Jacobson semisimple rings which are not primitive.

EXAMPLE 4.4.15. $\mathcal{Q} \subset \mathcal{P} \cap \mathcal{SJ}$: *The ring $\mathbb{Z}$ of integers is semiprimitive and prime, but not primitive.* Since $\mathbb{Z}$ is a subdirect sum of the prime fields, each of which is primitive, it follows that $\mathbb{Z}$ is semiprimitive and of course prime. Before Example 4.4.8 it has been shown that primitive commutative rings are fields, whence $\mathbb{Z}$ cannot be primitive.

We are going to show that the Jacobson radical is not a dual radical.

EXAMPLE 4.4.16. *There exists a nonzero ring T such that $\mathcal{J}(T) = 0$ and T cannot be mapped homomorphically onto a nonzero subdirectly irreducible prime ring.*

Let us consider the twisted polynomial ring $F[x; \sigma]$ over a field F with automorphism σ of infinite order, as given in Example 1.3.3. We shall show that $A = F[x; \sigma]$ is a primitive ring. Let us consider the left ideal

$$M = A(x - 1)$$

which is clearly not a right ideal. Thus $M \neq A$. M is maximal, for if L is any left ideal of A with $M \subseteq L$, then by ii) of Example 1.3.3, $L = Ag$ with an appropriate polynomial g of minimal degree. Since M contains polynomials of degree 1, g must be of degree 0 or 1. If g is constant, then $L = A$. If $\partial(g) = 1$, then $x - 1 \in L$ implies $x - 1 = hg$. Hence h must be constant. So $g = \frac{1}{h}(x - 1) \in M$, and also $L = Ag \subseteq M$. Thus M is a maximal left ideal in $A = F[x; \sigma]$. Since $1 \in F[x; \sigma]$, M is a maximal modular left ideal. Consider now

$$(M : A)_A = \{a \in A \mid aA \subseteq M\}.$$

As seen in (ii) of the proof of Theorem 4.4.9, $(M : A)_A \subseteq M$ and A/M is a faithful simple $A/(M : A)_A$-module. Thus A/M is an irreducible A-module, whence by Proposition 4.4.7 the ring $A = F[x; \sigma]$ is primitive. We claim that the ideal $T = Ax$ of $A = F[x; \sigma]$ is the desired ring. By Theorem 4.4.14 the primitivity of $F[x; \sigma]$ yields that of T, whence $\mathcal{J}(T) = 0$. Moreover, Ax^k is an ideal of T for each $k = 1, 2, \ldots$, and $\bigcap_k Ax^k = 0$. This latter relation shows that T is not subdirectly irreducible. Let $K \triangleleft T$, $K \neq T$ and $\overline{K}$ the ideal of $F[x; \sigma]$ generated by K. Using the Andrunakievich Lemma, we have

$$T/K \cong \frac{T/\overline{K}^3}{K/\overline{K}^3}.$$

Since $\overline{K} \subseteq T$, by iii) of Example 1.3.3 it follows that $\overline{K} = Ax^k$ and $\overline{K}^3 = Ax^{3k}$ with some $k \geq 1$. Hence $T/\overline{K}^3 = Ax/Ax^{3k}$ and also its homomorphic image T/K are nilpotent. Thus *the primitive ring T cannot be mapped homomorphically onto a nonzero subdirectly irreducible prime ring.* □

COROLLARY 4.4.17. *The Jacobson radical is not a supernilpotent dual radical, more precisely*

$$\mathcal{J} \neq \mathcal{J}_\varphi = \mathcal{U}s(\mathcal{Q}^*)$$

where $\mathcal{Q}^*$ denotes the class of all simple primitive rings.

Proof: The supernilpotent dual radical $\mathcal{J}_\varphi = \mathcal{U}s(\mathcal{Q}^*)$, where $s(\mathcal{Q}^*)$ is the class of all subdirectly irreducible primitive rings, contains the Jacobson radical class $\mathcal{J}$. Since the ring T of Example 4.4.16 is clearly in $\mathcal{J}_\varphi$ but not in $\mathcal{J}$, we get that $\mathcal{J} \subset \mathcal{J}_\varphi$. □

It may be of historical interest that in 1953 Kurosh raised the problem whether the Jacobson radical coincides with the upper radical $\mathcal{J}_s = \mathcal{U}\mathcal{Q}^*$ of simple primitive rings. A negative answer was given by Sąsiada and Suliński [1] in 1962, which follows from Corollary 4.4.17 trivially, for $\mathcal{J}_\varphi \subseteq \mathcal{J}_s$. In fact, Sąsiada and Suliński used the ring of Example 1.3.3 to solve the Kurosh problem.

Although the Jacobson radical is the most efficient radical with the mostly elaborated theory, for long no lower radical representation of $\mathcal{J}$ was known. Yu-lee Lee [2] proved that the Jacobson radical is the lower radical of the class $\mathcal{Z} \cup \mathcal{J}_2$ where $\mathcal{Z}$ is the class of all rings with zero multiplication and

$$\mathcal{J}_2 = \{(\mathcal{J}(A))^2 \mid A \text{ is any ring}\}.$$

More generally, for every supernilpotent radical γ it holds $\gamma = \mathcal{L}(\mathcal{Z} \cup \gamma_2)$ where

$$\gamma_2 = \{(\gamma(A))^2 \mid A \text{ is any ring}\}.$$

Puczyłowski and Zand [1] gave examples for rings which are in $\mathcal{J}$ but not in $\mathcal{J}_2$.

Jacobson's Conjecture is that

$$\bigcap_{n=1}^{\infty} (\mathcal{J}(A))^n = 0$$

for every noetherian ring A. The status of Jacobson's Conjecture for left and right noetherian rings is still open. More on this problem can be found in Krause [1] and Rowen [1].

4.5. Structure theorems for Jacobson semisimple rings

As already mentioned, the original aim of introducing radicals was to prove structure theorems for semisimple rings. In this respect the most successful radical turned out to be the Jacobson radical. In Corollary 4.4.11 we have already seen that Jacobson semisimple rings (that is, the semiprimitive rings) are exactly the subdirect sums of primitive rings, and so primitive rings can be regarded as building blocks of Jacobson semisimple rings. Primitive rings can be represented as dense subrings of linear transformations on a vector space over a division ring, this is the Jacobson–Chevalley Density Theorem.

Imposing descending chain condition for left ideals on semiprimitive rings, the subdirect decomposition becomes a finite direct sum and the subdirect components turn out to be full matrix rings over division rings (Wedderburn–Artin Structure Theorems). Requiring weaker finiteness conditions, interesting structure theorems can be obtained (for instance, the Litoff–Ánh Theorem).

Let A be a primitive ring, M a faithful irreducible A-module. Since M is simple, an A-module endomorphism is either 0 or an automorphism. Thus *the endomorphism ring of an irreducible A-module M is a division ring D.* This statement is known as *Schur's Lemma.* Considering the endomorphisms of the additive group M^+, that is, $\operatorname{End} M^+$, we have clearly

$$D \subseteq \operatorname{End} M^+.$$

For any element $a \in A$ the left multiplication

$$f_a \colon x \mapsto ax \qquad \forall x \in M$$

is in $\operatorname{End} M^+$, and $f_a = f_b$ means just

$$(f_a - f_b)(M) = (a-b)M = 0,$$

which implies $a = b$, for M is faithful. Thus A can be embedded into $\operatorname{End} M^+$. Moreover, for all elements $d \in D$ and $a \in A$ we have

$$df_a(x) = d(ax) = ad(x) = f_a d(x), \qquad \forall x \in M,$$

and so D is in the *centralizer*

$$Z_A(M^+) = \{g \in \operatorname{End} M^+ \mid gf_a = f_a g \text{ for } \forall a \in A\}$$

of A in $\operatorname{End} M^+$. In fact, $D = Z_A(M^+)$ because each element $z \in Z_A(M^+)$ is an A-module endomorphism:

$$z(ax) = zf_a(x) = f_a z(x) = az(x), \qquad \forall a \in A, \ x \in M.$$

Thus the irreducible A-module M may be viewed as a vector space over the division ring $D = Z_A(M^+)$, and the ring A as a subring of linear transformations of the vector space M. Hence in view of Proposition 4.4.7 we have got

THEOREM 4.5.1. *Every primitive ring A is isomorphic to a subring of all linear transformations of a vector space over a division ring.* □

The vector space in this theorem need not be finite dimensional. The next lemma provides some useful information on rings of linear transformations which will be crucial in proving the Jacobson–Chevalley Density Theorem. Let A denote a primitive ring, V an irreducible A-module which is a vector space over the centralizer D.

LEMMA 4.5.2. *If W_n is an n-dimensional subspace of the vector space V and $(0 : W_n)_A$ is the annihilator of W_n in A, then $(0 : W_n)_A v = V$ for every element $v \in V \setminus W_n$.*

Proof: By induction on the dimension n. In $n = 0$, then $(0 : W_0) = A$ and for every nonzero element $v \in V$ we have $Av = V$.

Let $n \geq 1$, and assume the validity of the statement for $n - 1$. Then $W_n = W_{n-1} + Dv_n$ where W_{n-1} is an $n - 1$ dimensional subspace and $v_n \in W_n$. It suffices to show that there exists an element $b \in (0 : W_{n-1})$ with $bv_n = 0$ and $bv \neq 0$ for all $v \in V \setminus W_n$, because V is irreducible. Suppose on the contrary that such an element b does not exist, that is, $v \in V \setminus W_n$ implies $cv = 0$ for every $c \in (0 : W_n)$. By the hypothesis $(0 : W_{n-1})v_n = V$, and so for every element $x \in V$ there is an element $a_x \in (0 : W_{n-1})$ such that $a_x v_n = x$. If $a'_x \in (0 : W_{n-1})$ is another element with $a'_x v_n = x$, then $(a_x - a'_x)v_n = 0$, implying $a_x - a'_x \in (0 : W_n)$. For any fixed element $z \in V \setminus W_n$, let us define the mapping $g : V \to V$ by $g(x) = a_x z$. Since $a_x - a'_x \in (0 : W_n)$, by the assumption it follows $a_x z - a'_x z = (a_x - a'_x)z = 0$, implying $a_x z = a'_x z$. Hence g is single valued. Since a_{x+y} is the element of $(0 : W_{n-1})$ such that

$$a_{x+y}v_n = x + y = a_x v_n + a_y v_n = (a_x + a_y)v_n,$$

so

$$g(x+y) = a_{x+y}z = (a_x + a_y)z = a_x z + a_y z = g(x) + g(y).$$

Further, by $a_{bx}v_n = bx = ba_x v_n$ we have

$$g(bx) = a_{bx}z = ba_x z = bg(x).$$

Thus g is in $\operatorname{End} V$. Moreover, g belongs to the centralizer D, for

$$f_b g(x) = bg(x) = g(bx) = g f_b(x).$$

Since for every element $a \in (0 : W_{n-1})$, $av_n = x \in V$, necessarily

$$az = a_x z = g(x) = g(av_n) = ag(v_n)$$

which implies $a(z - g(v_n)) = 0$ for every $a \in (0 : W_{n-1})$. By the induction hypothesis $(0 : W_{n-1})v = 0$ implies $v \in W_{n-1}$, and so for $v = z - g(v_n)$ we get

$$z - g(v_n) \in W_{n-1}.$$

By $g \in D$ it follows that $g(v_n) \in Dv_n$, whence

$$z \in W_{n-1} + Dv_n = W_n,$$

contradicting the choice of z. Thus the proof of the lemma has been completed. □

Applying this lemma we get easily the description of primitive rings.

A ring A is said to be a *dense subring of linear transformations on a vector space* V, if for any finitely many linearly independent elements $x_1, \ldots, x_n \in V$ and arbitrary not necessarily distinct elements $y_1, \ldots, y_n \in V$ there exists an element $a \in A$ such that $ax_i = y_i$ for $i = 1, \ldots n$. The term "dense" stems from the fact that $\operatorname{End} V$ can be endowed with the *finite topology* and a subring A is a dense subspace in $\operatorname{End} V$ if and only if A is a dense subring of $\operatorname{End} V$, (cf. for instance Kertész [1], Jacobson [5], Rowen [1]).

THEOREM 4.5.3 (Jacobson–Chevalley Density Theorem). *A ring A is primitive if and only if A is isomorphic to a dense subring of linear transformations on a vector space over a division ring.*

Proof: In view of Theorem 4.5.1 A can be considered as a subring of linear transformations on a vector space V over a division ring. Moreover, V is a simple faithful A-module. Take n linearly independent elements $x_1, \ldots, x_n \in V$ and arbitrary elements $y_1, \ldots, y_n \in V$. If $n = 1$, then by $Ax_1 = V$ there is an $a \in A$ such that $ax_1 = y_1$. Assume the statement true for $n - 1$. Then there exists an element $a \in A$ such that $ax_i = y_i$ for $i = 1, \ldots, n - 1$, but $ax_n = z$ may differ from y_n. Applying Lemma 4.5.2 there exists an element $b \in (0 : \{x_1, \ldots, x_{n-1}\})$ such that $bx_n = y_n - z$. The linear transformation $a + b \in A$ is then the desired one.

Conversely, let A be a dense subring of $\operatorname{End} V$ where V is a vector space over a division ring. Then V can be viewed as an A-module which is clearly faithful. Further, by the density of A in $\operatorname{End} V$ it follows $ax = y$ for any nonzero x and arbitrary y in V with an appropriate $a \in A$. Hence $Ax = V$, proving that V is a simple and also a faithful A-module. Thus by Proposition 4.4.7 A is a primitive ring. □

A ring A satisfies the *descending chain condition* (briefly *d.c.c.*) on left ideals, if every strictly descending chain of left ideals terminates in a finite number of steps. Rings with d.c.c. on left ideals are called also *(left) artinian rings*.

PROPOSITION 4.5.4. *A primitive ring A is artinian if and only if A is isomorphic to a full ring of linear transformations on a finite dimensional vector space over a division ring.*

Proof: Let the primitive ring A be artinian. By the Density Theorem A is a dense subring of $\operatorname{End} V$ for a vector space V. Let $\{x_1, \ldots, x_n, \ldots\}$ be a base for V, and let us consider the subspaces W_n generated by $x_1, \ldots, x_n$, for $n = 1, 2, \ldots$. Each annihilator $(0 : W_n)_A$ is clearly a left ideal of A and

$$(0 : W_1)_A \supset (0 : W_2)_A \supset \ldots$$

is a strictly descending chain by Lemma 4.5.2. Since A is artinian, this chain has to be finite, and so V is finite dimensional. Obviously, if A is dense in $\operatorname{End} V$, then by definition $A = \operatorname{End} V$ holds for V is finite dimensional.

Conversely, suppose that $A = \operatorname{End} V$ where V is an n dimensional vector space over a division ring D. Now A can be viewed also as an n^2 dimensional vector space over D, ($\operatorname{End} V$ is isomorphic to the $n \times n$ matrix ring $M_n(D)$), and the left ideals of A are certain subspaces. Hence every descending chain of left ideals is a chain of subspaces of decreasing dimension. $\square$

LEMMA 4.5.5. *Let A be a ring and L a left ideal of A. Then the following hold:*

(i) *The right multiplication $x \mapsto xa$, $x \in L$, induces an A-module homomorphism $L \to La$ for every $a \in A$,*

(ii) $K = L + \sum_{a \in A} La$ *is a two-sided ideal in A,*

(iii) *if $L^2 = 0$, then also $K^2 = 0$,*

(iv) *if L is a minimal left ideal and $L^2 \neq 0$, then there exists a nonzero idempotent e in L and $L = Le = Ae$.*

Proof: (i) and (ii) is straightforward. For (iii) we notice that for any products $\ell_1 a_1, \ell_2 a_2 \in K$

$$(\ell_1 a_1)(\ell_2 a_2) = \ell_1(a_1 \ell_2)a_2 \in L^2 a_2 = 0.$$

(iv) Since $L^2 \neq 0$ and L is minimal, there exists an element $c \in L$ such that $Lc = L$. Hence $c = ec$ with a suitable element $e \in L$. Further, also $ec = e(ec) = e^2 c$ holds, that is, $(e - e^2)c = 0$. Hence $e - e^2$ is in the left ideal $L \cap (0:c)_A$. Suppose $e \neq e^2$. Then $0 \neq e - e^2 \in L \cap (0:c)_A$ so by the minimality of L we have $L \cap (0:c)_A = L$. But then $lc = 0$ for all $l \in L$, while e is in L and $ec = c$. Thus we conclude that $e = e^2$ and e is an idempotent as desired. $\square$

An easy consequence of Proposition 4.5.4 and Lemma 4.5.5 is the

FIRST WEDDERBURN–ARTIN STRUCTURE THEOREM 4.5.6. *A ring A is simple, prime and artinian if and only if A is a full ring of linear transformations on a finite dimensional vector space over a division ring.*

Proof: Let A be simple, prime and artinian. Then A contains a minimal left ideal L. Since A is simple, by Lemma 4.5.5 (ii) $K = L + \sum_{a \in A} La$ must be A. Hence by Lemma 4.5.5 (iii) and (iv) it follows also that $L^2 \neq 0$ and there is an idempotent $e \in L$, $e \neq 0$. Thus by the simplicity of A and by Proposition 4.4.2 we get $\mathcal{J}(A) = 0$. Again by the simplicity of A it follows that A must be primitive. Hence Proposition 4.5.4 yields that A is isomorphic to $\operatorname{End} V$, $\dim V < \infty$.

Conversely, if $\dim V < \infty$, then $\operatorname{End} V$ is simple prime and artinian. □

From Proposition 4.5.4 and Theorem 4.5.6 we see that *a primitive artinian ring is simple.*

PROPOSITION 4.5.7. *If a ring A is semiprimitive and artinian, then A is a finite direct sum of simple artinian prime rings.*

Proof: In view of Theorem 4.4.9 $\mathcal{J}(A) = \bigcap_\alpha P_\alpha = 0$ where each P_α is a primitive ideal. Moreover, since a primitive artinian ring is simple, each P_α is a maximal ideal. We can also pick finitely many ideals $P_1, \ldots, P_n$ such that $\bigcap_{i=1}^n P_i = 0$ and $Q_1 = \bigcap_{i=2}^n P_i \neq 0$, because A is artinian. Hence by the maximality of P_1 we get $A = P_1 \oplus Q_1$ and Q_1 is a simple artinian ring. Further, by $P_1 \triangleleft A \in \mathcal{SJ}$ also P_1 is semiprimitive and by $P_1 \cong A/Q_1$, P_1 is artinian. Iterating the procedure the d.c.c. assures that in finitely many steps we arrive at a desired decomposition. □

LEMMA 4.5.8 (Hopkins [1]). *The Jacobson radical $\mathcal{J}(A)$ of an artinian ring A is nilpotent.*

Proof: Put $J = \mathcal{J}(A)$. In the descending chain

$$J \supseteq J^2 \supseteq \ldots$$

there exists an exponent n with $K = J^n = J^{n+1}$. Assume that $K \neq 0$, and consider the set $\mathcal{M}$ of all left ideals M of A for which $M \subseteq K$ and $KM \neq 0$. Clearly, $K \in \mathcal{M}$. Since A is artinian, we can choose a left ideal L which is minimal in $\mathcal{M}$. By $KL \neq 0$ there is an element $c \in L$ with $Kc \neq 0$. Further we have $K^2 = J^{2n} = J^n = K$ and hence $K \cdot Kc = K^2c = Kc \neq 0$. Therefore $Kc \in \mathcal{M}$ and by the choice of L it follows that $Kc = L$. Thus there exists an element $k \in K$ such that $kc = c$. Since $k \in K \subseteq J$, k has a quasi-inverse k': $k + k' - k'k = 0$. Hence

$$c = c - (k + k' - k'k)c = (c - kc) - k'(c - kc) = 0$$

holds, contradicting $L = Kc \neq 0$. □

An easy consequence is now the

SECOND WEDDERBURN–ARTIN STRUCTURE THEOREM 4.5.9. *If A is a semiprime artinian ring, then A is isomorphic to a finite direct sum of full rings of linear transformations on finite dimensional vector spaces over division rings, and conversely.*

Proof: Let A be semiprime and artinian. Then A is a subdirect sum of prime artinian rings B_α. By Lemma 4.5.8 $\mathcal{J}(B_\alpha)$ is nilpotent, and therefore $\mathcal{J}(B_\alpha) \subseteq \beta(B_\alpha) = 0$. Hence each B_α is semiprimitive and therefore so is A. Thus A is a subdirect sum of primitive rings, whence A is a semiprimitive artinian ring. Now Theorem 4.5.6 and Proposition 4.5.7 yields the assertion.

The converse is obvious. □

Let us mention that *in the decompositions of a semiprime artinian ring the direct summands are uniquely determined* (cf. for instance Jans [1, pp 21–22] or Kertész [1, Theorem 41.1]).

Imposing only the existence of a minimal left ideal in a simple prime ring, considerable generalizations of the First Wedderburn–Artin Structure Theorem can be derived from the Jacobson–Chevalley Density Theorem. In the rest of this section we give two neat characterizations for simple rings with minimal left ideals.

PROPOSITION 4.5.10 (McCoy [3]). *A prime ring A having a minimal left ideal L is primitive. In particular, every simple prime ring with minimal left ideal is primitive.*

Proof: Consider the ideal $I = L + LA$ of A generated by L. Since A is prime and $(0:L)_A$ is an ideal, the relation

$$(0:L)I = (0:L)(L+LA) = 0$$

yields $(0:L) = 0$. Hence L is a faithful simple A-module. □

LEMMA 4.5.11. *Let A be a dense ring of linear transformation on a vector space V over a division ring D, and let $t \neq 0$ be a finite valued linear transformation in A. Then to every finite dimensional subspace W of V there exists an element t^* in the principal ideal $(t)_A$ which is a projection of V onto W.*

Proof: We proceed exactly as in the proof of Example 1.2.11 (ii). Notice that by the density of A there exist elements $r, s \in A$ as required in Example 1.2.11 (ii). □

THEOREM 4.5.12. *A ring A is simple, prime and possesses a minimal left ideal if and only if A is a dense ring of finite valued linear transformations on a vector space over a division ring.*

Proof: Let A be a simple prime ring with a minimal left ideal L. By Proposition 4.5.10 the Jacobson–Chevalley Density Theorem 4.5.3 is applicable, and so A can be considered as a dense ring of linear transformations on a vector space V over a division ring D. Since L is an irreducible A-module, by Lemma 4.5.5 (iv) there exists an idempotent $e \in L$ such that $L = Ae$. We show that e is finite valued, more precisely, eV is a 1-dimensional subspace. Assume that there are two linearly independent elements ex and ey in eV. Since A is dense, there exists an $a \in A$ such that

$$aex = ex \quad \text{and} \quad aey = 0.$$

Now $ex \neq 0$ so $ae \neq 0$. But $ae \in (0:y)_A$ so $L \cap (0:y)_A \neq 0$ as $L = Ae$. By the minimality of L we then have $L \cap (0:y)_A = L$ and $L \subseteq (0:y)_A$.

But then $ey = 0$, a contradiction, as ex and ey are linearly independent. Therefore eV is 1-dimensional.

The ideal H of all finite valued linear transformations of $\operatorname{End} V$ contains e, and so by the simplicity of A also $A \subseteq H$ holds.

Conversely, let A be a dense ring of finite valued linear transformations on V over D, and $t \in A$ an arbitrary element. For any nonzero element $f \in A$, fV is finite dimensional, so by Lemma 4.5.11 there exists a projection e in the principal ideal $(t)_A$ which maps V onto fV. Then

$$(ef)x = e(fx) = fx$$

for all $x \in V$, and so $f = ef \in (t)_A$. Thus $A = (t)_A$ proving that A is a simple ring.

Finally, we show that A has a minimal left ideal. To this end, let us consider a 1-dimensional subspace $W = Dw$ of V. Lemma 4.5.11 is applicable, and so there exist an element $a \in A$ projecting V onto W. We claim that the left ideal $(a]$ of A generated by a, is minimal. Since a is a projection, we obtain $v - av \in \ker a$ for all $v \in V$. Hence $V = W \oplus \ker a$ holds. Take any nonzero element $b \in (a]$. Obviously $bu = 0$ is valid for all $u \in \ker a$. Hence for every element $v = dw + u$, $(d \in D, u \in \ker a)$, we have $bv = bdw$. As $b \neq 0$ and $b(\ker a) = 0$ we must have $bw \neq 0$. Since A is dense, there exists an element $c \in A$ such that $cbw = w$. Hence

$$av = adw = adcbw = ac(bdw) = acbv$$

holds for all $v \in V$, whence $a = acb \in (b]$. Thus $(a]$ is a minimal left ideal. □

Demanding the existence of a unity element in a simple ring is a fairly strong requirement, and Theorem 4.5.12 degenerates to the following

COROLLARY 4.5.13. *A ring A is simple, prime and has a unity element as well as a minimal left ideal if and only if A is isomorphic to the ring of all linear transformations of a finite dimensional vector space over a division ring.*

Proof: By Theorem 4.5.12 A is a dense ring of finite valued linear transformations, and in particular, the unity element is finite valued. Consequently the vector space has to be finite dimensional, and the necessity follows. The converse is trivial. □

LEMMA 4.5.14. *Let e be a nonzero idempotent in the simple ring A. Then the following conditions are equivalent:*

(i) *Ae is a minimal left ideal,*
(ii) *eAe is a division ring,*
(iii) *eA is a minimal right ideal.*

Proof: (i) $\Longrightarrow$ (ii) If $exe \neq 0$, then $Aexe \neq 0$ because $e(exe) = exe \neq 0$. Now $Aexe$ is a left ideal in A which is contained in Ae. Hence the minimality of Ae implies $Aexe = Ae$, whence $eAe \cdot exe = eAe$. Thus, if $z \in A$ is any element, then there exists an element $y \in A$ such that $eye \cdot exe = eze$, proving that eAe is a division ring.

(ii) $\Longrightarrow$ (iii) Let eAe be a division ring and R a right ideal of A such that $0 \neq R \subseteq eA$. We have to show that $R = eA$. Applying Lemma 4.5.5 (iii) for the *right* ideal R, it follows by $A^2 \neq 0$ that also $R^2 \neq 0$. Since $R^2 \subseteq ReA$, necessarily also $Re \neq 0$. Therefore, there exists an element $x \in R$ with $xe \neq 0$. The inclusion $R \subseteq eA$ implies $x = ey$ with a suitable $y \in A$ and so $ex = ey = x$. Thus $0 \neq xe = eye$. Since eAe is a division ring and $e \in eAe$, the equation $eye \cdot z = e$ has a solution $z \in eAe$. Moreover, since

$$e = (ey)(ez) \in Rz \subseteq R$$

we conclude that $eA \subseteq R$, proving the minimality of the right ideal eA.

The equivalence of (i) and (iii) follows immediately from the symmetry of (ii). $\square$

COROLLARY 4.5.15. *A simple prime ring has a minimal left ideal if and only if it has a minimal right ideal.* $\square$

An additive subgroup Q of a ring A is said to be a *quasi-ideal*, if $AQ \cap QA \subseteq Q$. This notion was introduced by Steinfeld [1] (see also Kertész [1]). Quasi-ideals are obviously subrings and every one-sided ideal is a quasi-ideal. An example of a quasi-ideal, which is perhaps the best illustration, is the subring $(D)_{ij}$ of the matrix ring $M_n(D)$ consisting of all matrices in which an arbitrary element of D may stand in the (i,j) position and 0 everywhere else. If D is a division ring, then $(D)_{ij}$ is a minimal quasi-ideal in $M_n(D)$ and clearly $M_n(D) = \sum_{i,j=1}^{n} (D)_{ij}$. Thus minimal quasi-ideals can be viewed as basic constituents of matrix rings.

We say that a ring A is a *strongly locally matrix ring* over a division ring D, if every finite subset C of A can be embedded into a quasi-ideal Q of A such that $Q \cong M_n(D)$ for some natural number n (which depends on the choice of C).

LITOFF–ÁNH THEOREM 4.5.16 (Ánh [1]). *A ring A is simple, prime and has a minimal one-sided ideal if and only if A is a strongly locally matrix ring over a division ring.*

Proof: Let A be a simple prime ring with a minimal one-sided ideal. Then by Corollary 4.5.15 A has a minimal left ideal as well as a minimal right ideal.

By Theorem 4.5.12 A is a dense ring of finite valued linear transformations on a vector space V. Let $C = \{t_1, \ldots, t_m\}$ be a finite subset in A. Since each t_i is finite valued, $W = t_1 V + \cdots + t_m V$ is a finite dimensional subspace in V. Thus by Lemma 4.5.11 there exists a projection $e \in A$ of V onto W such that $et_1 = t_1, \ldots, et_m = t_m$ and $e = e^2$. By symmetry it follows

Concrete Radicals and Structure Theorems

that to the finite set $C \cup \{e\}$ there exists an idempotent $f \in A$ such that $t_1 f = t_1, \ldots, t_m f = f$ and $ef = f$. Let us consider the linear transformation $a = e + f - fe$. As easily seen, $ae = e, fa = f$ and $a^2 = a$, and therefore

$$C = eCf = aeCfa \subseteq aAa.$$

Now $W^* = aV$ is a finite dimensional subspace in V. Clearly, each element $r \in aAa$ can be regarded as a linear transformation of W^*. Moreover, for any element $w \in W^*$ there exists an element $v \in V$ with $w = av$. Since $a^2 = a$, we have $aw = a^2 v = av = w$, and so a acts on W^* as the identical linear transformation. Hence every every linear transformation s of W^* can be written in the form $s = asa$, and therefore $aAa = \operatorname{End} W^*$.

The ring aAa is a quasi-ideal of A, for clearly

(∗) $$aAa \cdot A \cap A \cdot aAa \subseteq aAa.$$

Conversely, assume that A is a strongly locally matrix ring over a division ring D. First we prove that A is simple. Let I be any nonzero ideal of A and $0 \neq a \in I$. For any nonzero element $x \in A$ there exists a quasi-ideal Q of A such that $x, a \in Q$ and $Q \cong M_n(D)$ for some n. Hence Q is a simple ring, and so $a \in I \cap Q$ implies $Q \subseteq I$ and $x \in I$. Thus $I = A$, and A is a simple ring.

Since $Q \cong M_n(D)$, there exists a unity element e in Q, and we have

(∗∗) $$eAe \subseteq eA \cap Ae \subseteq QA \cap AQ \subseteq Q = eQe \subseteq eAe,$$

whence $Q = eAe$. Let f be an idempotent in Q such that Qf is a minimal left ideal. Since $Q \cong M_n(D)$, such an element f exists. Then by Lemma 4.5.14 we get

$$D \cong fQf = feAef = fAf,$$

and that Af is a minimal left ideal in A. □

Remark. In the definition of strongly locally matrix rings quasi-ideal can be replaced by biideal (defined at the end of Section 3.18) and the corresponding statement of Theorem 4.5.16 remains valid: in the proof we just change the lines (∗) and (∗∗) to

$$aAa \cdot A \cdot aAa \subseteq aAa,$$

and

$$eAe \subseteq QAQ \subseteq Q = eQe \subseteq eAe.$$

Litoff's Theorem (which has never been published but in Jacobson's book [5] and for which several proofs are known, e.g. Faith and Utumi [1], Kertész [1] and Steinfeld [2]) asserts that every simple ring with a minimal left ideal is a locally matrix ring, that is, every finite subset can be embedded into a *subring* S such that $S \cong M_n(D)$. Moreover, embedding of any finite subset into a left ideal L such that $L \cong M_n(D)$ cannot be achieved. Thus Theorem 4.5.16, which is due to Ánh [1], is the strong version of Litoff's Theorem.

The following example due to Ánh [1] shows that the converse of Litoff's Theorem is not true.

EXAMPLE 4.5.17. Let R_n denote a ring $R_n = M_{2^n}(D)$ of all $2^n \times 2^n$ matrices over a division ring D, and let us embed R_n into R_{n+1} along the diagonal by the correspondence

$$x \mapsto f_n(x) = \begin{pmatrix} x & 0 \\ 0 & x \end{pmatrix} \qquad \forall x \in R_n.$$

Define a relation $\equiv$ on the set $\bigcup_{n=1}^{\infty} R_n$ by $x \equiv y$ if and only if $x \in R_m$ and $y \in R_n$ for some m, n and there exists an integer $k > m, n$ such that

$$f_{k-1} \ldots f_m(x) = f_{k-1} \ldots f_n(y).$$

One can easily see that this relation is an equivalence relation. Let R denote the set of equivalence classes. In the usual way one can define addition and multiplication on R by using representatives of equivalence classes, and verify that R has become a ring such that each ring R_n can be embedded into R in a natural way, $g_n \colon R_n \to R$ (that is, R is a *direct limit* of the rings R_n, $n = 1, 2, \ldots$). Moreover, one can see that R *is a simple ring with unity element* e which is the image $e = g_n(e_n)$ of the unity element $e_n \in R_n$, $n = 1, 2, \ldots$. The ring R, however, *has no minimal one-sided ideals*, because otherwise by Corollary 4.5.13 R would be a matrix ring over a division ring, which is obviously not the case.

For more structure theorems the interested reader is referred to the books of Jacobson [5], Kertész [1] and Rowen [1], and to the series of papers of Xu [1], [2], [3], [4], [5], [6], [7].

4.6. One-sided primitivity and idempotent simple quasi-regular rings

As we have seen, the Jacobson radical can be defined in terms of left-handed or right-handed concepts: primitive rings, simple modules and so on. In particular, the Jacobson radical is defined as an upper radical by either the class of left primitive or the class of right primitive rings. How different are these classes?

We observe first of all that every left primitive ring, having zero Jacobson radical, is a subdirect sum of right primitive rings, and conversely. Thus a subdirectly irreducible ring is left primitive if and only if it is right primitive. In particular, this applies to simple rings. A (left or right) primitive PI-ring is a full matrix ring over a division ring and hence is primitive on both sides. A left primitive ring with a minimal left or right ideal is right

Concrete Radicals and Structure Theorems

primitive. The proof of this is based on the theory of dual vector spaces and would carry us too far afield. For details see Kaplansky [2], p. 146.

In general, however, the two kinds of primitivity are not equivalent, and we now present an example, due to Bergman [1] to show this.

Let us consider the field $\mathbb{Q}(x)$ of rational expressions and the skew polynomial ring $\mathbb{Q}(x)[y;\varphi]$ with endomorphism $\varphi : \mathbb{Q}(x) \to \mathbb{Q}(x)$ given by

$$\varphi(r(x)) = r(x^2) \quad \text{for all} \quad r(x) \in \mathbb{Q}(x)$$

and without derivation. We know from Theorem 1.3.1 that $yr(x) = r(x^2)y$ and that $A = \mathbb{Q}(x)[y;\varphi]$ is a left principal ideal domain with $\ker \varphi = 0$. Then φ is an injective endomorphism.

Our example of a ring which is right, but not left primitive is a subring of A. It is one of a collection of right primitive subrings of A, each of which has a part of $\mathbb{Q}(x)$ as a faithful simple module. The next few technicalities are necessary for the setting up of this module structure on $\mathbb{Q}(x)$.

LEMMA 4.6.1. *For every $r \in \mathbb{Q}(x)$ there is a unique $r^* \in \mathbb{Q}(x)$ such that $r^*(x^2) = \frac{1}{2}(r(x) + r(-x))$.*

Proof: First let r have the special form $ax^n/(b_0 + b_1x + b_2x^2 + \ldots)$. Then

$$\frac{1}{2}(r(x) + r(-x)) =$$

$$= \frac{1}{2}\left(\frac{ax^n}{(b_0 + b_2x^2 + b_4x^4 + \ldots) + (b_1x + b_3x^3 + \ldots)}\right.$$
$$+ \left.\frac{a(-x)^n}{(b_0 + b_2x^2 + b_4x^4 + \ldots) - (b_1x + b_3x^3 + \ldots)}\right)$$

$$= \begin{cases} \dfrac{ax^n(b_0 + b_2x^2 + b_4x^4 + \ldots)}{(b_0 + b_2x^2 + b_4x^4 + \ldots)^2 - (b_1x + b_3x^3 + \ldots)^2}, & n \text{ even} \\ \dfrac{-ax^n(b_1x + b_3x^3 + \ldots)}{(b_0 + b_2x^2 + b_4x^4 + \ldots)^2 - (b_1x + b_3x^3 + \ldots)^2}, & n \text{ odd}. \end{cases}$$

As all terms in both numerator and denominator involve even powers of x, we can write

$$\frac{1}{2}(r(x) + r(-x)) = r^*(x^2).$$

As x^2 is transcendental over $\mathbb{Q}$, r^* is unique. The result for general r is obtained by a simple linearity argument. □

Note that if $r(x) = x^n$ for an integer $n > 1$, then $r^*(x) = x^{n/2}$ if n is even and 0 if n is odd.

Let $\psi : \mathbb{Q}(x) \to \mathbb{Q}(x)$ be the function defined by
$$\psi(r) = r^* \text{ for all } r.$$

LEMMA 4.6.2. *For all $r, s \in \mathbb{Q}(x)$ we have*
$$\psi(r) s = \psi(r\varphi(s)).$$

Proof: We have
$$(\psi(r) s)(x^2) = \psi(r)(x^2) s(x^2) = \frac{1}{2}(r(x) + r(-x)) s(x^2)$$
$$= \frac{1}{2}\left(r(x) s(x^2) + r(-x) s\left((-x)^2\right)\right)$$
$$= \frac{1}{2}(r(x) \varphi(s)(x) + r(-x) \varphi(s)(-x))$$
$$= \psi(r\varphi(s))(x^2),$$
so as x^2 is transcendental, $\psi(r) s = \psi(r\varphi(s))$. □

For $r \in \mathbb{Q}(x)$ let $\rho_r : \mathbb{Q}(x) \to \mathbb{Q}(x)$ be the right multiplication map : $\rho_r(t) = tr$ for all $t \in \mathbb{Q}(x)$.

LEMMA 4.6.3. *For every $s \in \mathbb{Q}(x)$ we have $\rho_s \psi = \psi \rho_{\varphi(s)}$.*

Proof: For each $r \in \mathbb{Q}(x)$, we have
$$\rho_s \psi(r) = \psi(r) s = \psi(r\varphi(s)) = \psi \rho_{\varphi(s)}(r).$$ □

Now ψ and the ρ_s are $\mathbb{Q}$-linear transformations of $\mathbb{Q}(x)$. We want to define a function from A to the ring $\operatorname{End}_{\mathbb{Q}}(\mathbb{Q}(x))$ of all such linear transformations by sending r to ρ_r for all $r \in \mathbb{Q}(x)$ and y to ψ.

Ignoring y and ψ for the moment, we certainly have a well-defined function
$$\mathbb{Q}(x) \to \operatorname{End}_{\mathbb{Q}}(\mathbb{Q}(x)); \quad r \mapsto \rho_r.$$
Since $\rho_{rs}(t) = t(rs) = (tr)s = \rho_s \rho_r(t)$ for all t, we have $\rho_{rs} = \rho_s \rho_r$ and the function is a ring antihomomorphism. If we wish to extend this function by sending y to ψ then we need $ry \mapsto \psi \rho_r$, $yr \mapsto \rho_r \psi$ and, as $ys(x) = s(x^2) y = \varphi(s)(x) y$, we need
$$\rho_s \psi = \psi \rho_{\varphi(s)}$$
which we've just shown in Lemma 4.6.3.

Thus, taking account of uniqueness of representation in A, we get a

ring antihomomorphism from A to $\operatorname{End}_{\mathbb{Q}}(\mathbb{Q}(x))$ by requiring
$$\sum r_i y^i \mapsto \sum \psi^i \rho_{r_i}.$$

Using the more convenient "scalar multiplication" description of modules, we can summarize all this as

PROPOSITION 4.6.4. *A right A-module structure can be defined on $\mathbb{Q}(x)$ using its internal multiplication and the condition*
$$r(x) y = r^*(x) \quad \forall r \in \mathbb{Q}(x).$$

We note for later use that (in the module sense)
$$x^n y = \begin{cases} x^{n/2} & \text{if } n \text{ is even} \\ 0 & \text{if } n \text{ is odd,} \end{cases}$$

and, more generally,
$$x^n y^m = \begin{cases} x^{n/2^m} & \text{if } 2^m \mid n \\ 0 & \text{if } 2^m \nmid n. \end{cases}$$

We now find a family of right primitive subrings of A. The structure specified in Proposition 4.6.4 makes $\mathbb{Q}(x)$ a module over subrings of A and when these subrings are suitably restricted, a faithful simple submodule of $\mathbb{Q}(x)$ can be found.

PROPOSITION 4.6.5. *Let P be a $\mathbb{Q}$-subalgebra of A such that $x, y \in P$. Then P is right primitive.*

Proof: Let M be the P-submodule of $\mathbb{Q}(x)$ generated by x. If $0 \neq r/s \in M$, where $r = r(x)$, $s = s(x) \in \mathbb{Q}[x]$ (i.e. they are polynomials) let $\partial(r)$ denote the degree of r and let r have leading coefficient c. Then for $2^k > \partial(r)$ we have
$$\frac{r(x)}{s(x)} c^{-1} s(x) x^{2^k - \partial(r)} y^k = c^{-1} r(x) x^{2^k - \partial(r)} y^k.$$

Now the polynomial $c^{-1} r(x) x^{2^k - \partial(r)}$ has leading term x^{2^k} and no x-free term. Hence taking account of internal multiplication of Proposition 4.6.4,
$$c^{-1} r(x) x^{2^k - \partial(r)} y^k = x^{2^k / 2^k} = x.$$

Thus x belongs to the P-submodule generated by every nonzero element of M, so M is simple.

We next show that for every nonzero member of P there is a member of M which it does not annihilate. First consider a nonzero element

$\sum_{i \geq j} r_i(x) y^i \in P$, where each r_i is a polynomial in $\mathbb{Q}[x]$.

Choose a positive integer n such that
$$n > j \quad \text{and} \quad 2^n > \partial(r_j).$$

Then, again by Proposition 4.6.4,
$$\begin{aligned} x^{2^n - \partial(r_j)} r_j(x) y^j &= x^{2^n - \partial(r_j)} \left(\ldots + a x^{\partial(r_j)} \right) y^j \\ &= \left(\ldots + a x^{2^n} \right) y^j \\ &= \ldots + a x^{2^{n-j}}, \end{aligned}$$
where $0 \neq a \in \mathbb{Q}$.

For $i > j$ we have
$$x^{2^n - \partial(r_j)} r_i(x) y^i = \left(\ldots + b_i \, x^{2^n + \partial(r_i) - \partial(r_j)} \right) y^i$$

for some $b_i \neq 0$, so $\partial\left(\left(x^{2^n - \partial(r_j)} r_i(x)\right) y^i\right) \leq (2^n + \partial(r_i) - \partial(r_j))/2^i$ by Proposition 4.6.4. (To find the degree we work down through the powers of x till we find one with a coefficient divisible by 2^i or reach the x-free term. Also, we alert the reader to the fact that y is here acting as a scalar, so the term "degree" is unambiguous — it has to mean "x-degree".) We now make a further stipulation concerning n: $2^{n-j-1} \geq \partial(r_i) - \partial(r_j)$ for all i. (Note that this may involve no extra restriction at all: e.g. it is possible that $\partial(r_i) \leq \partial(r_j)$ for all i.) We now have by Proposition 4.6.4
$$\begin{aligned} \partial\left(x^{2^n - \partial(r_j)} r_i(x) y^i\right) &\leq 2^{n-i} + (\partial(r_i) - \partial(r_j))/2^i \\ &\leq 2^{n-i} + 2^{n-j-1-i} \\ &< 2^{n-j-1} + 2^{n-j-1} = 2^{n-j}. \end{aligned}$$

Since as we have seen above, $\partial\left(x^{2^n - \partial(r_j)} r_j(x) y^j\right) = 2^{n-j}$ it follows that
$$\partial\left(x^{2^n - \partial(r_j)} \sum_{i \geq j} r_i(x) y^i\right) = 2^{n-j}.$$

In particular, $x^{2^n - \partial(r_j)} \sum_{i \geq j} r_i(x) y^i \neq 0$, where $x^{2^n - \partial(r_j)} \in M$.

For a general nonzero element $\sum \frac{r_i(x)}{s_i(x)} y^i \in P$, $(r_i, s_i \in \mathbb{Q}[x])$, we have $\prod s_i(x) \in P$ (as $x \in P$ and P is a $\mathbb{Q}$-algebra closed under multiplication by scalars) so
$$0 \neq \prod s_i(x) \sum \frac{r_i(x)}{s_i(x)} y^i \in P.$$

As above, there exists $h \in M$ such that

$$0 \neq h \left(\prod s_i(x) \sum \frac{r_i(x)}{s_i(x)} y^i \right) = \left(h \prod s_i(x) \right) \sum \frac{r_i(x)}{s_i(x)} y^i.$$

Thus M is a faithful right P-module, and P is right primitive. □

We note in passing that by Proposition 4.6.5 A itself is right primitive, and so is the twisted polynomial ring over $\mathbb{Q}$ in x, y with $yx = x^2 y$. The ring we need is an intermediate one, defined in terms of polynomial divisibility, specifically divisibility by cyclotomic polynomials.

The *cyclotomic polynomial* $\Phi_m = \Phi_m(x)$ is $\prod (x - \xi)$ where ξ ranges over the set of primitive m-th roots of 1. If ξ_0 is a particular primitive m-th root of 1 then all such m-th roots are powers of ξ_0. For example

$$\Phi_3(x) = \left(x - \frac{-1 + i\sqrt{3}}{2} \right) \left(x - \frac{-1 - i\sqrt{3}}{2} \right) =$$
$$= \left(x + \frac{1}{2} \right)^2 - \left(\frac{i\sqrt{3}}{2} \right)^2 = x^2 + x + 1$$

and $\Phi_4(x) = (x - i)(x + i) = x^2 + 1$.

Each cyclotomic polynomial is irreducible in the principal ideal domain $\mathbb{Q}[x]$ so in $\mathbb{Q}[x]$ the divisibility of a product by Φ_m requires that some factor be divisible by Φ_m, there is a well-defined highest power of Φ_m dividing any nonzero non-unit, and so on. It is not immediately obvious how much of this works when y enters the picture: for instance if Φ_m divides the $\mathbb{Q}(x)$-coefficient of $r(x) y^m \cdot s(x) y^n$ where $r, s \in \mathbb{Q}[x]$, must Φ_m divide either r or s? We are only going to use Φ_m for odd m, and here everything goes smoothly because of the following result.

PROPOSITION 4.6.6. *Let m be an odd positive integer, $r = r(x) \in \mathbb{Q}[x]$. The following conditions are equivalent.*

(i) $\Phi_m(x) \mid r(x)$,
(ii) $\Phi_m(x^2) \mid r(x^2)$,
(iii) $\Phi_m(x) \mid r(x^2)$.

Proof: (after Herstein [2]). Both $\Phi_m(x)$ and $\Phi_m(-x)$ are irreducible and they are not associates as, m being odd, ξ and $-\xi$ can't both be (primitive) m-th roots of 1. Again because m is odd, if ξ is a primitive m-th root of 1 then so is ξ^2 so that ξ is a root of $\Phi_m(x^2)$ as well as $\Phi_m(x)$. Consequently $\Phi_m(x) \mid \Phi_m(x^2)$. But then $\Phi_m(-x) \mid \Phi_m((-x)^2) = \Phi_m(x^2)$ and thus $\Phi_m(x) \Phi_m(-x) \mid \Phi_m(x^2)$. Comparing degrees and noting that all three polynomials are monic, we see that $\Phi_m(x^2) = \Phi_m(x) \Phi_m(-x)$. Now to our assertions (i), (ii), (iii).

(i) $\Rightarrow$ (ii): This is clear.

(ii) $\Rightarrow$ (iii): $\Phi_m(x) \mid \Phi_m(x^2)$.

(iii) $\Rightarrow$ (i): If $\Phi_m(x) \mid r(x^2)$, then $\Phi_m(-x) \mid r\left((-x)^2\right) = r(x^2)$, so $\Phi_m(x^2) = \Phi_m(x)\Phi_m(-x) \mid r(x^2)$. If $r(x^2) = t(x^2)\Phi_m(x^2)$ then $r(x) = t(x)\Phi_m(x)$. □

COROLLARY 4.6.7. *If for some odd m $\Phi_m(x)^k \mid r(x) \in \mathbb{Q}(x)$ and $\Phi_m(x)^{k+1} \nmid r(x)$, then $\Phi_m(x)^k \mid r(x^2)$ and $\Phi_m(x)^{k+1} \nmid r(x^2)$.*

Proof: Let $r(x) = \Phi_m(x)^k t(x)$, where $\Phi_m(x) \nmid t(x)$. Then by the proof of Proposition 4.6.6 we have

$$r(x^2) = \Phi_m(x^2)^k t(x^2) = \Phi_m(x)^k \Phi_m(-x)^k t(x^2).$$

Since $\Phi_m(x) \nmid t(x)$, Proposition 4.6.6 says that $\Phi_m(x) \nmid t(x^2)$. □

Let

$$B = \left\{ \sum \frac{p_i(x)}{q_i(x)} y^i \mid p_i(x), q_i(x) \in \mathbb{Q}[x],\ \Phi_m(x) \nmid q_i(x)\ \forall\ \text{odd}\ m \right\}.$$

We assume always that $p_i(x)$ and $q_i(x)$ have no irreducible common factors. As $\mathbb{Q}[x]$ is a principal ideal domain, this presents no problems. We shall prove that B is a right primitive ring which is not left primitive. First we have to prove it is a ring.

PROPOSITION 4.6.8. *B is a $\mathbb{Q}$-subalgebra of A.*

Proof: Let $\sum_{i=0}^{n} \frac{p_i(x)}{q_i(x)} y^i$, $\sum_{i=0}^{n} \frac{r_i(x)}{t_i(x)} y^i$ be in B (with some of the indicated coefficients possibly zero). Then

$$\sum_{i=0}^{n} \frac{p_i(x)}{q_i(x)} y^i - \sum_{i=0}^{n} \frac{r_i(x)}{t_i(x)} y^i = \sum_{i=0}^{n} \frac{p_i(x) t_i(x) - q_i(x) r_i(x)}{q_i(x) t_i(x)} y^i$$

where none of the irreducibles $\Phi_m(x)$, m odd, is a factor of $q_i(x) t_i(x)$, so that

$$\sum_{i=0}^{n} \frac{p_i(x)}{q_i(x)} y^i - \sum_{i=0}^{n} \frac{r_i(x)}{t_i(x)} y^i \in B.$$

Also

$$\sum_{i=0}^{n} \frac{p_i(x)}{q_i(x)} y^i \sum_{i=0}^{n} \frac{r_i(x)}{t_i(x)} y^i = \sum_{i,j=0}^{n} \frac{p_i(x)}{q_i(x)} y^i \frac{r_j(x)}{t_j(x)} y^j = \sum_{i,j=0}^{n} \frac{p_i(x) r_j\left(x^{2^i}\right)}{q_i(x) t_j\left(x^{2^i}\right)} y^{i+j}.$$

No $\Phi_m(x)$ (m odd) is a factor of any $q_i(x)$ or $t_j(x)$ and by Proposition 4.6.6 and induction no such $\Phi_m(x)$ is a factor of any $t_j\left(x^{2^i}\right)$. Hence

$\sum_{i=0}^{n} \frac{p_i(x)}{q_i(x)} y^i \sum_{i=0}^{n} \frac{r_i(x)}{t_i} y^i \in B$. Clearly B is closed under multiplication by rationals, so it is a $\mathbb{Q}$-subalgebra of A. □

We can now show that left and right primitivity are independent properties.

THEOREM 4.6.9. *The ring B of Proposition 4.6.8 is right primitive but not left primitive.*

Proof: Since x and y are in B, it follows from Proposition 4.6.5 and Proposition 4.6.8 that B is right primitive.

Now let L be any maximal left ideal of B (so that B/L is in effect an arbitrary simple left B-module). We shall prove that the annihilator $(0 : B/L)_B$ is nonzero, so that B/L is not faithful and B is not left primitive.

Since left ideals of A are principal, there exists an element g of A such that $AL = Ag$. If $g = \sum_{i=0}^{n} s_i(x) y^i$ and $s_n(x) \neq 0$, then $s_n(x)^{-1} \in A$ and $Ag = A s_n(x)^{-1} g$, so we can assume that g has the form $s_0(x) + s_1(x) y + \cdots + s_{n-1}(x) y^{n-1} + y^n$.

Suppose $n > 0$. Then $AL = Ag \neq A$, so AL contains no units, i.e. no nonzero members of $\mathbb{Q}(x)$. In particular AL contains no cyclotomic polynomials. Let m be an odd positive integer such that $\Phi_m(x)$ divides neither the numerator nor the denominator of any coefficient of g. Then $\Phi_m(x) \notin AL$ so $\Phi_m(x) \notin L$. As $\Phi_m(x) \in B$ and L is a maximal left ideal, we therefore have

$$B = L + B\Phi_m(x),$$

whence it follows that $1 = \ell + b\Phi_m(x)$ for some $\ell \in L$, $b \in B$. Now $1 - b\Phi_m(x) = \ell \in L \subseteq AL$ and thus there is an element $\sum \frac{r_j(x)}{t_j(x)} y^j \in A$ such that

$$1 - b\Phi_m(x) = \sum \frac{r_j(x)}{t_j(x)} y^j g.$$

Let $b = \sum \frac{p_i(x)}{q_i(x)} y^i$ where the q_i are not divisible by any cyclotomic polynomials of odd degree. Then

$$1 - b\Phi_m(x) = 1 - \frac{p_0(x)}{q_0(x)} \Phi_m(x) + \sum_{i>0} \frac{p_i(x)}{q_i(x)} y^i \Phi_m(x).$$

Taking into account that A is a skew polynomial ring with endomorphism $\varphi(r(x)) = r(x^2)$, and therefore $yr(x) = r(x^2)y$ for all $r(x) \in \mathbb{Q}(x)$, we get

$$1 - b\Phi_m(x) = 1 - \frac{p_0(x)}{q_0(x)} \Phi_m(x) + \sum_{i>0} \frac{p_i(x)}{q_i(x)} \Phi_m(x^{2^i}) y^i,$$

so the coefficient of every power of y except for y^0 has numerator divisible by $\Phi_m(x)$. Hence when we expand $\sum \frac{r_j(x)}{t_j(x)} g$, not all of its coefficients have numerators divisible by Φ_m, so Φ_m does not divide every r_j. Let $j_0 = \max\{j : \Phi_m \nmid r_j\}$. The y^{j_0+n} term in $\sum \frac{r_j(x)}{t_j(x)} y^j g$ is

$$\frac{r_{j_0}(x)}{t_{j_0}(x)} y^{j_0} y^n + \frac{r_{j_0+1}(x)}{t_{j_0+1}(x)} y^{j_0+1} s_{n-1}(x) y^{n-1} + \ldots$$

$$= \frac{r_{j_0}(x)}{t_{j_0}(x)} y^{j_0+n} + \frac{r_{j_0+1}(x)}{t_{j_0+1}(x)} s_{n-1}\left(x^{2^{j_0+1}}\right) y^{j_0+n} + \ldots.$$

Since $\Phi_m(x) \nmid r_{j_0}(x)$, $\Phi_m(x) \mid r_{j_0+1}(x)$, $\Phi_m(x) \mid r_{j_0+2}(x)$, ... and since $\Phi_m(x)$ divides neither numerator nor denominator of $s_{n-1}(x)$, $s_{n-2}(x)$, it follows that the coefficient of y^{j_0+n} does not have a numerator divisible by $\Phi_m(x)$. Comparing this with $1 - b\Phi_m(x)$ we see that $j_0 + n = 0$, whence $n = 0$. But this means that $g = 1$ and so $AL = A$.

Let $1 = \sum a_k \ell_k$, $a_k \in A$, $\ell_k \in L$. Let Φ be a product of cyclotomic polynomials long enough to put every Φa_k in B. Then

$$\Phi = \Phi 1 = \Phi \sum a_k \ell_k = (\Phi \sum a_k) \ell_k \in BL = L,$$

so $B\Phi \subseteq L$.

Let $h = \frac{d_0(x)}{e_0(x)} + \frac{d_1(x)}{e_1(x)} y + \ldots + \frac{d_w(x)}{e_w(x)} y^w$ be in B. Then (calculating in A),

$$\Phi(x) h \Phi(x)^{-1} =$$
$$= \frac{d_0(x)}{e_0(x)} + \Phi(x) \frac{d_1(x)}{e_1(x)} \Phi(x^2)^{-1} y + \ldots + \Phi(x) \frac{d_w(x)}{e_w(x)} \Phi\left(x^{2^w}\right)^{-1} y^w.$$

By repeated use of Corollary 4.6.7 we see that no new cyclotomic factors are introduced to the coefficients, so $\Phi(x) h \Phi(x)^{-1}$ is in B. Similarly $\Phi(x)^{-1} h \Phi(x) \in B$. Since $\Phi(x) h = \left(\Phi(x) h \Phi(x)^{-1}\right) \Phi(x)$ and $h\Phi(x) = \Phi(x) \left(\Phi(x)^{-1} h \Phi(x)\right)$, we have $B\Phi = \Phi B$, so by what we've seen above, $\Phi B \subseteq L$. This means that Φ is in $(0 : B/L)_B$ so the simple left B-module B/L is not faithful. □

Other examples of rings which are primitive on one side only have been obtained by Irving [1] and Jategaonkar [1].

*

Every radical partitions the class of simple rings into two classes, the radical ones and the semi-simple ones. For the Jacobson radical it is easy to find simple rings in each class. The simple zero-rings are radical, but finding idempotent simple rings which are radical is a much more difficult task. The first example of a quasi-regular idempotent simple ring was obtained by

Sąsiada [1]; it had characteristic 2. A modification of Sąsiada's construction, which produced examples in all characteristics, was subsequently published by Sąsiada and Cohn [1]. We shall treat their construction in detail, and then take a cursory look at some other constructions of simple quasi-regular rings.

Let R be the ring of formal power series over a field F in two non-commuting indeterminates x and y. (We mean that x and y commute with the elements of F but not with each other.) Let I be the set of elements of R with x and y-free term zero. Then I is the set of non-units of R, I is an ideal and I is quasi-regular: every $f \in I$ has a quasi-inverse $-f - f^2 - \ldots$. Our simple idempotent quasi-regular ring will be the heart of a subdirectly irreducible homomorphic image I/M of I, where M is maximal with respect to the exclusion of x. We first need to find an ideal of I not containing x. Let $u = x - yx^2y$. The ideal (u) generated by u in R is suitable; in fact it contains no nonzero monomials. This will be established by means of a sequence of rather technical lemmas.

Every $b \in R$ is made up of a (possibly zero) term free of x and y and monomials with right-most factor x or y. Thus we can write $b = \lambda + b'x + b''y$, where $\lambda \in F$, $b', b'' \in R$, and this representation is unique. It will be used frequently in the sequel, and often, though not always, notation similar to that just described will be used without further comment. We shall frequently use the representation and its uniqueness in making deductions like the following.

If $dx = b + c = \lambda + b'x + b''y + \mu + c'x + c''y$, then $dx = b'x + c'x$ and hence $d = b' + c'$.

LEMMA 4.6.10. *If* $c = \sum_{i=1}^{n} a_i u b_i$, $a_i, b_i \in R$, *and* b_1 *is a unit, then there exist* $\widetilde{a}_1 \in R$, $\widetilde{b}_i \in I$ $(i = 1, 2, \ldots, n)$ *such that*

$$c = \widetilde{a}_1 u \left(1 + \widetilde{b}_1\right) + \sum_{i=2}^{n} a_i u \widetilde{b}_i.$$

Proof: For $i = 1, 2, \ldots, n$, let $b_i = \beta_i + \bar{b}_i$ where $\beta_i \in F$ and every term of $\bar{b}_i$ involves x or y. As b_1 is a unit, $\beta_1 \neq 0$. For $i > 1$ we have

$$b_i = \beta_i \beta_1^{-1} b_1 + b_i - \beta_i \beta_1^{-1} b_1,$$

so

$$\sum_{i=1}^{n} a_i u b_i = \beta_1 a_1 u \left(\beta_1^{-1} b_1\right) + \sum_{i=2}^{n} a_i u \left(\beta_i \beta_1^{-1} b_1 + b_i - \beta_i \beta_1^{-1} b_1\right)$$

$$= \beta_1 a_1 u \left(\beta_1^{-1} b_1\right) + \sum_{i=2}^{n} \beta_i a_i u \beta_1^{-1} b_1 + \sum_{i=2}^{n} a_i u \left(b_i - \beta_i \beta_1^{-1} b_1\right)$$

$$= \left(\sum_{i=1}^{n} \beta_i a_i\right) u \beta_1^{-1} b_1 + \sum_{i=2}^{n} a_i u \left(b_i - \beta_i \beta_1^{-1} b_1\right).$$

Let $\tilde{b}_1 = \beta_1^{-1} b_1 - 1 = \beta_1^{-1}(\beta_1 + \bar{b}_1) - 1 = \beta_1^{-1}\bar{b}$. Then $\tilde{b}_1 \in I$ and $\beta_1^{-1} b_1 = 1 + \tilde{b}_1$. For $i \geq 2$, let $\tilde{b}_i = b_i - \beta_i \beta_1^{-1} b_1 = \beta_i + \bar{b}_i - \beta_i \beta_1^{-1}(\beta_1 + \bar{b}_1) = \bar{b}_i - \beta_i \beta_1^{-1}\bar{b}_1$; this is in I. Finally, let $\tilde{a}_1 = \sum_{i=1}^{n} \beta_i a_i$. □

LEMMA 4.6.11. *If*

$$cx + dy = \sum_{i=1}^{n} a_i u b_i$$

where $c, d \in R$ and b_1 is a unit, then there exist $b_i^* \in R$ ($i = 1, 2, \ldots, n$) such that $b_1^* \in I$ and

$$cx + dy = cu(1 + b_1^*) + \sum_{i=2}^{n} a_i u b_i^*.$$

Proof: By Lemma 4.6.10 we can assume that $b_1 = 1 + b_1' x + b_1'' y$ and $b_i = b_i' x + b_i'' y$ for all $i \geq 2$. Then

$$cx + dy = a_1 u(1 + b_1' x + b_1'' y) + \sum_{i=2}^{n} a_i u(b_i' x + b_i'' y)$$

$$= a_1 (x - yx^2 y)(1 + b_1' x + b_1'' y) + \sum_{i=2}^{n} a_i u(b_i' x + b_i'' y).$$

Comparing terms ending in x and cancelling, we get

$$c = a_1 + a_1(x - yx^2 y) b_1' + \sum_{i=2}^{n} a_i u b_i'$$

$$= a_1 + a_1 u b_1' + \sum_{i=2}^{n} a_i u b_i'$$

$$= a_1(1 + u b_1') + \sum_{i=2}^{n} a_i u b_i'.$$

Now $u b_1' \in I$ so $1 + u b_1'$ is a unit. Also $(1 + u b_1')^{-1} = 1 + u g$ for some $g \in R$, so

$$c(1 + u g) = a_1 + \sum_{i=2}^{n} a_i u b_i'(1 + u g).$$

Let $d_i = b_i'(1 + u g)$, $i \geq 2$. Then

$$a_1 = c(1 + u g) - \sum_{i=2}^{n} a_i u b_i'(1 + u g),$$

so

$$cx + dy = a_1 u b_1 + \sum_{i=2}^{n} a_i u b_i$$

$$= \left(c(1 + u g) - \sum_{i=2}^{n} a_i u d_i \right) u b_1 + \sum_{i=2}^{n} a_i u b_i$$

$$= c(1+ug)ub_1 + \sum_{i=2}^{n} a_i u (b_i - d_i u b_1)$$
$$= cu(1+gu)b_1 + \sum_{i=2}^{n} a_i u (b_i - d_i u b_1),$$

since $(1+ug)u = u + ugu = u(1+gu)$.

Now

$$(1+gu)b_1 = (1+gu)(1+b_1'x + b_1''y)$$
$$= 1 + (gub_1 + b_1'x + b_1''y)$$

and the bracketed term is in I, so we can take b_1^* equal to this. □

LEMMA 4.6.12. *If* $dy = \sum_{i=1}^{n} a_i u b_i$ *and (for this d) n is as small as possible, then $b_i \in I$ for every i.*

Proof: If not, we may assume $b_1 \notin I$, i.e. b_1 is a unit. Then by Lemma 4.6.11 we have

$$dy = 0x + dy = 0 + \sum_{i=2}^{n} a_i u b_i^*$$

for suitable b_i^* in violation of the minimality of n. □

We can now consider directly the possibility of (u) containing a nonzero monomial. As $(u) \neq R$, any nonzero monomial in (u) must have (total) degree ≥ 1. It must also be a sum of elements of the form aub, $a,b \in R$. The next result is a sort of counterpart to Lemma 4.6.12 for x.

LEMMA 4.6.13. *Suppose (u) contains a nonzero monomial and let*

$$n = \min \left\{ k \,\middle|\, \begin{array}{l} (u) \text{ contains a nonzero monomial which is a sum} \\ \text{of } k \text{ elements of the form } aub \text{ with } a,b \in R \end{array} \right\}.$$

If for some $r \leq n$ there exist monomials $c, p_1, p_2, \ldots, p_r$ of positive degree such that

$$cp_1p_2 \ldots p_r x = \sum_{i=1}^{r} cp_1p_2 \ldots p_{i-1} u d_i + \sum_{j=r+1}^{n} a_j u d_j$$

and $d_{r+1}, d_{r+2}, \ldots, d_n \in I$, then also $d_1, d_2, \ldots, d_r \in I$.

(Note: the term in the first sum corresponding to $i = 1$ is just $c u d_1$.)

Proof: Suppose some of $d_1, d_2, \ldots, d_r \notin I$. Let $d_i = \lambda_i + d_i'x + d_i''y$ for all i (with $\lambda_i = 0$ for $i > r$). Then

$$cp_1p_2\ldots p_r x = \sum_{i=1}^{n} cp_1p_2\ldots p_{i-1}u\left(\lambda_i + d_i'x + d_i''y\right)$$
$$+ \sum_{j=r+1}^{n} a_j u\left(d_j'x + d_j''y\right)$$
$$= \sum_{i=1}^{n} cp_1p_2\ldots p_{i-1}\left(x - yx^2y\right)\lambda_i$$
$$+ \sum_{i=1}^{n} cp_1p_2\ldots p_{i-1}u\,d_i'x + \sum_{i=1}^{n} cp_1p_2\ldots p_{i-1}u\,d_i''y$$
$$+ \sum_{j=r+1}^{n} a_j u\,d_j'x + \sum_{j=r+1}^{n} a_j u\,d_j''y,$$

so matching terms ending in x and cancelling x we get

$$cp_1p_2\ldots p_r = \sum_{i=1}^{r} cp_1p_2\ldots \dot{p}_{i-1}\lambda_i + \sum_{i=1}^{r} cp_1p_2\ldots p_{i-1}u\,d_i'$$
$$+ \sum_{j=r+1}^{n} a_j u\,d_j'.$$

Let

$$s = \min\{i \mid \lambda_i \neq 0\} = \min\{i \mid d_i \notin I\}.$$

(Then $s \leq r$ by assumption.)

If $s \leq i \leq r$ and $\lambda_i \neq 0$, then

$$cp_1p_2\ldots p_{i-1}\lambda_i = cp_1\ldots p_{s-1}\lambda_s p_s\ldots p_{i-1}\lambda_s^{-1}\lambda_i$$
$$= (cp_1\ldots p_{s-1}\lambda_s)\left(p_s\ldots p_{i-1}\lambda_s^{-1}\lambda_i\right).$$

Let $f = \sum_{\substack{s < i \leq r \\ \lambda_i \neq 0}} p_s\ldots p_{i-1}\lambda_s^{-1}\lambda_i$. Then f is in I and we have

$$cp_1\ldots p_r = cp_1\ldots p_{s-1}\lambda_s(1+f) + \sum_{i=1}^{r} cp_1p_2\ldots p_{i-1}u\,d_i'$$
$$+ \sum_{j=r+1}^{n} a_j u\,d_j',$$

and transferring the first term on the right hand side and the sth term in the first sum to the left, we get

$$cp_1\ldots p_r - cp_1\ldots p_{s-1}(1+f)\lambda_s - cp_1\ldots p_{s-1}u\,d_s'$$
$$= \sum_{\substack{i=1 \\ i \neq s}}^{r} cp_1\ldots p_{i-1}u\,d_i' + \sum_{j=r+1}^{n} a_j u\,d_j'.$$

Concrete Radicals and Structure Theorems

Now the left hand side is equal to

$$c_1 p_1 \ldots p_{s-1} p_s \ldots p_r - c p_1 \ldots p_{s-1} (1+f) \lambda_s - c p_1 \ldots p_{s-1} u d'_s$$
$$= c_1 p_1 \ldots p_{s-1} (p_s \ldots p_r - (1+f) \lambda_s - u d'_s).$$

The bracketed expression has x and y-free term λ_s, which is nonzero, so the expression represents a unit. Let e be its inverse. Then

$$c p_1 \ldots p_{r-1} = \sum_{\substack{i=1 \\ i \neq s}}^{r} c p_1 \ldots p_{i-1} (d'_i e) + \sum_{j=r+1}^{n} a_i u (d'_j e).$$

But this is a contradiction, as the left hand side is a nonzero monomial and the right hand sum has $n-1$ terms. $\square$

LEMMA 4.6.14. *The ideal (u) contains no nonzero monomials.*

Proof: Suppose there are nonzero monomials in (u). As noted above, they cannot have degree 0. Let

$$c = \sum_{i=1}^{n} a_i u b_i,$$

where c is a nonzero monomial, there are no monomials in (u) expressible as a sum of fewer than n terms $a u b$ and among monomials expressible as a sum of n such terms, c has smallest possible degree.

Now c has either the form $c'x$ or the form $c''y$. But if $c''y = \sum_{i=1}^{n} a_i u b_i$ then by Lemma 4.6.12 and the minimality of n, $b_i \in I$ for each i, so we have

$$c''y = \sum_{i=1}^{n} a_i u b'_i x + \sum_{i=1}^{n} a_i u b''_i y,$$

whence $c'' = \sum_{i=1}^{n} a_i u b''_i \in (u)$, while $\partial(c'') < \partial(c)$ - a contradiction. We may therefore assume that $c = c'x$ for a suitable c'. If each $b_i \in I$ then arguing as in the latter part of the $c = c''y$ case, we get the conclusion $c' \in (u)$, but $\partial(c') < \partial(c)$, another contradiction.

We may now assume $c = c'x$ and some $b_i \notin I$. We lose no generality by supposing $b_1 \notin I$. Applying Lemma 4.6.11, we get, for some $d_i \in R$ with $d_1 \in I$,

$$c = c'x = a_1 u b_1 + \sum_{i=2}^{n} a_i u b_i$$
$$= c'u(1 + d_1) + \sum_{i=2}^{n} a_i u d_i$$
$$= c'u + c'u d_1 + \sum_{i=2}^{n} a_i u d_i.$$

Hence $c'yx^2y = c'(x-u) = c'ud_1 + \sum_{i=2}^{n} a_i u d_i$. By the minimality of n, Lemma 4.6.12 implies that each d_i is in I, and $d_i = d_i''y$. Writing e_i in place of d_i'', we get

$$(*) \qquad c'yx^2 = c'ue_1 + \sum_{i=2}^{n} a_i u e_i.$$

If all e_i are in I, then all $e_i = e_i'x$. Writing f_i for e_i', we get

$$(**) \qquad c'yx = c'uf_1 + \sum_{i=2}^{n} a_i u f_i.$$

If every $f_i = f_i'x$ is in I, then writing g_i for f_i', we get

$$c'y = c'ug_1 + \sum_{i=2}^{n} a_i u g_i,$$

and then all $g_i = g_i''y \in I$ by Lemma 4.6.12, so

$$c' = c'ug_1'' + \sum_{i=2}^{n} a_i u g_i''.$$

But then $c'(1 - ug_1'') = \sum_{i=2}^{n} a_i u g_i''$, where $1 - ug_1''$ is a unit, so

$$c' = \sum_{i=2}^{n} a_i u \left(g_i'' (1 - ug_1'')^{-1} \right)$$

contradicting the minimality of n.

Thus if a contradiction is to be avoided, then in $(*)$ or $(**)$ some e_i or some f_i must not belong to I. By Lemma 4.6.13 this must happen for some $i > 1$ and we may as well assume it happens for $i = 2$. Thus we have either

$$c'(yx)x = c'ue_1 + \sum_{i=2}^{n} a_i u e_i, \quad e_2 \notin I$$

or

$$c'(y)x = c'uf_1 + \sum_{i=2}^{n} a_i u f_i, \quad f_2 \notin I.$$

Let p_1 be an appropriate choice from yx, y. Then there exist $h_1, h_2, \ldots, h_n$ such that

$$c'p_1 x = c'uh_1 + \sum_{i=2}^{n} a_i u h_i, \quad h_2 \notin I.$$

We only used the minimality of $\partial(c)$ to show that some $b_i \notin I$. We have some $h_i \notin I$, so we can now repeat the argument above with the equation

$$c'x = a_1 u b_1 + \sum_{i=2}^{n} a_i u b_i$$

replaced by
$$c'p_1 x = a_2 u h_2 + \left(c' u h_1 + \sum_{i>2} a_i u h_i\right)$$
(since h_2, like b_1, doesn't belong to I). For a suitable choice of $p_2 \in \{yx, y\}$ we end up with an equation
$$c' p_1 p_2 x = (c' p_1) u k_2 + \left(c' u k_1 + \sum_{i>2} a_i u k_i\right)$$
$$= c' u k_1 + c' p_1 u k_2 + \sum_{i>2} a_i u k_i$$
where some $k_i \notin I$. By Lemma 4.6.13 we need $i > 2$ for this and may assume $i = 3$.

After some repetitions of this procedure, we arrive at an equation
$$c' p_1 p_2 \ldots p_{n-1} x = c' u \ell_1 + c' p_1 u \ell_2 + c' p_1 p_2 u \ell_3 + \ldots$$
$$+ c' p_1 p_2 \ldots p_{n-2} u \ell_{n-1} + a_n u \ell_n$$
where $p_1, p_2, \ldots, p_{n-1} \in \{yx, y\}$ and $\ell_n \notin I$.

We apply Lemma 4.6.11 one last time to get $q_1, q_2, \ldots, q_n$ such that
$$c' p_1 p_2 \ldots p_{n-1} x = c' u q_1 + c' p_1 u q_2 + c' p_1 p_2 u q_3 + \ldots$$
$$+ c' p_1 p_2 \ldots p_{n-2} u q_{n-1} + c' p_1 p_2 \ldots p_{n-1} u (1 + q_n)$$
and thus
$$c' p_1 p_2 \ldots p_{n-1} y x^2 y = c' p_1 p_2 \ldots p_{n-1} (x - u)$$
$$= c' u q_1 + c' p_1 u q_2 + c' p_1 p_2 u q_3 + \ldots$$
$$+ c' p_1 p_2 \ldots p_{n-2} u q_{n-1} + c' p_1 p_2 \ldots p_{n-1} u q_n.$$
By Lemma 4.6.12 (as each q_i is in I and so on), we have
$$c' p_1 p_2 \ldots p_{n-1} y x^2 = c' u q_1'' + c' p_1 u q_2'' + c' p_1 p_2 u q_3'' + \ldots$$
$$+ c' p_1 p_2 \ldots p_{n-2} u q_{n-1}'' + c' p_1 p_2 \ldots p_{n-1} u q_n''.$$
By Lemma 4.6.13, each q_i'' is in I so writing $q_i'' = s_i x + r_i y$ we have
$$c' p_1 p_2 \ldots p_{n-1} y x = c' u s_1 + c' p_1 u s_2 + c' p_1 p_2 u s_3 + \ldots$$
$$+ c' p_1 p_2 \ldots p_{n-2} u s_{n-1} + c' p_1 p_2 \ldots p_{n-1} u s_n.$$
But again by Lemma 4.6.13, each s_i is in I (since the second part of the hypothesis of that Lemma is satisfied vacuously) so
$$c' p_1 p_2 \ldots p_{n-1} y = c' u s_1' + c' p_1 u s_2' + c' p_1 p_2 u s_3' + \ldots$$
$$+ c' p_1 p_2 \ldots p_{n-2} u s_{n-1}' + c' p_1 p_2 \ldots p_{n-1} u s_n'.$$

Finally, let $s'_i = t_i x + v_i y$ for each i. Then by Lemma 4.6.12

$$c' p_1 p_2 \ldots p_{n-1} = c' u\, v_1 + c' p_1 u\, v_2 + c' p_1 p_2 u\, v_3 + \ldots$$
$$+ c' p_1 p_2 \ldots p_{n-2} u\, v_{n-1} + c' p_1 p_2 \ldots p_{n-1} u\, v_n.$$

But then

$$c' p_1 p_2 \ldots p_{n-1} (1 - u v_n) = c' u\, v_1 + c' p_1 u\, v_2 + c' p_1 p_2 u\, v_3 + \ldots$$
$$+ c' p_1 p_2 \ldots p_{n-2} u\, v_{n-1}.$$

As $v_n \in I$ (Lemma 4.6.12 or Lemma 4.6.13) $1 - uv_n$ is a unit so

$$c' p_1 p_2 \ldots p_{n-1} = \sum_{i=1}^{n-1} (c' p_1 p_2 \ldots p_{i-1}) u \left(v_i (1 - uv_n)^{-1} \right).$$

This violates the minimality of n. □

COROLLARY 4.6.15. $x \notin (u)$. □

Now $u \in I$, so $(u) \triangleleft I$. As $x \notin (u)$ we can choose an ideal M of I such that $(u) \subseteq M, x \notin M$ and M is maximal for this. Thus I/M is subdirectly irreducible.

THEOREM 4.6.16. *Let M be an ideal of I containing u which is maximal with respect to the exclusion of x. The heart of the subdirectly irreducible ring I/M is a quasi-regular idempotent simple ring.*

Proof: As I is the set of non-units and $R/I \cong F$, I is quasi-regular. Thus I/M and its heart are quasi-regular. Since $x \notin M$, the heart of I/M is $(x + M)$. But $x - yx^2y = u \in M$, so $x + M = (y + M)(x + M) \cdot (x + M)(y + M)$ and thus $(x + M)$ is idempotent. □

There are other known examples of quasi-regular idempotent simple rings. Morris [1] showed that the ring of compact operators on a separable infinite dimensional Hilbert space has a subdirectly irreducible homomorphic image whose heart is idempotent and quasi-regular. Dubrovin [1] found an example which arises in a different way. There is a certain linearly ordered group with positive cone P such that for every field F there exists a localization of the semigroup algebra $F[P]$ whose Jacobson radical is simple and idempotent. When $F = \mathbb{Z}_2$ the resulting simple ring has no zero divisors.

A natural question which now arises is: how do idempotent simple Jacobson radical rings behave towards radicals smaller than the Jacobson radical? To put it another way: can radical subclasses of the Jacobson radical class contain idempotent simple rings?

Somewhat trivially, a prime radical simple ring must be a zero-ring, as it must have a nonzero nilpotent ideal. The locally nilpotent case is less obvious.

THEOREM 4.6.17 (Szász [3], Herstein [1]). *If A is a locally nilpotent simple ring, then $A^2 = 0$.*

Proof: Let a be a nonzero element of A. Then $AaA \triangleleft A$, so $AaA = A$ or 0. Suppose $AaA = A$. Then $a = \sum_{i=1}^{n} r_i a s_i$ for some $r_i, s_i \in A$. Let T be the subring of A generated by $\{r_1, r_2, \ldots, r_n, s_1, s_2, \ldots, s_n\}$. As A is locally nilpotent, we have $T^m = 0$ for some m. Now

$$a = \sum_{i=1}^{n} r_i a s_i = \sum_{i=1}^{n} r_i \left(\sum_{j=1}^{n} r_j a s_j \right) s_i = \sum_{i,j=1}^{n} r_i r_j a s_j s_i$$
$$= \sum_{i,j=1}^{n} r_i r_j \left(\sum_{k=1}^{n} r_k a s_k \right) s_j s_i = \sum_{i,j,k=1}^{n} (r_i r_j r_k) a (s_k s_j s_i)$$
$$= \ldots = \sum t \, a \, v$$

for some elements t, v of $T^m = 0$. Hence $a = 0$ — contradiction. Thus $AaA = 0$ for all a ($\neq 0$ and therefore for all a). This means that $Aa \triangleleft A$ so $Aa = 0$ or A. If $Aa = 0$ then (as a was arbitrarily chosen) $A^2 = 0$. If $Aa = A$, then $A^2 = (Aa) A = AaA = 0$. □

Levitzki's problem asks whether or not there are any idempotent simple nil rings. Ryabukhin [5] proved that the Sąsiada–Cohn example is uncountable, but that it contains countable subrings which are quasi-regular, simple and idempotent. He noted that his argument also shows that if there are simple idempotent nil rings, there are countable ones. There are some further remarks on simple nil rings in a paper of McWorter [1]. Recently *Agata Smoktunowicz* [2] *solved Levitzki's problem: to every countable field K there exists a simple nil algebra over K* (see also L'vov [1]).

4.7. Weakly primitive rings

The success of the Jacobson radical as source of structure theorems is unquestionable. Nonetheless there are anomalies connected with $\mathcal{J}$ and its sorting of rings into "good" and "bad". For instance $\mathbb{Z}$, a "good" ring by most criteria, while semisimple, is not a fundamental building block of the semisimples (that is, $\mathbb{Z}$ is not primitive, see Example 4.4.14) but is a subdirect product of finite fields which are "good". Similar comments apply to matrix rings over $\mathbb{Z}$. Furthermore, the ring J^* of rationals with odd denominators, also a commutative integral domain, is not even semisimple (the even over odd Jacobson radical ring J is an ideal in J^*, cf. Example 3.2.13 (ii)).

In this section we introduce another radical – the *weak radical* – for which there is a generalized density theorem producing a class of rings – the *weakly primitive* rings – which contains the primitive rings and, for instance $\mathbb{Z}, J^*$, and $M_n(\mathbb{Z}), M_n(J^*)$ for all n. The weakly primitive rings are prime and their upper radical is the weak radical (which is special). As an application of the weak radical we obtain the important theorem of Goldie [1] concerning prime rings with artinian (simple) rings of quotients.

To get some preliminary insight into how weak primitivity works we'll examine the rings $M_2(\mathbb{Q})$ and $M_2(\mathbb{Z})$. Since multiplication by elements of $M_2(\mathbb{Q})$ gives all linear transformations of the two dimensional $\mathbb{Q}$-vector space $\mathbb{Q}^{(2)}$, $M_2(\mathbb{Q})$ is trivially (isomorphic to) a dense ring of linear transformations. As $M_2(\mathbb{Z}) \subseteq M_2(\mathbb{Q})$ we can view $M_2(\mathbb{Z})$ as a ring of linear transformations. It is not dense, but it comes close.

Let $\begin{pmatrix} x \\ y \end{pmatrix}$ and $\begin{pmatrix} z \\ w \end{pmatrix}$ be $\mathbb{Q}$-linearly independent elements of $\mathbb{Q}^{(2)}$. Choose a nonzero integer m such that mx, my, mz and mw are integers. Then $\begin{pmatrix} mx \\ my \end{pmatrix}$ and $\begin{pmatrix} mz \\ mw \end{pmatrix}$ are in $\mathbb{Z}^{(2)}$. As $\begin{pmatrix} mx \\ my \end{pmatrix}$ and $\begin{pmatrix} mz \\ mw \end{pmatrix}$ are $\mathbb{Q}$-linearly independent ($m \neq 0$) the matrix $\begin{pmatrix} mx & mz \\ my & mw \end{pmatrix}$ is invertible in $M_2(\mathbb{Q})$. As above we can find $n \in \mathbb{Z} \setminus \{0\}$ such that $n \begin{pmatrix} mx & mz \\ my & mw \end{pmatrix}^{-1}$ is in $M_2(\mathbb{Z})$.

Now let $\begin{pmatrix} p \\ q \end{pmatrix}$ and $\begin{pmatrix} r \\ s \end{pmatrix}$ be arbitrary elements of $\mathbb{Z}^{(2)}$ and let

$$\alpha = n \begin{pmatrix} p & r \\ q & s \end{pmatrix} \begin{pmatrix} mx & mz \\ my & mw \end{pmatrix}^{-1} \in M_2(\mathbb{Z}).$$

We have $\alpha \begin{pmatrix} mx \\ my \end{pmatrix} = \begin{pmatrix} np \\ nq \end{pmatrix}$ and $\alpha \begin{pmatrix} mz \\ mw \end{pmatrix} = \begin{pmatrix} nr \\ ns \end{pmatrix}$, so

$$\alpha \begin{pmatrix} x \\ y \end{pmatrix} = n/m \begin{pmatrix} p \\ q \end{pmatrix} \quad \text{and} \quad \alpha \begin{pmatrix} z \\ w \end{pmatrix} = n/m \begin{pmatrix} r \\ s \end{pmatrix}.$$

The rational number n/m depends on $\begin{pmatrix} x \\ y \end{pmatrix}$ and $\begin{pmatrix} z \\ w \end{pmatrix}$ but not on $\begin{pmatrix} p \\ q \end{pmatrix}$ and $\begin{pmatrix} r \\ s \end{pmatrix}$: if we take other elements of $\mathbb{Z}^{(2)}$ we can find another $\mathbb{Z}$-matrix to produce the same effect. Thus $M_2(\mathbb{Z})$, as a ring of linear transformations of $\mathbb{Q}^{(2)}$, is "dense up to scalars" to adapt a phrase of Zelmanowitz [1]. This example should be borne in mind when we come to our generalized density theorem. Note that the whole example goes through, *mutatis mutandis*, if we replace $\mathbb{Z}$ by J^*.

A nonzero module is *compressible* if it can be embedded in each of its nonzero submodules. A *critically compressible* module is a compressible module which cannot be embedded in any proper factor modules.

EXAMPLE 4.7.1. (i) *Simple modules are critically compressible.*

(ii) *A $\mathbb{Z}$-module is compressible if and only if it is cyclic of prime or infinite order.* For if G is compressible, it is embeddable in each of its nonzero cyclic subgroups and is therefore cyclic. If it is finite it is embeddable in each subgroup of prime order and thus has order p for some prime p. Conversely an infinite cyclic group is isomorphic to all its nonzero subgroups while a cyclic group of prime order has no proper subgroups. As all proper factor groups of an infinite cyclic group are finite, such a group is critically compressible. By (i) prime order cyclic groups are critically compressible too. Thus *compressible $\mathbb{Z}$-modules are critically compressible.*

(iii) By our introductory remarks, $\mathbb{Z}^{(2)}$ is a *compressible $M_2(\mathbb{Z})$-module*. But every proper homomorphic image of $\mathbb{Z}^{(2)}$ as a group has torsionfree rank ≤ 1 and hence can contain no subgroup isomorphic to $\mathbb{Z}^{(2)}$, which, accordingly is *critically* compressible as an $M_2(\mathbb{Z})$-module.

A useful alternative characterization of critically compressible modules is given by the next result. A *partial endomorphism* of a module M is a homomorphism to M from one of its submodules.

PROPOSITION 4.7.2. *For a compressible module M the following are equivalent.*

(i) *M is critically compressible.*

(ii) *Every nonzero partial endomorphism of M is a monomorphism.*

Proof: (i) $\Rightarrow$ (ii). Let $f : N \to M$ be a nonzero partial endomorphism (N being a submodule of M). Then there is an embedding $g : M \to f(N)$. Since the composite homomorphism $M \xrightarrow{g} f(N) \cong N/\ker f \subseteq M/\ker f$ is an embedding and M is critically compressible, we have $\ker f = 0$, i.e. f is an embedding.

(ii) $\Rightarrow$ (i): Let K be a nonzero submodule of M such that there is an embedding $h : M \to M/K$. Let $L/K = h(M)$. Then h induces an isomorphism $h^* : M \to L/K$ and this combines with the natural homomorphism $L \to L/K$ to give a partial endomorphism $L \to L/K \xrightarrow{(h^*)^{-1}} M$ whose kernel contains K and which is therefore not an embedding. □

Recall from Section 3.14 that a module is said to be uniform if each of its nonzero submodules is essential, i.e. $N \cap K \neq 0$ for all nonzero submodules N and K. A left ideal L of a ring A is said to be *uniform* if L is a uniform A-module.

PROPOSITION 4.7.3. *If a module M satisfies* (ii) *of Proposition 4.7.2 it is uniform.*

Proof: If N and K are submodules of M with $N \cap K = 0$, then $N \oplus K = N + K \subseteq M$ and the projection $N \oplus K \to N$ is a partial endomorphism. If $N \neq 0$ the projection is nonzero so its kernel, K, is zero. □

We now present some further examples related to compressibility.

PROPOSITION 4.7.4. *Let I be an ideal of a commutative ring A with unity element. Then A/I is a compressible A-module if and only if I is prime.*

Proof: Let I be a prime ideal. For $a \in A \setminus I$, define $f : A/I \to (Aa + I)/I$ by setting $f(r + I) = ra + I$ for all $r \in A$. If $r - s \in I$ then $ra - sa \in I$, so f is well-defined. If $f(r + I) = 0$, then $ra \in I$, so as a is not in I, r must be. Thus f is an embedding. This proves that A/I is compressible. Conversely, if A/I is compressible, $ba \in I$ and $a \notin I$, there is an embedding $g : A/I \to (Aa + I)/I$ Let $g(1 + I) = ua + I$. Then

$$g(b + I) = g(b(1 + I)) = bg(1 + I) = bua + I = uba + I = 0.$$

As g is an embedding, we have $b \in I$, whence I is prime. □

PROPOSITION 4.7.5. *A ring A with unity element is compressible as an A-module if and only if every nonzero left ideal L contains an element x with $(0 : x) = 0$.*

Proof: Let L be a left ideal of a ring A with unity element which is compressible as an A-module. Let $f(1) = x$ where $f : A \to L$ is an embedding. If $rx = 0$, then $0 = rf(1) = f(r)$ so $r = 0$. Thus $(0 : x) = 0$. Conversely, if A satisfies the stated conditions then for every left ideal L we can choose $x \in L$ with $(0 : x) = 0$ and then $A \cong A/(0 : x) \cong Ax \subseteq L$. □

Note that for the second implication in this proof we did not use the unity element. Thus we have

COROLLARY 4.7.6. *If a ring has no zero-divisors, it is a compressible module over itself.* □

Let A be a ring with unity element 1_A, S an ideal of A such that

$$A = \{s + n1_A \mid n \in \mathbb{Z}, s \in S\}$$

and $S \cap \mathbb{Z}1_A = 0$. If M is a unital A-module it is clearly an S-module by "restriction of scalars". Conversely, if M is an S-module and we define $(s + n1_A)x = sx + nx$ for all $s \in S$, $n \in \mathbb{Z}$, $x \in M$, this makes M into a unital A-module. If M_1 and M_2 are S-modules or, equivalently unital A-modules, a group homomorphism between them is an S-module homomorphism if and only if it is an A-module homomorphism. The same goes for partial endomorphisms (as "S-submodule" = "unital A-submodule"). Taking account of this, and of Proposition 4.7.2, we get

PROPOSITION 4.7.7. *Let A be a ring with unity element, S an ideal of A such that additively A is the direct sum of S and $\mathbb{Z}1_A$. Then M is a (critically) compressible S-module if and only if it is a (critically) compressible unital A-module.* □

The notion of noetherian rings can be extended to modules. A module with a.c.c. on submodules is called a *noetherian module*.

PROPOSITION 4.7.8. *Noetherian compressible modules are critically compressible.*

Concrete Radicals and Structure Theorems

Proof: Let M be noetherian and compressible. If there are nonzero submodules N such that M can be embedded in M/N, then there is a maximal such submodule N_0. Let $f : M \to M/N_0$ be an embedding, $f(M) = X/N_0$, $f(N_0) = Y/N_0$. Then $N_0 \subset Y$, but we have an embedding

$$M \to M/N_0 \cong (X/N_0)/(Y/N_0) \cong X/Y \subseteq M/Y$$

contrary to the maximality of N_0. Thus there are no such submodules N and M is critically compressible. □

COROLLARY 4.7.9. *If A is a noetherian ring, all compressible A-modules are critically compressible and finitely generated.*

Proof: If M is compressible and $m \in M \setminus \{0\}$, then M is embeddable in Aa, which is noetherian. Hence M is finitely generated. Again because A is noetherian, so is M, whence by Proposition 4.7.8 M is critically compressible. □

COROLLARY 4.7.10. *The following conditions are equivalent for a module M over a principal ideal domain A:*

(i) M *is compressible;*

(ii) M *is critically compressible;*

(iii) $M \cong A/I$ *where I is a prime ideal of A.*

Proof: (i) $\iff$ (ii) by Corollary 4.7.9. Every compressible module is embeddable in its nonzero cyclic submodules and is therefore cyclic. Hence we get (i) $\iff$ (iii) from Proposition 4.7.4. □

This result places the example of $\mathbb{Z}$-modules (Example 4.7.1) in context.

By Proposition 4.7.4 a commutative ring A with unity element is a compressible A-module if and only if it has no zero-divisors. We have seen some cases in which all compressibles are critically compressibles. We now present an example to show that in general the two concepts are not equivalent, even for modules over a domain. Recall that a left Ore domain A is a ring without zero-divisors in which $Aa \cap Ab \neq 0$ for all nonzero elements $a, b \in A$.

PROPOSITION 4.7.11. *If a ring A has no zero-divisors but is not a left Ore domain, then as an A-module A is compressible but not critically compressible.*

Proof: By Propositions 4.7.4 or 4.7.6, A is compressible. Since there exist $a, b \in A$ with $Aa \cap Ab = 0$, A is not a uniform module, so by Propositions 4.7.3 and 4.7.2 A is not critically compressible. □

We are now going to show that the non-trivial critically compressible modules satisfy conditions (M1), (M2), (SM3) and (SM4) of Section 3.14 and thus define a special radical. The first thing we need is

PROPOSITION 4.7.12. *If M is a compressible A-module and $AM \neq 0$, then M is a prime module.* (For *prime modules*, see Section 3.14.)

Proof: Suppose $AM \neq 0$. If $m \in M$, $J \triangleleft A$ and $Jm = 0$, then $J(Am + \mathbb{Z}m) = 0$. If $m \neq 0$ then $Am + \mathbb{Z}m \neq 0$. But as M is compressible, it can be embedded in $Am + \mathbb{Z}m$, whence $JM = 0$. Thus M is prime. □

Let $\mathcal{C}_A$ denote the class of critically compressible A-modules M such that $AM \neq 0$, and as in Section 3.14, let $\mathcal{C} = \bigcup \mathcal{C}_A$.

We shall prove that $\mathcal{C}$ satisfies (M1), (M2), (SM3) and (SM4).

If $M \in \mathcal{C}_{A/I}$ then M is a prime $\mathcal{C}_A$-module so $AM \neq 0$. In M, A/I-submodules and A-submodules coincide. If N is a nonzero submodule of M, there is an A/I-module embedding from M to N. This is also an A-module homomorphism so M is a compressible A-module. The partial A-endomorphisms are partial A/I-endomorphisms so they are embeddings and thus M is in $\mathcal{C}_A$ by Proposition 4.7.2. Thus $\mathcal{C}$ satisfies (M1).

Now let M be in $\mathcal{C}_A$ for some ring A and let I be an ideal of A with $IM = 0$. Then M is a prime A/I-module so $(A/I)M \neq 0$ and the submodules and partial endomorphisms of M are the same whether we use A or A/I as the ring of scalars. Thus M is in $\mathcal{C}_{A/I}$ and $\mathcal{C}$ satisfies (M2).

If $M \in \mathcal{C}_A$, $B \triangleleft A$ and $BM \neq 0$, then by Proposition 3.14.13, M is a prime B-module. Let N be a nonzero B-submodule of M. If $x \in N$, then $x \in M$, so as M is a prime A-module and $BM \neq 0$ we have $Bx = 0$ if and only if $x = 0$. (In particular, $BN \neq 0$.) Now BN is an A-module with respect to

$$a \sum b_i x_i = \sum (ab_i) x_i, \ a \in A, b_i \in B, x_i \in N.$$

The argument used to prove this is much like one used in the proof of Proposition 3.14.6 but we'll give it more or less completely. If

$$\sum b_i x_i = \sum b'_j x'_j \ (b'_j \in B, x'_j \in N)$$

then for all $b \in B$ we have

$$b \sum (ab_i) x_i = \sum b(ab_i) x_i = \sum (ba) b_i x_i$$
$$= ba \sum b_i x_i = ba \sum b'_j x'_j = \sum (ba) b'_j x'_j = \sum b \left(ab'_j\right) x'_j$$
$$= b \sum ab'_j x'_j$$

(the crucial point being that ab_i, ba and ab'_j are in B). Since therefore $B(\sum(ab_i)x_i - \sum(ab'_j)x'_j) = 0$ we have $\sum (ab_i) x_i = \sum \left(ab'_j\right) x'_j$ and scalar multiplication is well-defined. Clearly now BN is an A-submodule of M. As M is compressible, there is an A-module embedding $f : M \to BN$. This is clearly a B-module homomorphism, so combining it with the inclusion $BN \to N$ we have a B-module embedding of M in N, whence M is a compressible B-module.

Concrete Radicals and Structure Theorems 241

For N as above, let $g : N \to M$ be a B-module homomorphism. Let $g' = g\,|_{BN}$ the restriction of g to BN. Then for $a \in A, b_i \in B, x_i \in N$, we have

$$g'\left(a \sum b_i x_i\right) = g'\left(\sum (ab_i) x_i\right) = g\left(\sum (ab_i) x_i\right)$$
$$= \sum ab_i g(x_i) = a \sum b_i g(x_i) = a \sum g(b_i x_i)$$
$$= a \sum g'(b_i x_i) = ag'\left(\sum b_i x_i\right),$$

so g' is an A-module homomorphism, i.e. a partial endomorphism of M as an A-module. Hence g' is an embedding. If $y \in \ker g$, then $g'(by) = g(by) = bg(y) = 0$ for all $b \in B$, so $By \subseteq \ker g' = 0$. But then $y = 0$. Thus g is an embedding. As g was an arbitrary partial endomorphism of M as a B-module, we conclude that M is in $\mathcal{C}_B$. Thus $\mathcal{C}$ satisfies (SM3).

Finally we consider (SM4). Let A be a ring, $B \triangleleft A$, $M \in \mathcal{C}_B$. By Proposition 3.14.14, BM is a prime A-module with respect to

$$a \sum b_i x_i = \sum (ab_i x_i), \ a \in A, b_i \in B, x_i \in M.$$

If $x \in M \setminus \{0\}$ then as M is a prime B-module, $Bx \neq 0$. Hence $bx \neq 0$ for some $b \in B$. But then similarly $Bbx \neq 0$ whence $Abx = A(bx) \neq 0$. As $Abx \subseteq ABx \subseteq BM$ and the latter is a prime A-module, we have $A(Abx) \neq 0$.

For every $a, a' \in A$ we have

$$a'(a(bx)) = a'((ab)x) = (a'(ab))x = ((a'a)b)x$$

(where we have used the definition of the A-module structure on BM and the fact that $ab \in B$). Thus Abx is an A-submodule of BM. If BM viewed as an A-module, the induced B-module structure coincides with the original. Thus Abx is a B-module in a unique way.

Let N be a nonzero A-submodule of BM. If $x \in N \setminus \{0\}$ there exists an element b of B as above such that Abx is a nonzero A- and B-submodule of BM. But also $Abx \subseteq Ax \subseteq N$. Using the fact that M is a compressible B-module, we see that there is a B-module embedding $f : M \to Abx$. Now for $a \in A, b_i \in B, x_i \in M$ we have

$$f\left(a \sum b_i x_i\right) = f\left(\sum (ab_i) x_i\right) = \sum (ab_i) f(x_i) = a \sum b_i f(x_i)$$
$$= a \sum f(b_i x_i) = af\left(\sum b_i x_i\right),$$

so f induces an A-module embedding $f' : BM \to Abx \subseteq N$. This shows that BM is a compressible A-module.

If $h : N \to BM$ is an A-module homomorphism (N still being an arbitrary nonzero A-submodule of BM) then as the A-module structure on BM induces (i.e. extends) the B-module structure, h is a B-module homomorphism and hence is an embedding. This proves that $BM \in \mathcal{C}_A$ and completes the verification of (SM4) for $\mathcal{C}$. Now, adapting the notation of Section 3.14 and taking account of Theorem 3.14.22 we get

THEOREM 4.7.13. (Heinicke [1]). *For each ring A, let $\mathcal{C}_A$ denote the class of critically compressible A-modules with $AM \neq 0$. Let $\gamma_\mathcal{C} = \{A \mid \mathcal{C}_A = \emptyset\}$ and let*

$$F(\mathcal{C}) = \{A \mid \mathcal{C}_A \text{ contains a faithful } A\text{-module}\}.$$

Then

(i) *$F(\mathcal{C})$ is a special class and $\gamma_\mathcal{C}$ is its upper radical class,*
(ii) *$\gamma_\mathcal{C}(A) = 0$ if and only if A is a subdirect product of rings in $F(\mathcal{C})$,*
(iii) *$\gamma_\mathcal{C}(A) = \bigcap \{(0:M)_A \mid M \in \mathcal{C}_A\}$ for all A.* □

We shall use the simpler notation ζ for $\gamma_\mathcal{C}$ and call ζ the *weak radical class*.

This terminology is due to Koh and Mewborn [2] who first studied this class. Their approach was based on a generalization of the Density Theorem rather than modules. We shall now take up the density question, and show how the rings in $F(\mathcal{C})$ can be represented by linear transformations.

An important role in the subsequent discussion is played by the endomorphism ring of the quasi-injective hull of a module. We remind the reader that quasi-injective modules have been defined in Section 1.2. Quasi-injective modules include simple and injective modules. *Every module is contained in a minimal quasi-injective extension which is unique up to isomorphism.* It is convenient in what follows to have a fixed minimal quasi-injective extension at hand for a module. Thus we introduce the following notation, which will be preserved throughout: $\widehat{M}$ *will denote a fixed injective hull of M, Λ the endomorphism ring of $\widehat{M}$ and $\overline{M}$ will denote ΛM. Then $M \subseteq \overline{M} \subseteq \widehat{M}$ and $\overline{M}$ is a minimal quasi-injective extension of M.* (See Faith [1], Chapter 3 for a justification of all claims.)

To illustrate these ideas we consider $\mathbb{Z}$-modules. We have $\overline{\mathbb{Z}} = \mathbb{Q} = \widehat{\mathbb{Z}}$ and for a prime p and a positive integer n,

$$\mathbb{Z}(p^n) = \overline{\mathbb{Z}(p^n)} \subseteq \mathbb{Z}(p^\infty) = \widehat{\mathbb{Z}(p^n)}$$

(as $\mathbb{Z}(p^n)$ is a fully invariant subgroup of $\mathbb{Z}(p^\infty)$, being the set of elements x such that $p^n x = 0$).

We call a module *monoform* if each of its partial endomorphisms is an embedding. Thus by Proposition 4.7.2 critically compressible modules are monoform and by Proposition 4.7.3 monoform modules are uniform.

LEMMA 4.7.14. *Let M be a nonzero monoform module. Then every nonzero homomorphism f from M to an essential extension M^* is an embedding.*

Proof: By Proposition 4.2.3 M is uniform. If A, B are nonzero submodules of M^* then $A \cap M, B \cap M \neq 0$ so $A \cap B \cap M = (A \cap M) \cap (B \cap M) \neq 0$

and hence $A \cap B \neq 0$. Thus M^* is uniform too. Since $f(M) \neq 0$ we have $f(M) \cap M \neq 0$. Let f_1 be the restriction of f to $f^{-1}(M)$. Then $f^{-1}(M) = \{x \in M \mid f(x) \in M\} \neq 0$ and f_1 is a nonzero partial endomorphism of M so f_1 is an embedding. But

$$\ker f = \{x \in M \mid f(x) = 0\} = \{x \in M \mid f(x) = 0 \text{ and } f(x) \in M\} = \ker f_1 = 0.$$ □

PROPOSITION 4.7.15. (Zelmanowitz [1]). *If M is monoform then each endomorphism of M has a unique extension to an endomorphism of $\overline{M}$ and the ring $\mathrm{End}\left(\overline{M}\right)$ of endomorphisms of $\overline{M}$ is a division ring.*

Proof: Every endomorphism of M induces a partial endomorphism of $\overline{M}$ which has an extension in $\mathrm{End}\left(\overline{M}\right)$. The extension will be unique if for all $h_1, h_2 \in \mathrm{End}\left(\overline{M}\right)$ which agree on M we have $h_1 = h_2$ or, equivalently, if for all $h \in \mathrm{End}\left(\overline{M}\right)$ with $h(M) = 0$ we have $h = 0$. We shall prove this latter statement. Since $\overline{M} = \Lambda M$, this means showing, for such an h, that $hf(M) = 0$ for all $f \in \Lambda$.

Let f be in Λ. Since $\overline{M} = \Lambda M$ is a fully invariant submodule of $\widehat{M}$ (which is also an injective hull of $\overline{M}$) hf induces an endomorphism of $\overline{M}$. If $f(M) = 0$, then $hf.(M) = 0$ so we can assume that $f(M) \neq 0$. As $\widehat{M}$ is uniform (being an essential extension of the uniform module M) we have $f(M) \cap M \neq 0$. But $h(M) = 0$, so

$$f(M) \cap \ker h \supseteq f(M) \cap M \neq 0.$$

This means that there exists a nonzero $x \in M$ such that $f(x) \in f(M) \cap \ker h$ and thus $hf(x) = 0$. By Lemma 4.7.14, $hf(M) = 0$ as required.

Now let g be any nonzero endomorphism of $\overline{M}$. If $\ker g \neq 0$, then $M \cap \ker g \neq 0$ so $\ker g \mid_M \neq 0$ and then Lemma 4.7.14 says that $g \mid_M = 0$, i.e. $g(M) = 0$. By what we've seen already, this means that $g = 0$. Thus if $g \neq 0$ then g is an embedding.

Let g be an embedding and let g_1 be an extension of g to Λ. (Recall that $\widehat{M}$ is an injective hull of $\overline{M}$.) Then $\ker g_1 \cap \overline{M} = \ker g = 0$, so $\ker g_1 = 0$. Thus $\widehat{M} \cong g_1\left(\widehat{M}\right) \subseteq \widehat{M}$, so $g_1\left(\widehat{M}\right)$ is a direct summand of $\widehat{M}$. But M is uniform, so $\widehat{M}$ is indecomposable, whence $g_1\left(\widehat{M}\right) = \widehat{M}$ and g_1 is an isomorphism. Since $\overline{M} (= \Lambda M)$ is fully invariant in $\widehat{M}$, g_1^{-1} induces an inverse g^{-1} for g in $\mathrm{End}\left(\overline{M}\right)$. Thus $\mathrm{End}\left(\overline{M}\right)$ is a division ring. □

COROLLARY 4.7.16. *A critically compressible module M is simple if and only if it is quasi-injective.*

Proof: If M is quasi-injective then we can take $M = \overline{M}$ and then by Proposition 4.7.15 $\mathrm{End}(M)$ is a division ring so all nonzero endomorphisms of M are surjective. But for every nonzero submodule N of M there is an embedding of M into N which can be interpreted as an endomorphism of M, whence $N = M$ and M is simple. The converse is clear. □

Note that when M is simple, Proposition 4.7.15 is just Schur's Lemma; Proposition 4.7.15 is the first step in the proof of the generalized Density Theorem as Schur's Lemma is for the Density Theorem. We need two results involving quasi-injective modules before we can prove the generalized theorem.

PROPOSITION 4.7.17. (Anderson and Fuller [1]). *Let M be a quasi-injective module, $\{M_i \mid i \in I\}$ a set of isomorphic copies of M and let K be a submodule of $\oplus_{i \in I} M_i$. Then every homomorphism $h : K \to M$ has an extension $\overline{h} : \oplus_{i \in I} M_i \to M$.*

Proof: Let $\mathcal{F} = \{L \xrightarrow{f} M \mid K \subseteq L \subseteq \oplus_{i \in I} M_i \text{ and } f \mid_K = h\}$. Then $K \xrightarrow{h} M \in \mathcal{F}$, so $\mathcal{F} \neq \emptyset$. We can partially order $\mathcal{F}$ by defining $L_1 \xrightarrow{f_1} M \leq L_2 \xrightarrow{f_2} M$ to mean $L_1 \subseteq L_2$ and $f_2 \mid_{L_1} = f_1$. An application of Zorn's Lemma yields a maximal member $N \xrightarrow{\overline{h}} M$ of $\mathcal{F}$. Let $K_i = N \cap M_i$ for each i. As M is quasi-injective and $M \cong M_i$ there exists, for each i, a homomorphism $\overline{h}_i : M_i \to M$ such that $\overline{h}_i \mid_{K_i} = \overline{h} \mid_{K_i}$. Define $g_i : (M_i + N) \to M$ by setting

$$g_i(x_i + y) = \overline{h}_i(x_i) + \overline{h}(y) \text{ for all } x_i \in M_i, \ y \in N.$$

If $x_i + y = x_i' + y'$ ($x_i, x_i' \in M_i, y, y' \in N$) then $y - y' = x_i' - x_i \in N \cap M_i = K_i$ so $\overline{h}(y - y') = \overline{h}_i(y - y') = \overline{h}_i(x_i' - x_i)$ and thus $\overline{h}_i(x_i) + \overline{h}(y) = \overline{h}_i(x_i') + \overline{h}(y')$. Hence g_i is well-defined. Since $g_i \mid_N = \overline{h} \mid_N$ and hence $g_i \mid_K = \overline{h} \mid_K = h \mid_K$ we have $M_i + N \xrightarrow{g_i} M \in \mathcal{F}$. By the maximality of $\overline{h}$, M_i must be contained in N. This being so for all i, we have $N = \oplus_{i \in I} M_i$. □

PROPOSITION 4.7.18. (Cf. Johnson and Wong [1]). *Let M be a quasi-injective A-module satisfying the condition*

$$m \in M \text{ and } Am = 0 \Rightarrow m = 0.$$

Let $m_1, m_2, \ldots, m_t, m$ be in M. Then

$$(0 : m) \supseteq (0 : m_1) \cap (0 : m_2) \cap \ldots \cap (0 : m_t)$$

if and only if there exist $f_1, f_2, \ldots, f_t \in \mathrm{End}(M)$ such that

$$m = f_1(m_1) + f_2(m_2) + \ldots + f_t(m_t).$$

Concrete Radicals and Structure Theorems

Proof: If $(0:m) \supseteq \bigcap_{i=1}^{t}(0:m_i)$ then if we treat $(m_1, m_2, \ldots m_t)$ as a member of $M_1 \oplus M_2 \oplus \ldots M_t$, where each $M_i = M$, the function

$$h : A(m_1, m_2, \ldots m_t) \to Am; \ h(r(m_1, m_2, \ldots, m_t)) = rm$$

is well-defined and clearly a homomorphism. By Proposition 4.7.17 h has an extension $\overline{h} : M_1 \oplus M_2 \oplus \ldots \oplus M_t \to M$. For $i = 1, 2, \ldots, t$, let $\epsilon_i : M_i \to M_1 \oplus M_2 \oplus \ldots \oplus M_t$ and $\pi_i : M_1 \oplus M_2 \oplus \ldots \oplus M_t \to M_i$ be the natural injection and projection.

For every $r \in A$, we have

$$rm = h(rm_1, rm_2, \ldots, rm_t) = \overline{h}(rm_1, rm_2, \ldots rm_t)$$
$$= \overline{h}\sum_{i=1}^{t} \epsilon_i \pi_i (rm_1, rm_2, \ldots, rm_t)$$
$$= \overline{h}\sum_{i=1}^{t} \epsilon_i (rm_i) = \sum_{i=1}^{t} \overline{h}\epsilon_i (rm_i) = \sum_{i=1}^{t} r\overline{h}\epsilon_i (m_i) = r\sum_{i=1}^{t} \overline{h}\epsilon_i (m_i).$$

Let $f_i = \overline{h}\epsilon_i : M \to M$, $i = 1, 2, \ldots, t$. Then we have

$$r\left(m - \sum_{i=1}^{t} f_i(m_i)\right) = r\left(m - \sum_{i=1}^{t} \overline{h}\epsilon_i(m_i)\right) = 0$$

so that $m = \sum_{i=1}^{t} f_i(m_i)$.

The converse is clear. $\square$

We shall call a nonzero ring *weakly primitive* if it has a faithful critically compressible module.

Thus as simple modules are critically compressible, primitive rings are weakly primitive, while by Proposition 4.7.12 and Corollary 3.14.18 weakly primitive rings are prime.

By Theorem 4.7.13 we may call a ring A *weakly semiprimitive* if $\zeta(A) = 0$, that is, A is a subdirect sum of weakly primitive rings.

For a ring A, an *A-lattice* is a triple (D, V, M) where
 (i) D is a division ring,
 (ii) V is an A-module and a D-vector space,
 (iii) $d(rv) = r(dv)$ for all $r \in A$, $d \in D$, $v \in V$,
 (iv) M is a faithful A-submodule of V and
 (v) $DM = V$.

EXAMPLE 4.7.19. $(\mathbb{Q}, \mathbb{Q}^{(2)}, \mathbb{Z}^{(2)})$ is an $M_2(\mathbb{Z})$-lattice.

GENERALIZED DENSITY THEOREM 4.7.20 (Zelmanowitz [1]). *The following conditions are equivalent for a ring A.*
 (i) *A is weakly primitive.*

(ii) *There exists an A-lattice* (D, V, M) *such that if* $\{v_1, v_2, \ldots, v_k\}$ *is a D-linearly independent subset of V there exists* $d \in D \setminus \{0\}$ *such that for all* $x_1, x_2, \ldots, x_k \in M$ *there exists an* $r \in A$ *such that* $dx_i = rv_i$ *for all* i.

(iii) *There exists an A-lattice* (D, V, M) *such that for every* $\tau \in \mathrm{End}_D(V)$ *and for all finite dimensional subspaces* U', U *of* V *such that* $U' = \sum_{i=1}^{k} Dm_i$, $m_i \in M$, *there are elements* r, s *of* A *such that* $r\tau(u) = su$ *for all* $u \in U$ *and multiplication by* r *defines an automorphism of* U'.

Remark. (ii) says, informally, that if $v_1, v_2, \ldots, v_k$ are linearly independent and $x_1, x_2, \ldots, x_k$ arbitrary then there is an $r \in A$ which takes the v_i elements not to the x_i but to scalar multiples of them by a $d \in D$ which depends only on $v_1, v_2, \ldots, v_k$. See also the motivating example at the beginning of this section involving $M_2(\mathbb{Z})$.

Proof: (i) $\Rightarrow$ (ii): Let M be a faithful critically compressible A-module, $\overline{M} = \mathrm{End}\left(\widehat{M}\right) M$ a quasi-injective hull of M, $D = \mathrm{End}_A\left(\overline{M}\right)$. (Here, as before, $\widehat{M}$ is an injective hull of M.) By Proposition 4.7.15, D is a division ring and $\overline{M}$ a D-vector space. Let f be an R-endomorphism of $\widehat{M}$. Then $f\left(\overline{M}\right) \subseteq \overline{M}$ so the effect of f on M is the same as that of its restriction to $\overline{M}$. Hence $\overline{M} = \mathrm{End}\left(\widehat{M}\right) M = \mathrm{End}\left(\overline{M}\right) M = DM$. The other requirements are obviously met, so $(D, \overline{M}, M)$ is an R-lattice.

We wish to use Proposition 4.7.18; for this we have to show that if $v \in \overline{M} \setminus \{0\}$, then $Av \neq 0$. As $\overline{M}$ is an essential extension of M we have $M \cap (Av + \mathbb{Z}v) \neq 0$. As M is compressible there is an A-module embedding $f: M \to M \cap (Av + \mathbb{Z}v)$. Since M is a faithful A-module, if $r \in A \setminus \{0\}$ then $rm \neq 0$ for some $m \in M$. We then have $0 \neq f(m) = sv + \ell v$ for some $s \in R$, $\ell \in \mathbb{Z}$. But then $0 \neq f(rm) = rf(m) = r(sv + \ell v) = (rs + \ell r)v \in Rv$.

Let $v_1, v_2, \ldots v_k \in \overline{M}$ be linearly independent over D. For $i = 1, 2, \ldots, k$, let $A_i = \bigcap_{j \neq i} (0 : v_j)$. Then each A_i is a left ideal of A and by Proposition 4.7.18 (as $v_1, v_2, \ldots, v_k$ are linearly independent over $D = \mathrm{End}_A\left(\overline{M}\right)$) $A_i \not\subseteq (0 : v_i)$, i.e. $A_i v_i \neq 0$, for all i. As M is uniform by Propositions 4.7.2 and 4.7.3, so is its essential extension $\overline{M}$. Hence $\bigcap_{i=i}^{k} A_i v_i \cap M \neq 0$. As M is compressible, there is an A-module embedding $g: M \to \bigcap_{i=1}^{k} A_i v_i \cap M$. Thus for all $x_1, x_2, \ldots x_k \in M$ there exist $r_1 \in A_1, r_2 \in A_2, \ldots r_k \in A_k$ such that $g(x_i) = r_i v_i \in M$. Let $r = r_1 + r_2 + \ldots + r_k$. Then for $i = 1, 2, \ldots, k$, we have
$$rv_i = r_i v_i + \sum_{j \neq i} r_j v_i = r_i v_i = g(x_i).$$

Now g extends to an endomorphism of $\widehat{M}$; call this d. Then $d(x_i) = rv_i$ for $i = 1, 2, \ldots, k$.

(ii) $\Rightarrow$ (iii): There is no loss of generality in assuming $m_1, m_2, \ldots, m_k$ are D-linearly independent. Since $U \subseteq V = DM$ there exist $m_{k+1}, \ldots, m_\ell \in M$ such that $U \subseteq \sum_{i=k+1}^{\ell} Dm_i$. Let $V' = \sum_{i=k+1}^{\ell} Dm_i$. Then $U' + V'$ is spanned by $\{m_1, m_2, \ldots, m_k, m_{k+1}, \ldots, m_\ell\}$. Eliminating some of $m_{k+1}, \ldots, m_\ell$ if necessary, we can assume that $\{m_1, m_2, \ldots, m_k, m_{k+1}, \ldots, m_\ell\}$ is a basis for $U' + V'$. Call this space W.

Let $W_h = \sum_{i=1}^{h} Dm_i$. We shall prove by induction on h that there exist $r_h, s_h \in A$ such that $r_h \tau(w) = s_h w$ for all $w \in W_h$ and multiplication by r_h defines an automorphism of U'. The case $h = \ell$ will then give us (iii), as $U \subseteq W = W_\ell$.

We first consider $W_0 = 0$. By (ii) there exists $d \in D\setminus\{0\}$ such that for some $r_0 \in A$, $dm_i = r_0 m_i$ for $i = 1, 2, \ldots, k$. Trivially $\tau(r_0 w) - sw = 0$ for every $w \in W_0$ and for every $s \in A$. Also multiplication by r_0 is the same as multiplication by d in U' and this is an automorphism.

Now take $1 \leq h \leq \ell$ where there exist $r_{h-1}, s_{h-1} \in A$ such that $r_{h-1} \tau(w) = s_{h-1} w$ for all $w \in W_{h-1}$ and multiplication by r_{h-1} defines an automorphism of U'. Let $\tau' \in \operatorname{End}_D(V)$ be defined by

$$\tau'(v) = r_{h-1} \tau(v) - s_{h-1} v \quad \text{for all} \quad v \in V.$$

If we can find $r, s \in A$ such that $r\tau'(w) = sw$ for every $w \in W_h$ and multiplication by r defines an automorphism of U' we are finished, as then for all $w \in W_h$ we have

$$rr_{h-1} \tau(w) = r\tau'(w) + rs_{h-1} w = (s + rs_{h-1}) w$$

for all $w \in W_h$ and multiplication by rr_{h-1} defines an automorphism of U', so we can take

$$r_h = rr_{h-1} \quad \text{and} \quad s_h = s + rs_{h-1}.$$

The argument now splits to consider two cases.

(1) $\tau'(m_h) \notin U' + W_h$. In this case, if $h \leq k$ we have $\tau'(m_h) \notin U'$, i.e. $\tau'(m_h)$ is not in the span of $\{m_1, m_2, \ldots, m_k\}$ while if $h > k$, then $\tau'(m_h)$ is not in the span of $\{m_1, m_2, \ldots, m_k, m_{k+1}, \ldots, m_h\}$. Let $p = \max\{h, k\}$. Then by (ii), using the linearly independent set $\{m_1, m_2, \ldots, m_p, \tau'(m_h)\}$ and the sets

$$\{m_1, m_2, \ldots, m_h\} \quad \text{and} \quad \{0, 0, \ldots, 0, m_h, 0, 0, \ldots, 0\}$$

we can find $e \in D\setminus\{0\}$ and $r, s \in R$ such that

$$rm_1 = em_1, \quad rm_2 = em_2, \quad \ldots, \quad rm_p = em_p, \quad r\tau'(m_h) = em_h$$

and
$$sm_i = e0 = 0 \text{ for } i \neq h, \quad sm_h = em_h, \quad s\tau'(m_h) = e0 = 0.$$
On U' multiplication by r is the same as multiplication by e and is thus an automorphism. For $1 \leq i < h$, $sm_i = 0$. But $m_i \in W_{h-1}$ so $0 = r_{h-1}\tau(m_i) - s_{h-1}m_i = \tau'(m_i)$ and hence $r\tau'(m_i) = 0$. Also $r\tau'(m_h) = em_h$ and $sm_h = em_h$ so $r\tau'(m_h) = sm_h$. Thus $r\tau'(w) = sw$ for every $w \in W_h$ as required.

(2) $\tau'(m_h) \in U' + W_h$. If $\tau'(m_h) = 0$, then using the linearly independent set $\{m_1, m_2, \ldots, m_k\}$ we get nonzero $e \in D$, $r \in R$ such that $rm_i = em_i$ for $i = 1, 2, \ldots, k$. Then multiplication by r defines an automorphism of U'. If $j < h$, then $r\tau'(m_j) = r(r_{h-1}\tau(m_j) - s_{h-1}m_j) = 0$, while $r\tau'(m_h) = 0$. Thus we can take $s = 0$.

The remaining possibility is that there exist nonzero $c_{i_\mu} \in D$ with
$$0 \neq \tau'(m_h) = \sum_{\mu=1}^{\lambda} c_{i_\mu} m_{i_\mu} \text{ where } 1 \leq i_1 < \ldots < i_\lambda = p = \max\{h, k\}.$$
Again we use (ii). We associate $a \in D \setminus \{0\}$ with $\{m_1, m_2, \ldots, m_p\}$ and $b \in D \setminus \{0\}$ with $\{a^{-1}c_{i_1}m_{i_1}, \ldots, a^{-1}c_{i_\lambda}m_{i_\lambda}\} \cup \{m_i \mid i \neq i_\mu \forall \mu\}$. There exist $r, s, \in A$ such that $ra^{-1}c_{i_\mu}m_{i_\mu} = bm_{i_\mu}$ for each μ, $rm_i = bm_i$ for $i \notin \{i_1, \ldots, i_\lambda\}$, $sm_i = a0 = 0$ for $i \neq h$ and $sm_h = a\sum_{\mu=1}^{\lambda} bm_{i_\mu}$. Examining its effect on basis elements, we see that multiplication by r defines an automorphism of U'. For $1 \leq i < h$ we have $r\tau'(m_i) = 0 = sm_i$, while
$$r\tau'(m_h) = r\sum_{\mu=1}^{\lambda} c_{i_\mu} m_{i_\mu} = \sum_{\mu=1}^{\lambda} rc_{i_\mu} m_{i_\mu} = a\sum_{\mu=1}^{\lambda} ra^{-1}c_{i_\mu} m_{i_\mu} = a\sum_{\mu=1}^{\lambda} bm_{i_\mu} = sm_h.$$
Thus $r\tau'(w) = sw$ for all $w \in W_h$. This completes the proof that (ii) $\Rightarrow$ (iii).

(iii) $\Rightarrow$ (i): Let M be as in (iii), $0 \neq m \in M$. We'll show that $Am + \mathbb{Z}m$ is a faithful critically compressible A-module. Let t be in $A \setminus \{0\}$. As M is a faithful A-module, $tM \neq 0$. Let $x \in M$ be such that $tx \neq 0$. There exists $\tau \in \text{End}_D(V)$ such that $\tau(m) = x$. Using (iii) with $U = Dm$ and $U' = Dx$, we can find $r, s \in R$ such that $r\tau(\delta m) = s\delta m$ for all $\delta \in D$ and multiplication by r defines an automorphism of Dx. Then $tsm = tr\tau(m) = trx$. Now as r defines an automorphism on Dx, $rx \in Dx$, so $rx = \epsilon x$ for some $\epsilon \in D \setminus \{0\}$. This gives us $trx = t\epsilon x = \epsilon tx \neq 0$ as $tx \neq 0$, whence $tsm \neq 0$. Thus $t(Am + \mathbb{Z}m) \neq 0$ and so $Am + \mathbb{Z}m$ is a faithful A-module.

Let N be a nonzero A-submodule of $Am + \mathbb{Z}m$. If $y \in N \setminus \{0\}$ let $\sigma(y) = m$, $\sigma \in \text{End}_D(V)$. Choose $r^*, s^* \in A$ such that $r^*\sigma(\delta y) = s^*\delta y$ for all $\delta \in D$ and multiplication by r^* defines an automorphism of Dm (that is, take $U = Dy$ and $U' = Dm$ in (iii)). Let $r^*m = \gamma m$, $\gamma \in D \setminus \{0\}$. Then we have
$$0 \neq \gamma m = r^*m = r^*\sigma(y) = s^*y \in N.$$
Define $g : Am + \mathbb{Z}m \to N$ by setting
$$g(\ell m + km) = \gamma(\ell m + km) = \ell\gamma m + k\gamma m$$
$$= \ell s^*y + ks^*y$$

for all $\ell \in A$, $k \in \mathbb{Z}$. Then g is an embedding, so $Rm + \mathbb{Z}m$ is compressible.

We finally show that $Am + \mathbb{Z}m$ is critically compressible by showing that M is monoform. Let K be an A-submodule of M, $f : K \to M$ a nonzero A-module homomorphism and let $z \in N$ be such that $f(z) \neq 0$. Let W be in $K \setminus \{0\}$. Then there exists $\alpha \in \operatorname{End}_D(V)$ such that $\alpha(w) = z$. Using (iii) once more, we can find $\tilde{r}, \tilde{s} \in A$ such that $\tilde{r}\alpha(\delta w) = \tilde{s}\delta w$ for all $\delta \in D$, and multiplication by $\tilde{r}$ defines an automorphism of $Df(z)$. Then

$$\tilde{s}f(w) = f(\tilde{s}w) = f(\tilde{r}\alpha(w)) = f(\tilde{r}z) = \tilde{r}f(z) \neq 0.$$

Hence $f(w) \neq 0$. Thus f is an embedding. In particular if $K \subseteq Am + \mathbb{Z}m$ and f takes its values in $Am + \mathbb{Z}m$ then f is an embedding. Hence $Am + \mathbb{Z}m$ is monoform and therefore critically compressible by Proposition 4.7.2.

This completes the proof. □

Theorem 4.7.20 (in the form presented here) is due to Zelmanowitz [1]. Earlier versions were obtained by Heinicke [1], Koh and Mewborn [1] and Koh and Luh [1]. Critically compressible modules are also called *quasi-simple modules* in the literature. Further information about these modules and about connections between the module properties we have made use of and others (*rational extensions* and so on) can be obtained from the survey articles of Koh [1] and Storrer [1].

Now (preserving all the notation of Theorem 4.7.20) suppose V is finite dimensional over D. Then as $DM = V$, i.e. V is spanned by M, V has a basis in M. Thus in Theorem 4.7.20 (iii) we can take $U' = U = V$ and conclude that for every $\tau \in \operatorname{End}_D(V) \cong M_{\dim(V)}(D)$ there exist $r, s \in A$ such that multiplication by r defines an automorphism of V and $r\tau(u) = su$ for all $u \in V$. Identifying elements of A with the corresponding linear transformations in $\operatorname{End}_D(V)$ we can write $r\tau = s$ and thus $\tau = r^{-1}s$ and we have in view of Proposition 4.7.15 the following.

COROLLARY 4.7.21. (Koh and Luh [1], Zelmanowitz [1]). *If A has a faithful critically compressible module M whose quasi-injective hull $\overline{M}$ has finite dimension k over the division ring $\operatorname{End}_A(\overline{M})$, then A has the matrix ring $M_k(\operatorname{End}_A(\overline{M}))$ as a ring of left quotients.* □

We shall use Corollary 4.7.21 to prove an important theorem of Goldie [1] concerning prime rings with artinian quotient rings: it turns out that the relevant prime rings satisfy the condition of Corollary 4.7.21.

A ring A is called a *(left) Goldie ring* if
(i) A satisfies the ascending chain condition on annihilator left ideals (i.e. left ideals which are annihilators of subsets of A) and
(ii) A has no infinite set of left ideals whose sum is direct.

If a ring A has infinitely many left ideals $I_1, I_2, \ldots$ such that $\sum I_n = \oplus I_n$, then (assuming each $I_n \neq 0$)

$$I_1 \subset I_1 \oplus I_2 \subset I_1 \oplus I_2 \oplus I_3 \subset \ldots \subset I_1 \oplus I_2 \oplus \ldots \oplus I_n \subset \ldots$$

so A is not noetherian. Thus noetherian rings satisfy (ii). Clearly they satisfy (i) as well so noetherian rings are Goldie rings.

We need a bit more information about prime Goldie rings. The next few results provide it.

PROPOSITION 4.7.22. *Let A be any ring and*

$$Z(A) = \{x \in A \mid (0:x) \text{ is an essential left ideal}\}.$$

Then $Z(A)$ is an ideal, called the singular ideal *of A.*

Proof: As $(0:0) = A$ we have $0 \in Z(A)$ and $Z(A) \neq \emptyset$. If $x, y \in Z(A)$ then $(0:x-y) \supseteq (0:x) \cap (0:y)$ so $(0:x-y)$ is essential and $x-y \in Z(A)$. If $x \in Z(A)$ and $r \in A$, then $(0:xr) \supseteq (0:x)$ so $(0:xr)$ is essential and $xr \in Z(A)$. Finally we consider rx with $r \in A$ and $x \in Z(A)$. We have

$$(0:rx) = \{a \in A \mid ar \in (0:x)\}.$$

Let L be any nonzero left ideal. If $Lr = 0$ then $Lr \subseteq (0:x)$ so $L \subseteq (0:rx)$. If $Lr \neq 0$, then as Lr is a left ideal we have $Lr \cap (0:x) \neq 0$. Hence there exists $\ell \in L$ with $0 \neq \ell r \in (0:x)$, so $\ell \neq 0$ and $\ell \in (0:rx)$. Thus $L \cap (0:rx) \neq 0$, $(0:rx)$ is essential and $rx \in Z(A)$. □

PROPOSITION 4.7.23. *Every nonzero Goldie ring has a nonzero uniform left ideal.*

Proof: Let $A \neq 0$ be a ring with no nonzero uniform left ideals. We shall prove that A is not a Goldie ring. Since A itself is not uniform, there are left ideals $L_1, K_1 \neq 0$ such that $L_1 \cap K_1 = 0$. As L_1 is not uniform, there exist left ideals $L_2, K_2 \neq 0$ such that $L_2, K_2 \subseteq L_1$ and $L_2 \cap K_2 \neq 0$. Similarly there are nonzero $L_3, K_3 \subseteq L_2$ with $L_3 \cap K_3 = 0$ and in general for $n > 1$ there exist $L_n, K_n \subseteq L_{n-1}$ with $L_n \cap K_n = 0$. In particular, we have a chain $L_1 \supseteq L_2 \supseteq \ldots \supseteq L_n \supseteq \ldots$. Consider the sum $\sum_{n \geq 2} K_n$. If $x_{n_1} + x_{n_2} + \ldots + x_{n_m} = 0$, where $x_{n_1} \in K_{n_1}, x_{n_2} \in K_{n_2}, \ldots, x_{n_m} \in K_{n_m}$ and $n_1 < n_2 < \ldots < n_m$, then

$$x_{n_1} = -x_{n_2} - \ldots - x_{n_m} \in K_{n_1} \cap (K_{n_2} + \ldots + K_{n_m}) \subseteq$$

$$\subseteq K_{n_1} \cap (L_{n_2-1} + \ldots L_{n_m-1}) \subseteq K_{n_1} \cap L_{n_2-1} \subseteq K_{n_1} \cap L_{n_1} = 0.$$

By repetitions of this argument we show that similarly $x_{n_2} = 0, \ldots, x_{n_m} = 0$. Thus the sum $\sum_{n \geq 2} K_n$ is direct and hence (as each $K_n \neq 0$) R is not a Goldie ring. □

PROPOSITION 4.7.24. *Let A be a prime ring with $Z(A) = 0$. If A has a uniform left ideal $L \neq 0$, then L is a faithful critically compressible A-module.*

Proof: Let K be a nonzero left ideal of A with $K \subseteq L$, $f : K \to L$ a nonzero A-module homomorphism. Let $x \in K$ be such that $f(x) \neq 0$. As $Z(A) = 0$, $(0 : f(x))$ is not an essential left ideal. Let $J \neq 0$ be a left ideal with $J \cap (0 : f(x)) = 0$. Then

$$Jx \cap \ker f = \{jx \mid j \in J;\ jf(x) = f(jx) = 0\}$$
$$= \{J \cap (0 : f(x))\}\, x = 0.$$

Now $Jx \neq 0$, since otherwise we'd have $jf(x) = f(jx) = f(0) = 0$ for all $j \in J$, i.e. $J \subseteq (0 : f(x))$, whereas $J \cap (0 : f(x)) = 0 \neq J$. Hence, by the uniformity of L, we have $\ker f = 0$. This proves that

all nonzero partial A-module endomorphisms of L are embeddings.

Again let $K \neq 0$ be a left ideal with $K \subseteq L$. For $k \in K \setminus \{0\}$ the function

$$L \to K;\quad \ell \mapsto \ell k \text{ for all } \ell,$$

is a partial endomorphism, so by the italicized statement above it is an embedding. Hence L is compressible and, by the same italicized statement, critically compressible. Since A is prime, $(0 : L)_R = 0$, i.e. L is faithful. □

PROPOSITION 4.7.25. *If A is a prime Goldie ring, then $Z(A) = 0$.*

Proof: Suppose $Z(A) \neq 0$ and let $\mathcal{I} = \{(0 : x) \mid x \in Z(A) \setminus \{0\}\}$. Then $\mathcal{I}$ is a non-empty set of annihilator left ideals, so it has a maximal member, $(0 : x_0)$ $(0 \neq x_0 \in Z(A))$. Since $Z(A)$ is a nonzero ideal and A is prime, we have $(0 : Z(A))_A = 0$. In particular, $x_0 Z(A) \neq 0$ so there exists an $x \in Z(A)$ with $x_0 x \neq 0$. Clearly $(0 : x_0 x) \supseteq (0 : x_0) \in \mathcal{I}$. Again because A is prime we have $Ax_0 \neq 0$ so, as $(0 : x)$ is essential, $Ax_0 \cap (0 : x) \neq 0$. Hence there exists $a \in A$ with $ax_0 x = 0 \neq ax_0$. But then $(0 : x_0) \subset (0 : x_0 x)$, which contradicts the maximality of $(0 : x_0)$. Thus $Z(A) = 0$ as required. □

By Propositions 4.7.25, 4.7.23 and 4.7.24, every prime Goldie ring has a nonzero uniform left ideal, which is a faithful critically compressible module. Thus prime Goldie rings are weakly primitive. We can say more.

THEOREM 4.7.26. (Koh and Luh [1]). *Let A be a prime Goldie ring, L a nonzero uniform (whence faithful critically compressible) left ideal, $\overline{L}$ the quasi-injective hull of L, $D = \mathrm{End}_R(\overline{L})$. Then $\overline{L}$ is finite dimensional over D.*

Proof: Suppose not, and let $\{v_n \mid n = 1, 2, 3, \ldots\}$ be a D-linearly independent set in $\overline{L}$. Let $I_1 = (0 : v_1)$, $I_2 = (0 : v_1) \cap (0 : v_2)$, and in general let

$$I_n = (0 : v_1) \cap (0 : v_2) \cap \ldots \cap (0 : v_n).$$

By Proposition 4.7.18, we have

$$I_1 \supset I_2 \supset I_3 \supset \ldots \supset I_n \supset \ldots;$$

in particular, $I_n \neq 0$ for each n. We shall prove that for each n, there is a nonzero left ideal $J_n \subseteq I_n$ with $J_n \cap I_{n+1} = 0$.

We fix a value of n and a nonzero element w of L. Using Theorem 4.7.20 (ii), we associate with the linearly independent set $\{v_1, v_2, \ldots, v_{n+1}\}$ a nonzero $d \in D$ such that there exists an $r \in A$ with

$$rv_1 = d0 = 0, \; rv_2 = d0 = 0, \ldots, rv_n = d0 = 0 \text{ and } rv_{n+1} = dw.$$

(Note that $dw \neq 0$ as $w \neq 0$ and $d \in D \setminus \{0\}$.)

Since $Z(A) = 0$ by Proposition 4.7.23 there is a nonzero left ideal K such that $K \cap (0:w) = 0$. Let $J_n = Kr$. This is a left ideal. Since $K \cap (0:w) = 0 \neq K$, there exists $k \in K$ with $kw \neq 0$, whence

$$0 \neq d\,kw = k\,dw = k\,rv_{n+1}.$$

Thus $kr \notin (0:v_{n+1})$ so $J_n = Kr \nsubseteq (0:v_{n+1})$. In particular, $J_n \neq 0$. But for $i = 1, 2, \ldots n$, we have $J_n v_i = Kr\,v_i = 0$ so $J_n \subseteq (0:v_i)$ for such i and hence $J_n \subseteq I_n$ (but $J_n \nsubseteq I_{n+1}$). If $k'r \in J_n \cap I_{n+1}$ ($k' \in K$) then $k'r \in (0:v_{n+1})$ so $k'r\,v_{n+1} = 0$ whence

$$0 = d^{-1}k'r\,v_{n+1} = k'd^{-1}r\,v_{n+1} = k'w,$$

so $k' \in K \cap (0:w) = 0$. It follows that $J_n \cap I_{n+1} = 0$. Now we have

$$J_1, I_2 \subseteq I_1, J_1 \cap I_2 = 0;$$
$$J_2, I_3 \subseteq I_2, J_2 \cap I_3 = 0; \ldots,$$
$$J_n, I_{n+1} \subseteq I_n, J_n \cap I_{n+1} = 0.$$

Exactly as in the proof of Proposition 4.7.24 we can now prove that the sum $\sum J_n$ is direct. But this is not possible, as R is a Goldie ring. Thus indeed I is finite dimensional over D. □

Combining Theorem 4.7.26 with Corollary 4.7.21 we get

THEOREM 4.7.27. (Goldie [1]). *Every nonzero prime Goldie ring has a ring of left quotients of the form $M_n(D)$ where D is a division ring and $n \in \mathbb{Z}^+$.* □

COROLLARY 4.7.28. *Every nonzero noetherian prime ring has a ring of left quotients of the form $M_n(D)$, where D is a division ring and $n \in \mathbb{Z}^+$.* □

The explicit proof of Goldie's Theorem 4.7.27 using the ideas and concepts of this section appears to be due to Koh and Luh [1]. Our presentation has been greatly influenced by the account in the thesis of J. Horvath [1].

Let us mention that also the converse of Theorem 4.7.27 is true: *every ring whose ring of left quotients is simple and artinian, is a prime Goldie ring*. Moreover, in Theorem 4.7.27 and its converse the words "prime" and "simple artinian" may be changed to "semiprime" and "semiprime artinian", respectively. For details we refer to Herstein [3] and Kertész [1].

4.8. The Brown–McCoy radical

We recapitulate the statements of Examples 2.2.4, 3.2.17, 3.6.1 (i), 3.8.14 (i), 3.9.14, 3.16.1, and 3.17.8.

DEFINITION and THEOREM 4.8.1. *The Brown–McCoy radical $\mathcal{G}$ is the upper radical $\mathcal{G} = \mathcal{UM}$ of the class $\mathcal{M}$ of all simple rings with unity element. $\mathcal{M}$ is a special class and the Brown–McCoy radical $\mathcal{G}$ is a special radical. Moreover, $\mathcal{G}$ is a supernilpotent dual radical. A ring A is Brown–McCoy semisimple if and only if A is a subdirect sum of simple rings with unity element. Further, $\mathcal{G}(A) = \cap (I \triangleleft A \mid A/I \in \mathcal{M})$ for every ring A. The Brown–McCoy radical is neither left hereditary, nor left strong, and hence not a normal radical.* □

For characterizing the Brown–McCoy radical rings and the Brown–McCoy radical of a ring, we introduce the following notions. An element $a \in A$ is called *G-regular*, if

$$a \in G(a) = \{ax - x + \sum_{\text{finite}} (y_i a z_i - y_i z_i) \mid x, y_i, z_i \in A\},$$

and a ring A is said to be *G-regular*, if A consists of G-regular elements. Furthermore, an element $a \in A$ is said to be F_1-*regular*, if a is in the ideal generated by the set

$$\{ax - x + ya - y \mid x, y \in A\}.$$

Call a ring A F_1-*regular*, if each of its elements is F_1-regular.

THEOREM 4.8.2. *For a ring A the following three conditions are equivalent:*
  (i) $A \in \mathcal{G}$, *that is, A is a Brown–McCoy radical ring,*
  (ii) *A is G-regular,*
  (iii) *A is F_1-regular.*

Proof: (i) $\Longrightarrow$ (ii) If A is not G-regular, then it possesses an element $a \in A$ which is not G-regular, and so $a \notin G(a)$. Notice that $G(a)$ is a two-sided ideal of A. Using Zorn's Lemma, there exists an ideal I of A which contains $G(a)$ and is maximal with respect to the exclusion of the element a. The ideal I is a maximal ideal of A, for if K is any ideal of A properly containing I, then $a \in K$, and so by $G(a) \subseteq I \subset K$ we get that

$$x = ax - (ax - x) \in K + G(a) = K$$

for all $x \in A$, that is, $K = A$. Hence $B = A/I$ is a simple ring.

We claim that B has a unity element. Putting $b = a + I$, we have

$$G(a + I) = G(a) + I \subseteq I,$$

that is, $G(b) = 0$, and by $a \notin I$ also $b \neq 0$. Since $bx - x \in G(b) = 0$, for all $x \in A$, b is a left unity element of B. Thus the set

$$L = \{xb - x \mid x \in B\}$$

is obviously a two-sided ideal of B and $LB = 0$. Since B is a simple prime ring, we conclude that $L = 0$ and so $xb - x = 0$ for every $x \in B$. Thus b is also a right unity element of B.

So we have proved that A has a homomorphic image B which is a simple ring with unity element, and therefore $A \notin \mathcal{G}$.

(ii) $\Longrightarrow$ (iii) As $G(a)$ is contained in the ideal generated by $\{ax - x + ya - y \mid x, y \in A\}$ for each $a \in A$, every G-regular element is F_1-regular.

(iii) $\Longrightarrow$ (i) Let A be an F_1-regular ring and I be any ideal of A with $I \neq A$. We claim that the factor ring $B = A/I$ has no unity element. By way of contradiction, let us suppose that $f = e + I$, $e \notin I$, is a unity element of B. Since A is F_1-regular, e is contained in the ideal generated by the set

$$\{ex - x + ye - y \mid x, y \in A\} \subseteq I,$$

because e is a unity element modulo I. Thus it follows that $e \in I$, a contradiction.

Hence A cannot be mapped homomorphically onto a simple ring with unity element, and so $A \in \mathcal{UM} = \mathcal{G}$. □

We say that an ideal of a ring is G-regular (F_1-regular), if it is a G-regular (F_1-regular) ring. An obvious consequence of Theorem 4.8.2 is the following

COROLLARY 4.8.3. *The Brown–McCoy radical $\mathcal{G}(A)$ of a ring A is the sum of all G-regular (F_1-regular) ideals of A.* □

An ideal I of a ring A is said to be *small* in A, if $I \neq 0$ and $I + K \neq A$ for any other ideal $K \neq A$. Further, if $I \triangleleft A$ and A has a unity element, then A will be called a *unital extension* of I. There are interesting connections between small ideals and the Brown–McCoy radical.

PROPOSITION 4.8.4 (Gardner [16], Loi and Wiegandt [1]). *If $A \in \mathcal{G}$ and $A \neq 0$, then A is small in every unital extension B of A. If $A \triangleleft B$ and A is small in B, then $A \in \mathcal{G}$.*

Proof: Let B be a unital extension of $A \in \mathcal{G}$ and $A \neq 0$. Further, take an ideal K of B such that $\mathcal{G}(B) + K = B$. The factor ring B/K is either a ring with unity element or $B/K = 0$. Since

$$B/K \cong (\mathcal{G}(B) + K)/K \cong \mathcal{G}(B)/(\mathcal{G}(B) \cap K) \in \mathcal{G},$$

the first case is not possible. Hence $B = K$, showing that $\mathcal{G}(B)$ is small in B. Since $A \subseteq \mathcal{G}(B)$, also A is small in B.

For the second statement, assume that $A \notin \mathcal{G}$, and let $B \neq A$ be any extension of A with or without unity element. Since the Brown–McCoy radical is hereditary, $A \notin \mathcal{G}$ implies $B \notin \mathcal{G}$. Furthermore,

$$\mathcal{G}(B) = \cap(K_\alpha \triangleleft B \mid B/K_\alpha \in \mathcal{M}).$$

Since $A \notin \mathcal{G}$, we have $A \not\subseteq \mathcal{G}(B)$, and so $A \not\subseteq K_\alpha$ for some index α. The ideal K is maximal in B, so necessarily $A + K_\alpha = B$ which means that A is not small in B. □

PROPOSITION 4.8.5. *A nonzero ring A is Brown–McCoy semisimple if and only if every ideal I of A is not small in any unital extension of I.*

Proof: Suppose that $A \in \mathcal{SG}$. Then also $I \in \mathcal{SG}$ for every nonzero ideal I of A, and by Proposition 4.8.4 I is not small in any extension.

If $A \notin \mathcal{SG}$, then $I = \mathcal{G}(A) \neq 0$, and by Proposition 4.8.4 I is small in every unital extension. □

The Brown–McCoy radical is relatively large; we have namely,

PROPOSITION 4.8.6. *The Brown–McCoy radical class $\mathcal{G}$ properly contains the Jacobson radical class $\mathcal{J}$.*

Proof: In view of Corollary 4.4.3 every simple ring with unity element has to be Jacobson semisimple, that is, $\mathcal{M} \subseteq \mathcal{SJ}$. Hence it follows that $\mathcal{J} \subseteq \mathcal{UM} = \mathcal{G}$.

Let V be a vector space of countably infinite basis over a division ring, and T be the ring of all linear transformations of V. As we have seen already in Example 1.2.11, the subring

$$H = \{t \in T \mid t(V) \text{ is finite dimensional}\}$$

is an ideal of T and H is a simple ring without unity element. The ring H as well as T is Jacobson semisimple (cf. Example 4.4.8). Thus H is a ring which is in $\mathcal{G}$ but not in $\mathcal{J}$, proving $\mathcal{J} \neq \mathcal{G}$. □

An immediate consequence of Theorem 3.13.13 is

PROPOSITION 4.8.7. *The Brown–McCoy radical $\mathcal{G}$ is the smallest one for which every semisimple ring is a subdirect sum of simple rings with unity element.* □

As mentioned at the beginning of Section 3.13, Leavitt proved that if $\mathbb{M}$ is a class of simple prime rings and if every ring of the semisimple class of the radical class $\mathcal{U}\mathbb{M}$ is a subdirect sum of $\mathbb{M}$-rings, then $\mathbb{M}$ consists of simple rings with unity element. Thus *in Proposition 4.8.7 we may allow simple prime rings without unity element.*

We can follow the proof of Proposition 4.5.7 and get

THEOREM 4.8.8. *A ring A is Brown–McCoy semisimple and satisfies d.c.c. on ideals if and only if A is a finite direct sum of simple rings with unity element.* □

As we have seen in Theorem 3.9.15 and subsequent remarks, the Brown–McCoy radical class is the largest universal class of rings having no unity element. We have also the following characterization of the Brown–McCoy radical.

THEOREM 4.8.9 (de la Rosa and Wiegandt [1]). *The Brown–McCoy radical is the upper radical of the largest hereditary class of rings with unity element.*

Proof: Let $\mathbb{M}$ denote the class of all rings with unity element, and consider the class

$$\mathbb{K} = \{A \mid \text{if } 0 \neq I \triangleleft A \text{ then } I \text{ has a unity element}\}.$$

If $K \triangleleft I \triangleleft A \in \mathbb{K}$, then the ideal $\overline{K}$ of A generated by K has a unity element. Hence by the Andrunakievich Lemma $\overline{K} = \overline{K}^3 \subseteq K$ holds implying $K \triangleleft A$ and $K \in \mathbb{M}$. Thus $I \in \mathbb{K}$, proving that $\mathbb{K}$ is a hereditary class. It is obvious that $\mathbb{K}$ is the largest hereditary class consisting of rings with unity element.

Clearly $\mathbb{K}$ contains the class $\mathcal{M}$ of simple rings with unity element, so $\mathcal{U}\mathbb{K} \subseteq \mathcal{G}$. If $A \in \mathcal{G}$ then A cannot be mapped onto a (not necessarily simple) ring with unity element. Since each ring in $\mathbb{K}$ has a unity element, it follows that $A \in \mathcal{U}\mathbb{K}$. Thus $\mathcal{G} \subseteq \mathcal{U}\mathbb{K}$. □

For further study of Brown–McCoy semisimple rings the reader is referred to the results of McCoy [1] and Suliński [1], [2], [3] (cf. also Szász [6] Sections 40 and 41) to Gardner [12], Tumurbat and Wiegandt [3], Tumurbat and Zand [1]. The behavior of the Brown–McCoy radical with respect to products and certain ultraproducts was discussed by Olszewski [1].

4.9. Radicals of matrices and polynomials

It is natural to ask for connection between the radical of a ring A and that of a matrix ring, polynomial ring (skew polynomial ring, formal power series ring, etc.) over A. In this section we shall study the radical of a matrix ring and of a polynomial ring. For matrix rings we get easily general results which could claim their place in Chapter III. The case of polynomial rings is more subtle and less general. Also Köthe's Problem will be reformulated in terms of matrix rings and polynomial rings, respectively.

As usual, $M_n(A)$ will stand for the $n \times n$ matrix ring over the ring A, further $(A)_{ij}$ will denote the set of matrices which have elements from A at the i, j position and 0 everywhere else.

PROPOSITION 4.9.1 (Snider [1]). *If $I \triangleleft A$, then $M_n(I) \triangleleft M_n(A)$. If A is a ring with unity element and $K \triangleleft M_n(A)$, then $K = M_n(I)$ with some $I \triangleleft A$. If γ is a radical, then $\gamma(M_n(A)) = M_n(I)$ for some ideal I of A for every ring A.*

Proof: The first assertion is obvious.

For the second assertion one considers the set I of all elements of A which may occur in the matrices of K. Using the unity element of A one can show that $I \triangleleft A$, so $K \subseteq M_n(I)$; further, every element of $M_n(I)$ is a sum of elements in K, whence $M_n(I) \subseteq K$.

By the ADS-Theorem 3.1.2 and the first assertion we have $\gamma(M_n(A)) = M_n(I)$ where $I \triangleleft A^1$ and A^1 stands for the Dorroh extension of A. But $M_n(I) \subseteq M_n(A)$ so $I \subseteq A$ and thus $I \triangleleft A$. □

PROPOSITION 4.9.2 (Propes [3]). *Let γ be a radical. The following two statements are equivalent:*

(i) $A \in \gamma$ implies $M_n(A) \in \gamma$,

(ii) $M_n(\gamma(A)) \subseteq \gamma(M_n(A))$.

Also the following two conditions are equivalent:

(iii) $M_n(A) \in \gamma$ implies $A \in \gamma$,

(iv) $\gamma(M_n(A)) \subseteq M_n(\gamma(A))$.

Proof: (i) $\Longrightarrow$ (ii) Now $\gamma(A) \in \gamma$, so by (i) also $M_n(\gamma(A)) \in \gamma$. Since $M_n(\gamma(A)) \triangleleft M_n(A)$, it follows that $M_n(\gamma(A)) \subseteq \gamma(M_n(A))$.

(ii) $\Longrightarrow$ (i) Now $A \in \gamma$ implies $\gamma(A) = A$ and $M_n(\gamma(A)) = M_n(A)$. Hence by (ii) we have $M_n(A) \subseteq \gamma(M_n(A))$. Thus $M_n(A) \in \gamma$.

(iii) $\Longrightarrow$ (iv) Since $\gamma(M_n(A)) = M_n(I)$ for some ideal I of A, from (iii) we have $I \in \gamma$. Hence $I \subseteq \gamma(A)$ and so $\gamma(M_n(A)) = M_n(I) \subseteq M_n(\gamma(A))$.

(iv) $\Longrightarrow$ (iii) $M_n(A) \in \gamma$ means $M_n^*(A) = \gamma(M_n(A))$. Thus by (iv) $M_n(A) \subseteq M_n(\gamma(A))$, whence $A = \gamma(A) \in \gamma$. □

Following Leavitt [13], a class δ of rings is said *matric-extensible*, if for all natural numbers n, $A \in \delta$ if and only if $M_n(A) \in \delta$.

THEOREM 4.9.3 (Amitsur [3], Leavitt [14], Propes [3], Snider [1]). *The following statements are equivalent for a radical class γ:*

(1) *γ is matric-extensible,*

(2) *the semisimple class $S\gamma$ is matric-extensible,*

(3) *γ satisfies the matrix equation $\gamma(M_n(A)) = M_n(\gamma(A))$ for every natural number n.*

Proof: (1) $\Longrightarrow$ (3) Obvious by Proposition 4.9.2.

(2) $\Longrightarrow$ (1) Let $A \in \gamma$. If $M_n(A) \notin \gamma$, then $0 \neq M_n(A)/\gamma(M_n(A)) \in S\gamma$ and $\gamma(M_n(A)) = M_n(I)$ for some ideal I by Proposition 4.9.1. Hence

$$0 \neq M_n((A/I)) \cong M_n(A)/M_n(I) = M_n(A)/\gamma(M_n(A)) \in S\gamma.$$

Since $\mathcal{S}\gamma$ is matrix extensible, $A/I \in \mathcal{S}\gamma$ follows, a contradiction.

Let $M_n(A) \in \gamma$. If $A \notin \gamma$, then $0 \neq A/\gamma(A) \in \mathcal{S}\gamma$. Hence (2) implies

$$M_n(A)/M_n(\gamma(A)) \cong M_n(A/\gamma(A)) \in \mathcal{S}\gamma,$$

contradicting $M_n(A) \in \gamma$.

(3) $\implies$ (2) If $A \in \mathcal{S}\gamma$, then $\gamma(M_n(A)) = M_n(\gamma(A)) = 0$, that is, $M_n(A) \in \mathcal{S}\gamma$. Similarly, $M_n(A) \in \mathcal{S}\gamma$ implies $A \in \mathcal{S}\gamma$. $\square$

The matrix equation for lower and upper radicals as well as for the sum of two radical classes was investigated by Aslam and Zaidi [1]. Now we continue with some general results on matric-extensibility.

PROPOSITION 4.9.4. *If γ is a left (or right) hereditary radical, then*
(iii) $M_n(A) \in \gamma$ *implies* $A \in \gamma$.

Proof: We have

$$\sum_{i=2}^{n}(A)_{i1} \triangleleft \sum_{i=1}^{n}(A)_{i1} \triangleleft_l M_n(A) \in \gamma$$

and then

$$\left(\sum_{i=1}^{n}(A)_{i1}\right) \Big/ \left(\sum_{i=2}^{n}(A)_{i1}\right) \cong A \in \gamma.$$

Hence also $A \in \gamma$. $\square$

PROPOSITION 4.9.5. *If γ is a hereditary and left (or right) strong radical, then*
(i) $A \in \gamma$ *implies* $M_n(A) \in \gamma$.

Proof: Let $L_j = \sum_{i=1}^{n}(A)_{ij}$. Clearly $(A)_{11} \cong A$. If $A \in \gamma$, then by Lemma 3.16.6 also $A^0 \in \gamma$. Since $L_j^0 = \sum_{i=1}^{n}(A)_{ij}^0$ and $A^0 \cong (A)_{ij}^0 \triangleleft L_j^0$, we conclude that $L_j^0 \in \gamma$ for all $j = 1, \ldots, n$. Since

$$\sum_{j \neq i=1}^{n}(A)_{ij} \triangleleft L_j^0 \in \gamma,$$

by the hereditariness of γ it follows $\sum_{j \neq i=1}^{n}(A)_{ij} \in \gamma$. But also $\sum_{j \neq i=1}^{n}(A)_{ij} \triangleleft L_j$ and

$$L_j \Big/ \left(\sum_{j \neq i=1}^{n}(A)_{ij}\right) \cong A \in \gamma,$$

therefore $L_j \in \gamma$ for each $j = 1, \ldots, n$. Since γ is left strong and $M_n(A) = \sum_{j=1}^{n} L_j$, we get $M_n(A) \in \gamma$. $\square$

THEOREM 4.9.6 (Propes [3]). *If γ is a left (or right) hereditary and left (or right) strong radical, then γ is matric-extensible.*

Proof: Apply Propositions 4.9.2, 4.9.4 and 4.9.5. □

COROLLARY 4.9.7. *Every N-radical and every hereditary A-radical is matric-extensible.* □

EXAMPLES 4.9.8. *The Baer, Levitzki, Jacobson and antisimple radicals as well as the torsion and p-torsion radicals are matric-extensible.*

There are many more matric-extensible radicals. We present here some as upper radicals.

PROPOSITION 4.9.9 (Leavitt [13]). *Let ϱ be a homomorphically closed regular and matric-extensible class. If ϱ contains all nilpotent rings, then the upper radical $\mathcal{U}\varrho$ is matric-extensible.*

Proof: If $A \notin \mathcal{U}\varrho$, then there exists an ideal I of A such that $0 \neq A/I \in \varrho$. By the assumption also $M_n(A/I) \in \varrho$, and so $M_n(A)$ has a nonzero homomorphic image in ϱ, that is, $M_n(A) \notin \mathcal{U}\varrho$. This proves implication (iii).

Suppose that $A \in \mathcal{U}\varrho$ but $M_n(A) \notin \mathcal{U}\varrho$. Now $0 \neq M_n(A)/K \in \varrho$ with some ideal $K \triangleleft M_n(A)$. Also $M_n(A) \triangleleft M_n(A^1)$. Since A^1 is a ring with unity element, the ideal of $M_n(A^1)$ generated by K can be written by Proposition 4.9.1 as $M_n(I)$ where $I \triangleleft A$. By the Andrunakievitch Lemma $(M_n(I))^3 \subseteq K$. Then

$$M_n(A)/K \twoheadrightarrow M_n(A)/M_n(I) \cong M_n(A/I),$$

and so $M_n(A)/K \in \varrho$ implies $M_n(A/I) \in \varrho$. By (iii) we get $A/I \in \varrho$ contradicting $A \in \mathcal{U}\varrho$, unless $I = A$. But then $M_n(A)/K = M_n(I)/K$ is nilpotent. Since ϱ contains all nilpotent rings, $\mathcal{U}\varrho$ consists of idempotent rings. Thus A, $M_n(A)$ and also $M_n(A)/K$ are idempotent which is impossible. Thus $M_n(A) \in \mathcal{U}\varrho$. □

PROPOSITION 4.9.10 (Andrunakievich [1] and Leavitt [13]). *If ϱ is a regular and matric-extensible class of rings which contains no nonzero nilpotent rings, then $\mathcal{U}\varrho$ is a matric-extensible radical class. In particular, if ϱ is a (weakly) special matric-extensible class, then $\mathcal{U}\varrho$ is a matric-extensible (supernilpotent) special radical.*

Proof: As in the proof of Proposition 4.9.9 we see that (iii) of Proposition 4.9.1 is satisfied. For (i), suppose that $A \in \mathcal{U}\varrho$ but $0 \neq M_n(A)/K \in \varrho$ with some $K \triangleleft M_n(A)$. If $K = M_n(I)$ with a suitable ideal I of A, then

$$M_n(A/I) \cong M_n(A)/K \in \varrho$$

contradicts $A \in \mathcal{U}\varrho$. Hence the ideal of $M_n(A^1)$ generated by K has the form $M_n(I) \neq K$ with some $I \triangleleft A^1$ and $(M_n(I))^3 \subset K$. By the regularity of ϱ and $M_n(A)/K \in \varrho$ we know that $M_n(I)/K$ maps to some nonzero ring of ϱ. Since $M_n(I)/K$ is nilpotent, we have got a contradiction. Thus $M_n(A) \in \mathcal{U}\varrho$. □

EXAMPLES 4.9.11. (i) The class of all rings with zero multiplication satisfies the requirements of Proposition 4.9.9. Hence its *upper radical class $\mathcal{I}$ (Example 3.2.2 (ii)) consisting of all idempotent rings is matric-extensible.*

(ii) By Theorem 4.8.9 the Brown–McCoy $\mathcal{G}$ radical is the upper radical of the largest hereditary class $\mathcal{K}$ of rings with unity element. By Proposition 4.9.1 it is obvious that $\mathcal{K}$ is a matric-extensible class. Hence by Proposition 4.9.10 its upper radical, that is, *the Brown–McCoy radical $\mathcal{G}$ is matric-extensible.*

(iii) The upper radical β_s of the class of all simple prime rings is called the *Jenkins radical* (see Jenkins [1] and van Leeuwen [1]). De la Rosa ([1] and [2]) proved that an ideal I of a ring A is quasi-semiprime (see Section 3.6) if and only if $M_n(I)$ is so in $M_n(A)$. In particular, A is a simple prime ring if and only if $M_n(A)$ is a simple prime ring. Hence the class of all simple prime rings is trivially matric-extensible, and so by Proposition 4.9.10 *the Jenkins radical is matric-extensible.*

(iv) *The Thierrin radical $\mathcal{T}$,* as the upper radical of matrix rings over division rings, *is matric-extensible* by Proposition 4.9.10.

One more example.

PROPOSITION 4.9.12 (Brown and McCoy [3]). *The von Neumann regular radical ν is matric-extensible.*

Proof: We prove the validity of conditions (i) and (iii) of Proposition 4.9.2.

Let A be a von Neumann regular ring. For any element $a \in A$ we shall denote by a' an element of A such that $a = aa'a$. First, we shall show the validity of (ii) for $n = 2$. For any matrix $M = \begin{pmatrix} a & b \\ c & d \end{pmatrix} \in M_2(A)$, consider $K_1 = \begin{pmatrix} 0 & b' \\ 0 & 0 \end{pmatrix}$ and $K_2 = M - MK_1M$. A straightforward calculation shows that K_2 has the form $\begin{pmatrix} e & f \\ 0 & g \end{pmatrix}$. Now, if $K_3 = \begin{pmatrix} e' & 0 \\ 0 & g' \end{pmatrix}$ and

(u) $$K_4 = K_2 - K_2K_3K_2,$$

then K_4 has the form $K_4 = \begin{pmatrix} 0 & h \\ 0 & 0 \end{pmatrix}$. Similarly, let $K_5 = \begin{pmatrix} 0 & 0 \\ h' & 0 \end{pmatrix}$. Then

(v) $$K_4 = K_4K_5K_4,$$

as one readily checks. Substituting (u) and (v) into the right-hand side we arrive at

$$K_2 - K_2K_3K_2 = (K_2 - K_2K_3K_2)K_5(K_2 - K_2K_3K_2).$$

A direct verification — without using matrix multiplication — shows that $K_2 = K_2 K_6 K_2$ with

$$K_6 = K_5 - K_5 K_2 K_3 - K_3 K_2 K_5 + K_3 K_2 K_5 K_2 K_3 + K_3.$$

Hence by $K_2 = M - MK_1M$ we get

$$M - MK_1M = (M - MK_1M)K_6(M - MK_1M).$$

Similarly as above, there exists a matrix K_7 such that $M = MK_7M$. Thus $M_2(A)$ is von Neumann regular.

Since $M_{2^{k-1}}(M_2(A)) \cong M_{2^k}(A)$, we conclude by induction that also $M_{2^k}(A)$ is von Neumann regular for every $k = 2, 3, \ldots$. For $n \leq 2^k$ the matrix ring $M_n(A)$ is isomorphic to the subring B of $M_{2^k}(A)$ having zeros beyond the n-th column and n-th row. Since $M_{2^k}(A)$ is von Neumann regular, to every matrix $M \in M_n(B)$ there exists a matrix $M' \in M_{2^k}(A)$ such that $M = MM'M$. A moment's reflection on the matrix multiplication yields that also $M' \in M_n(B)$ can be achieved. Hence $M_n(A)$ is von Neumann regular, and condition (i) is satisfied.

It is straightforward to check that also condition (iii) is satisfied. $\square$

Having seen quite a few examples of matric-extensible radicals, it is a natural question to ask as whether the nil radical is matric-extensible. This problem turns out to be equivalent to Köthe's Problem.

THEOREM 4.9.13 (Krempa [1], Sands [1]). *The following statements are equivalent:*

(1) *the nil radical $\mathcal{N}$ is left strong,*
(2) $A \in \mathcal{N}$ *implies* $M_2(A) \in \mathcal{N}$,
(3) $A \in \mathcal{N}$ *implies* $M_n(A) \in \mathcal{N}$ *for every* $n = 1, 2, \ldots$.

Proof: (1) $\Longrightarrow$ (2) This follows immediately from Theorem 4.9.6.

(2) $\Longrightarrow$ (3) Note that for every ring A and natural number k we have $(M_2(A))_{2^{k-1}} \cong M_{2^k}(A)$. Hence, if $M_2(A)$ is nil for every nil ring A, then also $M_{2^k}(A)$ is nil for every $k = 2, 3, \ldots$, as one readily sees by induction. Now for $n \leq 2^k$ the matrix ring $M_n(A)$ is isomorphic to the subring of $M_{2^k}(A)$ of matrices with zero entries beyond the first n columns and first n rows. Hence $M_n(A)$ is nil.

(3) $\Longrightarrow$ (1) Suppose that L is a nil left ideal of a ring A. We show that for all elements $\ell_1, \ldots, \ell_n \in L$ and $r_1, \ldots, r_n \in A^1$ the element $a = \ell_1 r_1 + \cdots + \ell_n r_n$ is nilpotent. By (3) we know that also the matrix ring $M_n(L)$ is nil. Put $a_{ij} = r_i \ell_j$, and consider the matrix $M = (a_{ij}) \in M_n(L)$. Then there exists a natural number k such that $M^k = 0$. Observe that

$$a^{k+1} = \sum \ell_{i_1} r_{i_1} \ldots \ell_{i_{k+1}} r_{i_{k+1}}.$$

Now, for fixed indices i_1 and i_{k+1},

$$\sum\nolimits_{i_2,\ldots,i_k} \ell_{i_1} r_{i_1} \ldots \ell_{i_{k+1}} r_{i_{k+1}} = \ell_{i_1} \left(\sum\nolimits_{i_2,\ldots i_k} a_{i_1 i_2} a_{i_2 i_3} \ldots a_{i_k i_{k+1}} \right) r_{i_{k+1}}.$$

But $\sum_{i_2,\ldots i_k} a_{i_1 i_2} a_{i_2 i_3} \ldots a_{i_k i_{k+1}}$ is just the element at the (i_k, i_{k+1}) position of the matrix $M^k = 0$. Hence also $a^{k+1} = 0$.

The ideal of A generated by L is $L + LA = LA^1$, and by the previous considerations $L + LA$ is a nil ring. Thus $L \subseteq L + LA \subseteq \mathcal{N}(A)$, proving that $\mathcal{N}$ is a left strong radical. $\square$

Having in mind that the nil radical is strongly hereditary, an immediate consequence of Theorem 4.9.13 is

COROLLARY 4.9.14. *Köthe's problem is equivalent to the matric-extensibility of the nil radical.* $\square$

For radicals of infinite matrix rings we refer for instance to Patterson [1], [2], Sands [15], Slover [1], [2] and to the recent paper of Beidar, Ke and Puczyłowski [1].

Van Wyck [1], [2] introduced the notion of *structural matrix ring* as a generalization of full matrix rings and triangular matrix rings. The relationship between the radical of a ring and the radical of a structural matrix ring has been studied also by Groenewald and van Wyck [1], Sands [12], Veldsman [16] and van Wyck [3].

*

In the sequel we shall focus our attention to the relation between the radical of a ring A and of its polynomial ring $A[x]$. As we shall see, here the situation is more involved and less general than in the case of matrix rings.

For any radical γ Gardner [1] defined the class

$$\gamma_x = \{A \mid A[x] \in \gamma\}.$$

THEOREM 4.9.15. *For every radical γ the class γ_x is a radical class. If γ is hereditary (or supernilpotent), then so is γ_x.*

Proof: If B is a homomorphic image of $A \in \gamma_x$, then $B[x]$ is a homomorphic image of $A[x]$, so $B[x] \in \gamma$, that is, $B \in \gamma_x$. Let $\{I_\lambda \mid \lambda \in \Lambda\}$ be an ascending chain of ideals of a ring A such that $A = \cup I_\lambda$ and $I_\lambda \in \gamma_x$ for each $\lambda \in \Lambda$. Then $\{I_\lambda[x] \mid \lambda \in \Lambda\}$ is an ascending chain of ideals of $A[x]$ and $I_\lambda[x] \in \gamma$ for every $\lambda \in \Lambda$. Hence

$$A[x] = (\cup I_\lambda)[x] = \cup (I_\lambda[x]) \in \gamma,$$

that is, $A \in \gamma_x$. Suppose that I, $A/I \in \gamma_x$, that is, $I[x]$, $(A/I)[x] \in \gamma$. Since $I[x] \in \gamma$ and

$$A[x]/I[x] \cong (A/I)[x] \in \gamma,$$

it follows that $A[x] \in \gamma$, that is, $A \in \gamma_x$. Thus γ_x is a radical class.

The hereditariness of γ trivially implies that of γ_x. If A is a zero-ring, then so is $A[x]$. Hence if γ contains all zero-rings, so does γ_x. □

The radical γ_x may be very small.

PROPOSITION 4.9.16. *If γ is a subidempotent radical, then $\gamma_x = 0$. If $\gamma_x = 0$, then γ is hypoidempotent.*

Proof: If γ is subidempotent and $A \in \gamma_x$, then for any element $a \in A$ the monomial ax is in the square $(ax)^2$ of the principal ideal of $A[x]$ generated by ax. This can happen only when $a = 0$, so $\gamma = 0$.

Let $\gamma_x = 0$. If $A \in \gamma$, then $A/A^2 \in \gamma$. Since

$$A/A^2[x] \cong A/A^2 \oplus A/A^2 x \oplus A/A^2 x^2 \oplus \cdots$$

and each direct summand is in γ, we have $A/A^2[x] \in \gamma$, implying $A/A^2 \in \gamma_x = 0$. Thus $A = A^2$. □

Let γ be again any radical. Ortiz [1] assigned to every ring A the ideal

$$\varrho(A) = \gamma(A[x]) \cap A$$

and considered the class

$$\gamma_\varrho = \{A \mid \varrho(A) = A\}.$$

If ϱ is a Kurosh–Amitsur radical assignment (cf. 2.1), then γ_ϱ is just the radical class determined by ϱ.

PROPOSITION 4.9.17. *The assignment ϱ satisfies the following conditions:*
(i) *if $f: A \twoheadrightarrow f(A)$ then $f(\varrho(A)) \subseteq \varrho f(A)$,*
(ii) *ϱ is complete: if $\varrho(I) = I \triangleleft A$ then $I \subseteq \varrho(A)$,*
for all rings A. Moreover, $\gamma_\varrho = \gamma_x \subseteq \gamma$, $\gamma_x(A) \subseteq \varrho(A)$, $\varrho(A)[x] \subseteq \gamma(A[x])$ for every ring A. If $A[x] \in \gamma$ then also $A \in \gamma$.

Proof: For any homomorphisms $f: A \twoheadrightarrow f(A)$ we have

$$f(\varrho(A)) = f(\gamma(A[x]) \cap A) \subseteq \gamma(f(A)[x]) \cap f(A) = \varrho(f(A))$$

in view of Theorem 2.1.11.

Suppose that $\varrho(I) = I \triangleleft A$. Then making use of the ADS-Theorem 3.1.2 we have

$$I = \gamma(I[x]) \cap I \subseteq \gamma(A[x]) \cap A = \varrho(A),$$

proving that ϱ is complete.

If $A \in \gamma_x$, then $\varrho(A) = \gamma(A[x]) \cap A = A[x] \cap A = A$, proving $\gamma_x \subseteq \gamma_\varrho$.

Let $A \in \gamma_\varrho$. Then $A = \varrho(A) = \gamma(A[x]) \cap A$, whence $A \subseteq \gamma(A[x])$. The ideal of $A[x]$ generated by A is $A + A^2[x]$. Since $\gamma(A[x]) \triangleleft A[x]$, we have $A + A^2[x] \subseteq \gamma(A[x])$. So

$$\gamma(A[x]/(A + A^2[x])) = \gamma(A[x])/(A + A^2[x]),$$

whence

$$A[x]/\gamma(A[x]) \cong \frac{A[x]/(A + A^2[x])}{\gamma(A[x])/(A + A^2[x])} \cong \frac{A[x]/(A + A^2[x])}{\gamma(A[x]/(A + A^2[x]))}.$$

Now $A[x]/(A + A^2[x]) = x(A/A^2)[x]$. Let $K = \gamma(A[x]/(A + A^2[x]))$. Then we have surjective homomorphisms

$$A[x] \twoheadrightarrow (A/A^2)[x] \twoheadrightarrow x(A/A^2)[x] \twoheadrightarrow x(A/A^2)[x]/K.$$

These are ring homomorphisms – the second because $(A/A^2)[x]$ is a zero-ring. Now $x(A/A^2)[x]/K$ is γ-semisimple and $A + A^2[x] \subseteq \gamma(A[x])$. In particular, $A \subseteq \gamma(A[x])$. Thus for $a \in A$, under our homomorphisms, we have

$$a \mapsto 0 \qquad (\text{as } a \in A \subseteq A + A^2[x])$$

and

$$a \mapsto a + A^2 \mapsto (a + A^2)x \mapsto (a + A^2)x + K.$$

Hence we have

$$\begin{aligned}(A/A^2)x \subseteq K &= \gamma(A[x]/(A + A^2[x])) \\ &= \gamma((A/A^2)x \oplus (A/A^2)x^2 \oplus \cdots) \\ &= \gamma((A/A^2)x) \oplus \gamma((A/A^2)x^2 \oplus \cdots,\end{aligned}$$

as all rings are zero-rings. This means that $(A/A^2)x = \gamma((A/A^2)x)$, that is, $(A/A^2)x \in \gamma$. But $(A/A^2)x \cong A/A^2$ and

$$A[x]/(A + A^2[x]) \cong \bigoplus_{n=1}^{\infty} (A/A^2)x \cong \bigoplus_{n=1}^{\infty} A/A^2 \in \gamma.$$

Hence

$$A[x]/(A + A^2[x]) = \gamma(A[x]/(A + A^2[x])) = \gamma(A[x])/(A + A^2[x]),$$

so $A[x] = \gamma(A[x])$, that is, $A[x] \in \gamma$ proving $\gamma_\varrho \subseteq \gamma_x$.

If $I \triangleleft A$ and $I \in \gamma_x = \gamma_\varrho$, then by the completeness of ϱ we have $I \subseteq \varrho(A)$. In particular, $\gamma_x(A) \subseteq \varrho(A)$.

If $A \in \gamma_x$, then $A[x] \in \gamma$. Let A^1 be the Dorroh extension of A and (x) the principal ideal of $A[x]$ generated by x. Then $A \cong A[x]/(A[x] \cap (x)) \in \gamma$ which proves $\gamma_x \subseteq \gamma$ and also the implication $A[x] \in \gamma \Rightarrow A \in \gamma$.

Finally, we have obviously

$$\varrho(A)[x] = (\gamma(A[x]) \cap A)[x] \subseteq (\gamma(A[x]) \cap A)\mathbb{Z}[x] \subseteq$$
$$\subseteq \gamma(A[x])\mathbb{Z}[x] = \gamma(A[x]). \qquad \square$$

When γ is a hereditary radical, γ_x and ϱ coincide, providing a useful representation for γ_x.

THEOREM 4.9.18 (Gardner [1]). *If γ is a hereditary radical, then $\gamma_x(A) = \varrho(A) = \gamma(A[x]) \cap A$ for all rings A, whence ϱ is a Kurosh–Amitsur radical and $\gamma(A[x]) \cap A \subseteq \gamma(A)$.*

Proof: Let A^1 be the Dorroh extension of A. Since

$$\varrho(A^1) = \gamma(A^1[x]) \cap A^1 \subseteq \gamma(A^1[x]) \triangleleft A^1[x]$$

and $x \in A^1[x]$, we have $\varrho(A^1)[x] \subseteq \gamma(A^1[x])$. But $\varrho(A^1) \triangleleft A^1$, so $\varrho(A^1)[x] \triangleleft \gamma(A^1[x])$, and thus $\varrho(A^1)[x] \in \gamma$ by the hereditariness of γ. This means that $\varrho(A^1) \in \gamma_x$. Hence $\varrho(A^1) \subseteq \gamma_x(A^1)$. Furthermore, $\gamma_x(A^1)[x] \triangleleft A^1[x]$ and $\gamma_x(A^1)[x] \in \gamma$, so

$$\gamma_x(A^1) = \gamma_x(A^1)[x] \cap A^1 \subseteq \gamma(A^1[x]) \cap A^1 = \varrho(A^1).$$

The hereditariness of γ yields

$$\gamma_x(A) = \gamma_x(A^1) \cap A = \varrho(A^1) \cap A = \gamma(A^1[x]) \cap A^1 \cap A =$$
$$= \gamma(A^1[x]) \cap A = \gamma(A^1[x]) \cap A \cap A[x] =$$
$$= \gamma(A[x]) \cap A = \varrho(A).$$

The last assertion follows from $\gamma_x(A) = \varrho(A)$ and $\gamma_x \subseteq \gamma$. $\qquad \square$

LEMMA 4.9.19 (Ortiz [1]). *Let γ be any radical. If Q is a semiprime ideal of $A[x]$ such that $A[x]/Q$ is γ-semisimple, then $\varrho(A/(A \cap Q)) = 0$.*

Proof: Let $P = A \cap Q$. Since $P[x] \cdot P[x] \subseteq Q$ and Q is semiprime, $P[x] \subseteq Q$. Hence there is a surjective homomorphism

$$f: (A/P)[x] \twoheadrightarrow A[x]/Q.$$

If $a + P \in \varrho(A/P) \subseteq \gamma((A/P)[x])$, then

$$a + Q = f(a + P) \in \gamma(A[x]/Q) = 0.$$

Hence $a \in P$ and $\varrho(A/P) = 0$. $\qquad \square$

PROPOSITION 4.9.20 (Ortiz [1]). *If γ is a special radical, then so is γ_x.*

Proof: Since γ is hereditary, by Proposition 4.9.17 and Theorem 4.9.18 $\gamma_x = \gamma_\varrho$ and $\gamma_x(A) = \varrho(A)$ for all rings A.

Let $\mathcal{P}$ denote the class of all prime rings, and consider the ideal

$$T = \cap(P \triangleleft A \mid A/P \in \mathcal{P} \cap \mathcal{S}\gamma_x).$$

Clearly $\varrho(A) \subseteq T$. On the other hand, if Q is a prime ideal of $A[x]$ such that $A[x]/Q \in \mathcal{S}\gamma$, then by Lemma 4.9.19, $\varrho(A/(A \cap Q)) = 0$.

We claim that $Q \cap A$ is a prime ideal in A. Suppose that $Q \cap A = A$, that is, $A \subseteq Q$. Then $A \cdot A[x] \subseteq Q$, so

$$A[x]^2 = (A + \{\sum_{i>0} a_i x^i\})A[x] = AA[x] + \{\sum_{i>0} a_i x^i\}A[x] = AA[x] \subseteq Q.$$

Since Q is prime in $A[x]$, we therefore have $A[x] \subseteq Q$, a contradiction. Thus $A \not\subseteq Q$. If $c_0 + c_1 x + \cdots$ and $d_0 + d_1 x + \cdots$ are in $(Q \cap A)[x]$, then

$$(c_o + c_1 x + \cdots)(d_0 + d_1 x + \cdots) \in (Q \cap A)(Q \cap A)[x] \subseteq QA[x] \subseteq Q,$$

so that $(Q \cap A)[x]^2 \subseteq Q$ and so $(Q \cap A)[x] \subseteq Q$. Now if $I, J \triangleleft A$ and $IJ \subseteq Q \cap A$, then

$$I[x]J[x] \subseteq (IJ)[x] \subseteq (Q \cap A)[x] \subseteq Q,$$

so $I[x] \subseteq Q$ or $J[x] \subseteq Q$. But if, say, $I[x] \subseteq Q$, then $I = I[x] \cap A \subseteq Q \cap A$. Thus $Q \cap A$ is indeed a prime ideal of A.

Hence the factor ring $A/(A \cap Q)$ is a prime ring, so $A/(A \cap Q) \in \mathcal{P} \cap \mathcal{S}\gamma_x$, and therefore having in mind that γ is special, we have

$$T \subseteq \bigcap(A \cap Q \mid A[x]/Q \in \mathcal{P} \cap \mathcal{S}\gamma_x) = A \cap \bigcap(Q \mid A[x]/Q \in \mathcal{P} \cap \mathcal{S}\gamma_x) =$$
$$= A \cap \gamma(A[x]) = \varrho(A).$$

This proves $\gamma_x(A) = \varrho(A) = T$.

Clearly, $\gamma_x \subseteq \mathcal{U}(\mathcal{P} \cap \mathcal{S}\gamma_x)$. Suppose that there is a ring $A \in \mathcal{U}(\mathcal{P} \cap \mathcal{S}\gamma_x) \setminus \gamma_x$. Then the ring $A/\gamma_x(A) \neq 0$ is a subdirect sum of nonzero rings $B_\lambda \in \mathcal{P} \cap \mathcal{S}\gamma_x$. But B_λ, as a homomorphic image of A, is in $\mathcal{U}(\mathcal{P} \cap \mathcal{S}\gamma_x)$. This contradiction proves that $\gamma_x = \mathcal{U}(\mathcal{P} \cap \mathcal{S}\gamma_x)$. Now Corollary 3.8.9 is applicable, yielding that γ_x is a special radical. □

We might have applied also Theorem 3.8.11 to prove that γ_x is special.

Next, we shall look to the cases $\gamma(A) = \gamma_x(A)$ and $\gamma(A[x]) = \varrho(A)[x]$.

PROPOSITION 4.9.21. *For a radical γ the following conditions are equivalent:*

(i) $\gamma_x = \gamma$,
(ii) $A \in \gamma$ *implies* $A[x] \in \gamma$,
(iii) $A \in \gamma$ *if and only if* $A[x] \in \gamma$.

Concrete Radicals and Structure Theorems

If, in addition, γ is hereditary, then $\gamma_x = \gamma$ is equivalent to

($\diamond$) $$\gamma(A) = \gamma(A[x]) \cap A = \varrho(A).$$

Proof: (i) $\Longrightarrow$ (ii) Obvious.

(ii) $\Longrightarrow$ (iii) Trivial as A is a homomorphic image of $A[x]$.

(iii) $\Longrightarrow$ (i) If $A \in \gamma$, then by (iii) also $A[x] \in \gamma$, whence $A \in \gamma_x$. Thus $\gamma \subseteq \gamma_x$. Proposition 4.9.17 yields the inclusion $\gamma_x \subseteq \gamma$.

Let γ be a hereditary radical such that $\gamma_x = \gamma$. Then ($\diamond$) follows from Theorem 4.9.18. Conversely, if ($\diamond$) is true then by Theorem 4.9.18 for $A \in \gamma$ we have
$$A = \gamma(A) = \varrho(A) = \gamma_x(A),$$
proving $\gamma \subseteq \gamma_x$. $\square$

We say that *a radical γ has the Amitsur property*, if
$$\gamma(A[x]) = (\gamma(A[x]) \cap A)[x] = \varrho(A)[x].$$

Notice, that if a radical γ has the Amitsur property, then *the γ-radical of a polynomial ring is again a polynomial ring*.

THEOREM 4.9.22 (Krempa [2]). *For a radical γ to have the Amitsur property it is necessary and sufficient that it satisfies*

(*) $$\varrho(A) = \gamma(A[x]) \cap A = 0 \quad \text{implies} \quad \gamma(A[x]) = 0$$

for all rings A.

Proof: The necessity is obvious. For the sufficiency we recall that $\varrho(A)[x] \subseteq \gamma(A[x])$ by Proposition 4.9.17. Hence, since radical classes are closed under extensions, one gets
$$\gamma(A[x]/\varrho(A)[x]) = \gamma(A[x])/\varrho(A)[x].$$

So, using also the modularity law (see Section 1.1), we have

$$\gamma(A/\varrho(A)[x]) \cap A/\varrho(A) \cong \gamma(A[x]/\varrho(A)[x]) \cap (A + \varrho(A)[x])/\varrho(A)[x] =$$
$$= \frac{\gamma(A[x]) \cap (A + \varrho(A)[x])}{\varrho(A)[x]} = \frac{(\gamma(A[x]) \cap A) + \varrho(A)[x]}{\varrho(A)[x]} =$$
$$= \frac{\varrho(A) + \varrho(A)[x]}{\varrho(A)[x]} = 0.$$

Hence using condition (*), we have $\gamma(A/\varrho(A)[x]) = 0$. Taking account of the isomorphism
$$(A/\varrho(A))[x] \cong A[x]/\varrho(A)[x],$$

we conclude that $\gamma(A[x]/\varrho(A)[x]) = 0$. Consequently

$$\gamma(A[x]) \subseteq \varrho(A)[x],$$

whence $\gamma(A[x]) = \varrho(A)[x]$. □

PROPOSITION 4.9.23. *Let γ be a radical satisfying condition $(\diamond)$ and the implication*

$$A \in \mathcal{S}\gamma \quad \text{implies} \quad A[x] \in \mathcal{S}\gamma.$$

Then γ has the Amitsur property.

Proof: If $\varrho(A) = 0$, then by $(\diamond)$ also $\gamma(A) = 0$, that is, $A \in \mathcal{S}\gamma$. Hence by the assumption $A[x] \in \mathcal{S}\gamma$, that is, $\gamma(A[x]) = 0$. An application of Theorem 4.9.22 yields the assertion. □

EXAMPLE 4.9.24. *The torsion radical τ and the p-torsion radicals τ_p fulfil the requirements of $(*)$ and $(\diamond)$, and they have the Amitsur property.*

THEOREM 4.9.25 (Amitsur [5] and McCoy [4]). *The Baer radical β satisfies condition*

(ii) $A \in \beta$ *implies* $A[x] \in \beta$,

and hence also (iii) *and* $(\diamond)$. *Furthermore,*

(iv) $A \in \mathcal{S}\beta$ *if and only if* $A[x] \in \mathcal{S}\beta$,

and the Baer radical β has the Amitsur property.

Proof: Every nonzero homomorphism f of the ring $A[x]$ induces a surjective homomorphism $A \twoheadrightarrow B$ such that $f(A[x]) = B[f(x)]$. Since $A \in \beta$, also $B \in \beta$, and so B has a nonzero ideal I such that $I^2 = 0$. Hence $0 \neq I[f(x)] \triangleleft B[f(x)]$ and $(I[f(x)])^2 = 0$. Thus $A[x] \in \beta$, establishing (ii).

Suppose that $A[x] \in \mathcal{S}\beta$. Then by $(\diamond)$ we have

$$\beta(A) = \beta(A[x]) \cap A = 0,$$

that is, $A \in \mathcal{S}\beta$. Conversely, assume that $A \in \mathcal{S}\beta$, but $A[x] \notin \mathcal{S}\beta$. Then $\beta(A[x]) \neq 0$, and so there exists a nonzero ideal B of $\beta(A[x])$ such that $B^2 = 0$. As one readily sees, the set

$$I = \{0\} \cup \{\text{leading coefficients of polynomials in } B\} \triangleleft A,$$

and, of course, $I^2 = 0$. Hence $0 \neq I \subseteq \beta(A)$ follows, proving $A \notin \mathcal{S}\beta$. Thus also (iv) has been proved.

Finally, by Proposition 4.9.23, β has also the Amitsur property. □

THEOREM 4.9.26 (Amitsur [5]). *The Jacobson radical $\mathcal{J}$ has the Amitsur property.*

Proof: In view of Theorem 4.9.22 we have to prove that

$$\mathcal{J}(A[x]) \cap A = 0 \text{ implies } \mathcal{J}(A[x]) = 0$$

for all rings A. Assume that this is not true, and there exists a ring A such that $\mathcal{J}(A[x]) \neq 0$ though $\mathcal{J}(A[x]) \cap A = 0$. We shall carry out the proof in several steps.

Let A^1 be the Dorroh extension of A. Then $A[x] \triangleleft A^1[x]$ so $\mathcal{J}(A[x]) \triangleleft \mathcal{J}(A^1[x])$ and hence $0 \neq \mathcal{J}(A[x]) \subseteq \mathcal{J}(A^1[x])$. Also $A^1[x]/A[x] \cong (A^1/A)[x] \cong \mathbb{Z}[x]$ and this is semiprimitive, so $\mathcal{J}(A^1[x]) \subseteq A[x]$ and hence $\mathcal{J}(A^1[x]) \subseteq \mathcal{J}(A[x])$, so that $\mathcal{J}(A^1[x]) = \mathcal{J}(A[x])$. This gives us

$$\mathcal{J}(A^1[x]) \cap A^1 = \mathcal{J}(A[x]) \cap A^1 = \mathcal{J}(A[x]) \cap A = 0.$$

Thus if A has no unity element, we may replace A by A^1, or in other words we lose no generality by assuming that A has a unity element. We therefore make this assumption.

We choose a nonzero polynomial $f(x) \in \mathcal{J}(A[x])$ of minimal degree n. From $\mathcal{J}(A[x]) \cap A = 0$ it follows that $n \geq 1$. Observe that $\theta : g(x) \mapsto g(x+1)$ defines an automorphism of $A[x]$. Since the Jacobson radical, the unique largest quasi-regular ideal, is clearly invariant under automorphisms, it follows that $f(x+1) \in \mathcal{J}(A[x])$. Hence also

$$f_0(x) = f(x+1) - f(x) \in \mathcal{J}(A[x]).$$

Since the degree of $f_0(x)$ is less than that of $f(x)$, necessarily $f_0(x) = 0$. Thus

$$f(x+1) = f(x).$$

Let $f(x) = a_0 + a_1 + \cdots + a_n x^n$. Then

$$0 = f(x+1) - f(x) = \cdots + n x^{n-1},$$

so $n a_n = 0$. As $n > 1$ and $a_n \neq 0$, a_n has finite additive order m. Let p be a prime dividing m. Then $\frac{m}{p} a_n$ has additive order p. Let $g(x) = \frac{m}{p} f(x)$. Then $g(x) \in \mathcal{J}(A[x])$ so $pg(x) \in \mathcal{J}(A[x])$. But $pg(x)$ has leading coefficient $m a_n = 0$, so $pg(x)$ has degree less than n and hence $pg(x) = 0$.

Let A_p be the p-torsion ideal on A. Then the argument just given shows that $g(x) \in A_p[x]$ and $g(x)$ has degree n. Thus

$$0 \neq g(x) \in A_p[x] \cap \mathcal{J}(A[x]) = \mathcal{J}(A_p[x]).$$

But also

$$\mathcal{J}(A_p[x]) \cap A_p = A_p[x] \cap \mathcal{J}(A[x]) \cap A_p \subseteq \mathcal{J}(A[x]) \cap A_p \subseteq \mathcal{J}(A[x]) \cap A = 0.$$

Hence we may also assume without loss of generality that $pA = 0$ for some prime p. Thus the unity element 1 of A has additive order p and A is an algebra over the prime field of p-elements.

Next, we show that $g(x) = h(x^p - x)$ is a polynomial in $x^p - x$ with coefficients in A. We prove this by induction on the degree n of $g(x)$. First, let $n < p$. Since also $g(x+1) = g(x)$, it follows that $g(x+k) = g(x)$ for all integers k. Now

$$g(x+k) = g(k) + xg_1(k) + \cdots x^n g_n(k) = b_0 + xb_1 + \cdots + x^n b_n = g(x).$$

Hence $g(k1) = b_0$ for all integers k and so $g(x) - b_0 = g(x) - g(k1)$. Since the degree of $g(x)$ is less than p, it follows that $g(x) - b_0 = 0$, that is, $g(x) = b_0 \in A$.

Let $g(x)$ be of arbitrary degree. Then taking into consideration that the coefficients of $x^p - x$ are ± 1, there are polynomials $h(x) \in A[x]$ and $r(x) \in A$ such that

$$g(x) = h(x)(x^p - x) - r(x)$$

and the degree of $r(x)$ is less than p. Since we are working in characteristic p, we have $(x+1)^p = x^p + 1^p = x^p + 1$ and hence $(x+1)^p - (x+1) = x^p + 1$, so that

$$g(x+1) = h(x+1)(x^p - x) + r(x+1),$$

and since $g(x+1) = g(x)$, we obtain

$$(h(x+1) - h(x))(x^p - x) = r(x) - r(x+1).$$

Comparing the degrees of both sides, we conclude that

$$r(x+1) = r(x) \quad \text{and} \quad h(x+1) = h(x).$$

Since the degree of $r(x) \in A[x]$ is less than p, we know that $r(x) = r_0 \in A$. By the hypothesis we have $h(x) = h_0(x^p - x) \in A[x^p - x]$. Thus

$$g(x) = h_0(x^p - x) \cdot (x^p - x) + r_0 \in A[x^p - x].$$

We claim that if a polynomial $h(x^p - x)$ belongs to $\mathcal{J}(A[x])$ then $h(x^p - x) \in \mathcal{J}(A[x^p - x])$. Indeed, let $k(x) \in h(x^p - x)A[x^p - x]$. Then $k(x) \in \mathcal{J}(A[x])$ as $h(x^k - x) \in \mathcal{J}(A[x])$, whence $k(x+1) = k(x)$. Hence $k(x)$ has a quasi-inverse $k'(x)$. The quasi-inverse of $k(x+1)$ is readily seen to be $k'(x+1)$. Hence $k(x) = k(x+1)$ implies $k'(x) = k'(x+1)$. Taking into account that A has characteristic p, the same proof as for $g(x)$ yields that $k'(x) \in A[x^p - x]$. This shows that the right ideal $h(x^p - x)A[x^p - x]$ is quasi-regular in $A[x^p - x]$. Thus $h(x^p - x) \in \mathcal{J}(A[x^p - x])$ as claimed.

Summing up, we have proved that

$$\frac{m}{p} f(x) = g(x) = h(x^p - x) \in \mathcal{J}(A[x^p - x]).$$

The mapping $h(x) \mapsto h(x^p - x)$ induces an isomorphism between $A[x]$ and $A[x^p - x]$ such that $\mathcal{J}(A[x])$ and $\mathcal{J}(A[x^p - x])$ are corresponding ideals. In particular, since

$$h(x^p - x) = g(x) \in \mathcal{J}(A[x^p - x]),$$

it follows that $h(x) \in \mathcal{J}(A[x])$. Taking into account that the degree n of $g(x)$ is at least 1, the degree of $h(x)$ is less than n. This contradiction proves that $\mathcal{J}(A[x]) = 0$. □

Amitsur [5] proved that also the Levitzki and Köthe radicals have the Amitsur property. For the Brown–McCoy radical this was proved by Krempa [2].

Not every (special) radical has the Amitsur property. Krempa [2] gave the following easy example. Let F be a field of p elements, and γ the upper radical determined by F. Obviously, $F \in \mathcal{S}\gamma$, but $0 \neq x^p - x \in \gamma(F[x])$ as one easily checks. Hence in view of Proposition 4.9.22, γ does not have the Amitsur property.

Theorem 4.9.25 gives a decent relation between the Baer radical of a ring and of its polynomial ring. For the nil and Jacobson radicals the situation is more subtle and the nil and Jacobson radicals go hand in hand.

PROPOSITION 4.9.27. *If $A[x] \in \mathcal{J}$ then $A \in \mathcal{N}$.*

Proof: Embed $A[x]$ into the formal power series ring $A^1[[x]]$ where A^1 denotes the Dorroh extension of A. Let $a \in A$ be an arbitrary element. Since $ax \in A[x] \in \mathcal{J}$, the monomial ax has a quasi-inverse $\sum_{i=0}^{n} c_i x^i \in A[x]$. In $A^1[[x]]$ we can write

$$(1 - ax)(1 - c_0 - c_1 x - c_2 x^2 - \cdots - c_n x^n) = 1.$$

On the other hand, in $A^1[[x]]$ we have also

$$(1 - ax)(1 + ax + a^2 x^2 + \cdots + a^n x^n + \cdots) = 1.$$

Matching powers of x shows that

$$c_0 = 0, \ c_1 = -a, \ c_2 = -a^2, \ldots, c_n = -a^n, \ \text{and} \ a^{n+1} = 0.$$

Hence A is a nil ring. □

Observe that Proposition 4.9.27 is a substantially stronger statement than the implication $A[x] \in \mathcal{J} \Rightarrow A \in \mathcal{J}$ in Proposition 4.9.17.

PROPOSITION 4.9.28 (Amitsur [5]). *$\mathcal{J}(A[x]) \cap A$ is a nil ideal of A.*

Proof: To be consistent with the notation γ_x, we set $\mathcal{J}_x = \{A \mid A[x] \in \mathcal{J}\}$. As $\mathcal{J}$ is hereditary, Theorem 4.9.18 implies that $\mathcal{J}(A[x]) \cap A = \mathcal{J}_x(A) \in \mathcal{J}_x$, so $(\mathcal{J}(A[x]) \cap A)[x] \in \mathcal{J}$ and by Proposition 4.9.27 $\mathcal{J}(A[x]) \cap A \in \mathcal{N}$. □

PROPOSITION 4.9.29 (Amitsur [5]). *$A \in \mathcal{SN}$ implies $\mathcal{J}(A[x]) \cap A = 0$.* □

PROPOSITION 4.9.30 (Amitsur [5]). *If $A \in \mathcal{SN}$ then $A[x] \in \mathcal{SJ}$.*

Proof: Let us consider the set

$$I = \{0\} \cup \left\{ \begin{array}{l} a \in A \mid a \text{ is the leading coefficient} \\ \text{of some } a_1 x + a_2 x^2 + \cdots \in \mathcal{J}(A[x]) \end{array} \right\}.$$

The set I is readily seen to be an ideal of A. Let $a \in I$ be an arbitrary element, and $f \in \mathcal{J}(A[x])$ of degree n with leading coefficient $a_n = a \in I$ (that is, with zero x-free term). Let us embed $A[x]$ into the formal power series ring $A^1[[x]]$ over the Dorroh extension A^1 of A. Let h be the quasi-inverse of f in $\mathcal{J}(A[x])$. From $f + h - fh = 0$ we see that also h has zero x-free term. Further,

$$(1-f)(1-h) = 1,$$

and also

$$(1-f)(1 + f + f^2 + \cdots) = 1.$$

Hence

$$h = -f - f^2 - \cdots .$$

Since $h \in \mathcal{J}(A[x])$, necessarily the leading coefficient of h must be nilpotent, whence a is nilpotent. Consequently I is a nil ideal of A. But $A \in \mathcal{SN}$, therefore $I = 0$. Thus every nonzero element $f \in \mathcal{J}(A[x])$ has nonzero x-free term. Suppose that there is such an

$$f = b + b_1 x + b_2 x^2 \cdots \quad \text{with } b \neq 0.$$

Then for every $c \in A$ we have

$$(b + b_1 x + b_2 x^2 \cdots)cx = bcx + b_1 cx + b_2 cx^2 + \cdots \in \mathcal{J}(A[x]),$$

so $bc = 0$. Since c is arbitrary, so $b \in (0 : A)$ and $(0 : A)$ is a nil ideal in A. By $A \in \mathcal{SN}$ we conclude that $(0 : A) = 0$, so $b = 0$. This contradiction proves $\mathcal{J}(A[x]) = 0$. $\square$

Observe that we have deduced Proposition 4.9.30 from Proposition 4.9.27. Also the converse is true:

PROPOSITION 4.9.31. *The implications of Propositions 4.9.27 and 4.9.30 are logically equivalent.*

Proof: From Proposition 4.9.27 we have established Proposition 4.9.30 via Propositions 4.9.28 and 4.9.29.

Suppose that $A \notin \mathcal{N}$ but $A[x] \in \mathcal{J}$. Then

$$0 \neq B = A/\mathcal{N}(A) \in \mathcal{SN},$$

and by Proposition 4.9.30 we have $B[x] \in \mathcal{SJ}$. Moreover,

$$B[x] = (A/\mathcal{N}(A))[x] \cong A[x]/\mathcal{N}(A)[x] \in \mathcal{J}$$

in view of $A[x] \in \mathcal{J}$. This implies $B[x] = 0$, a contradiction. □

Next, we show that also the converse implications are equivalent.

PROPOSITION 4.9.32. *The implication*

(K1) $\qquad\qquad A \in \mathcal{N}$ *implies* $A[x] \in \mathcal{J}$

is equivalent to the implication

(K2) $\qquad\qquad A[x] \in \mathcal{SJ}$ *implies* $A \in \mathcal{SN}$.

Proof: (K1) $\Longrightarrow$ (K2). Let $A[x] \in \mathcal{SJ}$ and $A \notin \mathcal{SN}$. Then $0 \neq \mathcal{N}(A) \triangleleft A$, and so (K1) yields $\mathcal{N}(A)[x] \in \mathcal{J}$. Hence

$$0 \neq \mathcal{N}(A[x]) \subseteq \mathcal{J}(A[x]),$$

contradicting $A[x] \in \mathcal{SJ}$. Thus (K2) has been established.

(K2) $\Longrightarrow$ (K1) Let $A \in \mathcal{N}$. As proved in Theorem 4.9.26, the Jacobson radical has the Amitsur property. Hence we have

$$A[x]/\mathcal{J}(A[x]) = A[x]/(\mathcal{J}(A[x]) \cap A)[x] \cong (A/(\mathcal{J}(A[x]) \cap A))[x].$$

The ring $B = A/(\mathcal{J}(A[x]) \cap A)$ is nilpotent as A is so. But $B[x] \cong A[x]/\mathcal{J}(A[x]) \in \mathcal{SJ}$, so (K2) implies $B \in \mathcal{SN}$. Hence $B = 0$ and $A[x] = \mathcal{J}(A[x]) \in \mathcal{J}$. □

To decide as whether (K1) or (K2) is true is as difficult as to solve Köthe's Problem.

THEOREM 4.9.33 (Krempa [1], [3])*. The implication* (K1) *is true if and only if the nil radical* $\mathcal{N}$ *is left strong.*

Proof: In the proof we shall follow Puczyłowski [13] and shall apply Theorem 4.9.13: we shall prove the implications (K1) $\Longrightarrow$ Theorem 4.9.13 (1) and Theorem 4.9.13 (3) $\Longrightarrow$ (K1).

(K1) $\Longrightarrow$ (1) Let A be any ring and L a nil left ideal of A. Now $L[x] \triangleleft_l A[x]$, and so (K1) implies $L[x] \in \mathcal{J}$. The ideal of $A[x]$ generated by $L[x]$ is clearly $(L + LA)[x]$. Since the Jacobson radical is left strong (see Example 3.17.4), by Theorem 3.17.5 we have $(L + LA)[x] \in \mathcal{J}$. Hence Proposition 4.9.27 implies that $L + LA \in \mathcal{N}$. Applying Theorem 3.17.5 again, we get that $\mathcal{N}$ is a left strong radical.

(3) $\Longrightarrow$ (K1) We show, first, that if $A \in \mathcal{N}$ then $xA[x] \in \mathcal{J}$. Embedding $A[x]$ into the formal power series ring $A^1[[x]]$ over the Dorroh extension A^1 of A, we have $xA[x] \subseteq A[[x]]$. For any polynomial $xf \in xA[x]$ we can write the equation

$$(1 - xf)(1 + xf + x^2f^2 + \cdots) = 1$$

in $A^1[[x]]$. Hence $g = -xf - x^2f^2 - \cdots$ is a quasi-inverse of xf in $A[[x]]$. To avoid negative signs, in the sequel we write g in the form $g = \sum_{i=1}^{\infty}(-b_i)x^i$, and put $xf = \sum_{i=1}^{n} a_i x^i$. Now we have $xf - gxf = -g$, that is,

$$a_1 x + \cdots + a_n x^n + (b_1 x + b_2 x^2 + \cdots)(a_1 x + \cdots + a_n x^n) = b_1 x + b_2 x^2 + \cdots.$$

Matching powers of x for every $m \geq n$,

$$b_{m+1} = b_m a_1 + b_{m-1} a_2 + \cdots + b_{m-n+1} a_n$$
$$b_{m+2} = b_{m+1} a_1 + b_m a_2 + \cdots + b_{m-n+2} a_n$$
$$= (b_m a_1 + b_{m-1} a_2 + \cdots + b_{m-n+1} a_n) a_1 + b_m a_2 + \cdots + b_{m-n+2} a_n$$
$$= b_m c_{12} + b_{m-1} c_{22} + \cdots + b_{m-n+1} c_{n2}$$
$$\vdots$$
$$b_{m+n} = b_m c_{1n} + \cdots + b_{m-n+1} c_{nn}$$

for some $c_{ij} \in A$; $i,j \in \{1, \ldots, n\}$. Hence, putting

$$M = \begin{pmatrix} a_n & c_{n2} & \cdots & c_{nn} \\ \vdots & & & \\ a_1 & c_{12} & \cdots & c_{1n} \end{pmatrix},$$

for every $m \geq n$ we have

$$(b_{m+1}, \ldots, b_{m+n}) = (b_{m-n+1}, \ldots, b_m)M.$$

In particular,

$$(b_{n+1}, \ldots, b_{2n}) = (b_1, \ldots, b_n)M$$
$$(b_{2n+1}, \ldots, b_{3n}) = (b_{n+1}, \ldots, b_{2n})M = (b_1, \ldots, b_n)M^2$$
$$\vdots$$
$$(b_{kn+1}, \ldots, b_{(k+1)n}) = (b_1, \ldots, b_n)M^k$$

Now the implication (3) of Theorem 4.9.13 tells us that $M^k = 0$ for some k. Hence

$$b_{kn+1} = b_{kn+2} = \cdots = b_{(k+1)n} = 0.$$

Consequently the quasi-inverse g has the form $g = \sum_{i=1}^{kn}(-b_i)x^i$, and therefore $g \in xA[x]$. Thus $xA[x] \in \mathcal{J}$.

To finish the proof, we have to show that $A[x] \in \mathcal{J}$. Since $A \in \mathcal{N} \subseteq \mathcal{J}$ and $xA[x] \in \mathcal{J}$ from the isomorphism

$$A[x]/xA[x] = (A + xA[x])/xA[x] \cong A$$

we conclude that $A[x] \in \mathcal{J}$. □

COROLLARY 4.9.34. *Any of the implications* (K1) *and* (K2) *is equivalent to Köthe's Problem.* □

Recent investigations approximate the implication (K1). Puczyłowski and Smoktunowicz [1] proved that

$$A \in \mathcal{N} \text{ implies } A[x] \in {}^t\mathcal{G}.$$

Beidar, Fong and Puczyłowski [1] improved this to the implication

$$A \in \mathcal{N} \text{ implies } A[x] \in \mathcal{B}.$$

where $\mathcal{B}$ denotes the Behrens radical. Further approximation from above were given by Beidar, Puczyłowski and Wiegandt [1] and Tumurbat [2].

Agata Smoktunowicz [1] proved that the nil radical $\mathcal{N}$ does not satisfy the equivalent conditions of Proposition 4.9.21. She constructed a ring A such that

$$A \in \mathcal{N} \text{ but } A[x] \notin \mathcal{N},$$

answering a conjecture of Amitsur [5], [7] in the negative (at the time of Amitsur's conjecture Krempa's Theorem 4.9.33 was not yet known). For a survey of related questions we refer to Puczyłowski [17]. Smoktunowicz and Puczyłowski [1] constructed a polynomial ring which is Jacobson radical but not nil.

The study of radicals of polynomial rings and formal power series rings is challenging, as the results treated so far indicate.

At the end of this section we mention a sample of further results.

Jacobson rings (Brown–McCoy rings) are rings A in which the Baer radical $\beta(B)$ equals the Jacobson radical $\mathcal{J}(B)$ (the Brown–McCoy radical $\mathcal{G}(B)$, respectively) in all homomorphic images B of A. Watters [3], [4] proved that A is a Jacobson ring (Brown–McCoy ring) if and only if $A[x]$ is a Jacobson ring (Brown–McCoy ring, respectively). For further results, see e.g. Ferrero and Parmenter [1].

Let X be a set with cardinality $|X| \geq 2$, and $A[X]$ the ring of polynomials over A in commuting indeterminates in X. It goes back to Amitsur [5] that

$$\beta(A[X]) = \beta(A)[X]$$

and

$$\mathcal{J}(A[X]) = (\mathcal{J}(A[X]) \cap A)[X]$$

with $\mathcal{J}(A[X]) \cap A \in \mathcal{N}$. The corresponding result for the ring of polynomials $A\langle X \rangle$ in non-commuting indeterminates was proved by Puczyłowski [6] (see also Sierpińska [1]). For results on radicals and power series rings we refer e.g. to Puczyłowski [1], [7], and Puczyłowski and Smoktunowicz [2].

4.10. Radicals on artinian rings

As we have seen in Section 4.5, Jacobson semisimple artinian rings are described by the Wedderburn–Artin Structure Theorems and the Jacobson radical of an artinian ring is nilpotent. In this section we continue the study of radicals on artinian rings. In the first part the torsion radical of an artinian ring will be studied. Although the torsion radical as an A-radical looks suspicious in the context of structure theorems, we shall see that it has an important effect on artinian rings: the torsion radical is a direct summand in every artinian ring. In the second part we shall determine the radicals which have the same impact on artinian rings, and doing so we shall use the main result of the first part. As the torsion radical is an A-radical, it is no wonder that the behaviour of the additive group of rings plays a significant role in these investigations. Finally, the von Neumann radical ν will be studied on artinian rings.

PROPOSITION 4.10.1. *If A is an artinian ring, then the additive group A^+ is a direct sum $A^+ = D^+ \oplus B^+$ where D^+ is the maximal divisible subgroup in A^+ and B^+ is a torsion subgroup. In particular, if A is also torsionfree, then A^+ is divisible.*

Proof: Let $A^+ = D^+ \oplus B^+$ where D^+ is the maximal divisible subgroup and B^+ is reduced. Let us consider the set of left ideals

$$E = \{(b] \mid 0 \neq b \in B,\ o(b) = \infty\}.$$

We have to show that E is empty. Suppose that E is not empty. Since A is artinian, there exists a minimal left ideal $(c]$ in E. The choice of $(c]$ and $o(c) = \infty$ yield that

$$(c] = (2c] = \ldots = (nc],$$

whence $(c] = n(c]$ follows for $n = 2, 3, \ldots$. Hence $(c]^+$ is a divisible subgroup, so $(c] \subseteq D$. Thus $c \in B \cap D = 0$ follows, a contradiction. □

PROPOSITION 4.10.2. *If A is a torsionfree artinian ring, then every element $a \in A$ is a product $a = ba$ with a suitable element $b \in A$.*

Proof: Every descending chain

$$(a] \supseteq \ldots \supseteq (p^k a] \supseteq \ldots$$

terminates in a finite number k of steps for each prime p. Hence $(p^k a] = (p^{k+1} a]$ for some p, implying

$$p^k a = l(p^{k+1} a) + c(p^{k+1} a)$$

with a suitable integer l and an element $c \in A$. Now for the nonzero number $n = p^k - lp^{k+1}$ we have $na = p^{k+1} ca$. Since by Proposition 4.10.1 A is divisible, there exists an element $d \in A$ such that $nd = c$. Hence we get

$na = n(p^{k+1}da)$. Taking into account that A is torsionfree, we conclude that $a = p^{k+1}da$, and the assertion is proved with $b = p^{k+1}d$. □

PROPOSITION 4.10.3. *The class of all torsionfree artinian rings is homomorphically closed.*

Proof: Let A/I be a homomorphic image of a torsionfree artinian ring A, and let $a + I$ be an arbitrary coset with $a \notin I$. Suppose that $a + I$ is of finite order k in A/I, that is, $ka \in I$. Let us consider the descending chain

$$(ka] \supseteq (k^2 a] \supseteq \ldots \supseteq (k^n a] \supseteq \ldots$$

of left ideals. Since A is artinian, there exists an integer n such that $(k^n a] = (k^{n+1} a]$. Now we have $(k^n a] = k^n (ka]$. Thus $k^n a \in k^n (ka]$, that is, $k^n a = k^n x_0$ holds with some $x_0 \in (ka] \subseteq I$. Since A is torsionfree, it follows $a = x_0 \in I$, a contradiction. □

PROPOSITION 4.10.4 *Let A be an artinian ring with maximal divisible ideal D. If D^2 is torsionfree, then A is a direct sum of its torsion radical $\tau(A) = T$ and of a torsionfree ideal F.*

Proof: Firstly, we show that D^2 is a divisible subgroup. Let $d = \sum a_i b_i$ be any element of D^2, $a_i, b_i \in D$. For any nonzero integer n there exist elements $x_i \in D$ such that $nx_i = a_i$, and so

$$d = \sum (nx_i) b_i = n \sum x_i b_i \in D$$

for D is an ideal in A. Thus D^2 is divisible. Moreover, D^2 is obviously an additive subgroup.

Hence the maximal divisible subgroup D^+ decomposes as

$$D^+ = (T \cap D)^+ \oplus (D^2 \oplus C^+)$$

where C^+ is clearly torsionfree and divisible. In view of Proposition 4.10.1 there is a torsionfree subgroup B^+ such that

$$A^+ = D^+ \oplus B^+ = (T \cap D)^+ \oplus D^2 \oplus C^+ \oplus B^+.$$

Hence

$$T^+ = (T \cap D)^+ \oplus B^+$$

and then

$$A^+ = ((T \cap D)^+ \oplus B^+) \oplus (D^2 \oplus C^+) = T^+ \oplus D^2 \oplus C^+.$$

We claim that T is a direct summand of A as an ideal. Applying Proposition 1.2.2 we get

$$(D^2 + C)A = D^2(B + D) + C(B + D) = D^2 B + D^2 D + CB + CD \subseteq D^2,$$

and similarly also $A(D^2 + C) \subseteq D^2$, proving that $F = D^2 + C$ is an ideal in A. □

After these preparations we are ready to prove

SZÁSZ' THEOREM 4.10.5 ([2]). *Every artinian ring A is the direct sum $A = T \oplus F$ of its torsion radical $T = \tau(A)$ and of a uniquely determined torsionfree ideal F.*

Proof: In view of Proposition 4.10.4 we have to prove first that D^2 is torsionfree where D is the maximal divisible ideal of A. Let $a, b \in D$ be elements such that $ab \in T$. If $a, b \in D \cap T$, then by Proposition 1.2.1 we have $ab = 0$. If $a \in T$ and $o(b) = \infty$, then Proposition 1.2.2 yields $ab = 0$. So let $o(a) = o(b) = \infty$. By Proposition 4.10.1, Proposition 1.2.3 (iv) is applicable, and so $A/T \cong D/(D \cap T)$ holds showing that A/T is divisible and torsionfree. Proposition 4.10.2 assures the existence of an element $c \in A$ such that

$$a + T = (c + T)(a + T), \text{ that is, } a = ca + t, \ t \in T.$$

Clearly $o(c) = \infty$. Hence by Proposition 1.2.2 it follows that

$$ab = (ca + t)b = c(ab) + tb = 0.$$

This shows that $D^2 \cap T = 0$, that is, D^2 is torsionfree. An application of Proposition 4.10.4 yields that T is a direct summand in A, $A = T \oplus F$.

We still have to prove that the torsionfree ideal F of A is uniquely determined. Since $F \cong A/T$, F is artinian. Let K be any torsionfree ideal in A such that $A = T \oplus K$. Then we have

$$F/(F \cap K) \cong (F + K)/K \triangleleft A/K \cong T.$$

Since F is artinian and torsionfree, Proposition 4.10.3 implies that also $F/(F \cap K)$ is torsionfree, and so necessarily 0. Thus $F \subseteq K$. Analogous reasoning gives $K \subseteq F$. □

Theorem 4.10.5 features one of the important tasks of structure theoretic investigations: it chops (decomposes) artinian rings into two pieces of diverse properties (torsion and torsionfree ideals). F. A. Szász' Theorem 4.10.5 has been generalized to rings with d.c.c. on principal left ideals by Chr. Ayoub [1] and Dinh Van Huynh [1] (cf. also Kertész [1]), to alternative rings with d.c.c. on left ideals by Widiger [2] and to alternative rings with d.c.c. on principal left ideals by Beidar and Wiegandt [1].

In view of Hopkin's Lemma 4.5.8 and the Second Wedderburn-Artin Structure Theorem every artinian ring is the extension of a nilpotent ring (= its Jacobson radical) by a semiprime artinian ring. The so far best decomposition of artinian rings was achieved by Widiger [1] (see also Kertész [1]). Szele [1] described the structure of nilpotent artinian rings by subrings and natural numbers. Sometimes it is more convenient to have a representation of nilpotent artinian rings by homomorphic images and natural numbers (see Wiegandt [1]). Concerning the structure of nilpotent rings in general the reader is referred to the book of Kruse and Price [1].

Hopkins' Lemma 4.5.8 tells us that *for every artinian ring its Jacobson radical coincides with its Baer radical*. Since the Jacobson radical $\mathcal{J}$ is so successful on artinian rings, it is natural to focus our attention on those radicals γ which have the same effect on artinian rings as the Jacobson radical.

The Wedderburn–Artin Structure Theorems make it clear that the radicals we seek have to be contained in the upper radical $\mathcal{T}$ of all matrix rings over all division rings. Denoting the class of artinian rings by $\mathcal{A}$, we have obviously

(∗) $$\mathcal{T} \cap \mathcal{A} = \mathcal{J} \cap \mathcal{A},$$

that is, $\mathcal{T}(A) = A$ if and only if $\mathcal{J}(A) = A$ for an artinian ring A. However, (∗) is not enough for our purpose because the class $\mathcal{A}$ is not a universal class; it fails to be hereditary: the factor ring $\mathbb{Q}[x]/(x^2)$ of rational polynomials modulo the ideal (x^2) is a commutative artinian ring with ideals 0, $(x)/(x^2)$ and $\mathbb{Q}[x]/(x^2)$, but $(x)/(x^2)$ being isomorphic to the zero-ring built on the additive group $\mathbb{Q}^+$ is not artinian.

We say that *a radical γ coincides on artinian rings with the Jacobson radical $\mathcal{J}$*, if $\gamma(A) = \mathcal{J}(A)$ holds for every artinian ring A. Let us remind the reader that in Section 4.2 coincidence of the radicals β, $\mathcal{L}$ and $\mathcal{N}$ on certain classes of rings (commutative rings, noetherian rings) was meant in this sense. Divinsky [1], [2] calls this coincidence strong coincidence and (∗) weak coincidence.

PROPOSITION 4.10.6. *$\mathcal{T}(A) = \mathcal{J}(A)$ holds for every artinian ring A, that is, $\mathcal{T}$ coincides with $\mathcal{J}$ on $\mathcal{A}$.*

Proof: As matrix rings over division rings are primitive, we have $\mathcal{J} \subseteq \mathcal{T}$ so that $\mathcal{J}(A) \subseteq \mathcal{T}(A)$ for all rings A. Suppose there exists an artinian ring A such that $\mathcal{J}(A) \neq \mathcal{T}(A)$. Now

$$0 \neq \mathcal{T}(A)/\mathcal{J}(A) \triangleleft A/\mathcal{J}(A),$$

and $A/\mathcal{J}(A)$ is a finite direct sum of matrix rings over division rings. Hence $\mathcal{T}(A)/\mathcal{J}(A)$ can be mapped homomorphically onto at least one matrix ring over a division ring, which contradicts $\mathcal{T}(A)/\mathcal{J}(A) \in \mathcal{T}$. □

To get a lower bound for radicals coinciding with $\mathcal{J}$ on $\mathcal{A}$, we introduce *Divinsky's lower radical $\mathcal{D}$* of all nilpotent rings which are Jacobson radicals of artinian rings, more precisely, the lower radical $\mathcal{D} = \mathcal{LX}$ of the homomorphic closure $\mathcal{X}$ of nilpotent rings which are Jacobson radicals of artinian rings. We know from Theorem 3.4.12 that $\mathcal{D} = \mathcal{X}_3$.

PROPOSITION 4.10.7. $\mathcal{D}$ coincides with $\mathcal{J}$ on artinian rings.

Proof: Since $\mathcal{D} \subseteq \beta \subseteq \mathcal{J}$, it follows $\mathcal{D}(A) \subseteq \mathcal{J}(A)$ for all rings A. Moreover, by definition $\mathcal{J}(A) \subseteq \mathcal{D}(A)$ holds for every artinian ring A, proving the assertion. □

THEOREM 4.10.8 (Divinsky [1]). *A radical γ coincides with the Jacobson radical on artinian rings if and only if $\mathcal{D} \subseteq \gamma \subseteq \mathcal{T}$.*

Proof: If $\mathcal{D} \subseteq \gamma \subseteq \mathcal{T}$, then

$$\mathcal{J}(A) = \mathcal{D}(A) \subseteq \gamma(A) \subseteq \mathcal{T}(A) = \mathcal{J}(A)$$

for each $A \in \mathcal{A}$.

Conversely, if γ coincides with $\mathcal{J}$ on $\mathcal{A}$, then for every $A \in \mathcal{A}$ we have $\mathcal{D}(A) \in \gamma$ and so it follows $\mathcal{D} \subseteq \gamma$. Further, suppose that $\gamma \not\subseteq \mathcal{T}$. Then there exists a ring $B \in \gamma$ having a homomorphic image $M_n(D)$, a matrix ring over the division ring D. Hence $M_n(D) \in \gamma \cap \mathcal{A}$, which implies

$$\mathcal{J}(M_n(D)) = \gamma(M_n(D)) = M_n(D),$$

a contradiction. □

The position of $\mathcal{T}$ among the radicals is clear: $\mathcal{T}$ *properly contains the Brown–McCoy radical class $\mathcal{G}$, and $\mathcal{T}$ is properly contained in the Thierrin radical class $\mathcal{F}$.* Furthermore, $\mathcal{T}$ is a supernilpotent dual radical by Theorem 3.9.8, and so $\mathcal{T}$ is also a special radical.

Next, we are going to show that the radical $\mathcal{D}$ is not hereditary, and so it is properly contained in β.

PROPOSITION 4.10.9. *If L is a left ideal of an artinian ring A, then its torsion radical $\tau(L)$ is a direct summand of L.*

Proof: From Theorem 4.10.5 we know that $A = \tau(A) \oplus F$. Put briefly T for $\tau(A)$. Clearly $(L \cap T) \oplus (L \cap F) \subseteq L$. Let $a \in L$ be an arbitrary element: $a = t + f$ with $t \in T$ and $f \in F$. Applying Proposition 4.10.2 we get that $f = bf$ with a suitable element $b \in F$. Thus

$$ba = b(t+f) = bf = f,$$

and therefore $a \in L$ implies $f \in L \cap F$. Furthermore, also $t = f - a \in L \cap T$, proving

$$L \subseteq (L \cap T) \oplus (L \cap F) = \tau(L) \oplus (L \cap F). \qquad \square$$

PROPOSITION 4.10.10. *Every left ideal L of a torsionfree artinian ring A is divisible.*

Proof: A is divisible by Proposition 4.10.1. Let $a \in L$ be an arbitrary element. Proposition 4.10.2 is applicable, and so $a = ba$ with an appropriate element $b \in A$. Since A is divisible, for every nonzero integer n there exists a $y_0 \in A$ such that $ny_0 = b$ and so $ny_0 a = a$. Thus the equation $nx = a$ has a solution $x_0 = y_0 a \in L$. □

PROPOSITION 4.10.11. *The class $\mathcal{D}$ is contained in the class*

$$\mathcal{C} = \{A \in \beta \mid A/\tau(A) \text{ is divisible}\}.$$

Proof: By definition $\mathcal{D} = \mathcal{L}\mathcal{X}$ where

$$\mathcal{X} = \mathcal{X}_1 = \left\{ B \in \mathcal{J} \;\middle|\; \begin{array}{l} B \text{ is a homomorphic image of the} \\ \text{Jacobson radical of an artinian ring} \end{array} \right\},$$

and we know that $\mathcal{D} = \mathcal{X}_3$.

An arbitrary ring $B \in \mathcal{X}_1$ is obviously of the form $B = \mathcal{J}(A)/K$ where A is an artinian ring, and $A = \tau(A) \oplus F$ by Theorem 4.10.5. Moreover, applying Proposition 4.10.9 for $J = \mathcal{J}(A)$ and $T = \tau(A)$ we have

$$J = (J \cap T) \oplus (J \cap F).$$

By Proposition 4.10.3 along with F also $F/(F \cap K)$ is artinian and torsionfree, so from

$$((J \cap F) + K)/K \triangleleft (F + K)/K \cong F/(F \cap K)$$

we conclude that also $((J \cap F) + K)/K$ is torsionfree. In J/K, both $((J \cap T) + K)/K$ and $((J \cap F) + K)/K$ are ideals and their sum is J/K. As one is torsion, the other torsionfree, the sum is direct. Hence

$$\tau(J/K) = ((J \cap T) + K)/K,$$

and therefore

$$\frac{J/K}{\tau(J/K)} \cong \frac{(J \cap T) \oplus (J \cap F)}{(J \cap T) + K} \cong$$
$$\cong \frac{J \cap F}{((J \cap T) + K) \cap (J \cap F)} \triangleleft \frac{F}{(J \cap T) + K) \cap (J \cap F)}.$$

Since F is a torsionfree artinian ring, Propositions 4.10.3 and 4.10.10 yield that

$(J \cap F)/(((J \cap T) + K) \cap (J \cap F))$, as well as $(J/K)/\tau(J/K)$,

is divisible. Thus $\mathcal{X}_1 \subseteq \mathcal{C}$.

Next, let us consider an arbitrary element $B \in \mathcal{X}_2$, and put $C = B/\tau(B)$, and let D stand for the maximal divisible ideal of C. We want to show that $C = D$. Assume that $C \neq D$. Since $C \in \mathcal{X}_2$, $C/D \neq 0$ possesses a nonzero ideal $I/D \in \mathcal{X}_1 \subseteq \mathcal{C}$. But then, as C/D is torsionfree, so is I/D, where I/D is divisible. Since C/D is reduced, necessarily $I/D = 0$, a contradiction. Consequently $C = D$, which proves $\mathcal{X}_2 \subseteq \mathcal{C}$.

Repeating the above reasoning, we get that $\mathcal{D} = \mathcal{X}_3 \subseteq \mathcal{C}$. □

THEOREM 4.10.12 (Gardner [3] and Divinsky [1]). *Divinsky's radical $\mathcal{D}$ is not hereditary and so properly contained in the Baer radical β. The zero-ring $Z(\infty)$ is in β but not in $\mathcal{D}$.*

Proof: Let us consider the split-null extension A of the zero-ring $\mathbb{Q}^0$ over the rationals by the rationals $\mathbb{Q}$ (cf. Section 3.6): $A = \mathbb{Q} * \mathbb{Q}^0$. A is clearly a commutative ring, all the ideals of A are 0, $\mathbb{Q}^0$, A. Thus A is an artinian ring,
$$\mathcal{D}(A) = \beta(A) = \mathcal{J}(A) = \mathbb{Q}^0,$$
and so $\mathbb{Q}^0 \in \mathcal{D}$. Moreover, $Z(\infty) \triangleleft \mathbb{Q}^0$, $Z(\infty)$ is torsionfree and it is not divisible. Hence $Z(\infty) \notin \mathcal{C}$ and Proposition 4.10.11 implies that $Z(\infty) \notin \mathcal{D}$. Thus both statements have been proved. □

Let us mention that in Proposition 4.10.11 in fact equality holds, as it has been proved by Gardner [3]. For more details on the coincidence of radicals on artinian rings we refer to Divinsky [1], Gardner [3] and Szász [6].

Noetherian artinian rings can be characterized by the divisible torsion radical τ_D (see Example 3.2.2). An artinain ring A is noetherian if and only if $\tau_D(A) = 0$ (see Fuchs [1] and [2], Fuchs and Szele [1]).

Finally we discuss von Neumann regular rings in the context of artinian rings.

PROPOSITION 4.10.13. *Every von Neumann regular ring is Jacobson semisimple.*

Proof: Let A be a von Neumann regular ring, $a \in \mathcal{J}(A)$. Now there exists an $x \in A$ such that $a = axa$ and ax is an idempotent in $\mathcal{J}(A)$. Hence by Proposition 4.4.2 we conclude $ax = 0$ and so $a = axa = 0$. □

PROPOSITION 4.10.14. *Direct products and direct sums of von Neumann regular rings are von Neumann regular.*

Proof: Obvious. □

THEOREM 4.10.15. *Every artinian Jacobson semisimple ring is von Neumann regular.*

Proof: In view of the Wedderburn–Artin Structure Theorems 4.5.6, 4.5.9 and of Proposition 4.10.14 it is sufficient to prove that the ring $\text{End } V = M_n(D)$ of linear transformations on an n-dimensional vector space V over a division ring D is von Neumann regular. This has been shown already in Example 2.1.9. □

Although the radical class ν of von Neumann regular ring is not a semisimple class and the Jacobson semisimple rings do not form a radical class, their intersections with the class $\mathcal{A}$ of artinian rings coincide, as it follows immediately from Proposition 4.10.13 and Theorem 4.10.15.

COROLLARY 4.10.16. *An artinian ring A is von Neumann regular if and only if A is Jacobson semisimple:*

$$\nu \cap \mathcal{A} = \mathcal{SJ} \cap \mathcal{A}.$$ □

COROLLARY 4.10.17. *Every simple prime ring having a minimal left ideal, is von Neumann regular.*

Proof: Apply the Litoff–Ánh Theorem (Theorem 4.5.15) and Theorem 4.10.15. □

Notice that *a simple prime ring need not be von Neumann regular:* a simple prime Jacobson radical ring is not von Neumann regular by Proposition 4.10.13.

As the torsion radical also the von Neumann regular radical splits off in every artinian ring.

THEOREM 4.10.18 (Brown and McCoy [3], Mlitz, Sands and Wiegandt [1]). *The von Neumann regular radical $\nu(A)$ of any artinian ring A has a unity element, and hence $\nu(A)$ is a direct summand in A.*

Proof: Since $\nu(A)$ is hereditarily idempotent and $\mathcal{J}(A)$ is nilpotent for every $A \in \mathcal{A}$ (cf. Theorem 4.5.8), we have

$$\nu(A) \cap \mathcal{J}(A) = 0.$$

Hence

$$\nu(A) \cong \nu(A)/(\nu(A) \cap \mathcal{J}(A)) \cong (\nu(A) + \mathcal{J}(A))/\mathcal{J}(A) \triangleleft A/\mathcal{J}(A),$$

and $A/\mathcal{J}(A)$ is Jacobson semisimple and artinian. In view of the Second Wedderburn–Artin Structure Theorem 4.5.9 $\nu(A)$ as well as A is a finite direct sum of matrix rings over division rings. Thus $\nu(A)$ has a unity element. □

PROPOSITION 4.10.19. *On artinian rings the von Neumann regular radical ν coincides with the largest subidempotent radical χ, that is, $\nu(A) = \chi(A)$ for all $A \in \mathcal{A}$.*

Proof: A *verbatim* repetition of the proof of Theorem 4.10.18 applies to $\chi(A)$, so it is a direct summand and an artinian ring. Since

$$\mathcal{J}(\chi(A)) \subseteq \mathcal{J}(A) \cap \chi(A) = 0,$$

$\chi(A)$ is semisimple artinian, whence von Neumann regular. Thus $\chi(A) \subseteq \nu(A)$. □

For coincidence of the von Neumann regular radical with the Jacobson radical on larger classes than $\mathcal{A}$ we refer to Szász [5] and Wiegandt [2]. Coincidence of radicals in a more general setting has been treated by Mlitz [3], [4] and Sands [14].

4.11. Concrete hypernilpotent radicals

In previous sections some concrete special radicals, such as the Baer, the Levitzki, the Köthe, the Jacobson and the Brown–McCoy radicals, have already been treated in detail. We shall add a few more (special) radicals, and also position them in the lattice of radical classes.

Properties of the *Behrens radical* $\mathcal{B}$ have been given in Examples 3.6.8 (ii), 3.8.14 (iv), 3.9.14, 3.17.8 (ii) and 3.18.6 (iv). A reformulation of those is the following

DEFINITION AND THEOREM 4.11.1. *The Behrens radical $\mathcal{B} = \mathcal{U}\eta$ is the upper radical of the class η of all subdirectly irreducible rings each having a nonzero idempotent in its heart. The Behrens radical $\mathcal{B}$ is a special and a supernilpotent dual radical, which is not left strong and not normal.*
□

Puczyłowski and Zand [2] proved that *the Behrens radical is left hereditary.*

Let us mention that Behrens [1] introduced his radical of a ring A as

$$\mathcal{B}(A) = \bigcap \left(I \triangleleft A \;\middle|\; \begin{array}{l} \text{there is an element } e \in A \setminus I \text{ such that} \\ e^2 - e \in I \text{ and } e \in K \text{ for all } I \subset K \triangleleft A \end{array} \right).$$

This means exactly that each A/I is a subdirectly irreducible ring with an idempotent $e + I$ is its heart. This radical was introduced also by Nagata [1] three years before Behrens.

The following characterization of the Behrens radical is due to Propes [1].

PROPOSITION 4.11.2. *If $\wp$ denotes the class of all rings possessing a nonzero idempotent, then $\mathcal{B} = \mathcal{U}\wp$.*

Proof: Since $\wp$ contains all subdirectly irreducible rings with nonzero idempotents in their hearts, it follows that $\mathcal{U}\wp \subseteq \mathcal{B}$.

Assume that there exists a ring $A \in \mathcal{B} \setminus \mathcal{U}\wp$. Then A has a homomorphic image B possessing a nonzero idempotent e. Zorn's Lemma assures the existence of an ideal I of B which is maximal with respect to the exclusion of e. Then the ring $C = B/I$ is subdirectly irreducible because the coset $e + I$ is in the intersection of all nonzero ideals of C. Further, $e + I$ is a nonzero idempotent in the heart of C. Since C is a homomorphic image of A as well, we conclude that $A \notin \mathcal{B}$, a contradiction. □

Further characterizations of the Behrens radical can be found in de la Rosa and Wiegandt [2].

We remind the reader that beside the Baer radical β, the Levitzki radical $\mathcal{L}$, Köthe's nil radical $\mathcal{N}$, the Jacobson radical $\mathcal{J}$, the Behrens radical $\mathcal{B}$ and the Brown–McCoy radical $\mathcal{G}$, we have introduced also the *Thierrin radical* $\mathcal{F}$ (Examples 3.2.18, 3.8.14 (ii), 3.9.14) as the upper radical of all division rings and the upper radical $\mathcal{T}$ of all matrix rings over division rings (Proposition 4.10.6). These two latter are special and supernilpotent dual radicals.

PROPOSITION 4.11.3.
$$0 \subset \mathcal{D} \subset \beta \subset \mathcal{L} \subset \mathcal{N} \subset \mathcal{J} \subset \mathcal{J}_\varphi \subset \mathcal{B} \subset \mathcal{G} \subset \mathcal{T} \subset \mathcal{F}.$$

Proof: By Theorem 4.10.2 we have $0 \neq \mathcal{D} \subset \beta$. Examples 4.2.1 and 4.2.4 show $\beta \subset \mathcal{L}$ and $\mathcal{L} \subset \mathcal{N}$, respectively. The relation $\mathcal{N} \subset \mathcal{J}$ follows from Proposition 4.4.4 and Example 3.2.13 (ii), and $\mathcal{J} \subset \mathcal{J}_\varphi$ from Corollary 4.4.17.

Since a subdirectly irreducible ring with a nonzero idempotent in its heart, must be primitive (cf. Proposition 4.4.2), it follows $\mathcal{J}_\varphi \subseteq \mathcal{B}$.

In Example 1.3.2 the left ideal $W_1 f(x)$ of the Weyl algebra W_1 is a simple ring having no nonzero idempotents. Hence $L \in \mathcal{B}$. The ring A is also a simple ring but A has a unity element. Thus $A \notin \mathcal{J}_\varphi$. Assume that $L \in \mathcal{J}_\varphi$. Since L is simple, also $L \in \mathcal{J}$ holds. By Example 3.17.4 the Jacobson radical is left strong, and therefore $0 \neq L \subseteq \mathcal{J}(A)$ follows which implies the contradiction $A = \mathcal{J}(A) \in \mathcal{J}_\varphi$. This proves $\mathcal{J}_\varphi \neq \mathcal{B}$.

The ring H of finite valued linear transformations in Example 1.2.11 is a simple ring without unity element which has nonzero idempotents. Thus $H \in \mathcal{G} \setminus \mathcal{B}$, proving $\mathcal{B} \subset \mathcal{G}$.

The Weyl algebra W_1 in Example 1.3.2 is a simple ring with unity element, and A has no zero divisors. Moreover, non-constant elements of A have no multiplicative inverses, so A is not a division ring. Thus A is not a matrix ring over a division ring, proving $A \in \mathcal{T} \setminus \mathcal{G}$ and $\mathcal{G} \subset \mathcal{T}$.

The relation $\mathcal{T} \subset \mathcal{F}$ is obvious by definition. □

Next, we shall deal with hypernilpotent radicals which do not fit into the chain given in Proposition 4.11.3.

Summarizing the statements of Examples 3.6.8 (i), 3.8.14 (iii), 3.9.8 and 3.18.22 we arrive at

DEFINITION AND THEOREM 4.11.4. *Andrunakievich's antisimple radical β_φ is the upper radical of all subdirectly irreducible prime rings; β_φ is a normal and special radical and the smallest supernilpotent dual radical.*
□

Thus, rings being semisimple with respect to the antisimple radical, are subdirect sums of subdirectly irreducible prime rings. Also antisimple radical rings can be easily characterized.

PROPOSITION 4.11.5. *The following conditions on a ring A are equivalent:*

i) $A \in \beta_\varphi$,

ii) *no ideal of A can be mapped homomorphically onto a nonzero simple prime ring,*

iii) *every subdirectly irreducible homomorphic image of A has nilpotent heart.*

Proof: Straightforward, although for the implication i) $\implies$ ii) one needs the hereditariness of β_φ. □

More on the antisimple radical can be found in Andrunakievich [2], Szász [6].

To localize the position of the antisimple radical β_φ, first we reformulate Theorem 4.6.17.

PROPOSITION 4.11.6. *A simple prime ring is Levitzki semisimple. Otherwise expressed: a locally nilpotent simple idempotent ring does not exist.* □

PROPOSITION 4.11.7. $\mathcal{L} \subseteq \beta_\varphi$.

Proof: Let $A \neq 0$ be a Levitzki radical ring and B a subdirectly irreducible homomorphic image of A. Now both B and its heart $H(B)$ are in $\mathcal{L}$, and so by Proposition 4.11.6, $H(B)$ is nilpotent. Hence A cannot be mapped homomorphically onto a subdirectly irreducible prime ring, that is, $A \in \beta_\varphi$. This proves $\mathcal{L} \subseteq \beta_\varphi$. □

For characterizing the *Jenkins radical β_s which is the upper radical of all simple prime rings* (see Example 4.9.11 (iii)), a ring A is said to be an *m-ring*, if $A^2 \subseteq M$ for each maximal ideal M of A.

THEOREM 4.11.8. *The Jenkins radical class β_s consists of all m-rings.*

Proof: An m-ring A cannot be mapped homomorphically onto a simple prime ring, and so $A \in \beta_s$. If A is not an m-ring, then it has a maximal ideal M such that $A^2 \not\subseteq M$. Hence A/M is a simple prime ring, and therefore $A \notin \beta_s$. □

For further results on the Jenkins radical we refer to Loi [1].

Let $\mathcal{J}_\varphi$ and $\mathcal{J}_s$ denote *the upper radical of subdirectly irreducible primitive rings and simple primitive rings, respectively* (cf. Corollary 4.4.17).

PROPOSITION 4.11.9. $\beta_\varphi \subset \beta_s \subset \mathcal{J}_s \subset \mathcal{G}$, $\beta_\varphi \subset \mathcal{J}_\varphi \subset \mathcal{J}_s$, $\beta_\varphi \not\subseteq \mathcal{J}$ and so $\mathcal{L} \subset \beta_\varphi \neq \mathcal{N}$. Moreover, $\mathcal{J} \not\subseteq \beta_s \not\subseteq \mathcal{B} \not\subseteq \mathcal{J}_s$.

Proof: By definition the relations

$$\beta_\varphi \subseteq \beta_s \subseteq \mathcal{J}_s \subseteq \mathcal{G} \quad \text{and} \quad \beta_\varphi \subseteq \mathcal{J}_\varphi \subseteq \mathcal{J}_s$$

are obvious. Since a simple idempotent Jacobson radical ring (see Theorem 4.6.16) is in $\mathcal{J}_\varphi \subseteq \mathcal{J}_s$, but not in β_s, it follows that $\beta_s \neq \mathcal{J}_s$ and $\beta_\varphi \neq \mathcal{J}_\varphi$.

A simple prime Jacobson radical ring (cf. Theorem 4.6.16) is obviously in $\mathcal{J} \setminus \beta_s$ so $\mathcal{J} \not\subseteq \beta_s$.

The ring T of Example 4.4.16 is Jacobson semisimple and cannot be mapped homomorphically onto a subdirectly irreducible prime ring. This shows that $T \in \mathcal{SJ} \cap \beta_\varphi$, whence $\beta_\varphi \not\subseteq \mathcal{J}$.

We show that $\beta_\varphi \neq \beta_s$ and $\mathcal{J}_\varphi \neq \mathcal{J}_s \neq \mathcal{G}$. In fact, neither β_s nor $\mathcal{J}_s$ is supernilpotent because they are not hereditary: the ring $S = \langle t, H \rangle$ in Example 1.2.11 (v) is in β_s and so also in $\mathcal{J}_s$. Since S is subdirectly irreducible and its heart H is Jacobson semisimple, we have $H \not\in \mathcal{J}_s$ and $H \not\in \beta_s$. Hence $\beta_\varphi \neq \beta_s$, $\mathcal{J}_\varphi \neq \mathcal{J}_s$ and by $\mathcal{J}_s \subseteq \mathcal{G}$ also $\mathcal{J}_s \subset \mathcal{G}$, as claimed.

The ring $S = \langle t, H \rangle$ in Example 1.2.11 (v) is obviously in $\beta_s \setminus \mathcal{B}$ proving $\beta_s \not\subseteq \mathcal{B}$.

The simple ring $W_1 f(x)$ of Example 1.3.2, as seen in the proof of Proposition 4.11.3, is in $\mathcal{B} \setminus \mathcal{J}$ and hence also in $\mathcal{B} \setminus \mathcal{J}_s$. Thus $\mathcal{B} \not\subseteq \mathcal{J}_s$. $\square$

The question as whether $\mathcal{N} \subset \beta_\varphi$ is related to the existence of a simple idempotent nil ring.

PROPOSITION 4.11.10. *The existence of a simple idempotent nil ring is equivalent to the relation* $\mathcal{N} \not\subseteq \beta_\varphi$.

Proof: Suppose that a simple idempotent nil ring A does exist. Then $A \not\in \beta_\varphi$ and so $\mathcal{N} \not\subseteq \beta_\varphi$ follows.

Assume that $\mathcal{N} \not\subseteq \beta_\varphi$. Then there eixsts a nil ring A which is not in β_φ. Hence A has a homomorphic image B which is a subdirectly irreducible prime ring. Consequently, so is its heart $H(B)$, which is therefore a simple idempotent nil ring. $\square$

As mentioned at the end of Section 4.6, Agata Smoktunowicz [2] solved affirmatively the long standing and hard problem of Levitzki: *a simple idempotent ring exists.* Hence by Propositions 4.11.9 and 4.11.10 the nil radical $\mathcal{N}$ and the antisimple radical β_φ are not comparable.

Some other radicals lying between $\mathcal{B}$ and $\mathcal{T}$ but not comparable with $\mathcal{G}$, have been studied by Beidar and Wiegandt [4].

Another type of special radical is based on various notions of primeness.

Domains (that is, rings without zero divisors) are sometimes called *completely prime rings* (cf. Lemma 1.2.9), and an ideal I of a ring A is said to be a *completely prime ideal* if A/I is a domain. Summarizing Examples 3.8.16, 3.16.5 (ii), 3.16.11 (ii), 3.17.12 (ii) and Theorem 3.20.5 we have

DEFINITION AND THEOREM 4.11.11. *The generalized nil radical $\mathcal{N}_g$ of Andrunakievich and Thierrin is the upper radical $\mathcal{U}\varrho$ of the special class ϱ of all domains. The generalized nil radical $\mathcal{N}_g(A)$ of a ring A is the intersection of all completely prime ideals of A. $\mathcal{N}_g(A) = 0$ if and only if A is a subdirect sum of domains if and only if A contains no nonzero nilpotent* elements. $\mathcal{N}_g$ is a strict radical, whence left (and right) stable. $\mathcal{N}_g$ is not a left (or right) strong radical. □

PROPOSITION 4.11.12. *The generalized nil radical $\mathcal{N}_g$ is the upper radical $\mathcal{U}\varrho_0$ of the class ϱ_0 of all rings containing no nonzero nilpotent elements.*

Proof: Clearly $\varrho \subset \varrho_0$ and by Theorem 3.20.5 also $\varrho_0 = \mathcal{SN}_g$. Hence

$$\mathcal{N}_g = \mathcal{USN}_g \subseteq \mathcal{U}\varrho_0 \subseteq \mathcal{U}\varrho = \mathcal{N}_g$$

follows proving $\mathcal{N}_g = \mathcal{U}\varrho_0$. □

PROPOSITION 4.11.13. *The following relations hold:* $\mathcal{N} \subset \mathcal{N}_g \subset \mathcal{F}$, $\mathcal{J} \not\subseteq \mathcal{N}_g \not\subseteq \mathcal{T}$, $\beta_\varphi \not\subseteq \mathcal{N}_g$ *and* $\mathcal{T} \not\subseteq \mathcal{N}_g$.

Proof: The matrix ring $M_2(D)$ over a division ring D is in $\mathcal{N}_g$ but not in $\mathcal{N}$ yielding $\mathcal{N} \neq \mathcal{N}_g$.

The containment $\mathcal{N}_g \subseteq \mathcal{F}$ is obvious. The Weyl algebra W_1 in Example 1.3.2 is a simple domain with unity element which is not a division ring. Thus $W_1 \subset \mathcal{SN}_g \cap \mathcal{J}$ implying $\mathcal{N}_g \neq \mathcal{J}$. We have also $W_1 \in \mathcal{SN}_g \cap \mathcal{T}$, whence $\mathcal{T} \not\subseteq \mathcal{N}_g$.

The ring J of Example 3.2.13 (ii) is a Jacobson radical ring which is a domain. Thus $\mathcal{J} \not\subseteq \mathcal{N}_g$. The matrix ring $M_2(D)$ over a division ring D shows that $\mathcal{N}_g \not\subseteq \mathcal{T}$.

The ring T of Example 4.4.16 has no zero divisors, and it is an antisimple radical ring. Thus also $\beta_\varphi \not\subseteq \mathcal{N}_g$ has been established.

From $\mathcal{N}_g \not\subseteq \mathcal{T}$ and Theorem 4.10.8 we know that the generalized nil radical $\mathcal{N}_g$ does not coincide with the Jacobson radical on artinian rings, in fact $\mathcal{N}_g$ is too big. This is seen in the following

PROPOSITION 4.11.14. *Let A be an artinian ring. $\mathcal{N}_g(A) = 0$ if and only if A is a finite direct sum of division rings.*

Proof: Assume that $\mathcal{N}_g(A) = 0$. By $\mathcal{N} \subseteq \mathcal{N}_g$ it follows $\mathcal{N}(A) = 0$, and so the Second Wedderburn–Artin Structure Theorem 4.5.9 tells us that A is a finite direct sum of matrix rings over division rings. Moreover, by the hereditariness of the class $\mathcal{SN}_g$, these matrix rings must not contain zero divisors, so they are division rings.

The converse is trivial by definition. □

The Groenewald–Heyman (left) strongly prime radical s_ℓ was introduced in Example 3.16.14.

THEOREM AND DEFINITION 4.11.15 (Groenewald and Heyman [1], Desale and Varadarajan [1]). *The class of all (left) strongly prime rings is a special class. The upper radical s_ℓ of all (left) strongly prime rings is called the Groenewald–Heyman strongly prime radical. s_ℓ does not coincide with the upper radical s_r of right strongly prime rings.*

Proof: From Proposition 3.16.13 we know that the (left) strongly prime rings form a hereditary class of prime rings. To prove that it is a special class, we have to verify that it is closed under essential extensions. Let $I \triangleleft \cdot A$ and I a strongly prime ring. Now for an arbitrary nonzero ideal K of A also $K \cap I \neq 0$. By Lemma 3.16.12 there exists a finite subset $G \subseteq K \cap I \subseteq K$ such that $(0 : G)_I = 0$. Let $y \in A$ be any nonzero element. Since $I \triangleleft \cdot A$ and I is (strongly) prime, A is a prime ring (cf. Proposition 3.8.3). Thus, for the ideal (y) we have $I(y) \neq 0$, and so there exists an element $c \in I$ with $cy \neq 0$. Consequently $cy \notin (0 : G)_I$, whence $y \notin (0 : G)_A$. Since the choice of y was arbitrary, necessarily $(0 : G)_A = 0$, and so by Lemma 3.16.12, A is strongly prime. Thus (left) strongly prime rings form a special class.

The last statement has been proved in Example 3.16.14. □

PROPOSITION 4.11.16. $\mathcal{L} \subset s_\ell \subset \mathcal{G}$ and $s_\ell \not\subseteq \mathcal{B}$, $s_\ell \not\subseteq \mathcal{J}_s$.

Proof: Let A be a Levitzki radical ring, $0 \neq a \in A$ and $F = \{f_1, \ldots, f_n\}$ be a finite subset of A. Then the subring S generated by the set $\{f_1 a, \ldots, f_n a\}$ is nilpotent of degree, say k. Now, for any $0 \neq x \in S^{k-1}$ one has $xFa = 0$, and therefore A cannot be strongly prime. For A we could have taken any nonzero homomorphic image of A, so $A \in s_\ell$ follows implying $\mathcal{L} \subseteq s_\ell$.

Taking into account that $s_\ell \neq s_r$ and the symmetry of $\mathcal{L}$, we conclude $\mathcal{L} \neq s_\ell$.

As noted after Lemma 3.16.12, every simple ring with unity element is strongly prime. Hence $s_\ell \subseteq \mathcal{G}$ holds, so $s_\ell \neq s_r$ and the symmetry of $\mathcal{G}$ yield $s_\ell \neq \mathcal{G}$.

The ring H of Example 1.2.11 is a simple ring with nonzero idempotents, so $H \in \mathcal{SB}$. Further, since H consists of finite valued linear transformations of an infinite dimensional vector space, it is straightforward to check that any finite subset of H has a nonzero left annihilator. By the simplicity of H this means that $H \in s_\ell$. Thus $H \in s_\ell \setminus \mathcal{B}$, whence $s_\ell \not\subseteq \mathcal{B}$.

By the same token also $H \notin \mathcal{J}_s$, and so $s_\ell \not\subseteq \mathcal{J}_s$. □

PROPOSITION 4.11.17. $\mathcal{N} \not\subseteq s_\ell$.

Proof: We shall construct a strongly prime ring C which has a nonzero nil radical. Such a ring was given by L. Small (published in Goodearl, Handelman and Lawrence [1, Example 2.5]) and by Puczyłowski [9, Example 1.12]. Let us consider a finitely generated nil ring A which is not nilpotent. We remind the reader that such a ring does exists as seen in Theorem 4.2.4. Let B be the Dorroh extension of A which is again a finitely generated ring. Since for any natural number n, the ring A^n is finitely generated, by Zorn's Lemma we can find an ideal I of B such that I is maximal with respect to not containing any A^n.

First, we claim that I is a prime ideal in B. Let $K, L \triangleleft B$ such that $KL \subseteq I$ and $K, L \not\subseteq I$. Then by the choice of I there exist m and n such that $A^m \subseteq K + I$ and $A^n \subseteq L + I$. Hence

$$A^{m+n} \subseteq (K+I)(L+I) \subseteq I$$

follows, a contradiction. Thus the factor ring $C = B/I$ is prime and so is every ideal of C.

Next, we show that C is a strongly prime ring. In view of Lemma 3.16.12 we have to prove that every nonzero ideal J/I of C contains a finite subset G such that $(0 : G)_C = 0$. Since $I \subset J$, by the choice of I there exists an n such that $A^n \subseteq J$. Let $\{a_1, \ldots, a_k\}$ be a set of generators of A^n. Then for $G = \{a_1 + I, \ldots, a_k + I\}$ we have $(0 : G)_C = 0$, for C is a prime ring. Thus C is indeed a strongly prime ring. Moreover, $A \in \mathcal{N}$ implies that $\mathcal{N}(B) \neq 0$, and so also $\mathcal{N}(C) \neq 0$. This yields that $\mathcal{N}(C) \in \mathcal{S}s_\ell \cap \mathcal{N}$, that is, $\mathcal{N}(C) \in \mathcal{N} \setminus s_\ell$. □

A (left) strongly prime ring A is called *uniformly (left) strongly prime*, if the same insulator F can be chosen for every nonzero element of A. In this case F is said to be a *uniform insulator*. Clearly, every completely prime ring (that is, domain) is uniformly strongly prime, and any nonzero element is a uniform insulator. Furthermore, a uniformly right strongly prime ring can be defined analogously.

LEMMA 4.11.18 (Olson [1]). *For a ring A the following conditions are equivalent:*

i) *A is a uniformly left strongly prime ring,*

ii) *there exists a finite subset F in A such that $xFy \neq 0$ for any two nonzero elements $x, y \in A$.*

Proof: i) $\Longrightarrow$ ii) The ring A has a uniform insulator F such that for every nonzero element $y \in A$, $(0 : Fy)_A = 0$. Then, if x and y are nonzero elements in A, x cannot be in $(0 : Fy)_A$. Hence there exists an $f \in F$ such that $xfy \neq 0$.

ii) $\Longrightarrow$ i) It is easy to see that if ii) is satisfied, then for any nonzero element $y \in A$ no nonzero element annihilates Fy from the left. Hence A is uniformly (left) strongly prime.

Condition ii) is symmetric, hence we have

COROLLARY 4.11.19. *A ring is uniformly left strongly prime if and only if it is uniformly right strongly prime.*

THEOREM AND DEFINITION 4.11.20 (Olson [1]). *The class of all uniformly strongly prime rings is a special class. The corresponding special radical u is Olson's uniformly strongly prime radical.*

Proof: Obviously uniformly strongly prime rings are prime rings.

Let $0 \neq I \triangleleft A$ and A a uniformly strongly prime ring. Further, let us consider arbitrary nonzero elements $x, y \in I$, and let $c \in I$ be a fixed nonzero element. Since A is uniformly strongly prime, by Lemma 4.11.18 there exists a finite subset $F = \{f_1, \ldots, f_k\} \subseteq A$ such that $xf_ic \neq 0$ with a suitable $f_i \in F$, and there exists also an $f_j \in F$ such that

$$x(f_icf_j)y = (xf_ic)f_jy \neq 0.$$

Thus the subset
$$G = \{f_i c f_j \mid f_i, f_j \in F\} \subseteq I$$
satisfies condition ii) of Lemma 4.11.18, whence I is uniformly strongly prime. This proves that uniformly strongly prime rings form a hereditary class.

Assume that $I \triangleleft \cdot A$ and I is a uniformly strongly prime ring with a uniform insulator $F = \{f_1, \ldots, f_k\} \subseteq I$. If $x \in (0:F)_A$, then $xFy = 0$ for all $y \in I$, and hence $(ix)Fy = i(xFy) = 0$ for all $i \in I$. As F is a uniform insulator for I, $ix \in I$ and y need not be zero, we have $ix = 0$ for all $i \in I$, that is, $x \in (I:0)_A$. Since I is prime, so is A and therefore $(I:0)_A = 0$ which implies $x = 0$. Thus $xF \neq 0$ and similarly $Fz \neq 0$ for all nonzero elements $x, z \in A$. Hence for given nonzero elements $x, z \in A$ there exist elements $f_i, f_j \in F$ such that
$$0 \notin \{x f_i, f_j z\} \subseteq I.$$
Since F is a uniform insulator for I, by Lemma 4.11.18 (and its proof) there exists an element $f \in F$ such that $x f_i f f_j z \neq 0$. Hence the set
$$G = \{f_i f f_j \mid f_i, f_j \in F\}$$
satisfies condition ii) of Lemma 4.11.18, and therefore A is uniformly strongly prime.

Thus the uniformly strongly prime rings form a special class. $\square$

As shown by Beidar, Puczyłowski and Wiegandt [1], *the uniformly strongly prime radical u satisfies* condition (ii) of Proposition 4.9.21: $A \in u$ implies $A[x] \in u$.

Concerning the position of the uniformly strongly prime radical u we can state the following

PROPOSITION 4.11.21. $s_\ell \subset u \subset \mathcal{N}_g$ *and* $u \subset \mathcal{T}$.

Proof: The first containment is a trivial consequence of the definition, symmetry and Corollary 4.11.19.

For proving $u \subseteq \mathcal{T}$ it suffices to show that any matrix ring $M_n(D)$ over a division ring D is uniformly strongly prime. Clearly $F = \{1\} \subseteq D$ is a uniform insulator in D.

Let (a_{ij}) and (b_{kl}) arbitrary nonzero matrices in $M_n(D)$. Then $a_{ij} \neq 0$ and $b_{kl} \neq 0$ for some i, j, k, l. For the matrix unit e_{jk}, where in the j, k position the entry is 1 and everywhere else 0, we have $(a_{ij})e_{jk}(b_{kl}) \neq 0$ because in the i, l position of the product matrix the entry is $a_{ij} b_{kl} \neq 0$. Thus the set $G = \{e_{jk} \mid j, k = 1, \ldots, n\}$ is a uniform insulator in $M_n(D)$.

Since by Proposition 4.11.13 we have $u \subseteq \mathcal{N}_g \nsubseteq \mathcal{T}$ and $\mathcal{T} \nsubseteq \mathcal{N}_g$, we conclude $u \subset \mathcal{T}$ and $u \subset \mathcal{N}_g$. $\square$

PROPOSITION 4.11.22. $\mathcal{N} \subset u$ *and* $u \nsubseteq \mathcal{G}$.

Proof: Suppose that $\mathcal{N} \not\subseteq u$. Then there exists a nil ring A which is uniformly strongly prime with a uniform insulator

$$F = \{f_1, \ldots, f_k\}$$

For every pair (x, y) of nonzero elements of A we define $d(x, y)$ to be the number of nonzero members of the set xFy. Since A is uniformly strongly prime, $1 \leq d(x, y) \leq k$. Let

$$r = \min\{d(x, y) \mid x, y \in A \text{ and } yx \neq 0\}$$

The number r is well-defined because A is uniformly strongly prime and so $A^2 \neq 0$. Now $d(x_0, y_0) = r$ with some $x_0, y_0 \in A$, and without loss of generality we may assume that

$$x_0 f_1 y_0 \neq 0, \quad \ldots, \quad x_0 f_r y_0 \neq 0.$$

Since A is nil, there exists an $n \geq 2$ such that $(x_0 f_r y_0)^{n-1} \neq 0$ but $(x_0 f_r y_0)^n = 0$. If $x^* = (x_0 f_r y_0)^{n-1} x_0 \neq 0$, then $d(x^*, y_0) < r$, contradicting the minimality of r. If $(x_0 f_r y_0)^{n-1} x_0 = 0$ and $n \geq 3$, then $x' = (x_0 f_r y_0)^{n-2} x_0 \neq 0$ and also $y' = y_0 x_0 \neq 0$, but $d(x', y') < r$, a contradiction. If $n = 2$ and $(x_0 f_r y_0) x_0 = 0$ then $d(x_0, y_0 x_0) < r$, again a contradiction. Hence $\mathcal{N} \subseteq u$.

To prove $u \not\subseteq \mathcal{G}$, let us consider the matrix ring $M_n(D)$ over a *field* D. Since $M_n(D)$, being a simple ring with unity element, is strongly prime, the matrix unit e_{11} has an insulator

$$F = \{(f_{ij}^{(1)}), \ldots, (f_{ij}^{(k)})\},$$

and so the matrix equations

$$(x_{ij})(f_{ij}^{(r)}) e_{11} = 0, \qquad r = 1, \ldots, k$$

are fulfilled only with $(x_{ij}) = 0$. This means

$$x_{11} f_{11}^{(r)} + \cdots + x_{1n} f_{n1}^{(r)} = 0$$
$$\vdots$$
$$x_{n1} f_{11}^{(r)} + \cdots + x_{nn} f_{n1}^{(r)} = 0$$

for each $r = 1, \ldots, k$. Thus we have got nk equations with n^2 unknowns, and a necessary condition for having only trivial solutions is $nk \geq n^2$, that is $k \geq n$.

Next, let us consider the simple ring R with unity element as given in Example 4.5.17, into which the $2^n \times 2^n$ matrix rings $R_n = M_n(D)$ can be embedded canonically by $g_n : R_n \to R$, $n = 1, 2, \ldots$. Let us suppose that each R_n is a matrix ring over the same *field* D. The preceding considerations apply to the case $R_n = M_{2^n}(D)$, and so any insulator of the matrix unit $e_{11} \in R_n$ consists of at least 2^n elements, and this number does not decrease by the embedding $g_n(e_{11})$ into R. Hence no finite subset can be a uniform insulator, which proves that R is not uniformly strongly prime. Since R is a simple ring, necessarily $R \in u$. Furthermore, R as a simple ring with unity element, is not in $\mathcal{G}$, proving $u \not\subseteq \mathcal{G}$.

Since $\mathcal{N} \subset \mathcal{G}$ and $u \not\subseteq \mathcal{G}$, also $\mathcal{N} \neq u$ has been established. $\square$

There are many more special radicals based on various notions on primeness. Using van der Walt's superprimeness [1] Veldsman [4] introduced the (left) superprime radical. The insulators may be also of infinite cardinality or may have a finite or infinite upper bound. Radicals via constraints on insulators were investigated by Buys [1], Raftery [1], Raftery and van den Berg [1], [2] and van den Berg [1].

For a detailed positioning of all these radicals the reader is referred to van den Berg [2] and Wiegandt [9].

By the relationships proved in this section the position of the discussed radicals is given in the following diagram. Note that all these radicals but $\mathcal{D}$ are hypernilpotent, and with the exception of $\mathcal{D}$, β_s and $\mathcal{J}_s$ they are all special radicals.

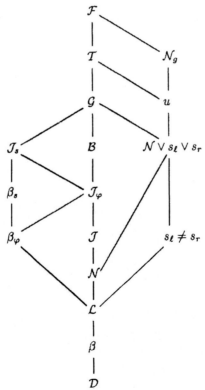

In section 4.7 at the definition of weak primitivity we have seen that primitive rings are weakly primitive and weakly primitive rings are prime. Further, the even over odd Jacobson radical ring J is weakly primitive. Hence we have $\beta \subseteq \zeta \subset \mathcal{J}$. The exact position of the week radical in the lattice of radicals in not known.

The uniform radical γ_Σ of Example 3.14.21 is a special radical and $\beta \subseteq \gamma_\Sigma \subset \mathcal{J}$.

4.12. Concrete hypoidempotent radicals

Restating the results of Examples 3.2.2 (ii) and 3.10.1 we have the following

THEOREM 4.12.1. *The class $\mathcal{I}$ of all idempotent rings is a non-hereditary radical class, in fact, $\mathcal{I}$ is the largest hypoidempotent radical class.*

The largest hereditary subclass χ of $\mathcal{I}$ is a radical class, the largest subidempotent radical class. χ is Blair's f-regular radical class, and χ is the dual subidempotent radical supplementing the anti-simple radical β_φ. χ also supplements the Baer radical β.

Recall that an element a of a ring A is called a λ-element, if $a \in AaA$. λ-rings and de la Rosa's λ-radical (that is, the class of all λ-rings) have been introduced in Example 3.15.2. Here we give also its position.

THEOREM 4.12.2. *The class λ of all λ-rings is a non-hereditary hypoidempotent radical class such that $\chi \subset \lambda \subset \mathcal{I}$.*

Proof: If $A \in \lambda$, then $a \in AaA$ holds for all elements $a \in A$, whence

$$A \subseteq A^3 \subseteq A^2 \subseteq A.$$

Thus λ consists of idempotent rings, that is, $\lambda \subseteq \mathcal{I}$.

The assertion that the class λ differs from $\mathcal{I}$ will be proved if we give a ring which is idempotent but not a λ-ring. To this end, let us consider the 2×2 matrices

$$a = \begin{pmatrix} 0 & 1 \\ 0 & 0 \end{pmatrix}, \qquad b = \begin{pmatrix} 1 & 0 \\ 0 & 0 \end{pmatrix}, \qquad c = \begin{pmatrix} 1 & 1 \\ 0 & 0 \end{pmatrix}.$$

over the two-element field, and the ring $A = \{0, a, b, c\}$. The ring A is clearly idempotent, because b is a left unity element in A. However, A is not a λ-ring, for

$$0 \neq a \notin AaA = 0,$$

as one readily checks.

If $R \in \chi$ and $a \in R$, then

$$a \in (a) = (a)^3 \subseteq R(a)R = RaR,$$

and so $\chi \subsetneq \lambda$.

Finally, the Dorroh extension of a nilpotent ring (say, of degree ≤ 3) is a λ-ring, but the nilpotent ring is not in λ. Therefore, λ is not hereditary, whence also $\chi \neq \lambda$ has been established. □

Let us recall that a ring A is said to be quasi-semiprime, if $0 \neq I \triangleleft A$ implies $AIA \neq 0$ (cf. the end of Section 3.6). An ideal Q of a ring A is a called quasi-semiprime ideal, if A/Q is a quasi-semiprime ring, that is, if $AIA \subseteq Q$ for an ideal I of A, then $I \subseteq Q$.

PROPOSITION 4.12.3. *A is a λ-ring if and only if every ideal of A is quasi-semiprime, that is, every homomorphic image of A is a quasi-semiprime ring.*

Proof: Let $0 \neq I \triangleleft A \in \lambda$ and $0 \neq i \in I$. Then $0 \neq i \in AiA \subseteq AIA$. Thus $AIA \neq 0$ and A is quasi-semiprime. So is every homomorphic image of A, as λ is homomorphically closed.

Conversely, assume that every ideal of a ring A is quasi-semiprime. For any element $a \in A$ and for the ideal $I = (a)$ we have

$$AIA = A((a) + AaA)A = A(a)A + A^2aA^2 = AaA.$$

Since I is quasi-semiprime, we conclude that

$$a \in I \subseteq AaA.$$

Thus $A \in \lambda$ follows. □

Remark. The intersection $q(A)$ of quasi-semiprime ideals is clearly contained in the Baer radical $\beta(A)$ and also $q(A) \neq \beta(A)$ may happen. For instance, let A be the Dorroh extension of a nilpotent ring B. Then $q(A) = 0$, but $q(B) = B \subseteq \beta(A)$. This example shows also that $B = q(B) \not\subseteq q(A)$, and so the assignment $A \to q(A)$ does not define a Kurosh–Amitsur radical: $q(A) = A$ for every zero-ring A and β is the lower radical of all zero-rings (cf. Example 2.2.2).

The λ-radical has the following interesting characterization, as proved in [1] and [2] of de la Rosa: $A \in \lambda$ if and only if every ideal of the matrix ring $M_n(A)$, $n > 1$, is of the form $M_n(I)$ where $I \triangleleft A$. This is by no mean contradicting Proposition 4.9.1 which claims only that for any radical γ the ideal $\gamma(M_n(A))$ is of that form.

THEOREM 4.12.4. *The class ν of all von Neumann regular rings is a subidempotent radical class which is properly contained in the class χ of hereditarily idempotent rings.*

Proof: The statements follow from Examples 2.1.9, 3.2.1, and for the last assertion we mention that a simple idempotent Jacobson radical ring cannot contain nonzero idempotents, whence it cannot be von Neumann regular. □

Next, we shall introduce a subidempotent radical, which is properly smaller than the von Neumann regular radical. A ring A is called *strongly regular*, if for every element $a \in A$ there exists an element $x \in A$ such that $a = a^2x$, or equivalently, $a \in a^2A$ (cf. Forsythe and McCoy [1], Arens and Kaplansky [1] and Kando [1]).

PROPOSITION 4.12.5. *Let A be a strongly regular ring. Then*
i)   *A does not contain nonzero nilpotent elements,*
ii)   *A is a von Neumann regular ring,*
iii)   *the idempotents of A are in the center of A.*
iv)   *If A is also simple, then A is a division ring.*

Moreover, a von Neumann regular ring need not be strongly regular, and a strongly regular ring is not necessarily a division ring.

Proof: i) For any element $a \in A$ we have
$$a = a^2x = a^3x^2 = \ldots = a^{n+1}x^n = \ldots.$$
Hence a nonzero element $a \in A$ cannot be nilpotent.

ii) Let $a = a^2x$. Then
$$(a - axa)^2 = a^2 - a^2xa - axa^2 + axa^2xa = a^2 - a^2 - axa^2 + axa^2 = 0,$$
and so by i) we conclude that $a = axa$.

iii) Let e be an idempotent in A. Then for any $x \in A$ we have
$$(exe - ex)^2 = exexe - exex - exexe + exex = 0,$$
which, in virtue of i), implies $exe = ex$. Similarly we can get $exe = xe$, whence $ex = xe$ follows.

iv) Let $a \in A$ be a nonzero element. The principal right ideal $[a)$ contains aA. Further, since $a = a^2x = axa$ by ii), and since xa is an idempotent, it follows by iii) that
$$aA = axaA \subseteq aAxa \subseteq Aa.$$
Analogous reasoning yields $Aa \subseteq aA$. Thus aA is a two-sided ideal, and therefore the simplicity of A implies $A = aA \subseteq [a)$. So the ring A has only trivial right ideals, whence A is a division ring.

The matrix ring $M_n(D)$, $n \geq 2$, over a division ring is von Neumann regular, but not strongly regular, because it contains nonzero nilpotent elements. The direct sum $D \oplus D$ of copies of the division ring D is clearly strongly regular, but not a division ring. □

THEOREM 4.12.6 *The class κ of all strongly regular rings is a subidempotent radical class which is properly contained in the class ν of all von Neumann regular rings.*

Proof: The class κ is clearly homomorphically closed and has the inductive property. Furthermore, let A be a ring having an ideal I such that both I and A/I are strongly regular. Then any element $a + I \in A/I$ is in $(a^2 + I)(A/I) = (a^2A + I)/I$. Thus $a \in a^2A + I$. Hence there exists an element $x \in A$ such that $a - a^2x \in I$. Since I is strongly regular, we have that

$$a - a^2x \in (a - a^2x)^2 I \subseteq (a^2 - a^3x - a^2xa + a^2xa^2x)I \subseteq a^2I \subseteq a^2A.$$

Hence also A is strongly regular.

Thus an application of Theorem 2.1.4 yields that κ is a radical class which is obviously hypoidempotent.

If A is a strongly regular ring and $I \triangleleft A$, then for each element $a \in I$ there exists an element $x \in A$ such that $a = a^2x = a^3x^2 = a^2(ax^2)$. Since $ax^2 \in I$, the element $a \in I$ and I itself must be strongly regular. Thus κ is hereditary, and so a subidempotent radical.

In view of Proposition 4.12.5 κ is properly contained in ν. □

The position of hypoidempotent radicals:

$$\kappa \subset \nu \subset \chi \subset \lambda \subset \mathcal{I}.$$

In addition, we know from Section 3.20 that the radical semisimple classes are all subidempotent radicals each contained in the class κ (cf. Proposition 4.12.5).

More concrete subidempotent radicals have been studied by Gardner [15] and Sands [13].

Chapter V
Special Features of the General Radical Theory

In this final chapter we shall see how radical theory degenerates for nonassociative rings, how the results of associative rings can be extended to bigger universal classes such as alternative rings, Jordan algebras, etc. Rings with involution are natural and important in algebra, so we shall look at their radical theory. Surprisingly, in the variety of associative rings with involution, the ADS-Theorem is not valid in general. For near-rings, radical theory degenerates in a very nice way, if we compare it with the associative radical theory.

5.1. Degeneracy and pathology of nonassociative radical theory

Moving from the variety or universal classes of associative rings to those of not necessarily associative rings, radical theory features pathological properties. We shall see that in the variety of all not necessarily associative rings torsion classes are scarce and degenerate inasmuch as torsion properties depend only on the additive group structure of rings. As expected, there is no possibility of introducing special radicals and getting structure theorems for semisimple rings. A bizarre situation is that smallest semisimple class containing a given class as well as smallest torsion class containing a given class need not exist.

In this section we shall work in *the variety $\mathbb{V}$ of all not necessarily associative rings*.

Leavitt and Armendariz [1] and Ryabukhin [4] gave examples of non-hereditary semisimple classes in $\mathbb{V}$. Hence in $\mathbb{V}$ a semisimple class need not be a torsionfree class. This is what was expected. In the sequel we are going to determine the torsion classes in $\mathbb{V}$. This results is the surprising degenerating Theorem 5.1.6. Let us remind the reader that torsion theories have been characterized already in Theorem 2.3.12. For the description of torsion classes in $\mathbb{V}$, we need two nonassociative ring construction $\Gamma(A)$ and

$\Lambda(A)$ for every ring $A \in \mathbb{V}$ in which addition is defined by

$$\Gamma(A)^+ = A^+ \oplus A^+ \oplus A^+ = \Lambda(A)^+$$

and the multiplication by

$$(x_1, y_1, z_1)(x_2, y_2, z_2) = (x_1 x_2 + z_1 y_2, z_2 x_1, 0) \in \Gamma(A)$$

and

$$(x_1, y_1, z_1)(x_2, y_2, z_2) = (x_1 x_2 + z_1 y_2, z_2 x_1, z_1 z_2) \in \Lambda(A),$$

respectively. It is a routine to check that both $\Gamma(A)$ and $\Lambda(A)$ are rings, though nonassociative even if A is associative. Moreover, denoting by A^0 the ring with zero-multiplication built on the additive group A^+, we have

$$A \oplus A^0 \cong (A, A, 0) \triangleleft \begin{cases} \Gamma(A) \\ \Lambda(A) \end{cases}$$

and we shall identify $A \oplus A^0$ with that ideal. Hence A represents the first component and A^0 the second one in both constructions. Furthermore, also

$$\Gamma(A)/(A, A, 0) \cong A^0$$

and

$$\Lambda(A)/(A, A, 0) \cong A$$

hold.

LEMMA 5.1.1. *The ideal of $\Gamma(A)$ generated by $(A, 0, 0)$ is $(A, A^2, 0)$.*

Proof: We have

$$(x, y, z) \cdot (a, \Sigma b_i c_i, 0) = (xa + \Sigma z(b_i c_i), 0, 0) \in (A, 0, 0)$$

and

$$(a, \Sigma b_i c_i, 0)(x, y, z) = (ax, za, 0) \in (A, A^2, 0).$$

Hence $(A, 0, 0) \subseteq (A, A^2, 0) \triangleleft \Gamma(A)$. Moreover, by

$$(u, 0, 0) + (w, 0, 0)(0, 0, v) = (u, vw, 0) \in (A, A^2, 0)$$

the subring $(A, 0, 0)$ generates the ideal $(A, A^2, 0)$ in $\Gamma(A)$. □

Similarly, one can prove the following.

LEMMA 5.1.2. *The ideal of $\Lambda(A)$ generated by $(0, A, 0)$ is $(A^2, A, 0)$.* □

LEMMA 5.1.3. *Let σ be a torsionfree class in $\mathbb{V}$ and $\gamma = \mathcal{U}\sigma$. If $A \in \gamma$ and $A^0 \in \sigma$, then $A = 0$.*

Proof: Since $\gamma(A \oplus A^0) = \gamma(A) \oplus \gamma(A^0)$, it follows that $\gamma(A \oplus A^0) = A$. Hence in view of Proposition 2.3.11 we have

$$A = \gamma((A, A, 0)) \subseteq \gamma(\Gamma(A)).$$

On the other hand,
$$\Gamma(A)/(A, A, 0) \cong A^0 \in \sigma,$$

and hence
$$\gamma(\Gamma(A)) \subseteq (A, A, 0)$$

holds implying
$$\gamma(\Gamma(A)) \triangleleft (A, A, 0).$$

From this we conclude

$$\gamma(\Gamma(A)) \subseteq \gamma((A, A, 0)) = (A, 0, 0),$$

and therefore we arrive at

$$\gamma(\Gamma(A)) = (A, 0, 0).$$

Thus
$$(A, 0, 0) = \gamma(\Gamma(A)) \triangleleft \Gamma(A),$$

and Lemma 5.1.1 yields

$$(A, 0, 0) = (A, A^2, 0)$$

that is, $A^2 = 0$, and so $A \cong A^0 \in \gamma \cap \sigma = 0$. $\square$

LEMMA 5.1.4. *Let σ be a torsionfree class in $\mathbb{V}$ and $\gamma = \mathcal{U}\sigma$. If $A^0 \in \gamma$ and $A \in \sigma$, then $A = 0$.*

Proof: Again by Proposition 2.3.11 we have

$$(0, A, 0) = \gamma((A, A, 0)) \subseteq \gamma(\Lambda(A)).$$

Further, by
$$\Lambda(A)/(A, A, 0) \cong A \in \sigma$$

it follows that
$$\gamma(\Lambda(A)) \triangleleft (A, A, 0).$$

Thus
$$\gamma(\Lambda(A)) \subseteq \gamma((A, A, 0) = (0, A, 0),$$

and so
$$(0, A, 0) = \gamma(\Lambda(A)) \triangleleft \Lambda(A).$$

Applying Lemma 5.1.2 we get

$$(0, A, 0) = (A^2, A, 0)$$

which implies $A^2 = 0$, that is, $A \cong A^0 \in \gamma \cap \sigma = 0$. □

THEOREM 5.1.5 (Gardner [9]). *If γ is a torsion class in $\mathbb{V}$, then γ is an A-radical.*

Proof: Let us recall from Proposition 3.19.2 that a radical γ is an A-radical if and only if $A \in \gamma \Leftrightarrow A^0 \in \gamma$. Further, by Lemma 3.12.10 the class

$$\gamma^0 = \{A \in \mathbb{V} \mid A^0 \in \gamma\}$$

is a radical class for every radical γ.

Suppose that γ is a torsion class and $A^0 \in \gamma$. Since

$$(A/\gamma(A))^0 \cong A^0/(\gamma(A))^0 \in \gamma,$$

and $A/\gamma(A) \in \mathcal{S}\gamma$, Lemma 5.1.4 yields $A = \gamma(A) \in \gamma$.

Assume that $A \in \gamma$. Since by Lemma 3.12.10 $\gamma^0(A) \triangleleft A$, we have $A/\gamma^0(A) \in \gamma$. Further, by $\gamma^0(A)^0 = \gamma^0(A^0)$ we have

$$(A/\gamma^0(A))^0 \cong A^0/\gamma^0(A)^0 \cong A^0/\gamma^0(A^0) \in \sigma.$$

Applying Lemma 5.1.3 we get $A/\gamma^0(A) = 0$. Hence $A/\gamma^0(A^0) = (A/\gamma^0(A))^0 = 0$, so $A^0 = \gamma^0(A^0) = \gamma(A^0) \in \gamma$. □

Let us mention that for hereditary torsion classes the assertion of Theorem 5.1.5 was proved by Andrunakievich and Ryabukhin [5].

Having proved Theorem 5.1.5, we get easily the following description of torsion classes in $\mathbb{V}$.

THEOREM 5.1.6 (Stewart [2], Gardner [9]). *In the variety $\mathbb{V}$ of all not necessarily associative rings the following three conditions are equivalent:*
   (I) *γ is a torsion class,*
   (II) *γ is an A-radical,*
   (III) *γ is a strict radical.*

Proof: (I) $\Longrightarrow$ (II) is the assertion of Theorem 5.1.5.

(II) $\Longrightarrow$ (III). By Proposition 3.19.2, $\gamma = \gamma^0$. Let U be a subring of a ring A such that $U \in \gamma$. Since γ is an A-radical, also $U^0 \in \gamma$, and so

$$U \subseteq \gamma^0(A) = \gamma(A).$$

Thus γ is a strict radical.

(III) $\Longrightarrow$ (I) is straightforward by Proposition 2.3.11. □

Theorem 5.1.6 tells us that, for instance, in the variety $\mathbb{V}$ the semisimple class of the Brown–McCoy radical $\mathcal{G}$ (cf. Example 2.2.4) is not hereditary because $\mathcal{G}$ is not an A-radical.

As an easy consequence we shall see — as expected — that in the variety $\mathbb{V}$ there is no possibility for introducing a (special) radical such that every semisimple ring should be a subdirect sum of reasonably good rings.

PROPOSITION 5.1.7. *If ϱ is a hereditary subclass of $\mathbb{V}$, then also its subdirect closure*

$$\overline{\varrho} = \{A \in \mathbb{V} \mid A = \sum_{\text{subdirect}} (A_\lambda \mid A_\lambda \in \varrho)\}$$

is hereditary.

Proof: Let $I \triangleleft A \in \overline{\varrho}$. Then there exist ideals K_λ of A such that $A/K_\lambda \in \varrho$ and $\cap K_\lambda = 0$. Hence by $I \cong I/\cap(K_\lambda \cap I)$, I is a subdirect sum of rings $I/(K_\lambda \cap I)$. Since

$$I/(K_\lambda \cap I) \cong (I + K_\lambda)/K_\lambda \triangleleft A/K_\lambda \in \varrho,$$

the hereditariness of ϱ yields $I/(K_\lambda \cap I) \in \varrho$. Thus $I \in \overline{\varrho}$. $\square$

COROLLARY 5.1.8 (Anderson, Kaarli, Wiegandt [1]). *If $\varrho \neq 0$ is a hereditary class of semiprime rings in $\mathbb{V}$, then the subdirect closure $\overline{\varrho}$ is not a semisimple class.*

Proof: Since ϱ consists of semiprime rings, the upper radical $\gamma = \mathcal{U}\varrho$ contains all rings with zero-multiplication, and therefore γ is not an A-radical as $\varrho \neq 0$ and $\gamma \neq \mathbb{V}$. Hence by Theorem 5.1.5 the semisimple class $\mathcal{S}\gamma$ is not hereditary, but by Proposition 5.1.7 the class $\overline{\varrho}$ is hereditary. Hence from $\varrho \subseteq \overline{\varrho} \subseteq \mathcal{S}\gamma$ it follows that $\overline{\varrho} \neq \mathcal{S}\gamma$. $\square$

In the light of these results we may observe the following interesting feature of the radical theory of rings: in the variety of associative rings (and also in some larger varieties) there is a "good" radical theory in which every semisimple class is hereditary and there are special radicals providing structure theorems by subdirect representations for semisimple rings. In the largest variety $\mathbb{V}$ of not necessarily associative rings the "good" radical theory collapses to that of abelian groups, for being an A-radical means that the radical property depends only on the additive group of rings.

Further degeneracy and pathology of nonassociative radical theory has been is explored by Gardner [14].

Concerning the lower radical construction in the variety $\mathbb{V}$, Ryabukhin [4] proved the following: *If δ is the class of all simple rings in $\mathbb{V}$, then the Kurosh lower radical construction does not terminate, that is, $\mathcal{L}\delta \neq \delta_\lambda$ for every ordinal number λ.* For the proof we refer to Gardner [18, 2.1.26 C Theorem].

Another pathological feature of nonassociative radical theory is due to Leavitt [3].

PROPOSITION 5.1.9. *In the variety $\mathbb{V}$ of all not necessarily associative rings there exists a subclass ϱ contained in two incomparable semisimple classes minimal relative to the inclusion of the subclass ϱ.*

Proof: Let us consider the ring A generated by nonassociative symbols $\{a, b, c\}$ subject to the relations

$$2a = 2b = 2c = 0,$$
$$a^2 = 0,$$
$$ab = ba = a,$$
$$ac = ca = bc = cb = b^2 = b,$$
$$c^2 = c.$$

It is clear that the only proper ideal of A is $I = \{0, a, b, a+b\}$, and that I has exactly one proper ideal $J = \{0, a\}$. Moreover, J is a two-element ring with zero-multiplication and $K = I/J$ is a two-element field. Let us consider the class $\varrho = \{0, A\}$ and also the classes $\varrho_1 = \{0, A, I, J\}$ and $\varrho_2 = \{0, A, K\}$. Obviously both ϱ_1 and ϱ_2 are regular classes and therefore $\sigma_1 = \mathcal{SU}\varrho_1$ and $\sigma_2 = \mathcal{SU}\varrho_2$ are semisimple classes containing ϱ. Clearly $J \in \mathcal{U}\varrho_2$ and so $J \notin \sigma_2$; further $K \in \mathcal{U}\varrho_1$ and $K \notin \sigma_1$. Hence σ_1 and σ_2 are incomparable.

Let σ be any semisimple class containing the class ϱ. If $\varrho_2 \subseteq \sigma$, then $\sigma_2 = \mathcal{SU}\varrho_2 \subseteq \mathcal{SU}\sigma = \sigma$, and hence $\sigma_2 \subseteq \sigma$. Thus, suppose that $\varrho_2 \not\subseteq \sigma$. Since $\varrho \subseteq \sigma$, it follows that $K \notin \sigma$. But the semisimple class σ is regular, so by $I \triangleleft A \in \varrho \subseteq \sigma$ and $K \notin \sigma$ we conclude $I \in \sigma$. Moreover, the regularity of σ yields also $J \in \sigma$, and hence $\varrho_1 \subseteq \sigma$. Consequently $\sigma_1 = \mathcal{SU}\varrho_1 \subseteq \mathcal{SU}\sigma = \sigma$, that is, $\sigma_1 \subseteq \sigma$. Thus both σ_1 and σ_2 are minimal semisimple classes with respect to the inclusion of the class ϱ.

Let us notice that Proposition 5.1.9 is in sharp contrast with Theorem 3.1.11 for associative rings. In the variety $\mathbb{V}$ the intersection of the regular classes ϱ_1 and ϱ_2 is ϱ which is not regular, and ϱ is not contained in a unique smallest regular class.

In this context we mention that Andrunakievich and Ryabukhin [5] have shown that in the universal class of all (not necessarily associative) algebras over a *field*, if $\{\sigma_i | i \in I\}$ is a collection of semisimple classes such that $\bigcap_{i \in I} \sigma_i$ is a semisimple class, then there exists an index i_0 such that $\bigcap_{i \in I} \sigma = \sigma_{i_0}$.

Theorem 2.2.1 tells us that the lower radical class of any subclass δ in a universal class $\mathbb{A}$ of not necessarily associative rings always exists. In particular, for associative rings the lower radical construction provides the smallest torsion class containing a given class. This latter is not so for not necessarily associative rings, as exhibited by Leavitt and Watters [1].

PROPOSITION 5.1.10. *In a universal class $\mathbb{A}$ of not necessarily associative rings there may exist a radical class δ which is not a torsion class and which is contained in two incomparable torsion classes τ_1 and τ_2 both minimal with respect to the inclusion of δ.*

Proof: Let us consider the ring A generated by nonassociative symbols $\{a, b, c, d\}$ satisfying
$$2a = 2b = 2c = 2d = 0$$
and determined by the multiplicative tableau

	a	b	c	d
a	a	a	a	b
b	a	0	0	a
c	a	0	c	d
d	b	a	0	c

Let I be the subring generated by a and b, and J the subring generated by a. Then one readily checks that I is the only proper ideal of A and J is the only proper ideal of I. Moreover, J is a two-element field, I/J is a two-element ring with zero-multiplication and A/I is a simple ring having more than two elements. Hence J, I/J and A/I are non-isomorphic simple rings. Let us choose the universal class as $\mathbb{A} = \{0, A, I, J, A/I, I/J\}$, and consider the subclass $\delta = \{0, J\}$. It can be easily checked that δ is a radical class with non-hereditary semisimple class $\mathcal{S}\delta = \{0, A, A/I, I/J\}$, further, $\tau_1 = \{0, I, J, I/J\}$ and $\tau_2 = \{0, J, A, A/I\}$ are incomparable radical classes in $\mathbb{A}$ both containing δ. The corresponding semisimple classes are $\sigma_1 = \mathcal{S}\tau_1 = \{0, A/I\}$ and $\sigma_2 = \mathcal{S}\tau_2 = \{0, I/J\}$, both hereditary and hence τ_1 and τ_2 are torsion classes. Since δ is not a torsion class, every torsion subclass μ of τ_1 which contains J must contain either I or I/J. But using homomorphic closure and extension closure, we see that then $I \in \mu$ if and only if $I/J \in \mu$, so $\mu = \tau_1$. Thus τ_1 is minimal. Further, $\{0, J, A/I\}$ is a radical class but not a torsion class as its semisimple class contains A but not I. Thus also τ_2 is minimal. □

For more positive and negative results on radical constructions we refer the reader also to Gardner [11] and Leavitt [2], [4], [6], [11].

5.2. Sufficient condition for a well-behaved radical theory: Terlikowska–Osłowska's approach

In the late 80's simultaneously and independently successful attempts were made to formulate conditions on universal classes of not necessarily associative rings (algebras over a commutative ring with identity, Ω-groups, groups, semigroups, objects of a category etc.) to ensure affirmative answer for three major issues of general radical theory:

  i) the validity of the ADS-Theorem (cf. Theorem 3.1.2),

ii) the validity of Sands' Theorem (cf. Theorem 3.1.8),

iii) the termination of the Kurosh lower radical construction at the first limit ordinal (cf. Theorem 3.4.5).

Puczyłowski [8], [15] and Veldsman [9], [10] followed the idea of Terlikowska–Osłowska [1], [2] (see also Krempa and Terlikowska [1]) and imposed her powerful category-theoretical condition (or modified versions) on the universal class considered which enabled them to derive positive answers for i), ii) and iii). Beidar [3] accomplished this project in a different way; he required purely nonassociative ring-theoretical conditions and obtained the same results.

In this section we shall discuss Terlikowska–Osłowska's approach mainly as presented by Puczyłowski [15]. We shall work in a *universal class* $\mathbb{A}$ *of not necessarily associative rings*.

A ring $A \in \mathbb{A}$ will be called *transitive*, if the relation $\triangleleft$ in A is transitive, that is, $J \triangleleft I \triangleleft A$ implies $J \triangleleft A$ for all $I \triangleleft A$ and $J \triangleleft I$. We say that an ideal J of A is *distinguished* and shall write $J \triangleleft_d A$, if for every ideals K and L of A such that $K \subseteq J \subseteq L$, $J/K \cong L/J$ and L/J is a transitive ring, then $K = J = L$. The universal class $\mathbb{A}$ is said to be *normal*, if

$$J \triangleleft_d I \triangleleft A \quad \text{implies} \quad J \triangleleft A$$

for every J, I and A.

EXAMPLE 5.2.1. *The variety of associative rings is normal.* Let $J \triangleleft_d I \triangleleft A$. We want to see that $J \triangleleft A$. Clearly, it suffices to show that $aJ \subseteq J$ for all elements $a \in A$, because the case $Ja \subseteq J$ is analogous. By Lemma 3.1.1 and its proof we know that

$$L = (aJ + J) \triangleleft I,$$
$$K = \{x \in J \mid ax \in J\} \triangleleft I,$$

and that

$$J/K \cong L/J.$$

Moreover, by Lemma 3.1.1 we have also $(L/J)^2 = 0$, and therefore L/J is a transitive ring. By the assumption $J \triangleleft_d I$ we conclude $K = J = L$, and therefore $aJ \subseteq J$ is valid. Thus $J \triangleleft A$ has been proved. □

It has been shown by Terlikowska–Osłowska [1] that also *the variety of alternative rings is normal*. Further examples of normal varieties are those of *autodistributive rings*, Γ-*rings*, *Nobusawa's rings*, as given by Veldsman [9]. Another example is the universal class of all *semigroups with* 0 in which one admits only Rees factor semigroups as homomorphic images (see Puczyłowski [15]).

Thus, from now onwards we shall assume that *the universal class* $\mathbb{A}$ *is normal*.

PROPOSITION 5.2.2. *For every radical class γ in $\mathbb{A}$, $\gamma(A) \triangleleft_d A$ for all $A \in \mathbb{A}$.*

Proof: If K and L are ideals of A such that $K \subseteq \gamma(A) \subseteq L$, $\gamma(A)/K \cong L/\gamma(A)$, then by $\gamma(A) \in \gamma$ it follows that $L/\gamma(A) \in \gamma$ and the extension property of the radical class γ (cf. Lemma 2.1.2) implies that $L \in \gamma$. This gives us $L \subseteq \gamma(A)$, and so $K = \gamma(A) = L$ follows, proving that $\gamma(A) \triangleleft_d A$. □

COROLLARY 5.2.3. *If $\mathbb{A}$ is a normal universal class, then the ADS-Theorem is valid: $I \triangleleft A$ implies $\gamma(I) \triangleleft A$ for every radical γ in $\mathbb{A}$.* □

Now we shall turn to the proof of the validity of Sands' Theorem in $\mathbb{A}$. We shall achieve it by proving some lemmas. For that purpose, let us consider a subclass ϱ of $\mathbb{A}$ which is *regular, coinductive, and is closed under extensions.*

LEMMA 5.2.4. *If $I \triangleleft A \in \mathcal{SU}\varrho$, and I is a transitive ring, then $I \in \varrho$.*

Proof: It is a trivial consequence of Corollary 5.2.3 that the semisimple class $\mathcal{SU}\varrho$ is hereditary, so $I \in \mathcal{SU}\varrho$. Hence I has a nonzero homomorphic image in ϱ and by the coinductivity of ϱ there exists an ideal J of I such that $I/J \in \varrho$ and J is minimal for this. Now $I \neq J$; suppose also $I \neq 0$. Then by the same token, there exists an ideal K of J such that $J/K \in \varrho$ and K is minimal for this. The transitivity of I implies $K \triangleleft I$. Since ϱ is closed under extensions, from $J/K \in \varrho$ and

$$\frac{I/K}{J/K} \cong I/J \in \varrho,$$

we conclude that $I/K \in \varrho$, which by $K \subset J$ contradicts the minimality of J. Thus $J = 0$, and hence $I \in \varrho$. □

LEMMA 5.2.5. *If $J \triangleleft A$, $A/J \in \varrho$ and J is minimal for this, then $J \triangleleft_d A$.*

Proof: Let K and L be ideals of A such that $K \subseteq J \subseteq L$, $J/K \cong L/J$ and L/J is a transitive ring. Since $A/J \in \varrho \subseteq \mathcal{SU}\varrho$, Lemma 5.2.4 is applicable to the transitive ring $L/J \triangleleft A/J$, implying $L/J \in \varrho$ and also $J/K \in \varrho$. Hence by

$$\frac{A/K}{J/K} \cong A/J,$$

and by A/J, $J/K \in \varrho$ the extension-closed property of ϱ yields $A/K \in \varrho$. This and the minimality of J implies $J \subseteq K$, that is, $K = J = L$, proving $J \triangleleft_d A$. □

Let us recall that in Theorem 2.3.9 a semisimple class σ has been characterized by regularity, coinductivity, being closed under extensions and by condition

(∗) If $I \triangleleft A$, $A/I \in \sigma$ and I is minimal for this, and if $K \triangleleft I$, $I/K \in \sigma$ and K is minimal for this, then $K \triangleleft A$.

LEMMA 5.2.6. ϱ satisfies condition (∗).

Proof: For an arbitrary ring $A \in \mathbb{A}$ the regularity and the coinductivity of ϱ and Zorn's Lemma guarantee the existence of an I and a K as required in condition (∗). If $I = 0$, then there is nothing to prove. So, assume that $I \neq 0$. Applying Lemma 5.2.5 we get $K \triangleleft_d I$. Since A is a normal universal class, we conclude $K \triangleleft A$. □

An immediate consequence of Theorem 2.3.9 and Lemma 5.2.6 is

COROLLARY 5.2.7. *If the universal class $\mathbb{A}$ is normal, then Sands' Theorem is valid, that is, the semisimple classes in $\mathbb{A}$ are precisely the subclasses of $\mathbb{A}$ which are regular, coinductive and closed under extensions.* □

Before proving that in $\mathbb{A}$ the lower radical construction terminates at the first limit ordinal ω, we remind the reader of the Kurosh lower radical construction (cf. § 3.3). Let $\delta = \delta_1$ be a homomorphically closed subclass of $\mathbb{A}$. The lower radical class $\mathcal{L}\delta$ of δ is the class $\mathcal{L}\delta = \cup \delta_\lambda$ where

$$\delta_\lambda = \left\{ A \;\middle|\; \begin{array}{l} \text{every nonzero homomorphic image of } A \\ \text{has a nonzero ideal in } \delta_\mu \text{ for some } \mu < \lambda \end{array} \right\}.$$

For a ring A we define inductively the ideals $\Delta^{(\lambda)}(A)$ depending on the class δ as follows:

$$\Delta^{(1)}(A) = \sum (I \triangleleft A \mid I \in \delta),$$

$$\Delta^{(\lambda+1)} = \sum (I \triangleleft A \mid I/\Delta^{(\lambda)}(A) \in \delta),$$

$$\Delta^{(\lambda)}(A) = \bigcup_{\mu < \lambda} \Delta^{(\mu)}(A) \text{ for limit ordinals } \lambda.$$

These ideals form an ascending chain, and we set $\Delta(A) = \bigcup \Delta^{(\lambda)}(A)$. Obviously $\Delta(A) = \Delta^{(\nu)}(A)$ for some ν and we may assume that ν is not a limit ordinal. Then a moment's reflection shows that $\Delta(A) \in \delta_2$.

LEMMA 5.2.8. *Let $A \in \delta_\lambda$. If A is transitive then $A \in \delta_2$.*

Proof: We proceed by induction on λ. For $\lambda = 2$ the assertion is trivial. Suppose that the statement is true for $\mu < \lambda$. If $\Delta(A) = A$ then, as already remarked, $A \in \delta_2$ and we are done. If $\Delta(A) \neq A$ then there exists an $I \triangleleft A$ such that $0 \neq I/\Delta(A) \in \delta_\mu$ for some $\mu < \lambda$. Since also $I/\Delta(A)$ is transitive, by the hypothesis we have $0 \neq I/\Delta(A) \in \delta_2$. Hence by the definition of δ_2 there exists a nonzero ideal J of I such that $J/\Delta(A) \in \delta$. But A is transitive, so $J \triangleleft A$. This, however, contradicts the definition of $\Delta(A)$ as $\Delta(A) = \Delta^{(\nu)}(A)$ for some ν, and then $J \subseteq \Delta^{(\nu+1)}(A) \neq \Delta^{(\nu)}(A)$. Thus $A \in \delta_2$. □

LEMMA 5.2.9. *If every transitive ring $R \in \mathcal{L}\delta$ is already in δ, the $\Delta(A) \triangleleft_d A$ for every ring $A \in \mathbb{A}$.*

Proof: Let K and L be ideals of A such that $K \subseteq \Delta(A) \subseteq L$, $L/\Delta(A) \cong \Delta(A)/K$, and $L/\Delta(A)$ is transitive. Obviously $\Delta(A)/K \in \mathcal{L}\delta$, so $L/\Delta(A) \in \delta$. Hence by the definition of $\Delta(A)$ it follows that $L = \Delta(A) = K$. □

THEOREM 5.2.10. *In a normal universal class* $\mathbb{A}$ *the Kurosh lower radical construction terminates at the first limit ordinal* ω: $\mathcal{L}\delta = \delta_\omega$ *for every homomorphically closed subclass* δ *of* $\mathbb{A}$.

Proof: Let A be a nonzero ring in $\delta_{\omega+1}$. Then there are nonzero rings I and J such that $J \triangleleft I \triangleleft A$ and $J \in \delta_k$ for some finite $k \geq 2$. Starting the construction of the ideals $\Delta^{(\lambda)}(A)$ from the class $\alpha = \delta_k$, we get the ideal $\Delta_k(A) = \bigcup \Delta^{(\lambda)}(A)$ of A. Since $\mathcal{L}\delta = \mathcal{L}\delta_2 = \mathcal{L}\delta_k = \mathcal{L}\alpha$, we may apply Lemmas 5.2.8 and 5.2.9 and get that $0 \neq J \subseteq \Delta_k(I) \triangleleft_d I$. Taking into account that the universal class $\mathbb{A}$ is normal, it follows that $\Delta_k(I) \triangleleft A$. Since $\Delta_k(I) \in (\delta_k)_2 = \delta_{k+1}$, the ring A has a nonzero ideal in δ_{k+1}, and clearly the same holds for every nonzero homomorphic image of A. Thus $A \in \delta_\omega$ has been proved. □

5.3. Sufficient conditions for a well-behaved radical theory: Beidar's approach

The results of this section are taken from Beidar's paper [3]. For short, a ring will always mean a *not necessarily associative ring*. For a ring A we define inductively

$$A^{(1)} = A \quad \text{and} \quad A^{(n+1)} = A^{(n)} \cdot A^{(n)}$$

for $n = 1, 2, \ldots$, that is, $A^{(n+1)}$ is the set of all finite sums $\Sigma a_i b_i$ with $a_i, b_i \in A^{(n)}$. Obviously $A^{(2)} = A^2$ is an ideal of A, and hence a subring, too. By induction we get that $A^{(n+1)}$ is an ideal of $A^{(n)}$ and a subring of A. We say that the ring A is *solvable of degree* $n > 1$, if $A^{(n)} = 0$ and $A^{(n-1)} \neq 0$.

Let us consider the functions

$$v_n \colon \bigoplus_{i=1}^{2^n} A \longrightarrow A^{(n+1)} \qquad n = 0, 1, 2, \ldots$$

defined on direct sums of copies of A and given by

$$v_0(x_1) = x_1$$

and

$$v_{n+1}(x_1, x_2, \ldots, x_{2^{n+1}}) = v_n(x_1, x_2, \ldots, x_{2^n}) \cdot v_n(x_{2^n+1}, \ldots, x_{2^{n+1}})$$

where $x_1, x_2, \ldots$ range through the elements of A. These functions $v_0, v_1, \ldots, v_n, \ldots$ can be considered also as monomials in the nonassociative polynomial ring $\mathbb{Z}\langle x_1,\ldots,x_n,\ldots\rangle$ with non-commutative indeterminates $x_1,\ldots,x_n,\ldots$ over the integers $\mathbb{Z}$. Let us mention two useful properties of these functions which can be readily verified by induction:

(5.3.1) $\quad v_{n+1}(x_1,\ldots,x_{2^{n+1}}) = v_n(x_1x_2, x_3x_4,\ldots, x_{2^{n+1}-1}x_{2^{n+1}}),$

(5.3.2) $\quad A^{(n)}$ *coincides with the additive subgroup of A generated by the subset* $v_{n-1}(A,\ldots,A)$ *for each* $n = 1, 2, \ldots$.

We shall denote by ann A the two-sided annihilator

$$\mathrm{ann}\, A = \{x \in A \mid xA = 0 = Ax\}$$

of the ring A. For the further developments the following statement will be crucial.

LEMMA 5.3.3. *Let A and B be rings and $f\colon A \to B$ a nonzero additive homomorphism such that $f(A^{(m)}) = 0$ for some $m > 1$. If $B^{(n)} = 0$ for some $n > 1$, then there exists an integer $k > 0$ and a nonzero ring homomorphism $h\colon A \to \mathrm{ann}\, B^{(k)}$ such that $h(A^2) = 0$.*

Proof: Without loss of generality we may assume that m is minimal, that is, $f(A^{(m-1)}) \neq 0$. By (5.3.2) it follows that

$$f(v_{m-2}(A,\ldots,A)) \neq 0.$$

Moreover, (5.3.1) yields

$$v_{m-2}(A^2,\ldots,A^2) = v_{m-1}(A,\ldots,A) \subseteq A^{(m)},$$

and therefore by $f(A^{(m)}) = 0$ we get

$$f(v_{m-2}(A^2,\ldots,A^2)) = 0.$$

Thus there exists a natural number $q \leq 2^{m-2}$ and elements $a_1,\ldots,a_q \in A$ such that

$$f(v_{m-2}(a_1,\ldots,a_{q-1},a_q,A^2,\ldots,A^2)) \neq 0,$$

but

$$f(v_{m-2}(a_1,\ldots,a_{q-1},A^2,A^2,\ldots,A^2)) = 0.$$

Let us choose elements $a_{q+1},\ldots,a_{2^{m-2}} \in A^2$ such that

$$f(v_{m-2}(a_1,\ldots,a_q,a_{q+1},\ldots,a_{2^{m-2}})) \neq 0.$$

Define the mapping $g\colon A \to B$ by

$$g(x) = f(v_{m-2}(a_1,\ldots,a_{q-1},x,a_{q+1},\ldots,a_{2^{m-2}})).$$

$g(x)$ is clearly a nonzero additive homomorphism such that $g(A^2) = 0$. Further, let us choose an integer p such that at least one of the following sets differs from the set $\{0\}$:

$$v_p(g(A), B, \ldots, B)$$
$$v_p(B, g(A), B, \ldots, B)$$
$$\vdots$$
$$v_p(B, \ldots, B, g(A)).$$

Since $g(A) \neq 0$ and $B^{(n)} = 0$, such an integer p does exist and $0 \leq p \leq n-2$. We may assume that p is maximal with respect to this property, and that

$$v_p(b_1, \ldots, b_{i-1}, g(A), b_{i+1}, \ldots, b_{2^p}) \neq 0$$

for some $b_1, \ldots, b_{i-1}, b_{i+1}, \ldots, b_{2^p} \in B$. Let us define now the mapping $h \colon A \to B$ by

$$h(x) = v_p(b_1, \ldots, b_{i-1}, g(x), b_{i+1}, \ldots, b_{2^p}).$$

Then $h(x)$ is clearly a nonzero additive homomorphism. Moreover, by the maximality of p we have

$$v_p(B, \ldots, B, g(A), B, \ldots, B) v_p(B, \ldots, B) =$$
$$= v_{p+1}(B, \ldots, B, g(A), B, \ldots, B) = 0$$

as well as

$$v_p(B, \ldots, B) v_p(B, \ldots, B, g(A), B, \ldots, B) =$$
$$= v_{p+1}(B, \ldots, B, g(A), B, \ldots, B) = 0.$$

Hence by (5.3.2) it follows that

$$h(A) = v_p(b_1, \ldots, b_{i-1}, g(A), b_{i+1}, \ldots, b_{2^p}) \subseteq \operatorname{ann} B^{(p+1)}$$

and $h(A^2) = 0$; in particular $h(xy) = 0$ for all $x, y \in A$. On the other hand, the maximality of p implies

$$h(x)h(y) =$$
$$= v_p(b_1, \ldots, b_{i-1}, g(x), b_{i+1}, \ldots, b_{2^p}) v_p(b_1, \ldots, b_{i-1}, g(y), b_{i+1}, \ldots, b_{2^p}) =$$
$$= v_{p+1}(b_1, \ldots, b_{i-1}, g(x) b_{i+1}, \ldots, b_{2^p}, b_1, \ldots, b_{i-1}, g(y), b_{i+1}, \ldots, b_{2^p}) \subseteq$$
$$\subseteq v_{p+1}(B, \ldots, B, g(A), B, \ldots, B) = 0.$$

Thus $h(x)$ is a ring homomorphism into $\operatorname{ann} B^{(k)}$ with $k = p+1$. $\square$

PROPOSITION 5.3.4 (Nikitin [2]). *Let γ be a radical in any universal class of rings. If $I \triangleleft A$ and I is solvable, then $\gamma(I) \triangleleft A$.*

Proof: We proceed by induction on the degree n of solvability of I. Let $n = 2$, that is, $I^2 = 0$. Then every additive subgroup of I is an ideal of I, and every left multiplication by an element $a \in A$ induces an endomorphism $f_a \colon I \to aI$ of the ring I. Hence $a\gamma(I)$ is a homomorphic image of $\gamma(I)$, and so $a\gamma(I) \in \gamma$. Since $a\gamma(I) \triangleleft I$, it follows that $a\gamma(I) \subseteq \gamma(I)$ for every $a \in A$. Thus $\gamma(I)$ is a left ideal of A. Similarly, $\gamma(I)$ is a right ideal of A, and therefore $\gamma(I) \triangleleft A$.

Next, assume that $n > 2$ and that the assertion holds for $2 \leq m < n$. By way of contradiction, let us suppose that $a\gamma(I) \not\subseteq \gamma(I)$ for some $a \in A$, and let us consider the mapping

$$f \colon \gamma(I) \to I/\gamma(I) = K$$

given by $f(x) = ax + \gamma(I)$ for all $x \in \gamma(I)$. Obviously $f(\gamma(I)) \neq 0$ and f is an additive homomorphism. Applying Lemma 5.3.3, there exists an integer $k > 0$ and a nonzero ring homomorphism $h \colon \gamma(I) \to \operatorname{ann} K^{(k)}$ such that $h((\gamma(I))^2) = 0$. Clearly $0 \neq h(\gamma(I)) \triangleleft K^{(k)}$, which implies $\gamma(K^{(k)}) \neq 0$ and hence $k < n$. Taking into account that

$$\gamma(K^{(1)}) = \gamma(K) = \gamma(I/\gamma(I)) = 0,$$

it follows that $k \geq 2$. Hence the number $m = n - k + 1$ is at least 2, but at most $n - 1$. Moreover, it is easy to see that

$$(K^{(k)})^{(m)} = K^{(k+m-1)} = K^{(n)} = 0.$$

Since $K^{(k)} \triangleleft K^{(k-1)}$, and $k < n$, the induction hypothesis implies

$$0 \neq \gamma(K^{(k)}) \triangleleft K^{(k-1)},$$

and consequently $\gamma(K^{(k-1)}) \neq 0$. If $k - 1 > 1$, then by the same reasoning we get that $\gamma(K^{(k-2)}) \neq 0$. Iterating this procedure we arrive at

$$0 \neq \gamma(K^{(k-(k-1))}) = \gamma(K) = \gamma(I/\gamma(I)) = 0,$$

a contradiction. Thus $a\gamma(I) \subseteq \gamma(I)$ for all $a \in A$. Analogously we get $\gamma(I)a \subseteq \gamma(I)$ for all $a \in A$, proving the assertion for an arbitrary degree n of solvability of I. □

PROPOSITION 5.3.5. *Let γ be a radical of rings, and B an accessible subring of a ring A. If $\gamma(A) = 0$ and B is solvable, then $\gamma(B) = 0$.*

Proof: Use induction and apply Proposition 5.3.4. □

So far the results have been achieved without imposing any constraint on the multiplication. Assuming the fulfilment of some conditions involving the multiplication, we can prove the ADS-Theorem and Sands' Theorem.

Therefore, in the sequel we shall require that the universal class $\mathbb{A}$ under consideration satisfies some of the following conditions:

(B1) if $K \triangleleft I \triangleleft A \in \mathbb{A}$, then to every element $a \in A$ there exists an integer $m > 0$ such that $aK^{(m)} + K^{(m)}a \subseteq K$,

(B2) if $K \triangleleft I \triangleleft A \in \mathbb{A}$, then to every element $a \in A$ there exists an integer $n > 0$ such that $((aK + Ka + K)_I)^{(n)} \subseteq K$,

(B3) if $K \triangleleft I \triangleleft A \in \mathbb{A}$, then there exists an integer $t > 0$ such that $(K^{(t)})_I \subseteq K^2$,

where $(\ldots)_I$ denotes the ideal of I generated by the subset $(\ldots)$.

THEOREM 5.3.6 (Nikitin [1]). *If the universal class $\mathbb{A}$ satisfies conditions* (B1) *and* (B2), *then the ADS-Theorem is valid in* $\mathbb{A}$: $I \triangleleft A \in \mathbb{A}$ *implies $\gamma(I) \triangleleft A$ for every radical γ.*

Proof: Suppose that $\gamma(I)$ is not an ideal in A, that is $a\gamma(I) + \gamma(I)a \not\subseteq \gamma(I)$ for some $a \in A$. We may confine ourselves to the case $a\gamma(I) \not\subseteq \gamma(I)$. Put
$$K = (a\gamma(I) + \gamma(I)a + \gamma(I))_I/\gamma(I)$$
and
$$J = I/\gamma(I),$$
and define a mapping $f\colon \gamma(I) \to K$ by the rule
$$f(x) = ax + \gamma(I), \qquad \forall\, x \in \gamma(I).$$

f is clearly an additive homomorphism. By virtue of (B1) there exists an integer $m > 0$ such that $a(\gamma(I))^{(m)} \subseteq \gamma(I)$, and since $a\gamma(I) \not\subseteq \gamma(I)$, necessarily $m > 1$. Hence $f((\gamma(I))^{(m)}) = 0$. Moreover, by condition (B2) there exists an integer $n > 0$ such that $K^{(n)} = 0$, and by the definition of K it follows that $n > 1$. Now, Lemma 5.3.3 is applicable, yielding the existence of an integer $k \geq 1$ and of a nonzero ring homomorphism
$$h\colon \gamma(I) \to \operatorname{ann} K^{(k)}$$
such that $h((\gamma(I))^2) = 0$. Obviously, $h(\gamma(I)) \triangleleft K^{(k)}$, and so by $0 \neq h((\gamma(I)) \in \gamma$ we conclude that $\gamma(K^{(k)}) \neq 0$. Since $\gamma(J) = \gamma(I/\gamma(I)) = 0$, $K \triangleleft J$ and $K^{(n)} = 0$, an application of Proposition 5.3.5 gives us $\gamma(K^{(k)}) = 0$, a contradiction. Thus $a\gamma(I) \subseteq \gamma(I)$ and similarly $a\gamma(I) + \gamma(I)a \subseteq \gamma(I)$ hold, proving the Theorem. $\square$

PROPOSITION 5.3.7. *Let ϱ be a subclass of a universal class $\mathbb{A}$ of rings such that ϱ is regular, coinductive and closed under extensions, and let $\gamma = \mathcal{U}\varrho$ be the upper radical of ϱ. If $\gamma(A) = 0$ for a ring $A \in \mathbb{A}$ and A is solvable, then $A \in \varrho$. Furthermore, if B is an accessible subring of A, then also $B \in \varrho$.*

Proof: Let n be the degree of solvability of $A \neq 0$. As the first step of the induction, assume that $n = 2$. Since ϱ is coinductive, there exists an ideal I of A with $A/I \in \varrho$, and minimal for this. Since $\gamma(A) = 0$, necessarily $I \neq A$. If $I = 0$, we are done. So, assume that $I \neq 0$. Employing Proposition 5.3.4 we get $\gamma(I) \triangleleft A$, and hence $\gamma(A) = 0$ implies $\gamma(I) = 0$. Again by the coinductivity of the class ϱ, there exists an ideal K of I with $I/K \in \varrho$ and minimal for this. By $I \neq 0$ and $\gamma(I) = 0$ it follows that $K \neq I$. Clearly $K \triangleleft A$ for $A^2 = 0$. Since the class ϱ is closed under extensions and $A/I \cong (A/K)/(I/K)$, we conclude that $A/K \in \varrho$. This and the minimality of I implies $I = K$, a contradiction. Hence $I = 0$ and $A \in \varrho$ for $n = 2$.

Since $A^2 = 0$, every subring B of A is an ideal of A, and so by Proposition 5.3.5 we obtain $\gamma(B) = 0$. Thus by the already proved assertion we get $B \in \varrho$.

Thus the assertion holds for $n = 2$.

Let us suppose that $n > 2$ and that the assertion holds for every k with $2 \leq k < n$. Further, let I be an ideal of A which contains A^2 and which is minimal with respect to $A/I \in \varrho$. (Since $A^2 \triangleleft A$, I is in fact an ideal of A such that I/A^2 is minimal in A/A^2 with respect to $(A/A^2)/(I/A^2) \in \varrho$.) Again, if $I = 0$, then $A \in \varrho$. So we assume that $I \neq 0$, and derive a contradiction. Now Proposition 5.3.5 implies that $\gamma(I) = 0$. Let k be the degree of solvability of I. Clearly $2 \leq k \leq n$. Let us consider the case $k = n$. Now we have

$$v_{n-2}(b_1, b_2, \ldots, b_{2^{n-2}}) \neq 0$$

for some elements $b_1, b_2, \ldots, b_{2^{n-2}} \in I$. In view of (5.3.1) and (5.3.2) we get that

$$v_{n-2}(A^2, \ldots, A^2) = v_{n-1}(A, \ldots, A) = 0.$$

Hence there exists an integer p with $1 \leq p \leq 2^{n-2}$ such that

$$v_{n-2}(b_1, \ldots, b_{p-1}, b_p, A^2, \ldots, A^2) \neq 0,$$

but

$$v_{n-2}(b_1, \ldots, b_{p-1}, A^2, A^2, \ldots, A^2) = 0.$$

Further, let $d_{p+1}, \ldots, d_{2^{n-2}} \in A^2$ be elements such that

$$v_{n-2}(b_1, \ldots, b_p, d_{p+1}, \ldots, d_{2^{n-2}}) \neq 0,$$

and define the mapping $f \colon A \to A^{(n-1)}$ by

$$f(x) = v_{n-2}(b_1, \ldots, b_{p-1}, x, d_{p+1}, \ldots, d_{2^{n-2}})$$

for every $x \in A$. Obviously, f is a nonzero additive homomorphism. Taking into account that $(A^{(n-1)})^2 = A^{(n)} = 0$, f is also a ring homomorphism. Again by $(A^{(n-1)})^2 = 0$ it follows that $f(A) \triangleleft A^{(n-1)}$ and that $(f(A))^2 = 0$. Thus Proposition 5.3.5 yields that $\gamma(f(A)) = 0$, and so by the induction hypothesis we get $f(A) \in \varrho$.

Put $\overline{A} = A/I$, and consider the direct sum $D = \overline{A} \oplus f(A)$. The mapping $g\colon A \to D$ defined by

$$g(x) = (x + I, f(x)) \in D, \qquad \forall\, x \in A$$

is a ring homomorphism with $g(A^2) = 0$, and $g(b_p) \neq 0$. Hence $A^2 \subseteq \ker g \subset I$. Since $\overline{A} = A/I \in \varrho$, $f(A) \in \varrho$ and ϱ is closed under extensions, we have also

$$A/\ker g \cong g(A) = D \in \varrho.$$

This and $\ker g \subset I$ contradict the choice of I. Thus the case $k = n$ is not possible, whence $k < n$, and the induction hypothesis implies $I \in \varrho$. Since ϱ is closed under extensions, by $A/I \in \varrho$ we get $A \in \varrho$, contradicting $I \neq 0$. Thus $A \in \varrho$ has been demonstrated.

Finally, let $B \rhd\!\!-\, A \in \varrho$ and $A^{(n)} = 0$. Then $\gamma(A) = 0$ and so Proposition 5.3.5 yields $\gamma(B) = 0$. Hence by the assertion already proved, we get $B \in \varrho$. $\square$

Let us remark that for Proposition 5.3.7 no constraint on the universal class $\mathbb{A}$ was demanded. For proving, however, the validity of Sands' Theorem, also condition (B3) is needed besides (B1) and (B2).

THEOREM 5.3.8. *Let the universal class $\mathbb{A}$ satisfy conditions* (B1), (B2), (B3). *A subclass ϱ of $\mathbb{A}$ is a semisimple class if and only if ϱ is regular, coinductive and closed under extensions.*

Proof: Only the sufficiency of these three properties needs demonstration. Let γ denote the upper radical $\gamma = \mathcal{U}\varrho$ of ϱ, and let $A \neq 0$ be any ring such that $\gamma(A) = 0$. The Theorem will be proved, if we exhibit that $A \in \varrho$. By way of contraposition, let us suppose that $A \notin \varrho$, and let I be an ideal of A with to $A/I \in \varrho$ and minimal for this. Since $\gamma(A) = 0$, we have $I \neq A$. Taking into consideration also the regularity of the semisimple class of γ, we get that $\gamma(I) \neq I$. Similarly, there exists an ideal K of I with $I/K \in \varrho$ and K is minimal for this. If $K \triangleleft A$, then by $(A/K) / (I/K) \cong A/I \in \varrho$ and since ϱ is closed under extensions we conclude that $A/K \in \varrho$ which contradicts the minimality of I, so $A \in \varrho$ and we are done. If K is not an ideal of A, then without loss of generality we may confine ourselves to the case $aK \not\subseteq K$ for some $a \in A$. Set $B = I/K$ and $C = (aK + Ka + K)_I/K$. The mapping $f\colon K \to C$ given by $f(x) = ax + K$, $x \in K$, is clearly a nonzero additive homomorphism, and condition (B1) implies $f(K^{(m)}) = 0$. Moreover, condition (B2) yields $C^{(n)} = 0$. Hence applying Lemma 5.3.3, there exists an integer $k \geq 1$ and a nonzero ring homomorphism

$$h\colon K \to \operatorname{ann} C^{(k)}$$

such that $h(K^2) = 0$. Since $B = I/K \in \varrho$, we have that $\gamma(B) = 0$. Since $C^{(n)} = 0$ and $C \triangleleft B$, Proposition 5.3.7 implies $\gamma(h(K)) = 0$. Since $h(K^2) = 0$, we get $K^2 \subseteq \ker h$ and so K/K^2 can be mapped homomorphically onto $K/\ker h \cong h(K)$. Moreover, by $\gamma(h(K)) = 0$ both $h(K)$ and K/K^2 can be mapped homomorphically onto a nonzero ring in ϱ, yielding $\gamma(K/K^2) \neq K/K^2$. By condition (B3) we have

$$L = (K^{(t)})_I \subseteq K^2$$

for some $t > 0$. Hence by the isomorphism $K/K^2 \cong (K/L) / (K^2/L)$ the ring K/L can be mapped homomorphically via K/K^2 onto a nonzero ring in ϱ, and therefore $\gamma(K/L) \neq K/L$. Putting $\overline{I} = I/L$ and $\overline{K} = K/L$, we have $\overline{K} \triangleleft \overline{I}$, $\overline{I}/\overline{K} \in \varrho$, and by the minimality property of K, $\overline{K}$ is likewise minimal with respect to having ist factor ring in ϱ. Moreover, by $L = (K^{(t)})_I$ and $\overline{K} = K/L$ it follows that $\overline{K}^{(t)} = 0$. Thus Proposition 5.3.7 applies to $\overline{K}/\gamma(\overline{K})$ yielding $\overline{K}/\gamma(\overline{K}) \in \varrho$. Furthermore, Proposition 5.3.4 gives $\gamma(\overline{K}) \triangleleft \overline{I}$. Hence by

$$\frac{\overline{I}/\gamma(\overline{K})}{\overline{K}/\gamma(\overline{K})} \cong \overline{I}/\overline{K} \cong I/K \in \varrho$$

and by the closure of ϱ under extensions, we get $\overline{I}/\gamma(\overline{K}) \in \varrho$. Thus the minimality of $\overline{K}$ and $\gamma(\overline{K}) \subseteq \overline{K}$ imply $\gamma(\overline{K}) = \overline{K}$, that is, $\gamma(K/L) = K/L$, a contradiction. Hence $A \in \varrho$ has been established. $\square$

It has been proved in Beidar [3] that the Kurosh lower radical construction terminates always at most at the first limit ordinal whenever the universal class $\mathbb{A}$ satisfies (B1) and a stronger version of (B2) in which the integer n depends only on the ring A but not on the choice of the element $a \in A$.

Though conditions (B1), (B2) and (B3) turned out to be quite powerful in proving strong results, we still have not seen universal classes which do fulfil these conditions. We shall prove that fairly many varieties of nonassociative rings, called Andrunakievich s-varieties, satisfy these conditions, beside many others.

For a ring A the power A^s is defined inductively as

$$A^1 = A \qquad \text{and} \qquad A^s = \sum_{i=1}^{s-1} A^i A^{s-i}$$

for $s = 2, 3, \ldots$. Clearly $A^{(n)} \subseteq A^{2^{n-1}}$. One can verify that $A^1 \supseteq A^2 \supseteq \cdots A^s \supseteq \cdots$ is a chain of ideals of A (see Zhevlakov, Slin'ko, Shestakov and Shirshov [1 p. 82]).

A subvariety $\mathbb{A}$ of not necessarily associative rings is called an *Andrunakievich s-variety of index k*, if the following two conditions hold:

(A) $K \triangleleft I \triangleleft A \in \mathbb{A}$ *implies that* $(K)_A^{(k)} \subseteq K$ *and k is minimal,* (i.e. $(K)_A/K$ is solvable of degree k);

(s) *there exists an integer* $s > 1$ *such that* $I^s \triangleleft A$ *whenever* $I \triangleleft A \in \mathbb{A}$.

It is clear that associative rings form an Andrunakievich 2-variety of index 3, more precisely, the degree of nilpotency is 3. Pchelintsev [1] proved that also alternative algebras (that is, algebras satisfying the identities $x^2y = x(xy)$ and $xy^2 = (xy)y$) over a commutative ring in which $6x = 0$ implies $x = 0$, form an Andrunakievich 2-variety, and Hentzel [2] showed that the degree of nilpotency is 4, improving substantially Pchelintsev's upper bound $4 \cdot 5^6$. Alternative algebras over a field of characteristic 3, and consequently alternative rings do not form an Andrunakievich s-variety, as Pchelintsev [2] proved. The notion of Andrunakievich s-variety was introduced by Anderson and Gardner [1] where also more examples were given (for instance, 4-permutable rings form an Andrunakievich 3-variety of index 3, autodistributive rings form an Andrunakievich 2-variety of index $k \leq 6$).

For a ring A we define

$$A^{(s,1)} = A \quad \text{and} \quad A^{(s,k)} = (A^{(s,k-1)})^s$$

for $k = 2, 3, \ldots$.

PROPOSITION 5.3.9. (i) $A^{(s,k)} \subseteq A^{(k)}$ *for* $k = 1, 2, \ldots$.

(ii) *If* $K \subseteq A$ *and* $k > 0$, *then there exists an integer* $t > 0$ *such that* $K^{(t)} \subseteq A^{(s,k)}$.

(iii) *If a universal class* $\mathbb{A}$ *satisfies* (s) *and* $I \triangleleft A \in \mathbb{A}$, *then* $I^{(s,k)} \triangleleft A$ *for every positive integer* k.

Proof: For $k = 1$ we have $A^{(s,1)} = A = A^{(1)}$. Assuming $A^{(s,k-1)} \subseteq A^{(k-1)}$ for $k > 1$ we get

$$A^{(s,k)} = (A^{(s,k-1)})^s = \sum_{i=1}^{s-1}(A^{(s,k-1)})^i(A^{(s,k-1)})^{s-i} \subseteq$$
$$\subseteq A^{(s,k-1)}A^{(s,k-1)} \subseteq A^{(k-1)}A^{(k-1)} = A^{(k)}.$$

(ii) For $k = 1$ we may choose $t = 1$ getting

$$K^{(1)} = K \subseteq A = A^{(s,1)}.$$

Assume that $K^{(p)} \subseteq A^{(s,k-1)}$ for some $p > 0$, and choose an integer $q > 0$ such that $s \leq 2^{q-1}$. As already mentioned, $A^s \subseteq A^t$ for $t \leq s$, so we have

$$(K^{(p)})^{(q)} \subseteq (K^{(p)})^{2^{q-1}} \subseteq (K^{(p)})^s \subseteq (A^{(s,k-1)})^s = A^{(s,k)}.$$

As one easily sees, $(K^{(p)})^{(q)} = K^{(p+q-1)}$. Hence for $t = p + q - 1$ we get

$$K^{(t)} = (K^{(p)})^{(q)} \subseteq A^{(s,k)}.$$

(iii) If $k = 1$, then $I^{(s,1)} = I \triangleleft A$. Supposing the assertion for $k - 1$, and using condition (s) we get

$$I^{(s,k)} = (I^{(s,k-1)})^s \triangleleft A.$$

□

PROPOSITION 5.3.10. *An Andrunakievich s-variety* $\mathbb{A}$ *of index* k *satisfies conditions* (B1), (B2), (B3) *and the integers* m, n, t *depend neither on* $a \in A$ *nor on* A *but only on the index* k *of* $\mathbb{A}$.

Proof: Let $K \triangleleft I \triangleleft A \in \mathbb{A}$ and set $L = (K)_A$. Now by (A) we have $L^{(k)} \subseteq K$. From the inclusion

$$(aK + Ka + K)_I \subseteq (K)_A = L$$

we infer that

$$(aK + Ka + K)_I^{(k)} \subseteq L^{(k)} \subseteq K,$$

and hence (B2) is satisfied with $n = k$.

By $K \subseteq L$ Proposition 5.3.9 (ii) yields the existence of a $t > 0$ with $K^{(t)} \subseteq L^{(s,k+1)}$, and so Proposition 5.3.9 (i) implies that

$$K^{(t)} \subseteq L^{(s,k+1)} \subseteq L^{(k+1)} = L^{(k)} L^{(k)} \subseteq K^2.$$

Proposition 5.3.9 (iii) gives $L^{(s,k+1)} \triangleleft A$, so as $K^{(t)} \subseteq L^{(s,k+1)}$ we get

$$(K^{(t)})_I \subseteq L^{(s,k+1)} \subseteq K^2,$$

that is,

$$(K^{(t)})_I \subseteq K^2$$

holds which is just (B3), and t depends only on the index k.

Using again $L^{(s,k+1)} \triangleleft A$, and the containment $K^{(t)} \subseteq L^{(s,k+1)}$, we get

$$aK^{(t)} + K^{(t)}a \subseteq L^{(s,k+1)} \subseteq K^2 \subseteq K,$$

and therefore (B1) is fulfilled with $m = t$. □

We end this section with some remarks. Beidar [3] proved his results also for not necessarily associative *algebras* over a commutative ring R with unity element without extra effort. He has shown that Jordan R-algebras with $1/2 \in R$ and (γ, δ)-algebras with $1/6 \in R$ satisfy conditions (B1), (B2), (B3), though Jordan algebras do not satisfy condition (A), but only (s) with $s = 3$ (cf. Slin'ko [1], Medved'ev [1], Beidar [5]).

5.4. On the radical theory of associative rings with involution

An associative ring A is said to be a *ring with involution*, if on the ring A there is defined also a unary operation $*$, called involution, subject to the familiar identities

$$(x + y)^* = x^* + y^*$$

Special Features of the General Radical Theory

$$(xy)^* = y^*x^*$$
$$x^{**} = x$$

for all $x, y \in A$. We shall write A^* if we wish to emphasize that A is a *ring with involution* $*$. Rings with involution are very common, and in fact the most important and most appreciated rings as, for instance, complex numbers, Gaussian integers, real and complex matrix rings, are endowed with a natural involution. Besides, on every commutative ring the identical mapping trivially induces an involution. Because of their importance, rings with involution deserve a radical theory. In this section we shall deal with some special features of this theory.

Considering and working in the variety or category of rings with involution, it is natural and consistent to demand that all homomorphisms preserve also involution besides the ring operations. Thus we discard many ring homomorphisms, exactly those which do not preserve involution. Consequently we may consider only subrings which are closed under involution. Such subrings, in particular ideals, will be referred to as *-subrings* and *-ideals*. The fact that I is a $*$-ideal of A will be denoted by $I \triangleleft^* A$.

EXAMPLE 5.4.1. Let A be any ring without involution, and let A^{op} denote its opposite ring. On the direct sum $B = A \oplus A^{\mathrm{op}}$ one may define the *exchange involution* $*$ by

$$(x, y)^* = (y, x) \qquad \forall\, (x, y) \in B.$$

If A happens to be simple with $A^2 \neq 0$, then B has no other $*$-ideals than 0 and itself, that is, B is $*$-*simple*, but of course, not simple as a ring.

A ring theorist may expect an even better radical theory for rings with involution than for associative rings, because one has more calculation rules. Nonetheless, a universal algebraist would warn that working in another category, the situation may change drastically. And he is right! An impact of discarding some ring homomorphisms is that, for instance, the ADS-Theorem is no longer valid for rings with involution, as exhibited by Salavová [1]. In the sequel we shall work *in the category of all associative rings with involution and all homomorphisms preserving also involutions*. Thus, in defining radical classes and semisimple classes, we must consider only $*$-ideals and involution preserving homomorphisms. We say that a radical γ of rings with involution has the *ADS-property*, if $\gamma(I) \triangleleft^* A$ for every $I \triangleleft^* A$. The main objective of this section is to prove easily testable criteria for a radical to have the ADS-property.

Having a ring A with zero-multiplication and involution $*$, it is straightforward to see that also the unary operations $-* \colon x \mapsto x^{-*} = -x^*$ and id $\colon x \mapsto x^{\mathrm{id}} = x$ are involutions on A.

THEOREM 5.4.2 (Loi and Wiegandt [2]). *For a radical γ of associative rings with involution the following conditions are equivalent:*

(i) γ has the ADS-property;
(ii) $A^{\text{id}} \in \gamma$ if and only if $A^{-\text{id}} \in \gamma$ whenever $A^2 = 0$;
(iii) if $A^* \in \gamma$ and $A^2 = 0$, then $A^\circ \in \gamma$ for any involution $\circ$ built on A;
(iv) if $A^* \in \gamma$ and $A^2 = 0$, then $A^{-*} \in \gamma$.

The proof will be decomposed into several lemmas and propositions.

Let A be a ring such that $A^2 = 0$, and let us consider the additive group
$$E^+ = \mathbb{Z}^+ \oplus \mathbb{Z}^+ \oplus A^+ \oplus A^+.$$
On E^+ we define a multiplication and an involution $*$ by the following rules:
$$(m, n, x, y)(k, l, u, v) = (mk - nl, ml + nk, mu - nv + kx - ly, mv + nu + ky + lx)$$
$$(m, n, x, y)^* = (m, -n, x, -y),$$
and we denote by E^* the structure obtained in this way.

LEMMA 5.4.3. E^* is a ring with involution and the subset $I = (0, 0, A, A)$ is a $*$-ideal of E^*. The subsets $K = (0, 0, A, 0)$ and $L = (0, 0, 0, A)$ are $*$-ideals of I, $K^* \cong A^{\text{id}}$, $L^* \cong A^{-\text{id}}$ and $I^* = K^* \oplus L^*$. Neither K nor L is an ideal of E.

Proof: A tedious verification confirms that E^* is a ring with involution. The further assertions are straightforward. For the last statements we notice that
$$(0, 1, 0, 0)(0, 0, x, 0) = (0, 0, 0, x)$$
and
$$(0, 1, 0, 0)(0, 0, 0, y) = (0, 0, y, 0)$$
yield $EK \not\subseteq K$ and $EL \not\subseteq L$. □

PROPOSITION 5.4.4. Condition (i) implies condition (ii).

Proof: Suppose that (ii) is not satisfied, that is, there exists a ring A such that $A^2 = 0$ and either $A^{\text{id}} \in \gamma$, $A^{-\text{id}} \notin \gamma$ or $A^{-\text{id}} \in \gamma$, $A^{\text{id}} \notin \gamma$. Obviously all ideals of A are closed under both involutions id and $-$id. Hence without loss of generality we may confine ourselves to the cases $A^{\text{id}} \in \gamma$, $A^{-\text{id}} \in \mathcal{S}\gamma$ and $A^{-\text{id}} \in \gamma$, $A^{\text{id}} \in \mathcal{S}\gamma$. In the first case by Lemma 5.4.3 we have $I^* \cong A^{\text{id}} \oplus A^{-\text{id}}$ and so $\gamma(I^*) = K^* \cong A^{\text{id}}$. Moreover, again by Lemma 5.4.3 $\gamma(I^*)$ is not an ideal of E, although $I \triangleleft^* E$. In the second case a similar argument yields $\gamma(I^*) = L^* \cong A^{-\text{id}}$, and that $\gamma(I^*)$ is not an ideal of E. Hence γ does not possess the ADS-property, contradicting (i). □

PROPOSITION 5.4.5. Condition (ii) implies condition (iii).

Proof: Let us assume that $A^* \in \gamma$ and $A^2 = 0$. Firstly we show that $A^{\text{id}} \in \gamma$. The set
$$D = \{x + x^* \mid x \in A\}$$

is clearly a $*$-ideal of A, moreover, $D^* = D^{\mathrm{id}}$ holds. The mapping

$$g\colon A^* \to D^*$$

defined by $g(x) = x + x^*$, $x \in A$, is obviously a homomorphism onto D^*, and as $A^* \in \gamma$ it follows that $D^{\mathrm{id}} = D^* \in \gamma$. In the factor ring $(A/D)^*$ the relations

$$x + D = -x^* + D$$

and

$$(x + D)^* = (-x^* + D)^* = -x + D$$

are valid. Hence we have

$$(A/D)^{-\mathrm{id}} = (A/D)^* = A^*/D^* \in \gamma.$$

Applying (ii) we get
$$A^{\mathrm{id}}/D^{\mathrm{id}} = (A/D)^{\mathrm{id}} \in \gamma.$$

Since γ is closed under extensions, by $D^{\mathrm{id}} \in \gamma$ and $A^{\mathrm{id}}/D^{\mathrm{id}} \in \gamma$ we conclude that $A^{\mathrm{id}} \in \gamma$.

An application of (ii) yields also $A^{-\mathrm{id}} \in \gamma$. Let $\circ$ denote an arbitrary involution on A, and let us consider the set

$$C = \{x + x^\circ \mid x \in A\}.$$

As above, we get that $C^{-\circ} = C^{-\mathrm{id}}$ is a homomorphic image of $A^{-\mathrm{id}} \in \gamma$, whence $C^{-\mathrm{id}} \in \gamma$ follows and so in view of (ii) we conclude $C^{\mathrm{id}} \in \gamma$. Moreover, we have also

$$A^\circ/C^{\mathrm{id}} = A^\circ/C^\circ = (A/C)^{-\mathrm{id}} = A^{-\mathrm{id}}/C^{-\mathrm{id}} \in \gamma,$$

and therefore we get $A^\circ \in \gamma$ because γ is closed under extensions. Thus the validity of (iii) has been proved. $\square$

Let A be a ring with zero-multiplication and involution $*$. As one readily sees, on the direct sum $B = A \oplus A$ one can define an involution $\square$ by the rule

$$(x, y)^\square = (y^*, x^*) \qquad \forall\, (x, y) \in B.$$

LEMMA 5.4.6. *Let γ be a radical of rings with involution such that γ fulfils condition (iii). Further, let I be a $*$-ideal of a ring A^* with involution, and put $L = \gamma(I^*)$. Then the following hold:*

a) $(L^2/L^3)^* \in \gamma$,

for every element $a \in A$

b) $aL^2 a^* \subseteq L$,

c) $aL + La^* + L \vartriangleleft^* I$,

d) *the mapping*

$$f: (L/L^2 \oplus L/L^2)^\square \longrightarrow ((aL + La^* + L)/L)^*$$

defined by $f(x+L^2, y+L^2) = ax + ya^* + L$ *is a homomorphism onto* $((aL + La^* + L)/L)^*$.

Proof: a) Let us consider an element $a \in L$ and the mapping

$$g_a: L/L^2 \to L^2/L^3$$

defined by $g_a(x+L^2) = ax + L^3$ for all $x \in L$. The mapping g_a is obviously a homomorphism of $(L/L^2)^{\mathrm{id}}$ to $(L^2/L^3)^{\mathrm{id}}$, and $g_a(L/L^2)^{\mathrm{id}} \triangleleft (L^2/L^3)^{\mathrm{id}}$. Since $L^* \in \gamma$, condition (iii) implies that $(L/L^2)^{\mathrm{id}} \in \gamma$. Thus also $g_a(L/L^2)^{\mathrm{id}} \in \gamma$. Hence by

$$(L^2/L^3)^{\mathrm{id}} = \sum_{a \in L} g_a(L/L^2)^{\mathrm{id}} \subseteq \gamma((L^2/L^3)^{\mathrm{id}}) \subseteq (L^2/L^3)^{\mathrm{id}}$$

we have that $(L^2/L^3)^{\mathrm{id}} \in \gamma$, and so condition (iii) ensures that $(L^2/L^3)^* \in \gamma$.

b) Let us define a mapping

$$h: (L^2/L^3)^* \to ((aL^2a^* + L)/L)^*$$

by $h(\Sigma x_i y_i + L^3) = a(\Sigma x_i y_i)a^* + L$. It is easy to check that h is a homomorphism onto $((aL^2a^* + L)/L)^*$. As $(L^2/L^3)^* \in \gamma$, which has already been proved, we get

$$((aL^2a^* + L)/L)^* \in \gamma.$$

Furthermore,

$$I(aL^2a^*) = (Ia)(L^2a^*) \subseteq IL^2a^* \subseteq L^2a^* = L(La^*) \subseteq LI \subseteq L$$

and similarly $(aL^2a^*)I \subseteq L$. Thus $(aL^2a^* + L)/L \triangleleft I/L$, and a moment's reflection reveals that $(aL^2a^* + L)/L \triangleleft^* I^*/L^*$. Hence we have

$$((aL^2a^* + L)/L)^* \subseteq \gamma(I^*/L^*) = \gamma(I^*/\gamma(I^*)) = 0.$$

Thus $aL^2a^* \subseteq L$ holds, proving b).

c) One can easily verify that

$$(aL + La^* + L) \triangleleft^* I.$$

Statement d) is a straightforward calculation. □

Lemma 5.4.6 is crucial in proving the next implication.

PROPOSITION 5.4.7. *Condition* (iii) *implies condition* (i).

Proof: Let $I \triangleleft^* A$ be arbitrary and set $L = \gamma(I)$. Lemma 5.4.6 d) tells us that $((aL + La^* + L)/L)^*$ is a homomorphic image of $(L/L^2 \oplus L/L^2)^\square$ which is in γ in view of condition (iii) and by $L = \gamma(I^*) \in \gamma$. Hence we get $((aL + La^* + L)/L)^* \in \gamma$. Applying Lemma 5.4.6 c) gives us

$$(aL + La^* + L)/L \triangleleft^* I/L = I/\gamma(I) \in \mathcal{S}\gamma.$$

Thus we get
$$(aL + La^* + L)/L \subseteq \gamma(I/L) = 0,$$

that is, $aL + La^* + L \subseteq L$. Hence

$$ax = ax + 0a^* \in L \quad \text{and} \quad xa^* = a0 + xa^* \in L$$

for all $x \in L$. Since the choice of $a \in A$ was arbitrary, we thus have $AL \cup LA^* \subseteq L$, implying $\gamma(I) = L \triangleleft^* A$. $\square$

Proof of Theorem 5.4.2: Propositions 5.4.4, 5.4.5 and 5.4.7 yield the equivalence of conditions (i), (ii) and (iii). The implications (iii) $\Longrightarrow$ (iv) $\Longrightarrow$ (ii) are trivial. $\square$

Defining hypernilpotency and hypoidempotency analogously for rings with involution, it is an immediate consequence of Theorem 5.4.2 that *every hypernilpotent and every hypoidempotent radical of rings with involution has the ADS-property*.

In Loi and Wiegandt [2] example of a non-hereditary semisimple class of rings with involution has been given.

Given a radical γ of associative rings the question arises whether γ induces a radical for rings with involution just by designating $\gamma(A)$ as a candidate for the radical of the ring A with involution. Obviously the ring radical γ induces a radical for rings with involution if and only if $\gamma(A)$ is closed under involution for all rings A with involution.

Next we shall give easily testable criteria for a ring radical γ to induce a radical for rings with involution.

THEOREM 5.4.8 (Lee and Wiegandt [1]). *Let γ be a radical of associative rings. The following three conditions are equivalent:*
  (i) *$\gamma(A)$ is closed under involution for every ring A with involution,*
  (ii) *$A \in \gamma$ implies $A^{\mathrm{op}} \in \gamma$ for every ring A,*
  (iii) *$A \in \mathcal{S}\gamma$ implies $A^{\mathrm{op}} \in \mathcal{S}\gamma$ for every ring A.*

Proof: (i) $\Longrightarrow$ (ii). Given a ring $A \in \gamma$, let us consider the direct sum $B = A \oplus A^{\mathrm{op}}$ endowed with the exchange involution $*$ (cf. Example 5.4.1). Since $A = \gamma(A) \triangleleft B$, it follows that $A \subseteq \gamma(B)$. Hence (i) yields $A^{\mathrm{op}} \subseteq (\gamma(B))^* = \gamma(B)$. Thus $B = A \oplus A^{\mathrm{op}} \subseteq \gamma(B)$ holds, implying $B \in \gamma$. Since A^{op} is a homomorphic image of B, we conclude $A^{\mathrm{op}} \in \gamma$, proving (ii).

(ii) $\implies$ (iii). Suppose that $A \in \mathcal{S}\gamma$. Now we have $\gamma(A^{\mathrm{op}}) \in \gamma$, and so (ii) implies $(\gamma(A^{\mathrm{op}}))^{\mathrm{op}} \in \gamma$. Since $(\gamma(A^{\mathrm{op}}))^{\mathrm{op}} \triangleleft A$, it follows that

$$(\gamma(A^{\mathrm{op}}))^{\mathrm{op}} \subseteq \gamma(A) = 0,$$

which implies $A^{\mathrm{op}} \in \mathcal{S}\gamma$.

(iii) $\implies$ (i). Let A be a ring with involution $*$. For a subset S of A, let $S^{(*)}$ denote the involutive image of S. The correspondence

$$x + (\gamma(A))^{(*)} \mapsto x^* + \gamma(A) \qquad \forall \, x \in A$$

establishes an isomorphism between $A/\gamma(A)^{(*)}$ and $(A/\gamma(A))^{\mathrm{op}}$, as one can readily check. Since $A/\gamma(A) \in \mathcal{S}\gamma$, condition (iii) implies $(A/\gamma(A))^{\mathrm{op}} \in \gamma$. Hence also $A/\gamma(A)^{(*)} \in \mathcal{S}\gamma$, which implies $\gamma(A) \subseteq \gamma(A)^{(*)}$, whence

$$(\gamma(A))^{(*)} \subseteq (\gamma(A))^{(*)(*)} = \gamma(A).$$

Thus $\gamma(A)$ is closed under involution. $\square$

COROLLARY 5.4.9. *Let ϱ be a regular class of associative rings. If ϱ satisfies condition*

(iv) $A \in \varrho$ *implies* $A^{\mathrm{op}} \in \mathcal{SU}\varrho$,

then the upper radical $\mathcal{U}\varrho$ satisfies condition (i) *of Theorem 5.4.8.*

Proof: Let B be a ring such that $B^{\mathrm{op}} \in \mathcal{SU}\varrho$. Assume that $B \notin \mathcal{SU}\varrho$. Then its $\mathcal{U}\varrho$-radical is a nonzero ideal I of B. Since $B^{\mathrm{op}} \in \mathcal{SU}\varrho$, the ideal I^{op} of B^{op} has a nonzero homomorphic image C^{op} in ϱ. Applying (iv), we get $C \in \mathcal{SU}\varrho$. But the ring C is a homomorphic image of I which contradicts $I \in \mathcal{U}\varrho$. Thus $B^{\mathrm{op}} \in \mathcal{SU}\varrho$ implies $B \in \mathcal{SU}\varrho$, and so Theorem 5.4.8 (iii) is satisfied. $\square$

COROLLARY 5.4.10. *If a homomorphically closed class δ of associative rings satisfies condition* (ii) *of Theorem 5.4.8, then the lower radical $\mathcal{L}\delta$ satisfies condition* (i) *of Theorem 5.4.8.*

Proof: Let $B \neq 0$ be a ring such that $B^{\mathrm{op}} \in \mathcal{L}\delta$. Suppose that $B \notin \mathcal{L}\delta$. Then from the Yu-lee Lee radical construction (see Theorem 3.3.2 and Corollary 3.3.4) we know that B has a nonzero homomorphic image C which has no nonzero accessible subring in δ. But then C^{op} is a nonzero homomorphic image of $B^{\mathrm{op}} \in \mathcal{L}\delta$, and therefore C^{op} has a nonzero accessible subring D^{op} in δ. Applying condition (ii), we get $0 \neq D \in \delta$, and D is an accessible subring of C. This contradiction proves $B \in \mathcal{L}\delta$. $\square$

We can harvest on Theorem 5.4.8 and its Corollaries, obtaining that every left-right symmetric radical of associative rings induces a radical for rings with involution. Thus with the exception of left (and right) strongly prime radical all the other *concrete* radicals treated in this book induce radicals for rings with involution.

Special Features of the General Radical Theory   325

For further results on radicals of rings with involution we refer to the papers of Aburawash [4], Booth and Groenewald [2], [4], and Loi [3], [4], [6], [7].

Recently also the structure theory of rings with involution has been developed. In this theory *-biideals play a dominant role (here one-sided ideals make no sense). A *-*biideal* B is a biideal (cf. the end of section 3.18) of a ring A with involution such that $B^* \subseteq B$. Imposing d.c.c. on *-biideals is a strong requirement, for instance, if a ring A with involution has d.c.c. on *-biideals, then A is an artinian ring and its Jacobson radical $\mathcal{J}(A)$ satisfies d.c.c. on subgroups, and hence $\mathcal{J}(A)$ is an artinian ring (cf. Beidar and Wiegandt [3]). For details the reader is referred to the works of Aburawash, [2], [3], [5], [6], Domokos [1], Loi [5], Loi and Wiegandt [3], and to the survey paper of Aburawash [7] and Wiegandt [7].

5.5. On the radical theory of near-rings

An algebraic structure N endowed with an addition + and multiplication · is called a *right near-ring*, if N is a not necessarily commutative group with respect to addition and a semigroup with respect to multiplication, and these operations are linked with the right distributive law:

$$(x+y)z = xz + yz \qquad \forall\ x, y, z \in N.$$

In the sequel we shall write just *near-ring* for right near-ring. Thus a near-ring is "nearly" a ring, but addition need not be commutative and left distributivity is missing. Near-rings have strong and natural links with groups. But anyone who thinks that near-ring theory is nearly ring theory, is absolutely mistaken. Their theory started with investigations of L. E. Dickson (1905), H. Wielandt and H. Zassenhaus (in the 30's) and has become a well established flourishing branch of algebraic researches (cf. Clay [1], Meldrum [1], Pilz [1]) with many applications. Sometimes, however, there is an analogy with ring theory inasmuch as many ring theoretic notions, such as ideal, primeness, radical, etc, make sense also for near-rings.

There are two very natural and fundamental examples of near-rings:

i) The set $M(G)$ of all mappings of an additive group G into itself forms a near-ring with operations

$$(f+g)(x) = f(x) + g(x)$$

and

$$(f \circ g)(x) = f(g(x))$$

for all $f, g \in M(G)$ and $x \in G$.

ii) Polynomials of one indeterminate over a commutative ring with unity element form a near-ring with respect to addition + and substitution ∘.

The axioms imposed on near-rings imply

$$0x = 0 \quad \forall\, x \in N,$$

in view of $0x = (0+0)x = 0x + 0x$. Nevertheless, $x0 \neq 0$ is possible in a near-ring. If, however, $x0 = 0$ for all $x \in N$, then we talk about a *0-symmetric near-ring*. A *constant near-ring* is a near-ring such that $xy = x$ holds for all $x, y \in N$. Every near-ring N is the sum of two subnear-rings

$$N_0 = \{x \in N \mid x0 = 0\}$$

and

$$N_c = \{x \in N \mid xy = x,\ \forall y \in N\} = \{x \in N \mid x0 = x\},$$

as every element $x \in N$ decomposes uniquely into the sum

$$x = (x - x0) + x0$$

where $(x - x0) \in N_0$ and $x0 \in N_c$. Clearly $N_0 \cap N_c = 0$. Notice that N_c, called the *constant part* of N, is a *right invariant subnear-ring*, that is, $N_c N \subseteq N_c$ as well as a *left invariant subnear-ring*, that is $NN_c \subseteq N_c$, but it need not be a normal subgroup.

A subnear-ring I of a near-ring N is called an *ideal* of N if

i) I^+ is a normal subgroup of N^+,

ii) I is right invariant: $IN \subseteq I$,

iii) $x(y + i) - xy \in I \quad \forall x, y \in N, \forall i \in I$.

If I is an ideal of a 0-symmetric near-ring, then I is left invariant, although a left and right invariant normal subgroup of a 0-symmetric near-ring need not be an ideal. The subnear-ring N_0, called the *0-symmetric part* of N, is a *left ideal* of N, that is, it satisfies i) and iii), but not necessarily ii).

The ideals of a near-ring are just the kernels of near-ring homomorphisms. Having a one-to-one correspondence between surjective homomorphisms and ideals, and in the possession of the isomorphism theorems, the basic facts of the general radical theory of nonassociative rings carry over to near-rings (cf. Chapter II). In the development of structural theory for near-rings in the 60's, concrete radical counterparts of established concrete ring radicals were introduced and investigated. The study of the general radical theory of near-rings started only in the 80's, and brought unexpectedly interesting results. A sort of degeneracy of the near-ring radical theory was anticipated. Indeed, it degenerates but in a nice way. The main features of this kind of degeneracy are that a nontrivial radical class with hereditary semisimple class contains all near-rings with zero-multiplication as well as all constant near-rings; further, a nontrivial radical class contains all near-rings with zero-multiplication if and only if its semisimple class is weakly homomorphically closed (cf. Theorem 3.6.11).

Before turning our attention to the special features of the general radical theory of near-rings we compile a brief account of concrete radicals.

In the sequel $\mathbb{N}$ will denote the variety of all near-rings and $\mathbb{N}_0$ that of all 0-symmetric near-rings.

In the variety $\mathbb{N}$ the subvarieties $\mathbb{N}_0$ and $\mathbb{N}_c$ of all constant near-rings are radical classes, although $\mathbb{N}_0(N) \neq N_0$ and $\mathbb{N}_c(N) \neq N_c$ whenever N_0 and N_c are not ideals of the near-ring N. Observe that both $\mathbb{N}_0$ and $\mathbb{N}_c$ are extension closed varieties, but no semisimple classes (cf. Theorem 2.3.9). These examples provide a nice contrast with Theorem 3.20.3.

The notion of primeness makes sense also for near-rings, so the Baer radical β of near-rings is defined as the upper radical of all prime near-rings, but β fails to be a Kurosh–Amitsur radical even in the variety $\mathbb{N}_0$ of 0-symmetric near-rings. Kaarli and Kriis [1] proved that $I \triangleleft N$ and $\beta(I) = I$ does not imply $I \subsetneq \beta(N)$.

Köthe's nil radical $\mathcal{N}$ is a hereditary Kurosh–Amitsur radical in the variety $\mathbb{N}_0$ but its semisimple class is not hereditary (Kaarli [2]).

Generalizing the notion of primitivity to near-rings one gets four variants which are called ν-primitivity, $\nu = 0, 1, 2, 3$. ν-primitive near-rings are defined via annihilators of various types of near-ring modules (called N-groups), which in the case of *rings* all coincide with the irreducibles. ν-primitivity gives rise to Jacobson type radicals $\mathcal{J}_\nu$, $\nu = 0, 1, 2, 3$, as upper radicals, and the strict containment

$$\mathcal{J}_0 \subset \mathcal{J}_1 \subset \mathcal{J}_2 \subset \mathcal{J}_3$$

holds among these radical classes in $\mathbb{N}_0$. However, $\mathcal{J}_0$ and $\mathcal{J}_1$ are not Kurosh–Amitsur radicals as exhibited by Kaarli [4]: there are 0-symmetric near-rings N and M such that $\mathcal{J}_0(\mathcal{J}_0(N)) \neq \mathcal{J}_0(N)$ and $\mathcal{J}_1(\mathcal{J}_1(M)) \neq \mathcal{J}_1(M)$. $\mathcal{J}_2$ and $\mathcal{J}_3$ are Kurosh–Amitsur radicals in the variety $\mathbb{N}_0$, moreover, both are *right invariantly strong*, that is $\mathcal{J}_2(N)$ and $\mathcal{J}_3(N)$ contain all right invariant subnear-rings which are in the class $\mathcal{J}_2$ and $\mathcal{J}_3$, respectively, and each of them has a hereditary semisimple class (cf. Anderson, Kaarli and Wiegandt [2], Holcombe [1], Holcombe and Walker [1], Kaarli [1]). There are two more Jacobson type radicals between $\mathcal{J}_0$ and $\mathcal{J}_1$. $\mathcal{J}_{1/2}$ was introduced by Betsch [2] (see also Pilz [1]). $\mathcal{J}_{1/2}(N)$ is not necessarily an ideal of N, but $\mathcal{J}_0(N)$ is the largest ideal of N contained in $\mathcal{J}_{1/2}(N)$, (Ramakotaiah [1]). The radical $\mathcal{J}_s$ was defined by Laxton [1] and $\mathcal{J}_s(N)$ lies between $\mathcal{J}_{1/2}(N)$ and $\mathcal{J}_1(N)$, (see also Hartney [1], [2]).

The Brown–McCoy radical $\mathcal{G}$, as the upper radical of all simple near-rings with unity element, is a Kurosh–Amitsur radical in the variety of $\mathbb{N}_0$ and it shares the good properties of its ring theoretical counterpart (cf. Anderson, Kaarli and Wiegandt [1]). It is a bit surprising that there exists a near-ring $N \in \mathbb{N}$ with unity element such that $\mathcal{J}_2(N) = N$, and so $\mathcal{J}_2 \nsubseteq \mathcal{G}$, as pointed out in the paper [1] of P. Fuchs, C. J. Maxson, A. P. J. van der Walt and K. Kaarli.

The notion of primeness can be interpreted for near-rings in a more subtle way as done by Booth, Groenewald and Veldsman [1]. Call a near-ring N *equiprime*, if for every nonzero element $a \in N$ and $x, y \in N$, $anx = any$ for all $n \in N$ implies $x = y$. For rings the notions of primeness and equiprimeness coincide. An equiprime near-ring is always 0-symmetric. The upper radical $\mathcal{E}$ of all equiprime near-rings is referred to as the *equiprime radical* which is a hereditary Kurosh–Amitsur radical with hereditary semisimple class in the variety $\mathbb{N}_0$. So equiprimeness seems to be the right definition of primeness for near-rings, although there is one shortcoming: the equiprime radical is not comparable with $\mathcal{J}_2$.

Notice that all the positive results of the above mentioned concrete radicals are valid for the variety $\mathbb{N}_0$ of 0-symmetric near-rings. The question arises naturally: how to extend a radical from the variety of all 0-symmetric near-rings to that of all near-rings? There may be several possibilities. There is always an easy way: if γ is a radical class in the variety $\mathbb{N}_0$, then γ is a radical class also in the variety $\mathbb{N}$ of all near-rings (it is crucial here that the extension of a 0-symmetric near-ring by a 0-symmetric near-ring remains a 0-symmetric one). Nevertheless, in the variety $\mathbb{N}$ the semisimple class of γ becomes too big and so no description of the semisimple near-rings can be expected. To illustrate other possibilities, we consider generalizations of the Brown–McCoy radical:

$$\mathcal{G} = \left\{ N \in \mathbb{N}_0 \;\middle|\; \begin{array}{l} \text{no nonzero homomorphic image of} \\ N \text{ is a simple near-ring with unity} \end{array} \right\},$$

$$\mathcal{G}_1 = \left\{ N \in \mathbb{N} \;\middle|\; \begin{array}{l} \text{no nonzero homomorphic image of} \\ N \text{ is a simple near-ring with unity} \end{array} \right\},$$

$$\mathcal{G}_2 = \left\{ N \in \mathbb{N} \;\middle|\; \begin{array}{l} \text{no nonzero homomorphic image of } N \text{ is a} \\ \text{0-symmetric simple near-ring with unity} \end{array} \right\}.$$

$\mathcal{G}$ is a Kurosh–Amitsur radical class in $\mathbb{N}_0$ as well as in $\mathbb{N}$. Since $\mathcal{G}_1$ and $\mathcal{G}_2$ are upper radical classes of simple near-rings, also they are Kurosh–Amitsur radicals. Since $\mathcal{G}_1$ contains non-0-symmetric near-rings, and $\mathcal{G}_2$ contains non-0-symmetric simple near-rings with unity, the containments

$$\mathcal{G} \subset \mathcal{G}_1 \subset \mathcal{G}_2$$

are proper.

After this brief summary on concrete near-ring radicals we are going to discuss some special features of the general radical theory of near-rings. Unless specifically mentioned otherwise, *all considerations will take place in the variety $\mathbb{N}$ of all near-rings*.

To attack the problem of hereditariness of semisimple classes of near-rings via the ADS-Theorem, is a hopeless enterprise because in proving the ADS-Theorem one uses *in concerto* the associativity of multiplication, both left and right distributivity and also the commutativity of addition.

Special Features of the General Radical Theory

A method for proving specific results in the general radical theory of near-rings is to employ near-ring constructions the properties of which force some near-rings into the radical or semisimple class considered.

As for rings, in the sequel N^0 will denote the near-ring on the group N^+ with zero-multiplication.

LEMMA 5.5.1. *Let N and M be near-rings, and let us define a multiplication on the additive group $N^+ \oplus N^+ \oplus M^+$ by the rule*

$$(x, y, m)(u, v, n) = \begin{cases} (y, 0, 0) & \text{if } n \neq 0 \\ (0, 0, 0) & \text{if } n = 0 \end{cases}$$

for all $x, y, u, v \in N$ and $m, n \in M$. Then we obtain a 0-symmetric near-ring $\Sigma = \Sigma(N, M)$ such that
   i) $(N, 0, M) \triangleleft \Sigma$ *and* $\Sigma/(N, O, M) \cong N^0$,
   ii) $M^0 \cong (0, 0, M) \triangleleft (N, 0, M)$ *and* $(N, 0, M)/(0, 0, M) \cong N^0$.

A straightforward verification gives the proof. □

PROPOSITION 5.5.2. γ *be a radical of near-rings with hereditary semisimple class σ. If there exists a near-ring $N = N^0 \neq 0$ such that $N \in \sigma$, then σ contains all near-rings with zero-multiplication.*

Proof: Let $M = M^0$ be any near-ring with zero-multiplication, and let us construct the near-ring $\Sigma = \Sigma(N, M)$ of Lemma 5.5.1. Then we have

$$(0, 0, M) \cong M^0 = M,$$

and we want to show that $(0, 0, M) \in \sigma$. Again by Lemma 5.5.1

$$\Sigma/(N, 0, M) \cong N^0 \in \sigma,$$

and therefore $\gamma(\Sigma) \subseteq (N, 0, M)$ follows, implying that $\gamma(\Sigma) \subseteq \gamma(N, 0, M)$. Similarly, we get $\gamma(N, 0, M) \subseteq \gamma(0, 0, M)$. Thus $\gamma(\Sigma) \subseteq (0, 0, M)$. Hence every element of $\gamma(\Sigma)$ has the form $(0, 0, m)$, $m \in M$. Since $\gamma(\Sigma) \triangleleft \Sigma$, for the elements $(0, y, 0), (0, 0, 0) \in \Sigma$ with $y \neq 0$ and for any $(0, 0, m) \in \gamma(\Sigma)$ we have

$$(0, y, 0)((0, 0, 0) + (0, 0, m)) - (0, y, 0)(0, 0, 0) \in \gamma(\Sigma).$$

Since

$$(0, y, 0)((0, 0, 0) + (0, 0, m)) - (0, y, 0)(0, 0, 0) = \begin{cases} (y, 0, 0) & \text{if } m \neq 0 \\ (0, 0, 0) & \text{if } m = 0, \end{cases}$$

we conclude that $m = 0$, whence $\gamma(\Sigma) = 0$, that is, $\Sigma \in \sigma$. Since σ is hereditary, the relation

$$(0, 0, M) \triangleleft (N, 0, M) \triangleleft \Sigma$$

implies $M = M^0 \cong (0, 0, M) \in \sigma$. □

Consistently with section 3.11, we call a near-ring radical *hypernilpotent*, if it contains all near-rings with zero-multiplication. Mlitz and Oswald [1] proved that *a hypernilpotent near-ring radical contains also all nilpotent near-rings*.

LEMMA 5.5.3. *For any near-ring N, let a multiplication be defined on the additive group*

$$\theta^+(N) = N^+ \oplus N^+ \oplus N^+$$

by the rule

$$(x, y, z)(u, v, w) = \begin{cases} (xu, z, 0) & \text{if } u \neq 0 \text{ and } w \neq 0 \\ (xu, 0, 0) & \text{otherwise} \end{cases}$$

for all $x, y, z, u, v, w \in N$. Then $\theta = \theta(N)$ becomes a near-ring such that
  i) $(N, N, 0) \triangleleft \theta$ *and* $\theta/(N, N, 0) \cong N^0$,
  ii) $N \cong (N, 0, 0) \triangleleft (N, N, 0)$ *and* $(N, N, 0)/(N, 0, 0) \cong N^0$.
Moreover, θ is 0-symmetric if and only if so is N.

The proof is straightforward, so we omit it. □

The next theorem was proved by Betsch and Kaarli [1].

THEOREM 5.5.4. *If $\gamma \neq 0$ is a torsion class of near-rings, that is, γ is a radical of near-rings with hereditary semisimple class σ, then γ is hypernilpotent.*

Proof: Let us suppose that γ is not hypernilpotent. Then by virtue of Proposition 5.5.2 σ contains all near-rings with zero-multiplication. Let N be a near-ring such that $0 \neq N \in \gamma$. Now $N^0 \in \sigma$ holds. Construct the near-ring $\theta(N)$ of Lemma 5.5.3. Then by

$$\theta/(N, N, 0) \cong N^0 \in \sigma$$

it follows that $\gamma(\theta) \subseteq (N, N, 0)$, whence $\gamma(\theta) \subseteq \gamma(N, N, 0)$. Since σ is hereditary, also $\gamma(N, N, 0) \subseteq \gamma(\theta)$, so

$$\gamma(N, N, 0) = \gamma(\theta).$$

The relation
$$(N, N, 0)/(N, 0, 0) \cong N^0 \in \sigma$$
yields
$$\gamma(\theta) = \gamma(N, N, 0) \subseteq (N, 0, 0).$$
Further, $(N, 0, 0) \triangleleft (N, N, 0)$ and $(N, 0, 0) \cong N \in \gamma$ imply

$$(N, 0, 0) \subseteq \gamma(N, N, 0) = \gamma(\theta).$$

Thus
$$(N,0,0) = \gamma(\theta) \triangleleft \theta,$$
and therefore for any nonzero element $x \in N$ we have
$$(0,0,x)((0,0,x) + (x,0,0)) - (0,0,x)(0,0,x) \in \gamma(\theta).$$
On the other hand,
$$(0,0,x)((0,0,x) + (x,0,0)) - (0,0,x)(0,0,x) = (0,x,0).$$
Hence $(0,x,0) \in (N,0,0)$ so $x = 0$, a contradiction. Thus $\gamma = 0$ which case has been excluded in the theorem, and consequently γ must be hypernilpotent. □

Remarks: 1) Theorem 5.5.4 remains true also in the smaller variety $\mathbb{N}_0$, since the near-ring $\theta(N)$ of Lemma 5.5.3 is 0-symmetric whenever N is such a near-ring.

2) Theorem 5.5.4 is not reversible. As has been mentioned, the nil radical $\mathcal{N}$ has non-hereditary semisimple class even in $\mathbb{N}_0$.

We shall see that hereditary semisimple classes of near-rings consist of 0-symmetric near-rings. For this purpose we need an ingenious construction due to Barua [1]. Let K and L be near-rings. The cartesian power K^L of L copies of K can be viewed as the set of all functions $L \to K$. Then K^L is a near-ring with respect to the componentwise operations
$$(f+g)(x) = f(x) + g(x)$$
and
$$(f \cdot g)(x) = f(x) \cdot g(x)$$
for all $x \in L$ and $f, g \in K^L$.

Let α be a mapping designating to each $x \in L$ an element α_x of $M((K^L)^+)$, (see i) at the beginning of this section), such that $f^x(y) = f(xy)$, where $f^x = \alpha_x(f)$, for all $x, y \in L$. It is straightforward to verify that α_x is a near-ring endomorphism of K^L. It is routine to check that a near-ring $\Xi = \Xi(K, L)$ can be defined on the cartesian product $L \times K^L$ with addition
$$(x, f) + (y, g) = (x+y, f+g)$$
and multiplication
$$(x, f) \cdot (y, g) = (xy, f^y g)$$
for all $x, y \in L$ and $f, g \in K^L$. The near-ring $\Xi(K, L)$ is called the *wreath-sum* of K with L.

LEMMA 5.5.5. *The mapping*
$$\varphi \colon (0, K^L) \to K$$

defined by $\varphi(0,f) = f(0)$ is a near-ring homomorphism onto K. Moreover,
$$\ker \varphi \lhd (0, K^L) \lhd \Xi,$$
$$\Xi/(0, K^L) \cong L \quad \text{and} \quad (0, K^L)/\ker \varphi \cong K,$$
$\ker \varphi$ is a left ideal in Ξ and $(\ker \varphi)^2 = 0$. $\ker \varphi \lhd \Xi$ if and only if $k(x0)f(0) = 0$ for all $k \in \ker \varphi$, $x \in L$ and $f \in K^L$.

Proof: φ is clearly an additive homomorphism. Furthermore,
$$\varphi((0,f)(0,g)) = \varphi(0, f^0 g) = (f^0 g)(0) =$$
$$= f^0(0)g(0) = f(0)g(0) = \varphi(0,f)\varphi(0,g).$$

Let $k \in K$, and define $f: L \to K$ by $f(x) = k$ for all $x \in L$. Then $(0, f) \in (0, K^L)$ and $\varphi(0, f) = f(0) = k$. Hence φ is a near-ring homomorphism onto K and $(0, K^L)/\ker \varphi \cong K$ where
$$\ker \varphi = \{(0, k) \in (0, K^L) \mid k(0) = 0\}.$$

Since for any $k_1, k_2 \in \ker \varphi$ we have
$$(k_1^0 k_2)(x) = k_1^0(x) k_2(x) = k_1(0x) k_2(x) = k_1(0) k_2(x) = 0 k_2(x) = 0 = 0(x)$$
for all $x \in L$, we get
$$(0, k_1)(0, k_2) = (0, k_1^0 k_2) = 0,$$
proving that $(\ker \varphi)^2 = 0$.

It is also routine to check that
$$(0, K^L) \lhd \Xi \quad \text{and} \quad \Xi/(0, K^L) \cong L,$$
and that $\ker \varphi$ is a left ideal in Ξ.

Finally, $\ker \varphi \lhd \Xi$ if and only if
$$(0,k)(x,f) = (0, k^x f) \in \ker \varphi,$$
that is
$$0 = (k^x f)(0) = k^x(0) f(0) = k(x0)f(0)$$
for all $(0, k) \in \ker \varphi$, $f \in K^L$ and $x \in L$. □

LEMMA 5.5.6. *Let K be a near-ring with $K^2 \neq 0$. If L is a non-0-symmetric near-ring, then $\ker \varphi$ is not an ideal in $\Xi(K, L)$.*

Proof: Let $s, t \in K$ with $st \neq 0$. Define mappings k and f in K^L by
$$k(x) = \begin{cases} s & \text{if } x \neq 0 \\ 0 & \text{if } x = 0 \end{cases}$$

and $f(x) = t$ for all $x \in L$. Then $(0,k) \in \ker \varphi$, and for any $0 \neq c \in L_c$, (L_c is the constant part of L), we get that

$$k(c0)f(0) = k(c)f(0) = st \neq 0.$$

Hence $\ker \varphi$ is not an ideal in Ξ by Lemma 5.5.5. □

THEOREM 5.5.7 (Veldsman [15]). *Let $\sigma_1 \neq \mathbb{N}$ be a semisimple class in the variety $\mathbb{N}$ of all near-rings. If σ is hereditary, then σ consists entirely of 0-symmetric near-rings.*

Proof: If $\sigma = 0$, the statement is true. Thus, let us choose a near-ring $K \neq 0$ from σ, and suppose that $K_c \neq 0$. We are going to apply the wreath-sum $\Xi(K, L)$ in the case when $K = L$. Since $K_c \neq 0$, also $K^2 \neq 0$.

Let $\gamma = \mathcal{U}\sigma$ be the corresponding radical class. In view of $(\ker \varphi)^2 = 0$, (cf. Lemma 5.5.5) and of the hereditariness of σ, Theorem 5.5.4 yields that $\ker \varphi = \gamma(\ker \varphi)$. By Proposition 2.3.11 the hereditariness of σ implies

$$\ker \varphi = \gamma(\ker \varphi) \subseteq \gamma(0, K^K) \subseteq \gamma(\Xi).$$

By Lemma 5.5.5 we have

$$(0, K^K)/\ker \varphi \cong K \in \sigma \quad \text{and} \quad \Xi/(0, K^K) \cong K \in \sigma,$$

so we conclude that

$$\gamma(0, K^K) \subseteq \ker \varphi \quad \text{and} \quad \gamma(\Xi) \subseteq (0, K^K).$$

The latter yields $\gamma(\Xi) \subseteq \gamma(0, K^K)$, and the former gives us $\ker \varphi = \gamma(0, K^K)$. Hence $\ker \varphi = \gamma(\Xi)$ which is impossible, for $\ker \varphi$ is not an ideal in Ξ by Lemma 5.5.6. Thus $K_c = 0$, that is K is 0-symmetric. □

Theorem 5.5.7 together with Theorem 5.5.4 can be reformulated in the following

COROLLARY 5.5.8. *If a near-ring radical $\gamma \neq \{0\}$ has a hereditary semisimple class σ, then γ contains all nilpotent near-rings and σ consists of 0-symmetric near-rings, whence all the constant near-rings are in γ.* □

The statement that radicals with hereditary semisimple classes contain all constant near-rings was proved by Velsman [11] using a near-ring construction on the cartesian product $N \times N \times N$ with non-componentwise addition. This was, however, not enough to show that hereditary semisimple classes consist of 0-symmetric near-rings.

Not every 0-symmetric near-ring may be a member of a hereditary semisimple class of $\mathbb{N}$. Veldsman [8] proved that in $\mathbb{N}$ a hereditary semisimple class with hereditary radical class must not contain the two-element field.

(The two-element field is a dubious character, it is also the two-element "0-symmetric" constant near-ring, that is, $xy = x$ for $y \neq 0$ and $x0 = 0$.) Thus, for instance, no reasonable extensions of the radicals $\mathcal{J}_2$ and $\mathcal{J}_3$ from $\mathbb{N}_0$ to $\mathbb{N}$ will have a hereditary semisimple class or will remain itself hereditary.

Exploiting the properties of the construction $\Xi(K, L)$ Veldsman [15] proved also that in the variety $\mathbb{N}$ if a radical γ has the ADS-property then $\gamma = \{0\}$ or $\mathbb{N}$, and if a class ϱ satisfies condition (F) then $\varrho = \{0\}$, (cf. Propositions 3.6.4 and 3.6.13 and Theorem 3.6.15).

THEOREM 5.5.9 (Veldsman [11] and Anderson, Kaarli and Wiegandt [2]). *Let $\gamma \neq 0$ be a right invariantly strong radical of near-rings. Then γ has a hereditary semisimple class if and only if γ contains all constant near-rings. In particular, in the variety $\mathbb{N}_0$ of 0-symmetric near-rings the semisimple class of any right invariantly strong radical is hereditary.*

Proof: Suppose that the semisimple class $\sigma = \mathcal{S}\gamma$ is hereditary. Then by Corollary 5.5.8 γ contains all constant near rings.

Conversely, assume that γ contains all constant near-rings. Since the constant part of a near-ring is a right invariant subnear-ring, and since γ is right invariantly strong, the semisimple class $\sigma = \mathcal{S}\gamma$ consists of 0-symmetric near-rings. Further, let us consider any $I \triangleleft N \in \sigma$, and put $K = \gamma(I)$. Since $N \in \sigma$, N is 0-symmetric, and hence I is left invariant in N. We claim that K is right invariant in N. Let $a \in N$ be an arbitrary element. The mapping

$$\varphi\colon K \to (Ka + K)/K$$

defined by

$$\varphi(k) = ka + K \qquad \forall\, k \in K$$

is a near-ring homomorphism, as one can readily check by using the left invariance of I. Since $K = \gamma(I) \in \gamma$, also $(Ka+K)/K \in \gamma$ holds. $(Ka+K)/K$ is clearly a right invariant subnear-ring in I/K and so by the assumption on γ we conclude that

$$(K\alpha + K)/K \subseteq \gamma(I/K) = L/K \triangleleft I/K$$

with an appropriate ideal L of I. Hence the extension property of radical classes yields $L \in \gamma$, and therefore

$$\gamma(I)a + \gamma(I) = Ka + K \subseteq L \subseteq \gamma(I)$$

follows. This implies $\gamma(I)a \subseteq \gamma(I)$ for all $a \in N$, that is, $\gamma(I)$ is a right invariant subnear-ring of N whence

$$\gamma(I) \subseteq \gamma(N) = 0$$

as $N \in \sigma$. Consequently $I \in \sigma$, proving the Theorem . □

Special Features of the General Radical Theory   335

Call a radical *hyperconstant*, if it contains all constant near-rings. Then an immediate consequence of Theorems 5.5.4 and 5.5.9 is

COROLLARY 5.5.10. *Every right invariantly strong radical $\gamma \neq 0$ with hereditary semisimple class is hypernilpotent and hyperconstant.* □

For the following characterization of hypernilpotent radicals of near-rings, one more construction is required.

LEMMA 5.5.11. *For any two near-rings N and M, let a multiplication be defined on the additive group $\Psi^+(N,M) = N^+ \oplus N^+ \oplus M^+$ by the rule*

$$(x,y,m)(u,v,n) = \begin{cases} (y,0,mn) & \text{if } v \neq 0 \text{ and } n \neq 0 \\ (0,0,mn) & \text{otherwise.} \end{cases}$$

Then $\Psi = \Psi(N,M)$ is a near-ring such that
  i) $(N,0,M) \triangleleft \Psi$ and $\Psi/(N,0,M) \cong N^0$,
  ii) $(0,0,M) \triangleleft (N,0,M)$ and $(N,0,M)/(0,0,M) \cong N^0$
  iii) $N^0 \cong (N,0,0) \triangleleft \Psi$ and $\Psi/(N,0,0) \cong N^0 \oplus M$
  iv) $N \neq 0$, $C \subseteq (0,0,M)$ and $C \triangleleft \Psi$ imply $C = 0$.
Moreover, $\Psi(N,M)$ is 0-symmetric if and only if so is M.

Proof: Only iv) needs demonstration, as everything else is a straightforward verification.

Let us consider a nonzero element $x \in N$ and an arbitrary element $(0,0,m) \in C$. If $m \neq 0$, then by $C \triangleleft \Psi$ we have

$$(x,0,0) = (0,x,0)(0,x,m) =$$
$$= (0,x,0)((0,x,0) + (0,0,m)) - (0,x,0)(0,x,0) \in C \subseteq (0,0,M).$$

Hence $x = 0$ follows, a contradiction. Thus $m = 0$ and also $C = 0$. □

The next Theorem due to Veldsman [6] characterizes hypernilpotent radicals as radicals having weakly homomorphically closed semisimple class (cf. Section 3.6).

THEOREM 5.5.12. *A radical $\gamma \neq \{0\}$ of near-rings is hypernilpotent if and only if the semisimple class $\sigma = S\gamma$ is weakly homomorphically closed.*

Proof: Suppose that σ is weakly homomorphically closed. Further, let us assume that γ is not hypernilpotent, and hence there exists a near-ring N such that $0 \neq N = N^0 \in \sigma$. We have to show that every near-ring M is in σ. To this end we construct the near-ring $\Psi = \Psi(N,M)$ of Lemma 5.5.11. Then i) and ii) of that Lemma give

$$\Psi/(N,0,M) \cong N \in \sigma \qquad \text{and} \qquad (N,0,M)/(0,0,M) \cong N \in \sigma.$$

Hence
$$\gamma(\Psi) \subseteq (N,0,M) \qquad \text{and} \qquad \gamma(N,0,M) \subseteq (0,0,M)$$

hold, implying
$$\gamma(\Psi) \subseteq \gamma(N,0,M) \subseteq (0,0,M).$$

Applying iv) of Lemma 5.5.11 we get that $\gamma(\Psi) = 0$, that is, $\Psi \in \sigma$, and so by Lemma 5.5.11 iii), we get
$$N \cong (N,0,0) \triangleleft \Psi \in \sigma.$$

Since σ is weakly homomorphically closed and $N^2 = 0$, in view of Lemma 5.5.11 iii) we conclude that
$$N \oplus M \cong \Psi/(N,0,0) \in \sigma.$$

Since every ideal of the direct component M is also an ideal in the direct sum $N \oplus M$, from $N \oplus M \in \sigma$ we conclude that $M \in \sigma$. Thus σ consists of all near-rings and so $\gamma = \{0\}$. This contradicts the assumption $\gamma \neq \{0\}$, and therefore γ has to be hypernilpotent.

The converse is clear. □

Remarks: 1) Theorem 5.5.1 remains valid also in $\mathbb{N}_0$ because $\Psi(N,M)$ is 0-symmetric whenever M is such a near-ring.

2) Comparing Theorem 5.5.12 with Theorem 3.6.11, the degeneracy here is that the weak homomorphically closed property of σ forces the nilpotent near-rings to its radical class γ without demanding that γ be hereditary.

Combining Theorems 5.5.12 and 2.3.9 we get

COROLLARY 5.5.13. *For classes γ and σ of near-rings the following are equivalent:*

(i) *$\gamma \neq 0$ is hypernilpotent radical and $\sigma = \mathcal{S}\gamma$,*

(ii) *σ is regular, closed under subdirect sums and extensions, weakly homomorphically closed, $((N)\sigma)\sigma \triangleleft N$ for every near-ring N, σ is not the class of all near-rings and $\gamma = \mathcal{U}\sigma$.* □

Semisimple classes of radicals containing all nilpotent and all constant near-rings were characterized by Mlitz and Wiegandt [1], [3].

To build a decent theory for supernilpotent and special radicals of near-rings, one obstacle is that the corresponding assertion to Proposition 3.6.4 is not true for near-rings, that is, $K \triangleleft I \triangleleft N$ and I/K prime does not imply $K \triangleleft N$ (cf. Veldsman [15]). Nevertheless, for 0-symmetric near-rings, if I/K is 2-primitive or simple with unity element or equiprime, then $K \triangleleft A$ follows. The above mentioned shortcoming of prime (and semiprime) near-rings and the examples for which $K \triangleleft A$ holds true has led to the delicate theory of *overnilpotent radicals* of near-rings (cf. Veldsman [13]).

Special radicals of near-rings were studied by Birkenmeier, Heatherly and Lee [1], Booth and Groenewald [1], [4] and Groenewald [2]. For further developments in the radical theory of near-rings one can consult the survey papers of Kaarli [3] and Veldsman [14].

References

A. Abian

[1] Rings without nilpotent elements, *Mat. Časopis.* 25 (1975), 289–291.

U. A. Aburawash

[1] Characterizations of semisimple classes by means of ring elements, *Periodica Math. Hungar.* 22 (1991), 1–13.

[2] Semiprime involution rings with chain conditions, *Contr. General Alg.* 7, Hölder–Pichler–Tempsky, Wien & B. G. Teubner, Stuttgart, 1991, pp. 7–11.

[3] The structure of ∗-simple involution rings with minimal ∗-biideals, *Beitr. Alg. und Geom., Halle* 33 (1992), 77–83.

[4] On the structure of involution rings, *Math. Japonica* 37 (1992), 987–994.

[5] On ∗-simple involution rings with minimal ∗-biideal, *Studia Sci. Math. Hungar.* 32 (1996), 455-458.

[6] ∗-biideals in the radical theory of involution rings, *Quaest. Math.* 19 (1997), 517–525.

[7] On involution rings, *East-West J. Math.* 2 (2000), 109–126.

S. A. Amitsur

[1] Nil PI-rings, *Proc. Amer. Math. Soc.* 2 (1951), 538–540.

[2] A general theory of radicals I, Radicals in complete lattices, *Amer. J. Math.* 74 (1952), 774–786.

[3] A general theory of radicals II, *Amer. J. Math.* 76 (1954), 100–125.

[4] A general theory of radicals III, Applications, *Amer. J. Math.* 76 (1954), 126–136.

[5] Radicals of polynomial rings, *Canad. J. Math.* 8 (1956), 355–361.

[6] Rings of quotients and Morita contexts, *J. Algebra* 17 (1971), 273–298.

[7] Nil radicals. Historical notes and some new results, *Coll. Math. Soc. J. Bolyai 6, Rings, Modules and Radicals, Keszthely 1971*, North-Holland 1973, pp. 47–65.

S. A. Amitsur and J. Levitzki
 [1] Maximal identities for algebras, *Proc. Amer. Math. Soc.* 1 (1950), 449–463.

F. W. Anderson and K. R. Fuller
 [1] *Rings and categories of modules*, Springer-Verlag, 1992.

T. Anderson, N. Divinsky and A. Suliński
 [1] Hereditary radicals in associative and alternative rings, *Canad. J. Math.* 17 (1965), 594–603.

T. Anderson and B. J. Gardner
 [1] Semisimple classes in a variety satisfying an Andrunakievich Lemma, *Bull. Austral. Math. Soc.* 18 (1978), 187–200.

T. Anderson, K. Kaarli and R. Wiegandt,
 [1] Radicals and subdirect decompositions, *Comm. in Algebra* 13 (1985), 479–494.
 [2] On left strong radicals of near-rings, *Proc. Edinburgh Math. Soc.* 31 (1988), 447–456.

T. Anderson and R. Wiegandt
 [1] Weakly homomorphically closed semisimple classes, *Acta Math. Acad. Sci. Hungar.* 34 (1979), 329–336.
 [2] On essentially closed classes of rings, *Annales Univ. Sci. Budapest.* 24 (1981), 107–111.
 [3] Semisimple classes of alternative rings, *Proc. Edinburgh Math. Soc.* 25 (1982), 21–26.

V. A. Andrunakievich (= V. A. Andrunachievici = В. А. Андрунакиевич)
 [1] Radicals of associative rings I, (Russian), *Mat. Sb.* 44 (1958), 179–212; English transl.: *Amer. Math. Soc. Transl.* (2) 52 (1966), 95–128.
 [2] Radicals of associative rings II, (Russian), *Mat. Sb.* 55 (1961), 329–346; English transl.: *Amer. Math. Soc. Transl.* (2) 52 (1966), 129–150.
 [3] Prime modules and the Baer radical, (Russian), *Sibirsk. Mat. Zh.* 2 (1961), 801–806.
 [4] On strongly regular rings, (Russian), *Izv. Akad. Nauk Moldav. SSR* 11 (1963), 75–77.

V. A. Andrunakievich and Yu. M. Ryabukhin
 [1] Special modules and special radicals, (Russian), *Dokl. Akad. Nauk SSSR* 147 (1962), 1274–1277.
 [2] Special modules and radicals, (Russian), *Izdat. Kazan. Univ., Kazan* 1964, 7–17.
 [3] Modules and radicals, (Russian), *Dokl. Akad. Nauk SSSR* 156 (1964), 991–999, English translation: *Soviet Math. Dokl.* 3 (1964), 728–732.
 [4] Rings without nilpotent elements and completely prime ideals, (Russian), *Dokl. Akad. Nauk SSSR* 180 (1968), 9–11.
 [5] Torsion and Kurosh chains in algebras, (Russian), *Trudy Mosk. Mat. Obshch.* 29 (1973), 19–49, English transl.: *Trans. Moscow. Math. Soc.* 29 (1973), 17–47.
 [6] *Radicals of algebras and structure theory*, (Russian), Nauka, Moscow, 1979.

References

R. R. Andruszkiewicz
 [1] On accessible subrings of associative rings, *Proc. Edinburgh Math. Soc.* 35 (1992), 101–107.

R. R. Andruszkiewicz and E. R. Puczyłowski
 [1] Kurosh's chains of associative rings, *Glasgow Math. J.* 32 (1990), 67–69.
 [2] Accessible subrings and Kurosh's chains of associative rings, *Algebra Coll.* 4 (1997), 79–88.

P. N. Ánh
 [1] On Litoff's theorem, *Studia Sci. Math. Hungar.* 18 (1983), 153–157.

P. N. Ánh, N. V. Loi and R. Wiegandt
 [1] On the radical theory of Andrunakievich varieties, *Bull. Austral. Math. Soc.* 31 (1985), 257–269.

P. N. Ánh and R. Wiegandt
 [1] Semisimple classes of nonassociative rings in Jordan algebras, *Comm. in Algebra* 13 (1985), 2669–2690.

R. F. Arens and I. Kaplansky
 [1] Topological representation of algebras, *Trans. Amer. Math. Soc.* 63 (1948), 457–481.

A. V. Arhangel'skiĭ (= А. В. Архангельский) and R. Wiegandt
 [1] Connectednesses and disconnectednesses in topology, *General Topology & Appl.* 5 (1975), 9–33.

E. P. Armendariz
 [1] Closure properties in radical theory, *Pacific J. Math.* 26 (1968), 1–7.

E. P. Armendariz and W. G. Leavitt
 [1] The heredity property in the lower radical construction, *Canad. J. Math.* 20 (1968), 474–476.

V. I. Arnautov (= В. И. Арнаутов)
 [1] General theory of radicals of topological rings (Russian), *Bul. Acad. Ştiinţe Moldoven, Matematica* 2(21) 1996, 5–45.
 [2] The theory of radicals of topological rings, *Math. Japonica* 47 (1998), 439–544; correction: *ibidem* 48 (1999), 310–312.

M. Aslam and A. M. Zaidi
 [1] Matrix equation in radicals, *Studia Sci. Math. Hungar.* 28 (1993), 447–452.

Christine Ayoub
 [1] Conditions for a ring to be fissile, *Acta Math. Acad. Sci. Hungar.* 30 (1977), 233–237.

A. M. Babich (= А. М. Бабич)
 [1] On the radical of Levitzki, (Russian), *Dokl. Akad. Nauk SSSR* 126 (1959), 242–243.

R. Baer
 [1] Radical ideals, *Amer. J. Math.* 65 (1943), 537–568.

M. N. Barua
 [1] *Near-rings and near-ring modules — Some special types*, Ph. D. Thesis, Gauhati University, Assam, India.

E. A. Behrens
 [1] Nichtassoziative Ringe, *Math. Ann.* 127 (1954), 441–452.

K. I. Beidar (= K. I. Beĭdar = К. И. Бейдар)
 [1] A chain of Kurosh may have an arbitrary finite length, *Czechosl. Math. J.* 32 (1982), 418–422.
 [2] Examples of rings and radicals, *Coll. Math. Soc. J. Bolyai 38, Radical Theory, Eger 1982*, North-Holland 1985, pp. 19–46.
 [3] Semisimple classes and the lower radical, (Russian), *Mat. Issled., Kishinev* 105 (1988), 13–29.
 [4] The intersection property for radicals, (Russian), *Usp. Mat. Nauk* 44 (1989), 1 (265), 187–188.
 [5] The Andrunakievich lemma and Jordan algebras, (Russian), *Usp. Mat. Nauk* 45 (1990), no 4 (274), 137–138; English transl.: *Russian Math. Surveys* 45 (1990), no 4, 159.
 [6] On essential extensions, maximal essential extensions and iterated maximal essentail extensions in radical theory, *Coll. Math. Soc. J. Bolyai 61, Theory of Radicals, Szekszárd 1991*, North-Holland, 1993, pp. 17–26.
 [7] On questions of B. J. Gardner and A. D. Sands, *J. Austral. Math. Soc.* 56 (1994), 314–319.
 [8] On principally hereditary radicals, *Comm. in Algebra* 26 (1998), 3899–3912.

K. I. Beidar, Y. Fong and W. F. Ke
 [1] On complemented radicals, *J. Algebra* 201 (1998), 328–356.

K. I. Beidar, Y. Fong, W. F. Ke and K. P. Sum
 [1] On radicals with semisimple essential covers, *Preprint*, 1995.

K. I. Beidar, Y. Fong and E. R. Puczyłowski
 [1] A polynomial ring over a nil ring cannot be homomorphically mapped onto a ring with nonzero idempotent, *J. Algebra* 238 (2001), 389–399.

K. I. Beidar, Y. Fong and C. S. Wang
 [1] On the lattice of strong radicals, *J. Algebra* 180 (1996), 334–340.

K. I. Beidar, W.-F. Ke and E. R. Puczyłowski
 [1] On subhereditary radicals and reduced rings, *Proc. Royal Soc. Edinburgh* 132 A (2002), 255–266.
 [2] On matrix rings and subhereditary radicals, *Preprint*, 2002.

K. I. Beidar, E. R. Puczyłowski and R. Wiegandt
 [1] Radicals and polynomial rings, . *J. Austral. Math. Soc.* 71 (2002), 1–7.

K. I. Beidar and K. Salavová

[1] Some examples of supernilpotent nonspecial radicals, *Acta Math. Hungar.* 40 (1982), 109–112.

[2] On the lattice of N-radicals, left strong radicals and left hereditary radicals, (Russian), *Acta Math. Hungar.* 42 (1983), 81–95.

K. I. Beidar and R. Wiegandt

[1] Splitting theorems for nonassociative rings, *Publ. Math. Debrecen* 38 (1991), 121–143.

[2] Rings with involution and chain conditions on biideals, (Russian), *Usp. Mat. Nauk* 485 (293) (1993), 159–160. English transl.: *Russian Math. Surv.* 48 no 5 (1992), 161–162.

[3] Rings with involution and chain conditions, *J. Pure & Appl. Algebra* 87 (1993), 205–220.

[4] Radicals induced by the total of rings, *Beitr. Alg. und Geom.* 38 (1997), 149–159.

[5] Radical assignments and radical classes, *Bul. Acad. Ştiinţe Rep. Moldova, Matematica*, 2(30) 1999, 17–27.

H. E. Bell

[1] Duo rings, some applications to commutativity theorems, *Canad. Math. Bull.* 11 (1968), 375–380.

J. E. van den Berg

[1] On uniformly strongly prime rings, *Math. Japonica* 38 (1993), 1157–1166.

[2] *On chain domains, prime rings and torsion preradicals*, Ph. D. Thesis, University of Natal, Pietermaritzburg 1995.

G. M. Bergman

[1] A ring primitive on the right but not on the left, *Proc. Amer. Math. Soc.* 15 (1964), 473–475.

[2] Radicals, tensor products, and algebraicity, *Ring Theory, Israel Math. Conf. Proc.* 1 (1989), 150–192.

G. Betsch

[1] *Struktursätze für Fastringe*, Dissertation, Universität Tübingen, 1963.

[2] Ein Radikal für Fastringe, *Math. Zeitschr.* 78 (1962), 86–90.

[3] Some structure theorems on 2-primitive near-rings, *Coll. Math. Soc. J. Bolyai 6, Rings, Modules and Radicals, Keszthely 1971*, North-Holland 1973, pp. 73–102.

[4] Primitive near-rings, *Math. Zeitschr.* 130 (1973), 351–361.

[5] Embedding of a near-ring into a near-ring with identity, *Near-rings and Near-fields, Proc. Conf. Tübingen 1985*, Elsevier 1987, pp 37–40.

G. Betsch and K. Kaarli

[1] Supernilpotent radicals and hereditariness of semisimple classes of near-rings, *Coll. Math. Soc. J. Bolyai 38, Radical Theory, Eger 1982*, North-Holland, 1985, pp. 47–58.

G. Betsch and R. Wiegandt
[1] Non-hereditary semisimple classes of near-rings, *Studia Sci. Math. Hungar.* 17 (1982), 69–75.

G. F. Birkenmeier
[1] Rings which are essentially supernilpotent, *Comm. in Algebra* 22 (1994), 1063–1082.
[2] Radicals whose essential covers are semisimple classes, *Comm. in Algebra* 22 (1994), 6239–6258.

G. F. Birkenmeier, H. E. Heatherly and E. K. S. Lee
[1] Special radicals for near-rings, *Tamkang J. Math.* 27 (1996), 281–288.

G. F. Birkenmeier and R. Wiegandt
[1] Essential covers and complements of radicals, *Bull. Austral Math. Soc.* 53 (1996), 261–266.

R. L. Blair
[1] Ideal lattices and structure of rings, *Trans. Amer. Math. Soc.* 75 (1953), 135–153.
[2] A note on f-regularity in rings, *Proc. Amer. Math. Soc.* 6 (1955), 511–515.

G. L. Booth and N. J. Groenewald
[1] Special radicals of near-rings, *Math. Japonica* 37 (1992), 701–706.
[2] Special radicals in rings with involution, *Publ. Math. Debrecen* 48 (1994), 241–251.
[3] Special radicals of Ω-groups, *Nearrings, Nearfields and K-loops, Proc. Conf. Hamburg 1995*, Kluwer 1997, pp 211–218.
[4] Radicals of involution rings, *Algebra Coll.* 5 (1998), 277–289.

G. L. Booth, N. J. Groenewald and S. Veldsman
[1] A Kurosh–Amitsur prime radical for near-rings, *Comm. in Algebra* 18 (1990), 3111–3122.

R. Brauer
[1] On the nilpotency of the radical of a ring, *Bull. Amer. Math. Soc.* 48 (1942), 752–758.

B. Brown and N. H. McCoy
[1] Radicals and subdirect sums, *Amer. J. Math.* 69 (1947), 46–58.
[2] The radical of a ring, *Duke Math. J.* 15 (1948), 495–499.
[3] The maximal regular ideal of a ring, *Proc. Amer. Math. Soc.* 1 (1950), 165–171.

Annemarie Buys
[1] Uniformly strongly prime radicals, *Quaest. Math.* 21 (1998), 149–156.

Annemarie Buys and G. Gerber
[1] Special classes of Ω-groups, *Annales Univ. Sci. Budapest.* 29 (1986), 73–85.

Cai Chuan Ren
[1] Dual radicals and F. A. Szász Problem 21, (Chinese), *Acta Math. Sinica* 32 (3) (1989), 394–400.

References

Chen Shen Can
 [1] Supplementing, dual and quasi-hereditary radicals, *Northeastern Math. J.* 8 (3) (1992), 357–363.

J. R. Clay
 [1] *Nearrings: Genesis and Applications*, Oxford University Press, 1992.

P. M. Cohn
 [1] *Universal algebra*, Harper & Row, 1965.

J. Dauns
 [1] Prime modules, *J. reine angew. Math.* 298 (1978), 156–181.

G. B. Desale and K. Varadarajan
 [1] SP-modules and related topics, *University of Calgary, Research Paper 463*, 1980.

S. E. Dickson
 [1] A torsion theory for abelian categories, *Trans. Amer. Math. Soc.* 121 (1966), 223–235.
 [2] A note on hyper-nilpotent radical properties for associative rings, *Canad. J. Math.* 19 (1967), 447–448.

Dinh Van Huynh
 [1] Die Spaltbarkeit von MHR-Ringen, *Bull. Acad. Polon. Sci.* 25 (1977), 939–941.

N. Divinsky (= N. J. Divinsky)
 [1] General radicals that coincide with the classical radicals on rings with d.c.c., *Canad. J. Math.* 13 (1961), 639–644.
 [2] *Rings and radicals*, Allen & Unwin, London, 1965.
 [3] Duality between radical and semisimple classes of associative rings, *Scripta Math.* 29 (1973), 409–416.
 [4] Unequivocal rings, *Canad. J. Math.* 24 (1975), 679–690.

N. Divinsky, J. Krempa and A. Suliński
 [1] Strong radical properties of alternative and associative rings, *J. Algebra* 17 (1971), 369–388.

M. Domokos
 [1] Goldie's theorems for involution rings, *Comm. in Algebra* 22 (1994), 371–380.

J. L. Dorroh
 [1] Concerning adjunctions to algebras, *Bull. Amer. Math. Soc.* 38 (1932), 85–88.

M. P. Drazin and M. L. Roberts,
 [1] Polynomial, multiplicative and special radicals, *Comm. in Algebra* 28 (2000), 3073–3093.

I. N. Dubrovin (= И. Н. Дубровин)
[1] Chain domains, (Russian), *Vestnik Moskov. Univ. Ser. I. Mat. Mekh.* no 2 (1980), 51–54.

G. G. Emin (= Г. Г. Емин)
[1] Modules and strict radicals, (Russain), *Dokl. Akad. Nauk Armyanskoi SSR* 81 (1985), 3–6.

P. O. Enersen and W. G. Leavitt
[1] The upper radical construction, *Publ. Math. Debrecen* 20 (1973), 219–222.

C. Faith
[1] *Lecutes on injective modules and quotient rings*, Springer-Verlag, 1967.

C. Faith and Y. Utumi
[1] On a new proof of Litoff's theorem, *Acta Math. Acad. Sci. Hungar.* 14 (1963), 369–372.

Sh. Feigelstock
[1] *Additive groups of rings*, Pitman Res. Notes in Math. 83, Longman Sci. & Tech., 1983.
[2] *Additive groups of rings II*, Pitman Res. Notes in Math. 169, Longman Sci. & Tech., 1988.
[3] Additive groups of unequivocal rings, *Acta Math. Hungar.* 79 (1998), 31–34.

M. Ferrero and M. M. Parmenter
[1] A note on Jacobson rings and polynomial rings, *Proc. Amer. Math. Soc.* 105 (1989), 281–286.

I. Fischer and Ruth R. Struik
[1] Nil algebras and periodic groups, *Amer. Math. Monthly* 75 (1968), 611–623.

Y. Fong, F. K. Huang and R. Wiegandt
[1] Radical theory for group semiautomata, *Acta Cybernetica* 11 (1994), 169–188.

A. Forsythe and N. H. McCoy
[1] On the commutativity of certain rings, *Bull. Amer. Math. Soc.* 52 (1946), 523–526.

Halina France-Jackson (= Halina Korolczuk)
[1] On prime essential rings, *Bull. Austral. Math. Soc., Ser. A* 47 (1993), 287–290.

E. Fried and R. Wiegandt
[1] Connectednesses and disconnectednesses of graphs, *Algebra Universalis* 5 (1975), 411–428.
[2] Abstract relational structures, I (General theory), *Algebra Universalis* 15 (1982), 1–21; II (Torsion theory), *ibidem* 15 (1082), 22–39.

L. Fuchs
[1] Wann folgt die Minimalbedingung aus der Maximalbedingung? *Archiv Math.* 8 (1957), 317–319.

[2] *Infinite abelian groups* vol. II, Academic Press, 1973.

P. Fuchs, C. J. Maxson, A. P. J. van der Walt and K. Kaarli
  [1] Centralizer near-rings determined by PID-modules, II. *Periodica Math. Hungar.* 26 (1993), 111–114.

B. J. Gardner
  [1] A note on radicals and polynomial rings, *Math. Scand.* 31 (1972), 83–88.
  [2] Radicals of abelian groups and associative rings, *Acta Math. Acad. Sci. Hungar.* 24 (1973), 259–268.
  [3] Some remarks on radicals of rings with chain conditions, *Acta Math. Acad. Sci. Hungar.* 25 (1974), 263–268.
  [4] Some aspects of T-nilpotence, *Pacific J. Math.* 53 (1974), 117–130.
  [5] Some radical constructions for associative rings, *J. Austral. Math. Soc.* 18 (1974), 442–446.
  [6] Semi-simple classes of algebras and attainability of identities, *Pacific J. Math.* 61 (1975), 401–416.
  [7] Sub-prime radical classes determined by zerorings, *Bull Austral. Math. Soc.* 12 (1975), 95–97.
  [8] Radicals and left ideals, *Bull. Acad. Polon. Sci.* 24 (1976), 943–945.
  [9] Some degeneracy and pathology in non-associative radical theory, *Annales Univ. Sci. Budapest.* 22–23 (1979/1980), 65–74.
  [10] Extension closed varieties need not have attainable identities, *Bull. Malaysian Math. Soc.* 2 (2) (1979), 37–39.
  [11] The lower radical construction for non-associative rings: examples and counterexamples, *Bull. Austral. Math. Soc.* 20 (1979), 259–271.
  [12] Radicals related to the Brown–McCoy radical in some varieties of algebras, *J. Austral. Math. Soc., Ser. A* 28 (1979), 283–294.
  [13] Extension-closure and attainability for varieties of algebras with involution, *Comment. Math. Univ. Carol.* 21 (1980), 285–292.
  [14] Some degeneracy and pathology in non-associative radical theory II, *Bull. Austral. Math. Soc.* 23 (1981), 423–428.
  [15] Radical classes of regular rings with artinian primitive images, *Pacific J. Math* 99 (1982), 337–349.
  [16] Small ideals in radical theory, *Acta Math. Hungar.* 43 (1984), 287–294.
  [17] Some recent results and open problems concerning special radicals, *Radical Theory, Proc. 1988 Sendai Conf.* Uchida Rokakuho Tokyo, 1989, pp. 25–56.
  [18] *Radical theory*, Pitman Res. Notes in Math., 198, Longman Sci. & Tech., 1989.
  [19] Strong semisimplicity, *Periodica Math. Hungar.* 24 (1992), 23–35.

B. J. Gardner and A. V. Kelarev
  [1] Invariant radicals, *Proc. Royal Soc. Edinburgh* 127 A (1997), 773–780.

B. J. Gardner and P. N. Stewart
  [1] On semi-simple radical classes, *Bull. Austral. Math. Soc.* 13 (1975), 349–353.
  [2] Prime essential rings, *Proc. Edinburgh Math. Soc.* 34 (1991), 241–250.

B. J. Gardner and R. Wiegandt
  [1] Characterizing and constructing special radicals, *Acta Math. Acad. Sci. Hungar.* 40 (1982), 73–83.

A. W. Goldie
  [1] The structure of prime rings under accending chain conditions, *Proc. London Math. Soc.* (3) 8 (1958), 589–608.

E. S. Golod (= Е. С. Голод)
  [1] On nil-algebras and finitely approximable p-groups, (Russian), *Izv. Akad. Nauk SSSR, Ser. Mat.* 28 (1964), 273–276.

E. S. Golod and I. M. Safarevich (= И. М. Шафаревич)
  [1] On class field towers, (Russian), *Izv. Akad. Nauk SSSR, Ser. Mat.* 28 (1964), 261–272.

K. R. Goodearl, D. Handelman and J. Lawrence
  [1] *Strongly prime and completely torsion-free rings*, Carleton Math. Series, 109 (1974).

N. J. Groenewald
  [1] Strongly prime group rings, *Quaest. Math.* 3 (1949), 241–247.

N. J. Groenewald and G. A. P. Heyman
  [1] Certain classes of ideals in group rings II, *Comm. in Algebra* 9 (1981), 137–148.

N. J. Groenewald and W. A. Olivier
  [1] Regularities and complementary radicals, *Qaest. Math.* 18 (1995),427–434.

N. J. Groenewald and L. van Wyck
  [1] Polynomial regularities in structural matrix rings, *Comm. in Algebra* 22 (1994), 2101–2113.

Guo Jin-yun
  [1] On the termination of the construction of the lower radical class for a class of associative rings, (Chinese), *Chinese Ann. Math. Ser. A* 8 (1987), 433–444; English summary: *Chinese Ann. Math. Ser. B* 8 (1987), 488.

D. Handelman and J. Lawrence
  [1] Strongly prime rings, *Trans. Amer. Math. Soc.* 211 (1975), 209–223.

J. F. T. Hartney
  [1] On the radical theory of a distributively generated near-ring, *Math. Scand.* 23 (1968), 214–220.
  [2] *Radicals and antiradicals in near-rings*, Ph. D. Thesis, University of Nottingham, 1979.

U. Hebisch and H. J. Weinert
  [1] Radical theory for semirings, *Quaest. Math.* 20 (1997), 647–661.
  [2] On the interrelation between radical theories for semirings and for rings, *Comm. in Algebra* 29(2001), 109–129.
  [3] Semisimple classes of semirings, *Algebra Coll.* 9 (2002), 177–196.

References

A. G. Heinicke
[1] Some results in the theory of radicals of associative rings, *Ph. D. Thesis, University of British Columbia, Vancouver*, 1968.
[2] A note on lower radical constructions for associative rings, *Canad. Math. Bull.* 11 (1968), 23–30.
[3] Hereditary radical ideals in a ring, *J. London Math. Soc.* 2 (1970), 539–543.

I. R. Hentzel
[1] A note on modules and radicals, *Proc. Amer. Math. Soc.* 19 (1968), 1385–1386.
[2] The Andrunakievich lemma for alternative rings, *Algebras, Groups and Geometries* 6 (1989), 55–64.

I. N. Herstein
[1] *Theory of rings*, Univ. Chicago Math. Lecture Notes, 1961.
[2] *Topics in ring theory*, Univ. Chicago Math. Lecture Notes, 1965.
[3] *Noncommutative rings*, Carus Math. Monogr. 15, MAA New York, 1968.

G. A. P. Heyman and C. Roos
[1] Essential extensions in radical theory for rings, *J. Austral. Math. Soc.* 23 (1977), 340–347.

H.-J. Hoehnke
[1] Radikale in allgemeinen Algebren, *Math. Nachr.* 32 (1966), 347–383.

A. E. Hoffman and W. G. Leavitt
[1] Properties inherited by the lower radical, *Portugal Math.* 27 (1968), 63–66.
[2] A note on the termination of the lower radical construction, *J. London Math. Soc.* 43 (1968), 617–618.

M. Holcombe
[1] A hereditary radical for near-rings, *Studia Sci. Math. Hungar.* 17 (1982), 453–456.

M. Holcombe and R. Walker
[1] Radicals in categories, *Proc. Edinburgh Math. Soc.* 24 (1978), 111–128.

M. Hongan
[1] Some remarks on normal classes of semiprime rings, *Math. J. Okayama Univ.* 25 (1983), 139–143.

Ch. Hopkins
[1] Nilrings with minimal conditions for admissible left ideals, *Duke J. Math.* 4 (1938), 664–667.

J. Horvath
[1] *Density theorems and applications*, M. Sc. Thesis, University of British Columbia, Vancouver, 1977.

Hsieh Pang Chieh (= Xie Bang Jie)
[1] Different Köthe radicals, Köthe semisimplicities and quasi-nil radicals of non-asscociative rings, *Acta Sci. Natur.*, 1 (1957), 19–26.

Huang Wen Ping
[1] A characterization of dual pairs of supplementing radicals, (Chinese), *Pure Appl. Math.* 7 No 1 (1991), 122–123.

R. S. Irving
[1] On the primitivity of certain Ore extensions, *Math. Ann.* 242 (1979), 177–192.

N. Jacobson
[1] *The theory of rings*, Amer. Math. Soc., New York 1943.
[2] The radical and semisimplicity of arbitrary rings, *Amer. J. Math.* 67 (1945), 300–320.
[3] Structure theory of simple rings without finiteness assumptions, *Trans. Amer. Math. Soc.* 57 (1945), 228–245.
[4] Structure theory for algebraic algebras of bounded degree, *Annals of Math.* 46 (1945), 695–707.
[5] *Structure of rings*, Amer. Math. Soc. Coll. Publ. 37, Providence, 1968.
[6] *Finite-dimensional division algebras over fields*, Springer-Verlag, 1996.

M. Jaegermann
[1] Morita contexts and radicals, *Bull. Acad. Polon. Sci.* 20 (1972), 619–623.
[2] Normal radicals of endomorphism rings of free and projective modules, *Fund. Math.* 86 (1975), 237–250.
[3] Normal radicals, *Fund. Math.* 95 (1977), 147–155.

M. Jaegermann and A. D. Sands
[1] On normal radicals and normal classes of rings, *J. Algebra* 50 (1978), 337–349.

G. Janelidze and L. Márki
[1] Radicals of rings and pullbacks, *J. Pure & Appl. Algebra* 97 (1994), 29–36.
[2] A combinatorial approach to radicals, *Comm. in Algebra*, to appear.

G. Janelidze, L. Márki and W. Tholen
[1] Locally semisimple coverings, *J. Pure & Appl. Algebra* 128 (1998), 281–289.

J. P. Jans
[1] *Rings and homology*, Holt, Rinehart and Winston, 1964.

A. V. Jategaonkar
[1] A counter-example in ring theory and homological algebra, *J. Algebra* 12 (1969), 418–440.

T. L. Jenkins
[1] A maximal ideal radical class, *J. Nat. Sci. & Math., Lahore* 7 (1967), 191–195.

R. E. Johnson and E. T. Wong
 [1] Quasi-injective modules and irreducible rings, *J. London Math. Soc.* 36 (1961), 260–268.

S. J. Joubert and M. J. Schoeman
 [1] Torsion theories and corresponding radical and semisimple classes, *Karachi J. Math.* 3 (1985), 9–17.

K. Kaarli (= К. Каарли)
 [1] Radicals in near-rings, (Russian), *Tartu Riikl. Ül. Toimetised* 390 (1976), 134–171.
 [2] Classification of irreducible R-groups over a semiprimary near-ring, (Russian), *Tartu Riikl. Ül. Toimetised* 556 (1981), 47–63.
 [3] Survey on the radical theory of near-rings, *Contr. General Algebra 4, Proc. Krems Conf. 1985*, Hölder–Pichler–Tempsky, Wien & B. G. Teubner, Stuttgart, 1987, pp. 45–62.
 [4] On Jacobson type radicals of near-rings, *Acta Math. Hungar.* 50 (1987), 71–78.

K. Kaarli and Tiina Kriis
 [1] Prime radical near-rings, *Tartu Riikl. Ül. Toimetised* 764 (1987), 23–29.

T. Kando
 [1] Strong regularity in arbitrary rings, *Nagoya Math. J.* 4 (1952), 51–63.

I. Kaplansky
 [1] Rings with polynomial identity, *Bull. Amer. Math. Soc.* 54 (1948), 575–580.
 [2] *Fields and rings*, Univ. Chicago Press, Chicago and London, 1969.

A. V. Kelarev (= А. В. Келарев)
 [1] A general approach to the structure of radicals in some ring constructions, *Coll. Soc. Math. J. Bolyai 61, Theory of Radicals, Szekszárd 1991*, North-Holland, 1993, pp. 131–144.

A. Kertész
 [1] *Lectures on artinian rings*, (ed. R. Wiegandt), Akadémiai Kiadó, Budapest 1987.

A. Kertész and A. Widiger
 [1] Artinsche Ringe mit artinschen Radikal, *J. reine angew. Math.* 242 (1970), 8–15.

S. P. Kogalovskiĭ (= С. П. Когаловский)
 [1] Structural characteristics of universal classes, (Russian), *Sibirsk. Mat. Zh.* 4 (1963), 97–119.

K. Koh
 [1] Quasisimple modules and other topics in ring theory, *Lectures on ring and modules*, Lecture Notes in Math. 246, Springer-Verlag 1972, 323–428.

K. Koh and J. Luh
 [1] On a finite dimensional quasi-simple module, *Proc. Amer. Math. Soc.* 25 (1970), 801–807.

K. Koh and A. C. Mewborn
[1] A class of prime rings, *Canad. Math. Bull.* 9 (1966), 63–72.
[2] The weak radical of a ring, *Proc. Amer. Math. Soc.* 18 (1967), 554–559.

G. Köthe
[1] Die Struktur der Ringe, deren Restklassenring nach dem Radikal vollständig reduzibel ist, *Math. Zeitschr.* 32 (1930), 161–186.

K. K. Krachilov (= К. К. Крачилов)
[1] Complementedness in lattices of radicals I, (Russian), *Mat. Issled. Kishinev* 62 (1981), 76–88.
[2] Complementedness in lattices of radicals II, (Russian), *Mat. Issled. Kishinev* 62 (1981), 89–111.

G. Krause
[1] Some recent developments in the theory of Noetherian rings, *Springer Lecture Notes in Math.* 641 (1978), 209–219.

J. Krempa
[1] Logical connections among some open problems in non-commutative rings, *Fund. Math.* 76 (1972), 121–130.
[2] On radical properties of polynomial rings, *Bull. Acad. Polon. Sci.* 20 (1972), 545–548.
[3] Radicals of semi-group rings, *Fund. Math.* 85 (1974), 57–71.
[4] Lower radical properties for alternative rings, *Bull. Acad. Polon. Sci.* 23 (1975), 139–142.
[5] Radicals and derivations of algebras, *Coll. Math. Soc. J. Bolyai 38, Radical Theory, Eger 1982*, North-Holland, 1985, pp. 195–227.

J. Krempa and Barbara Terlikowska
[1] Theory of radicals in self-dual categories *Bull. Acad. Polon. Sci.* 22 (1974), 367–373.

R. L. Kruse and D. T. Price
[1] *Nilpotent rings*, Gordon and Breach, New York– London–Paris, 1969.

C. Kuratowski
[1] Une methode d' élimination des nombres transfinis des raisonnements mathématiques, *Fund. Math.* 3 (1922), 76–108.

A. G. Kurosh (= A. G. Kuroš, = А. Г. Курош)
[1] Radicals of rings and algebras, *Mat. Sb.* 33 (1953), 13–26, (Russian), English translation: *Coll. Math. Soc. J. Bolyai 6, Rings, Modules and Radicals, Keszthely 1971*, North-Holland, 1973, pp. 297–312.

S. Lajos and F. Szász
[1] Biideals in associative rings, *Acta Sci. Math. Szeged* 32 (1971), 185–193.

J. Lawrence
[1] *Primitive group rings*, Thesis, McGill Univ., 1973.

R. R. Laxton
[1] A radical and its theory for distributively generated near-rings, *J. London Math. Soc.* 38 (1963), 40–49.

W. G. Leavitt
[1] Sets of radical classes, *Publ. Math. Debrecen* 14 (1967), 321–324.
[2] Strongly hereditary radicals, *Proc. Amer. Math. Soc.* 21 (1969), 703–705.
[3] Hereditary semisimple classes, *Glasgow Math. J.* 11 (1970), 7–8.
[4] Radical and semisimple classes with specified properties, *Proc. Amer. Math. Soc.* 24 (1970), 680–687.
[5] *The general theory of radicals*, University of Nebraska, Lincoln.
[6] General radical theory in rings, *Rocky Mountain J. Math.* 3 (1973), 431–439.
[7] The intersection property of an upper radical, *Archiv Math.* 24 (1973), 486–492.
[8] A minimally embeddable ring, *Period. Math. Hungar.* 12 (1981), 129–140.
[9] Hereditary upper radicals, *Studia Sci. Math. Hungar.* 16 (1981), 15–23.
[10] Upper radicals of regular classes, *Acta Math. Hungar.* 42 (1983), 213–220.
[11] Cogenerators of radicals, *Acta Math. Hungar.* 51 (1988), 79–83.
[12] A note on radicals and torsion theories, *Publ. Math. Debrecen* 37 (1990), 47–53.
[13] Matric-extensible radicals, *Rings and Radicals, Proc. Int. Conf., Shijiazhuang 1994*, Pitman Res. Notes in Math., 346, Longman 1996, pp. 211–216.
[14] A note on matric-extensibility and the ADS-condition, *Studia Sci. Math. Hungar.* 32 (1996), 407–414.

W. G. Leavitt and E. P. Armendariz
[1] Non-hereditary semi-simple classes, *Proc. Amer. Math. Soc.* 18 (1967), 1111–1117.

W. G. Leavitt and T. L. Jenkins
[1] Non-hereditariness of the maximal ideal radical class, *J. Nat. Sci. & Math., Lahore* 7 (1967), 202–205.

W. G. Leavitt and Lee Yu-lee
[1] A radical coinciding with the lower radical in associative and alternative rings, *Pacific J. Math.* 30 (1969), 459–462.

W. G. Leavitt and L. C. A. van Leeuwen
[1] Multiplier algebras and minimal embeddability, *Publ. Math. Debrecen* 29 (1982), 95–99.

W. G. Leavitt and J. F. Watters
[1] On smallest radical and semisimple classes, *Glasgow Math. J.* 12 (1971), 98–104.
[2] Special closure, M-radicals and relative complements, *Acta Math. Acad. Hungar.* 28 (1976), 55–67.

W. G. Leavitt and R. Wiegandt
[1] Torsion theory for not necessarily associative rings, *Rocky Mountain J. Math.* 9 (1979), 259–271.

P. H. Lee and R. Wiegandt
[1] On radicals and rings with involution, *PU. M. A. Ser A* 3 (1992), 219–224.

Lee Yu-lee (= Yu-lee Lee)
[1] On the construction of lower radical properties, *Pacific J. Math.* 28 (1969), 393–395.
[2] A note on the Jacobson radical, *Proc. Amer. Math. Soc.* 118 (1993), 337–338.

L. C. A. van Leeuwen
[1] A generalization of the Jenkins radical, *Archiv Math.* 22 (1971), 155–160.
[2] Properties of semisimple classes, *J. Nat. Sci. & Math. Lahore* 15 (1975), 59–67.
[3] Complements of radicals in the class of hereditarily artinian rings, *Acta Sci. Math. Szeged* 39 (1977), 313–318.

L. C. A. van Leeuwen and T. L. Jenkins
[1] A supernilpotent non-special radical class, *Bull. Austral. Math. Soc.* 9 (1973), 343–348.
[2] On the herditariness of the upper radical, *Archiv Math.* 25 (1974), 135–137.

L. C. A. van Leeuwen, C. Roos and R. Wiegandt
[1] Characterizations of semisimple classes, *J. Austral. Math. Soc.* 23 (1977), 172–182.

L. C. A. van Leeuwen and R. Wiegandt
[1] Radicals, semisimple classes and torsion theories, *Acta Math. Acad. Sci. Hungar.* 36 (1980), 37–47.
[2] Semisimple and torsionfree classes, *Acta Math. Acad. Sci. Hungar.* 38 (1981), 73–81.

J. Levitzki
[1] On the radical of a general ring, *Bull. Amer. Math. Soc.* 49 (1943), 462–466.
[2] On multiplicative systems, *Comp. Math.* 8 (1950), 76–80.
[3] A theorem on polynomial identities, *Proc. Amer. Math. Soc.* 1 (1950), 334–341.

W. Lex and R. Wiegandt
[1] Torsion theory for acts, *Studia Sci. Math. Acad. Hungar.* 16 (1981), 263–280.

Liu Shao-xue
[1] Recent research work on radicals in China, *Contr. to General Algebra 4, Proc. Krems Conf.* 1985, Hölder–Pichler–Tempsky, Wien and B. G. Teubner, Stuttgart, 1987, pp. 85–97.

Liu Shao-xue, Luo Yun-lun, Tang Ai-ping, Xiao Jie and Guo Jin-yun
[1] Some recent results on modules and rings, *Bull. Soc. Math. Belgique Sèr. B* 39 (1987), 181–193.

N. V. Loi
[1] On the Jenkins radical and semisimple rings, *Annales Univ. Sci. Budapest.* 26 (1983), 205–211.
[2] Essentially closed radical classes, *J. Austral. Math. Soc., Ser. A* 35 (1983), 132–142.
[3] On the radical theory of involution algebras, *Acta Univ. Carol. — Math. & Phys.* Prague, 27 (1986), 29–40.
[4] The ADS-property for radicals of involution K^*-algebras, *Archiv Math.* 49 (1987), 196–199.
[5] On the structure of semiprime involution rings, *General Alg., Proc. Krems Conf. 1988*, North-Holland, 1990, 155–161.
[6] A note on the radical theory of involution algebras, *Studia Sci. Math. Hungar.* 23 (1988), 157–160.
[7] Semisimple radical classes of involution algebras, *Proc. Edinburgh Math. Soc.* 32 (1989), 1–9.

N. V. Loi and R. Wiegandt
[1] Small ideals and the Brown–McCoy radical, *Coll. Math. Soc. J. Bolyai 38, Radical Theory, Eger 1982*, North-Holland 1985, pp. 253–263.
[2] Involution algebras and the Anderson–Divinsky–Suliński property, *Acta Sci. Math. Szeged* 50 (1986), 5–14.
[3] On involution rings with minimum condition, *Ring Theory, Israel Math. Conf. Proc.* 1 (1989), 203–214.

I. V. L'vov (= И. А. Львов)
[1] The existence of a simple nil ring, *Preprint*, Sobolev Inst. Math., Novosibirsk, March 2003, pp 85.

I. A. L'vov and A. V. Sidorov (= А. В. Сидоров)
[1] On the stabilization of Kurosh chains, (Russian), *Mat. Zametki* 36 (1984), 815–821.

S. Majumdar
[1] Semi-simple radical classes, *J. Bangladesh Acad. Sci.* 1 no 2 (1977), 39–44.

L. Márki, R. Mlitz and R. Wiegandt
[1] A general Kurosh–Amitsur radical theory, *Comm. in Algebra* 16 (1988), 249–305.

L. Márki, P. N. Stewart and R. Wiegandt
[1] Radicals and decomposability of semigroups and rings, *Annales Univ. Sci. Budapest.* 18 (1975), 27–36.

J. Martinez
[1] Torsion theory for lattice-ordered groups, *Czechosl. J. Math.* 25 (1975), 284–299.

J. McConnell and J. C. Robson
[1] *Noncommutative noetherian rings*, John Wiley & Sons, 1987.

N. R. McConnell
[1] Hereditary and strict domains for radical classes of associative rings, *Bull. Austral. Math. Soc.* 41 (1990), 255–269.

[2] Radical ideal of Dedekind domains and their extensions, *Comm. in Algebra* 19 (1991), 559–583.

N. R. McConnell and T. Stokes
[1] Rings having simple adjoint semigroup, *Contemporary Mathematics* 273 (2001), 203–208.

N. H. McCoy
[1] Subdirect sums of rings, *Bull. Amer. Math. Soc.* 53 (1947), 856–877.
[2] *Rings and ideals*, Carus Math. Monogr. 8, Math. Assoc. America, 1948.
[3] Prime ideals in general rings, *Amer. J. Math.* 71 (1949), 823–833.
[4] The prime radical of a polynomial ring, *Publ. Math. Debrecen* 4 (1956), 161–162.

J. D. McKnight and G. L. Musser
[1] Special (p,q)-radicals, *Canad. J. Math.* 24 (1972), 38–44.

W. A. McWorter
[1] Some simple properties of simple nil rings, *Canad. Math. Bull.* 9 (1966), 197–200.

Yu. A. Medvedev (= Ю. А. Медведев)
[1] An analogue of the Andrunakievich lemma for linear Jordan rings, (Russian), *Sibirsk. Mat. Zh.* 28 no 6 (1987), 81–89; English transl.: *Siberian Math. J.* 28 (1987), 928–936.

J. D. P. Meldrum
[1] *Near-rings and their links with groups*, Pitman Res. Notes in Math., 134 London, 1985.

D. Isabel C. Mendes
[1] Some characterizations of supernilpotent and hereditarily idempotent radical classes, *Quaest. Math.* 13 (1990), 1–15.
[2] Left hereditary supernilpotent radical classes, *Acta Math. Hungar.* 71 (1996), 123–129.
[3] A note on left strong supernilpotent radical classes, *Acta Math. Hungar.* 76 (1997), 351–355.

D. Isabel C. Mendes and S. Tumurbat
[1] Left subhereditary supernilpotent radical classes, *Math. Notes, University Miskolc* 2 (2000), 75–80.

R. Mlitz
[1] Ein Radikal für universale Algebren und seine Anwendung auf Polynomringe mit Komposition, *Monatshefte für Math.* 75 (1971), 144–152.
[2] Radicals and semisimple classes of Ω-groups, *Proc. Edinburgh Math. Soc.* 23 (1980), 37–41.
[3] Extending radicals of Ω-groups, *Rings and Radicals, Proc. Int. Conf., Shijiazhuang 1994*, Pitman Res. Notes in Math. 346, Longman, 1996, pp. 230–248.
[4] Smallest extensions of radicals with given properties, *Proc. Conf. General Alg. and Appl. in Discr. Math., Potsdam 1996*, Shaker Verlag, Aachen, 1997, pp. 153–157.

R. Mlitz and A. Oswald
[1] Hypersolvable and supernilpotent radicals of near-rings, *Studia Sci. Math. Hungar.* 24 (1989), 239–258.

R. Mlitz, A. D. Sands and R. Wiegandt
[1] Radicals coinciding with the von Neumann regular radical on artinian rings, *Monatshefte für Math.*, 125 (1998), 229–239.

R. Mlitz and S. Veldsman
[1] Radicals and subdirect decompositions of Ω-groups, *J. Austral Math. Soc., Ser. A* 48 (1990), 171–198.

R. Mlitz and R. Wiegandt
[1] Semisimple classes of hypernilpotent and hyperconstant near-ring radicals, *Archiv Math.* 63 (1994), 414–419.
[2] Near-ring radicals depending only on the additive groups, *Southeast Asian Bull. Math.* 22 (1998), 171–177.
[3] Semisimple classes containing no trivial near-rings, *Studia Sci. Math. Hungar.* 38 (2201), 331–337.

Bettina Morak
[1] On the radical theory for semirings, *Beiträge Alg. und. Geom.* 40 (1999), 533–549.

K. Morita
[1] Duality for modules and its applications to the theory of rings with minimum condition, *Sci. Rep. Tokyo Kyoiku Daigaku, Sect. A* 6 (1958), 83–142.
[2] Category-isomorphisms and endomorphism rings of modules, *Trans. Amer. Math. Soc.* 103 (1962), 451–469.

D. Morris
[1] A new example of a Jacobson radical simple ring, *Indiana Univ. Math. J.* 23 (1974), 591–592.

G. L. Musser
[1] On linear semiprime (p,q)-radicals, *Pacific J. Math.* 37 (1971), 749–757.

M. Nagata
[1] On the theory of radicals of a ring, *J. Math. Soc. Japan* 3 (1951), 330–344.

W. K. Nicholson and J. F. Watters
[1] Normal radicals and normal classes of rings, *J. Algebra* 59 (1979), 5–15.

A. A. Nikitin (= А. А. Никитин)
[1] On heredity of radicals in Jordan rings, (Russian), *Alg. i Logika* 17 (1978), 303–315.
[2] On lower radicals of rings, (Russian), *Alg. i Logika* 17 (1978), 596–610; English transl.: *Alg. and Logic* 17 (1978), 392–401.

D. M. Olson
[1] A uniformly strongly prime radical, *J. Austral. Math. Soc.* 43 (1987), 95–102.

D. M. Olson and T. L. Jenkins
  [1] Upper radicals and essential ideals, *J. Austral. Math. Soc.* 30 (1981), 385–389.
  [2] Radical theory for hemirings, *J. Nat. Sci. & Math., Lahore* 23 (1983), 23–32.

D. M. Olson and S. Veldsman
  [1] Some remarks on uniformly strongly prime radicals, *Math. Japonica* 33 (1988), 445–455.

J. Olszewski
  [1] The Brown–McCoy radical class is not closed under products, *Comm. in Algebra* 22 (1994), 117–122.

A. H. Ortiz
  [1] A construction in general radical theory, *Canad. J. Math.* 22 (1970), 1097–1100.

B. Osłowski and E. R. Puczyłowski
  [1] On strong radical properties of alternative algebras, *Bull. Acad. Polon. Sci.* 25 (1977), 845–850.

Th. Palmer
  [1] Banach algebras and the general theory of *-algebras I, Algebras and Banach algebras, *Encyclopedia Math. & Appl.* 49, Cambridge Univ. Press, 1994.

M. M. Parmenter, D. S. Passman and P. N. Stewart
  [1] The strongly prime radical of crossed products, *Comm. in Algebra* 12 (1984), 1099–1113.

M. M. Parmenter, P. N. Stewart and R. Wiegandt
  [1] On the Groenewald–Heyman strongly prime radical, *Quaest. Math.* 7 (1984), 225–240.

D. S. Passman
  [1] A new radical for group rings?, *J. Algebra* 28 (1974), 556–572.
  [2] *The algebraic structure of group rings*, John Wiley & Sons, 1977.

E. M. Patterson
  [1] On the radicals of certain infinite matrices, *Proc. Royal Soc. Edinburgh* 65 A (1960), 263–271.
  [2] On the radicals of row-finite matrices, *Proc. Royal Soc. Edinburgh* 66 A (1961/62), 42–46.

S. V. Pchelintsev (= С. В. Пчелинцев)
  [1] On metaideals of alternative algebras, (Russian), *Sibirsk. Mat. Zh.* 24 (1983), 142–148.
  [2] Prime algebras and absolute zero divisors (Russian) *Izv. Akad. Nauk SSSR* 50 (1986), 79–100.

S. Perlis
  [1] A characterization of the radical of an algebra, *Bull. Amer. Math. Soc.* 48 (1942), 128–132.

G. Pilz
  [1] *Near-rings*, North-Holland, 1983.

B. I. Plotkin (= Б. И. Плоткин)
  [1] On functorials, radicals and coradials in groups, (Russian), *Mat. Zapiski* 3 (1970), 150–182.
  [2] Radicals in groups operations on classes of groups, and radical classes, (Russian), *Selected Questions of Algebra and Logic*, Nauka, Novosibirsk 1973, pp. 205–245.
  [3] *Groups of automorphisms of algebraic systems*, Wolters–Nordhooff, Groningen, 1972.

G. Preuß
  [1] Eine Galois-Korrespondenz in der Topologie, *Monatshefte für Math.* 75 (1971), 447–452.

C. Procesi
  [1] *Rings with polynomial identities*, Marcel Dekker, 1973.

R. E. Propes
  [1] A characterization of the Behrens radical, *Kyungpook Math. J.* 10 (1970), 49–52.
  [2] Radicals of PID's and Dedekind domains, *Canad. J. Math.* 24 (1972), 566–572.
  [3] The radical equation $P(A_n) = (P(A))_n$, *Proc. Edinburgh Math. Soc.* 19 (1974/75), 257–259.

E. R. Puczyłowski
  [1] Radicals of polynomial rings, power series rings and tensor products, *Comm. in Algebra* 8 (1980), 1699–1709.
  [2] On unequivocal rings, *Acta Math. Acad. Sci.* 36 (1980), 57–62.
  [3] Remarks on stable radicals, *Bull. Acad. Polon. Sci.* 28 (1980), 11–16.
  [4] A note on hereditary radicals, *Acta Sci. Math. Szeged.* 44 (1982), 133–135.
  [5] Hereditariness of strong and stable radicals, *Glasgow Math. J.* 23 (1982), 85–90.
  [6] Radicals of polynomial rings in non-commutative indeterminates, *Math. Zeitschr.* 182 (1983), 63–67.
  [7] Nil ideals of power series rings, *J. Austral. Math. Soc.* 34 (1983), 287–292.
  [8] On semisimple classes of associative and alternative rings, *Proc. Edinburgh Math. Soc.* 27 (1984), 1–5.
  [9] The behaviour of radical properties of rings under some algebraic constructions, *Coll. Math. J. Bolyai 38, Radical Theory, Proc. Conf. Eger 1982*, North-Holland, 1985, pp. 449–480.
  [10] On Sands question concerning strong and hereditary radicals, *Glasgow Math. J.* 28 (1986), 1–3.
  [11] On questions concerning strong radicals of associative rings, *Quaest. Math.* 10 (1987), 321–338.
  [12] On normal classes of rings, *Comm. in Algebra* 20 (1992), 2999–3013.

[13] On Koethe's Problem, *Inst. Mat. UFRGS, Série A*, No 36, Porto Alegre 1993.
[14] Some questions concerning radicals of associative rings, *Coll. Soc. Math. J. Bolyai 61, Theory of Radicals, Szekszárd 1991*, North-Holland, 1993, pp. 209–227.
[15] On general theory of radicals, *Alg. Universalis* 30 (1993), 53–60.
[16] Radicals of a ring, *Comm. in Algebra.*, 22 (1994), 5419–5436.
[17] Some results and questions on nil rings, *Mat. Contemp.* 16 (1999), 265–280.

E. R. Puczyłowski and Agata Smoktunowicz
[1] On maximal ideals and the Brown–McCoy radical of polynomial rings, *Comm. in Algebra* 26 (1998), 2473–2482.
[2] The nil radical of power series rings, *Israel J. Math.* 110 (1999), 317–324.

E. R. Puczyłowski and H. Zand
[1] On squares of Jacobson radical rings, *Bull. Austral. Math. Soc.* 53 (1996), 299–303.
[2] The Brown–McCoy radical and one-sided ideals, *Quaest. Math.* 19 (1996), 47–58.
[3] Subhereditary radicals of associative rings, *Algebra Coll.* 6 (1999), 215–223.

J. G. Raftery
[1] On some special classes of prime rings, *Quaest. Math.* 10 (1987), 257–263.

J. G. Raftery and J. E. van den Berg
[1] On a classification of prime rings, *Quaest. Math.* 15 (1992), 139–150.
[2] On prime rings and the radicals associated with their degrees of primeness, *Quaest. Math.* 18 (1995), 437–475.

D. Ramakotaiah
[1] Radicals for near-rings, *Math. Zeitschr.* 97 (1967), 45–66.

M. A. Rashid and R. Wiegandt
[1] The hereditariness of the upper radical, *Acta Math. Acad. Sci. Hungar.* 24 (1973), 343–347.

D. I. Riley
[1] Infinitesimally PI radical algebra, *Israel J. Math.* 123 (2001), 365–379.

J. C. Robson
[1] Do simple rings have unity elements? *J. Algebra* 7 (1967), 140–143.

C. Roos
[1] *Regularities of rings*, Doctoral Dissertation, Technische Hogeschool Delft, 1975.
[2] A class of regularities for rings, *J. Austral. Math. Soc.* 27 (1979), 437–453.

B. de la Rosa
[1] *Ideals and radicals*, Doctoral Dissertation, Technische Hogeschool Delft, 1970.

[2] A radical class which is fully determined by a lattice isomorphism, *Acta Sci. Math. Szeged* 33 (1972), 337–341.

[3] On a refinement in the classification of the nil radical, *Portugal. Math.* 36 (1977), 49–60.

B. de la Rosa, Y. Fong and R. Wiegandt
[1] Complementary radicals revisited, *Acta Math. Hungar.* 65 (1994), 253–264.

B. de la Rosa, Julia S. van Niekerk and R. Wiegandt
[1] A concrete analysis of the radical concept, *Math. Pannonica* 3/2 (1992), 3–15, corrigendum: *ibidem* 4/1 (1993), 151.

B. de la Rosa, S. Veldsman and R. Wiegandt
[1] On the theory of Plotkin radicals, *Chinese J. Math.* 21 (1993), 33–54.

B. de la Rosa and R. Wiegandt
[1] Characterizations of the Brown–McCoy radical, *Acta Math. Hungar.* 46 (1985), 129–132.
[2] On dual radicals and ring elements, *J. Austral. Math. Soc., Ser. A* 44 (1988), 164–170.

R. F. Rossa
[1] More properties inherited by the lower radical, *Proc. Amer. Math. Soc.* 33 (1972), 247–249.
[2] Radical properties involving one-sided ideals, *Pacific J. Math.* 49 (1973), 467–471.
[3] On stability and radical theory, *unpublished manuscript.*

R. F. Rossa and R. Tangeman
[1] General heredity for radical theory, *Proc. Edinburgh Math. Soc.* 20 (1976–77), 333–337.

H. J. le Roux, G. A. P. Heyman and T. L. Jenkins
[1] Essentially closed classes of rings and upper radicals, *Acta Math. Acad. Sci. Hungar.* 38 (1981), 63–68.

L. H. Rowen
[1] *Ring theory I, II,* Academic Press, 1988.
[2] Koethe's conjecture, *Israel Math. Conf. Proc.* 1 (1989), 193–202.

R. A. Rubin
[1] Absolutely torsionfree rings, *Pacific J. Math.* 46 (1973), 503–514.

Yu. M. Ryabukhin (= Ju. M. Rjabuhin = Ю. М. Рябухин)
[1] On overnilpotent and special radicals, (Russian), *Issled. Alg. Mat. Anal., Kishinev* (1965), 65–72.
[2] Radicals in categories, (Russian), *Mat. Issled. Kishinev* 2 (1967), No. 3, 107–165.
[3] On lower radicals of rings, (Russian), *Mat. Zametki* 2 (1967), 239–244.; English transl.: *Math. Notes* 2 (1968), 631–633.
[4] The theory of radicals in non-associative rings, (Russian), *Mat. Issled. Kishinev* 3. vyp. 1 (7) (1968), 86–99.
[5] A countable simple quasi-regular ring, (Russian), *Mat. Zametki* 4 (1968), 399–403; English transl.: *Math. Notes* 4 (1968), 731–733.

[6] Supernilpotent and special radicals, (Russian), *Mat. Issled. Kishinev* 48 (1978), 80–83.
[7] Private communication, 1985.

Yu. M. Ryabukhin and K. K. Krachilov
[1] Lattice complementary torsions in algebras, (Russian), *Mat. Issled. Kishinev* 48 (1978), 94–111.

Ju. M. Rjabuhin and R. Wiegandt
[1] On special radicals, supernilpotent radicals and weakly homomorphically closed classes, *J. Austral. Math. Soc.* 31 (1981), 152–162.

I. Sakhajev (= И. И. Сахаев)
[1] Negative solution hypothesis of Koethe and some problems connected with it, *Algebra Conference Venezia 2002, Abstract of Talks*, pp 37–38.

Katarina Salavová
[1] Radicals of rings with an involution I, (Russian), *Comment. Math. Univ. Carol.* 16 (1977), 367–381.

A. D. Sands
[1] Radicals and Morita contexts, *J. Algebra* 24 (1973), 335–345.
[2] On normal radicals, *J. London Math. Soc.* 11 (1975), 361–365.
[3] Strong upper radicals, *Quart. J. Math. Oxford* 27 (1976), 21–24.
[4] Relations among radical properties, *Glasgow Math. J.* 18 (1977), 17–23.
[5] A characterization of semisimple classes, *Proc. Edinburgh Math. Soc.* 24 (1981), 5–7.
[6] Weakly homomorphically closed semisimple classes of subidempotent radicals, *Acta Math. Acad. Sci. Hungar.* 39 (1982), 179–183.
[7] On M-nilpotent rings, *Proc. Royal Soc. Edinburgh* 93 A (1982), 63–70.
[8] Radical properties and one-sided ideals, *Contr. General Algebra 4, Proc. Krems Conf. 1985*, Hölder–Pichler–Tempsky, Wien & B. G. Teubner, Stuttgart, 1987, pp. 151–171.
[9] Radicals and one-sided ideals, *Proc. Royal Soc. Edinburgh* 103 A (1986), 241–251.
[10] On ideals in over-rings, *Publ. Math. Debrecen* 35 (1988), 245 and 274–279.
[11] On dependence and independence among radical properties, *Rings, Modules and Radicals, Proc. Hobart Conf. 1987*, Pitman Res. Notes in Math., 204, Longman Sci. & Tech., 1989, pp. 125–141.
[12] Radicals of structural matrix rings, *Quaest. Math.* 13 (1990), 77–81.
[13] Some subidempotent radicals, *Coll. Math. Soc. J. Bolyai 63, Theory of Radicals, Szekszárd 1991*, North-Holland 1993, pp. 239–248.
[14] Radicals which coincide on a class of rings, *Math. Pannonica*, 8 (1997), 261–267.
[15] On radicals of infinite matrix rings, *Proc. Edinburgh Math. Soc.* 16 (1968/69), 195–203.

A. D. Sands and P. N. Stewart
[1] General heredity and strength for radical classes *Canad. J. Math.* 40 (1988), 1410–1421.

A. D. Sands and S. Tumurbat
  [1] On the semisimple classes of left hereditary radicals, *Bul. Acad. Ştiinţe Rep. Moldova, Matematica* 2(36) (2001), 97–100.

E. Sąsiada
  [1] Solution of the problem of existence of a simple radical ring, *Bull. Acad. Polon. Sci.* 9 (1961), 257.

E. Sąsiada and P. M. Cohn
  [1] An example of a simple radical ring, *J. Algebra* 5 (1967), 373–377.

E. Sąsiada and A. Suliński
  [1] A note on the Jacobson radical, *Bull. Acad. Polon. Sci.* 10 (1962), 421–423.

E. G. Shul'geĭfer (= Е. Г. Шульгейфер)
  [1] General theory of radicals in categories, (Russian), *Mat. Sb.* 51 (1960), 487–500.

A. V. Sidorov (= А. В. Сидоров)
  [1] On the stabilization of Kurosh chains in the class of semigroups with zero, (Russian), *Sibir. Mat. Zh.* 29 (3) (1988), 131–136.

Anna Sierpińska
  [1] Radicals of rings of polynomials in non-commutative indeterminates, *Bull. Acad. Polon. Sci.* 21 (1973), 805–808.

A. M. Slińko (= А. М. Слинько)
  [1] On radicals of Jordan algebras, (Russian), *Alg. i Logika* 11 (1972), 206–215.

Rebecca Slover
  [1] The Jacobson radical of row-finite matrices, *J. Algebra* 12 (1969), 345–359.
  [2] A note on the radical of row-finite matrices, *Glasgow Math. J.* 13 (1972), 80–81.

Agata Smoktunowicz
  [1] Polynomial rings over nil rings need not be nil, *J. Algebra* 233 (2000), 427–436.
  [2] A simple nil ring exists, *Comm. in Algebra* 30 (2002), 27–59.

Agata Smoktunowicz and E. R. Puczyłowski
  [1] A polynomial ring that is Jacobson radical and not nil, *Israel J. Math.* 124 (2001), 317–325.

R. L. Snider
  [1] Complemented hereditary radicals, *Bull. Austral. Math. Soc.* 4 (1971), 307–320.
  [2] Lattices of radicals, *Pacific J. Math.* 42 (1972), 207–220.

A. G. Sokolskiĭ (= А. Г. Соколский)
  [1] N-radicals of rings of matrices of the Rees type, (Russian), *Izv. Vyssh. Uchebn. Zaved. Mat.* 1992 no. 3, 56–57. English transl.: *Russian Mathematics (Iz. VUZ)* 36 (1992), 54–55.

O. Steinfeld
 [1] Über Quasiideale von Ringen, *Acta Sci. Math. Szeged* 17 (1956), 170–180.
 [2] *Quasi-ideals in rings and semigroups*, Akadémiai Kiadó, Budapest, 1978.

P. N. Stewart
 [1] Semi-simple radical classes, *Pacific J. Math.* 32 (1970), 249–254.
 [2] Strict radical classes of associative rings, *Proc. Amer. Math. Soc.* 39 (1973), 273–278.
 [3] On the lower radical construction, *Acta Math. Acad. Sci. Hungar.* 25 (1974), 31–32.

H. H. Storrer
 [1] On Goldman's primary decomposition, *Lectures on ring and modules*, Lecture Notes in Math. 246, Springer-Verlag 1972, 618–661.

A. Suliński
 [1] On subdirect sums of simple rings with unity, *Bull. Acad. Polon. Sci.* 8 (1960), 223–228.
 [2] A classification of semisimple rings, *Bull. Acad. Polon. Sci.* 9 (1961), 1–6.
 [3] Certain questions in the general theory of radicals (Russian), *Mat. Sb.* 44 (1958), 273–286.

A. Suliński, T. Anderson and N. Divinsky
 [1] Lower radical properties for associative and alternative rings, *J. London Math. Soc.* 41 (1966), 417–424.

F. A. Szász (= F. Szász)
 [1] Über Ringe mit Minimalbedingung für Hauptrechtsideale III, *Acta Math. Acad. Sci. Hungar.* 14 (1963), 447–461.
 [2] Über artinsche Ringe, *Bull. Acad. Polon. Sci.* 11 (1963), 351–354.
 [3] Bemerkungen über Rechtsockel und Nilringe, *Monatsh. Math.* 67 (1963), 359–362.
 [4] On simple Jacobson radical rings, *Math. Japonica* 18 (1973), 225–228.
 [5] A solution of a problem on upper radicals of rings, *Math. Nachr.* 66 (1975), 401.
 [6] *Radicals of rings*, Akadémiai Kiadó, Budapest and John Wiley & Sons, 1981; German original: *Radikale der Ringe*, Akadémiai Kiadó, Budapest and Deutscher Verlag der Wissenschaften, Berlin, 1975.

T. Szele
 [1] Nilpotent artinian rings, *Publ. Math. Debrecen* 4 (1955), 71–78.

T. Szele and L. Fuchs
 [1] On artinian rings, *Acta Sci. Math. Szeged* 17 (1956), 30–40.

J. Szendrei
 [1] On rings admitting only direct extensions, *Publ. Math. Debrecen* 3 (1953), 180–182.

R. Tangeman and D. Kreiling
 [1] Lower radicals in nonassociative rings, *J. Austral. Math. Soc.* 14 (1972), 419–423.

Barbara Terlikowska-Osłowska (= B. Terlikowska)
[1] Category with a selfdual set of axioms, *Bull. Acad. Polon. Sci.* 25 (1977), 1207–1214.
[2] Radical and semisimple classes of objects in categories with self-dual set of axioms, *Bull. Acad. Polon. Sci.* 26 (1978), 7–13.

G. Thierrin
[1] Sur les idéaux complétement premiers d'un anneau quelconque, *Acad. Royal Belgique Bull.* 43 (1957), 124–132.
[2] Sur le radical corpoïdal d'un anneau, *Canad. J. Math.* 12 (1960), 101–106.

S. Tumurbat
[1] On principally left strong radicals, *Acta Math. Hungar.* 93 (2001), 99–108.
[2] The radicalness of polynomial rings over nil rings, *Math. Pannonica* 13 (2002), 191–199.
[3] On special radicals coinciding on simple rings and on polynomial rings, *J. Algebra & Appl.* to appear.
[4] A note on normal radicals and principally left and right strong radicals, *Southeast Asian Bull. Math.* to appear.
[5] Some issues in the theory of supernilpotent radicals, *Bul. Acad. Ştiinţe Rep. Moldova* 1(38) (2002), 3–43.

S. Tumurbat and R. Wiegandt
[1] Principally left hereditary and principally left strong radicals, *Algebra Coll.* 8 (2001), 409–418.
[2] A note on special radicals and partitions of simple rings, *Comm. in Algebra* 30 (2002), 1769–1777.
[3] On subhereditary radicals and Brown-McCoy semisimple rings, *Bul Acad. Ştiinţe Rep. Moldova, Matematica* 3(34) (2000), 11–20.
[4] On polynomial and multilpicative radicals, *Preprint*, 2002.
[5] On radicals with Amitsur property, *Preprint*, 2002.

S. Tumurbat and H. Zand
[1] Hereditariness, strongness and relationship between Brown–McCoy and Behrens radicals, *Beiträge Alg. und Geom.* 42 (2001), 275–280.
[2] On subhereditary and principally left strong radicals, *East-West J. Math.* 3 (2001), 101–108.
[3] On principally left hereditary radicals and A-radicals, *Kyungpook Math. J.* 43 (2003), 19–26.

Y. Utumi
[1] A theorem on Levitzki, *Amer. Math. Monthly* 70 (1963), 286.

S. Veldsman
[1] *A general radical theory in categories*, Ph. D. Thesis, Univ. Port Elizabeth, S. A., 1980.
[2] On characterizations of radical and semisimple classes in categories, *Comm. in Algebra* 10 (1982), 913–938.
[3] The elements in the strongly prime radical of a ring, *Math. Japonica* 31 (1986), 283–285.

[4] The superprime radical, *Contributions to general Algebra 4, Proc. Krems Conf. 1985*, Hölder–Pichler–Tempsky, Wien & B. G. Teubner, Stuttgart, 1987, pp. 181–188.
[5] Properties of ring elements that determine supernilpotent and special radicals, *Acta Math. Hungar.* 50 (1987), 249–252.
[6] Supernilpotent radicals of near-rings, *Comm. in Algebra* 15 (1987), 2497–2509.
[7] Subidempotent radical classes, *Quaest. Math.* 11 (1988), 361–370.
[8] On the non-hereditariness of radical and semisimple classes of near-rings, *Studia Sci. Math. Hungar.* 24 (1989), 315–323.
[9] Sufficient condition for a well-behaved Kurosh–Amitsur radical theory, *Proc. Edinburgh Math. Soc.* 32 (1989), 377–394.
[10] Sufficient condition for a well-behaved Kurosh–Amitsur radical theory II, *Rings, Modules and Radicals, Proc. Conf. Hobart 1987*, Pitman Res. Notes in Math. 204, Longman Sci. & Tech., 1989, pp 142–152.
[11] Near-ring radicals with hereditary semisimple classes, *Archiv Math.* 54 (1990), 443–447.
[12] Extensions of ideals of rings, *Publ. Math. Debrecen* 38 (1991), 297–309.
[13] An overnilpotent radical theory for near-rings, *J. Algebra* 144 (1991), 248–265.
[14] To the abstract theory of radicals: a contribution from near rings, *Coll. Math. Soc. J. Bolyai 61, Theory of Radicals, Szekszárd 1991*, North-Holland, 1993, pp. 275–296.
[15] The general radical theory of near-rings — answers to some open problems, *Alg. Universalis* 36 (1996), 185–189.
[16] On the radical of structural matrix rings, *Monatshefte für Math.* 122 (1996), 227–238.
[17] The general radical theory of incidence algebras, *Comm. in Algebra* 27 (1999), 3659–3673.
[18] Connectednesses and disconnectednesses of Petri nets, *Applied Categorical Structures* 8 (2000), 485–504.

A. P. J. van der Walt
[1] Prime rings — the strong and not so strong, *Proc. 3rd Alg. Symp., Pretoria 1981*, pp. 91–97.

Wang Xion Hau
[1] On Köthe semisimple rings, (Chinese), *Bull. North-East People's University, Ji-Lin* 1 (1955), 143–147.

J. F. Watters
[1] Lower radicals in associative rings, *Canad. J. Math.* 21 (1969), 466–476.
[2] On internal constructions of radicals in rings, *J. London Math. Soc.* 1 (1969), 500–504.
[3] Polynomial extensions of Jacobson rings, *J. Algebra* 36 (1975), 302–308.
[4] The Brown–McCoy radical and Jacobson rings, *Bull. Acad. Sci. Polon.* 24 (1976), 91–99.
[5] Essential cover and closure, *Annales Univ. Sci. Budapest.* 25 (1982), 279–280.
[6] On the lower radical construction for algebras, *Preprint*, 1985.

[7] A note on strongly prime radicals, *Acta Math. Hungar.* 69 (1995), 169–174.

J. H. M. Wedderburn
[1] On hypercomplex numbers, *Proc. London Math. Soc.* (2) 6 (1908), 77–118.

H. J. Weinert and R. Wiegandt
[1] A Kurosh–Amitsur radical theory for proper semifields, *Comm. in Algebra* 20 (1992), 2419–2458.

A. Widiger
[1] A general decomposition for artinian rings, *Studia Sci. Math. Hungar.* 12 (1977), 29–36.
[2] Decompostion of artinian alternative rings, *Beitr. Alg. und Geom., Halle DDR* 12 (1982), 57–71.

R. Wiegandt
[1] Über transfinit nilpotente Ringe, *Acta Math. Acad. Sci. Hungar.* 17 (1966), 101–114.
[2] Radicals coinciding with the Jacobson radical on linearly compact rings, *Beitr. Alg. und Geom., Halle DDR* 1 (1971), 195–199.
[3] Homomorphically closed semisimple classes, *Studia Univ. Babeş–Bolyai, Cluj, Ser. Math.-Mech.* 17 (1972), no 2, 17–20.
[4] *Radical and semisimple classes of rings*, Queen's Papers in Pure & Appl. Math. 37, Kingston Ontario, 1974.
[5] Radicals of rings defined by means of elements, *Sitzungsber. Österr. Akad. d. Wiss., Mathem.-naturw. Klasse* 184 (1975), 117–125.
[6] Near-rings and radical theory, *Near-rings and Near-fields*, Proc. Conf. San Benedetto del Tronto 1981, pp. 49–58.
[7] On the structure of involution rings with chain condition, *Tap Chí Toán Hoc (J. Math. Vietnam Math. Soc.)* 21 (1993), 1–12.
[8] A note on supplementing radicals, *Northeastern Math. J.* 11 (1995), 476–482.
[9] Rings distinctive in radical theory, *Quaest. Math.* 22 (1999), 303–328.

Wu Tong-suo
[1] On essentially supernilpotent rings and their dual, *Comm. in Algebra* 23 (1995), 4473–4479.

L. van Wyck
[1] *Subrings of matrix rings*, Ph. D. Thesis, Univ. Stellenbosch, 1986.
[2] Maximal left ideals in structural matrix rings, *Comm. in Algebra* 16 (1988), 399–419.
[3] Special radicals in structural matrix rings, *Comm. in Algebra* 16 (1988), 421–435.

Xu Yong Hua (= Hsü Yong-hua)
[1] A theory of rings that are isomorphic to complete rings of linear transformations I, Chinese), *Acta Math. Sinica* 22 (1979), 204–218.
[2] A theory of rings that are isomorphic to complete rings of linear transformations II, (Chinese), *Acta Math. Sinica* 22 (1979), 303–315.

- [3] A theory of rings that are isomorphic to complete rings of linear transformations III, (Chinese), *Acta Math. Sinica* 22 (1979), 389–403.
- [4] A theory of rings that are isomorphic to complete rings of linear transformations IV, (Chinese), *Acta Math. Sinica* 22 (1979), 556–568.
- [5] A theory of rings that are isomorphic to complete rings of linear transformations V, (Chinese), *Acta Math. Sinica* 23 (1980), 547–553.
- [6] A theory of rings that are isomorphic to complete rings of linear transformations VI, (Chinese), *Acta Math. Sinica* 23 (1980), 646–657.
- [7] On the strucure of primitive rings having minimal one-sided ideals, *Sci. Sinica* 24 (1981), 1056–1065.

E. I. Zel'manov (= Е. И. Зельманов)
- [1] An example of a finitely generated prime ring, (Russian), *Sibirsk. Mat. Zh.* 20 (1979), 423.

J. Zelmanowitz
- [1] Weakly primitive rings, *Comm. in Algebra* 9 (1981), 23–45.

K. A. Zhevlakov, A. M. Slin'ko, I. P. Shestakov and A. I. Shirshov (= К. А. Жевлаков, А. М. Слинько, И. П. Шестаков and А. И. Ширшов)
- [1] *Rings that are nearly associative*, Academic Press, 1982; Russian original: Кольца близкие к ассоциативным, Наука, Москва, 1978.

Zhou Yiqiang
- [1] On the supplementing radial of a radical class, (Chinese), *Natural Science Journal of Hunan Normal Univ.* 12 (1989), 197–199.

M. Zorn
- [1] A remark on a method in transfinite algebra, *Bull. Amer. Math. Soc.* 41 (1935), 667–670.

List of Symbols

$\subseteq$	containment
$\subset$	proper containment
$\vert\ \vert$	cardinality
$(A, +, \cdot)$	set A with operation $+$ and
$\mapsto$	designating an element by a mapping
$\{\ \}$	set of elements
A^+	additive group of a ring A
A^0	zero-ring on the additive group A^+
$\mathbb{Z}$	ring of integers
$\triangleleft_l$	left ideal
$\triangleleft_r$	right ideal
$\triangleleft$	two-sided ideal
A^2	subring consisting of finite sums of products of elements of A
$\langle a \rangle$	subring generated by the element a
$(a]$	principal left ideal generated by the element a
$[a)$	principal right ideal generated by the element a
(a)	principal ideal generated by the element a
$\to$	mapping
$\twoheadrightarrow$	surjection ($=$ surjective homomorphism)
$\cong$	isomorphism
$\ker f$	kernel of the homomorphism f
Π	direct product
$(\ldots, a_\lambda, \ldots)$	element of a cartesian product
$\oplus$	direct sum

$\sum_{\text{subdirect}}$	subdirect sum
$\mathbb{A}$	universal class
$H(A)$	heart of the ring A
$M_n(A)$	$n \times n$ matrix ring over A
$(a_{ij})_{n \times n}$	$n \times n$ matrix
$(S_{ij})_{n \times n}$	$n \times n$ matrices with elements $a_{ij} \in S_{ij}$ at the (i,j) position
s_n	standard polynomial of degree n
$C(\infty)$	infinite cyclic group
$Z(\infty)$	zero-ring on $C(\infty)$
$C(p^\infty)$	quasi-cyclic or Prüfer group
$Z(p^\infty)$	zero-ring on $C(p^\infty)$
$o(a)$	order of the group element a
$\mathbb{Q}$	field of rational numbers
A_p	p-component of the ring A
ann A	annihilator of the ring A
A^1	Dorroh extension of the ring A
$A[x]$	polynomial ring with coefficients from A
$\overline{K}$	the ideal generated by the subring K
End (V,V)	ring of linear transformations of the vector space V
$\partial(f)$	degree of the polynomial f
$A[x; \sigma, \delta]$	skew (or twisted) polynomial ring with endomorphism σ and derivation δ
W_1	first Weyl algebra
F^σ	fixed subring of the automorphisms σ in a field F
$A[X]$	polynomials with commuting indeterminates from the set X
$A\langle X \rangle$	polynomials with non-commuting indeterminates from the set X
$A[[x]]$	formal power series ring over the ring A
$A[[X]]$	formal power series ring with commuting indeterminates from X
$A\langle\langle X \rangle\rangle$	formal power series ring with non-commuting indeterminates from X
$A[x, x^{-1}]$	Laurent polynomial ring over the ring A
$A[x, x^{-1}, \sigma]$	twisted Laurent polynomial ring
$\gamma(A)$	γ radical of the ring A
$\mathcal{N}$	Köthe's nil radical
$\mathcal{J}$	Jacobson radical
$\mathcal{L}$	Levitzki radical and also the lower radical operator

List of Symbols

ν	von Neumann regular radical
β	Baer or prime radical
$\mathcal{Z}$	class of zero-rings
$\mathcal{U}$	upper radical operator
$\mathcal{G}$	Brown–McCoy radical
$\mathcal{S}$	semisimple operator
$(A)\sigma$	intersection of ideals I such that $A/I \in \sigma$
(τ, φ)	torsion theory
$\succ\!\!-$	accessible subring
τ_D	radical of divisible torsion rings
$\triangleleft\cdot$	essential ideal
$\hat{\gamma}$	the largest hereditary subclass in γ
$\mathcal{F}$	Thierrin radical
$\mathcal{Y}$	Yu-lee Lee radical operator
$\mathcal{W}$	Watters' radical
G	Gaussian integers
$\mathbb{Q}(i)$	rational complex numbers
β_φ	antisimple radical
$\mathcal{B}$	Behrens radical
$A * M$	split-null extension of M by A
$\mathcal{I}$	radical class of idempotent rings
$\mathcal{E}$	essential cover operator
$\mathcal{H}$	hereditary closure operator
$\mathcal{P}$	class of prime rings
$\mathcal{L}_{\text{sp}}$	lower special radical operator
I^*	annihilator of I in a ring A
I^l	left annihilator of I in a ring A
I^r	right annihilator of I in a ring A
$\mathcal{N}_g$	generalized nil radical
$\gamma^\perp$	supplementing radical to γ
s	class of subdirectly irreducible rings
h	largest homomorphically closed subclass of
τ	torsion radical
$s(\alpha)$	class of subdirectly irreducible rings with heart in α
$t(\alpha)$	class of subdirectly irreducible rings with heart not in α
χ	radical class of hereditarily idempotent rings

List of Symbols

Δ	divisible radical
τ_p	p-torsion radical
γ^0	radical class of rings A with $A^0 \in \gamma$
$\mathcal{M}$	a class of simple prime rings
$\mathcal{R}$	upper radical of $\mathcal{M}$
$(0:M)_A$	annihilator of the A-module M
L_a	left multiplication by a
R_a	right multiplication by a
Σ_A	class of A-modules M with $AM \neq 0$
Σ	union of the classes Σ_A
$F(\Sigma)$	class of rings A for which Σ_A contains a faithful module
γ_Σ	class of rings A for which Σ_A is empty
$\neg \mathbb{P}$	not $\mathbb{P}$
$\gamma_\mathbb{P}$	class of $\mathbb{P}$-rings
λ	de la Rosa's λ-radical
s_ℓ	Groenewald–Heyman (left) strongly prime radical
s_r	right strongly prime radical
(R,V,W,S)	Morita context
$\Delta(P)$	ideal of S assigned to the ideal P of R in a Morita context (R,V,W,S)
$\nabla(Q)$	ideal of R assigned to the ideal Q of S in a Morita context (R,V,W,S)
$\mathcal{V}(A)$	sum of nilpotent ideals relative to certain factor rings of A in a transfinite procedure
$N^*(A)$	Passman's N^*-radical
$\partial_x(w)$	degree of w in x
$\mathcal{Q}$	class of (left) primitive rings
$\mathcal{Q}^*$	class of simple primitive rings
$\mathcal{J}_\varphi$	upper radical of subdirectly irreducible primitive rings
Φ_m	cyclotomic polynomial
$\mathcal{C}_R$	class of critically compressible R-modules with $RM \neq 0$
ζ	weak radical
(Δ, V, M)	R-lattice for a ring R
$Z(R)$	singular ideal of the ring R
$(A)_{ij}$	matrices having elements from A at the (i,j) position and 0 elsewhere
β_s	Jenkins radical

List of Symbols

γ_x	class of rings A for which $A[x]$ is in γ
$\varrho(A)$	intersection of A and the γ-radical of the polynomial ring $A[x]$
γ_ϱ	class of rings A with $\varrho(A) = A$
$\mathcal{T}$	upper radical of matrix rings over division rings
$\mathcal{A}$	class of artinian rings
$\mathcal{D}$	Divinsky's lower radical
$\wp$	class of rings having a nonzero idempotent
$\mathcal{U}$	Olson's uniformly strongly prime radical
κ	radical of strongly regular rings
$\mathbb{V}$	variety of nonassociative rings
$\Gamma(A)$	nonassociative ring construction on $A^+ \oplus A^+ \oplus A^+$
$\Lambda(A)$	nonassociative ring construction on $A^+ \oplus A^+ \oplus A^+$
$\triangleleft_d$	distinguished ideal
$A^{(n)}$	power of powers of the nonassociative ring A
A^s	power of the nonassociative ring A
$A^{(s,k)}$	certain power of the nonassociative ring A
$*$	involution
$\triangleleft^*$	$*$-ideal
A^{op}	opposite ring of A
N_0	0-symmetric part of the near-ring N
N_c	constant part of the near-ring N
$\mathbb{N}$	variety of near-rings
$\mathbb{N}_0$	variety of 0-symmetric near-rings
$\Sigma(N, M)$	near-ring construction on $N^+ \oplus N^+ \oplus M^+$
$\Theta(N)$	near-ring construction on $N^+ \oplus N^+ \oplus N^+$
$\Phi(N)$	near-ring construction on $N \times N \times N$
$\Xi(K, L)$	wreath-sum of K with L
$\Psi(N, M)$	near-ring construction on $N^+ \oplus N^+ \oplus M^+$

List of Standard Conditions

(a)	γ is homomorphically closed
(b)	for $\forall$ rings A, $\gamma(A) = \Sigma(I \triangleleft A \mid I \in \gamma) \in \gamma$
(c)	$\gamma(A/\gamma(A)) = 0$ for $\forall$ rings A
$(\bar{c})$	γ is closed under extensions
$(\bar{b})$	γ has the inducitve property
$(R1)$	if $A \in \gamma$, then for $\forall A \longrightarrow B \neq 0$ $\exists C \triangleleft B$ such that $0 \neq C \in \gamma$
$(R2)$	if $A \in \mathbb{A}$ and for $\forall A \longrightarrow B \neq 0$ $\exists C \triangleleft B$ such that $0 \neq C \in \gamma$, then $A \in \gamma$
$(S1)$	σ is regular
$(S2)$	if $A \in \mathbb{A}$ and for $\forall 0 \neq B \triangleleft A$ $\exists B \longrightarrow C$ such that $0 \neq C \in \sigma$, then $A \in \sigma$
$(R1^\circ)$	if $A \in \gamma$, then for $\forall A \longrightarrow B \neq 0$ $\exists C \succ\!\!- B$ such that $0 \neq C \in \gamma$
$(R2^\circ)$	if $A \in \mathbb{A}$ and for $\forall 0 \neq B \triangleleft A$ $\exists B \longrightarrow C$ such that $0 \neq C \in \gamma$, then $A \in \gamma$
$(S1^\circ)$	if $A \in \sigma$, then for $\forall 0 \neq B \succ\!\!- A$ $\exists B \longrightarrow C$ such that $0 \neq C \in \sigma$
$(S2^\circ)$	if $A \in \mathbb{A}$ and for $\forall 0 \neq B \succ\!\!- A$ $\exists B \longrightarrow C$ such that $0 \neq C \in \sigma$, then $A \in \sigma$
(λ)	ϱ is closed under essential extensions
(F)	if $K \triangleleft I \triangleleft A$ and $I/K \cong S$, then $K \triangleleft A$
(R)	if $I \triangleleft A$, $I \in \varrho$ and $I^* = 0$, then $A \in \varrho$
(A)	if $I \triangleleft A$ and $I \in \varrho$, then $A/I^* \in \varrho$
$(M1)$	if $M \in \Sigma_{A/I}$, then $M \in \Sigma_A$
$(M2)$	if $M \in \Sigma_A$, $I \triangleleft A$ and $I \subseteq (0:M)_A$, then $M \in \Sigma_{A/I}$
$(M3)$	if $\ker(\Sigma_A) = 0$, then $\Sigma_B \neq \emptyset$ for $\forall 0 \neq B \triangleleft A$
$(M4)$	if $\Sigma_B \neq \emptyset$ for $\forall 0 \neq B \triangleleft A$, then $\ker(\Sigma_A) = 0$
$(SM3)$	if $M \in \Sigma_A$, $B \triangleleft A$ and $BM \neq 0$, then $M \in \Sigma_B$
$(SM4)$	if $B \triangleleft A$ and $M \in \Sigma_B$, then $BN \in \Sigma_A$
$(l-s)$	$L \in \gamma$ implies $L \subseteq \gamma(A)$ for $\forall L \triangleleft_l A$
$(N1)$	if $L_l \triangleleft K_r \triangleleft R \in \varrho$ and L is a prime ring, then $L \in \varrho$
$(N2)$	if $L \triangleleft_l K \triangleleft_r R$, $L \in \varrho$ and R is a prime ring, then $R \in \varrho$
$(K1)$	$A \in \mathcal{N}$ implies $A[x] \in \mathcal{J}$
$(K2)$	$A[x] \in \mathcal{SJ}$ implies $A \in \mathcal{SN}$

Author Index

Abian, A., 176, 337
Aburawash, U. A., 134, 324, 325, 337
Amitsur, S. A., vi, 6, 21, 142, 161, 198, 200, 257, 268, 271, 275, 337, 338
Anderson, F. W., 244, 338
Anderson, T., vi, 40, 45, 51, 56, 57, 60, 67, 69, 71, 73, 102, 303, 317, 327, 334, 338, 362
Andrunakievich, V. A., vi, vii, 11, 78, 80, 87, 92, 97, 117, 119, 120, 124, 126, 127, 128, 134, 176, 185, 259, 286, 302, 304, 338
Andruszkiewicz, R. R., 65, 339
Ánh, Pham Ngoc, 102, 216, 218, 339
Arens, R. F., 296, 339
Arhangel'skiĭ, A. V., vii, 339
Armendariz, E. P., 47, 59, 174, 299, 339, 351
Arnautov, V. I., v, 339
Aslam, M., 258, 339
Ayoub, Christine, 278, 339

Babich, A. M., 185, 339
Baer, R., 184, 188, 340
Barua, M. N., 331, 340
Behrens, E. A., 284, 340
Beidar, K. I., 60, 64, 81, 83, 88, 101, 142, 149, 156, 164, 181, 186, 262, 275, 278, 287, 291, 306, 309, 316, 318, 325, 340, 341

Bell, H. E., 178, 341
Berg, J. E. van den, 219, 341
Bergman, G. M., 219, 341
Betsch, G., 327, 330, 341, 342
Birkenmeier, G. F., 98, 181, 336, 342
Birkhoff, G., 5
Blair, R. L., 99, 342
Booth, G. L., 89, 324, 327, 336, 342
Brauer, R., 197, 342
Brown, B., 260, 283, 342
Buys, Annemarie, 89, 102, 293, 342

Cai Chuan Ren, 89, 90, 342
Chen Shen Can, 89, 90, 343
Clay, J. R., 325, 343
Cohn, P. M., 175, 227, 343, 361
Connell, E. H., 80

Dauns, J., 124, 343
Desale, G. B., 288, 343
Dickson, L. E. 325
Dickson, S. E., 36, 57, 343
Dinh Van Huynh, 278, 343
Divinsky, N., vi, viii, 40, 43, 45, 51, 56, 57, 60, 103, 104, 145, 146, 149, 279, 280, 282, 338, 343, 362
Domokos, M., 325, 343
Dorroh, J. K., 10, 343
Drazin, M. P., 134, 343
Dubrovin, I. N., 234, 344

Author Index

Emin, G. G., 131, 344
Enersen, P. O., 31, 50, 344
Erő, Zsuzsa, viii

Faith, C., 218, 242, 344
Feigelstock, Sh., 10, 105, 106, 344
Ferrero, M., 275, 344
Fisher, I., 190, 344
Fong Yuen, vii, 98, 149, 181, 275, 340, 344, 358
Forsythe, A., 7, 296, 344
France-Jackson, Halina, 88, 344
Fried, E., vii, 344
Fuchs, L., 282, 344, 362
Fuchs, P., 327, 345
Fuller, K. R., 244, 338

Gardner, B. J., vi, vii, 21, 37, 54, 59, 84, 85, 88, 89, 96, 102, 103, 108, 138, 139, 164, 165, 168, 171, 174, 177, 178, 179, 181, 254, 256, 262, 265, 282, 297, 302, 303, 305, 317, 338, 345, 346
Gerber, G., 89, 102, 342
Goethe, J. W., v
Goldie, A. W., 236, 249, 252, 346
Golod, E. S., 190, 346
Goodearl, K. R., 289, 346
Groenewald, N. J., 89, 98, 141, 262, 288, 324, 327, 336, 342, 346
Guo Jin-yun, 65, 346, 352

Handelman, D., 140, 142, 289, 346
Hartney, J., 327, 346
Heatherly, H. E., 336, 342
Hebisch, U., vii, 346
Heinicke, A. G., 56, 64, 108, 127, 242, 249, 347
Hentzel, I. R., 119, 130, 316, 347
Herstein, I. N., 8, 223, 235, 252, 347
Heyman, G. A. P., 74, 75, 76, 77, 78, 141, 288, 346, 347, 359
Hoehnke, H.-J., 28, 347
Hoffman, A. E., 49, 57, 58, 347

Holcombe, M., 327, 347
Hongan, M., 162, 163, 347
Hopkins, Ch., 213, 347
Horvath, J., 252, 347
Hsieh Pang Chieh, 134, 188, 348
Huang, Feng-Kuo, vii, 344
Huang, Wen Ping, 98, 348

Irving, R. S., 226, 348

Jacobson, N., 7, 8, 16, 117, 211, 218, 348
Jaegermann, M., 137, 150, 153, 154, 155, 156, 164, 168, 170, 348
Janelidze, G., viii, 348
Jans, J. P., 8, 214, 348
Jategaoncar, A. V., 226, 348
Jenkins, T. L., vii, 74, 75, 88, 108, 109, 260, 348, 351, 352, 356, 359
Johnson, R. E., 244, 349
Joubert, S. J., 36, 349

Kaarli, K., 71, 303, 327, 330, 334, 336, 338, 341, 345, 349
Kando, T., 296, 349
Kaplansky, I., 8, 200, 201, 219, 296, 339, 349
Ke, Wen-Fong, 142, 181, 262, 340
Kelarev, A. V., 164, 345, 349
Kertész, A., 8, 205, 211, 214, 216, 218, 252, 278, 349
Kogalovskiĭ, S. P., 173, 174, 349
Koh, Kwangil, 242, 249, 251, 252, 349, 350
Köthe, G., vi, 21, 144, 187, 350
Krachilov, K. K., 98, 350, 360
Krause, G., 208, 350
Kreiling, D., 28, 49, 362
Krempa, J., vii, 37, 55, 137, 144, 145, 146, 149, 261, 267, 271, 273, 306, 343, 350
Kriis, T., 327, 349
Kruse, R. L., 279, 350
Kuratowski, C., 1, 350

Author Index

Kurosh, A. G., vi, 21, 51, 208, 350

Lajos, S., 164, 350
Lawrence, J., 140, 142, 289, 346, 350
Laxton, R. R., 327, 351
Leavitt, W. G., vi, 31, 36, 37, 49, 50, 52, 53, 57, 58, 59, 76, 89, 107, 108, 109, 111, 112, 115, 117, 257, 259, 299, 304, 305, 339, 344, 347, 351
Lee, Enoch K. S., 336, 342
Lee Pjek-Hwee, 329, 352
Lee, Yu-lee, 52, 53, 208, 351, 352
Leeuwen, L. C. A. van, 34, 40, 41, 45, 48, 88, 98, 107, 108, 109, 111, 112, 115, 260, 351, 352
Levitzki, J., 6, 184, 196, 197, 200, 201, 338, 352
Lex, W., vii, 352
Litoff, 218
Liu Shao-xue, 65, 352
Locke, J., v
Loi, Nguyen Van, 102, 179, 254, 286, 319, 323, 324, 325, 339, 353
Luh, Jiang, 249, 251, 252, 349
Luo Yun-lin, 65, 352
L'vov, I. A., 65, 235, 353

Majumdar, S., 45, 181, 353
Márki, L., vii, viii, 21, 81, 348, 353
Martinez, J., vii, 353
Maxson, C. J., 327, 345
McConnell, J., 16, 353
McConnell, N. R., 102, 108, 353, 354
McCoy, N. H., 7, 8, 11, 214, 256, 260, 268, 283, 296, 342, 344, 354
McKnight, J. D., 80, 354
McWorter, W. A., 235, 354
Medved'ev, Yu. A., 318, 354
Meldrum, J. D. P., 325, 354
Mendes, D. Isabel C., 101, 142, 354
Mewborn, A. C., 242, 249, 350
Mlitz, R., vii, 21, 28, 34, 35, 283, 284, 330, 336, 353, 354, 355

Morak, Bettina, vii, 355
Morita, K., 149, 355
Morris, D., 234, 355
Musser, G. L., 80, 354, 355

Nagata, M., 284, 355
Nicholson, W. K., 158, 161, 162, 163, 355
Niekerk, Julia S. van, 28, 358
Nikitin, A. A., 311, 313, 355

Olivier, W. A., 98, 346
Olson, D. M., vii, 290, 355, 356
Olszewski, J., 256, 356
Ortiz, A. H., 263, 265, 356
Osłowski, B., 149, 356
Oswald, A., 330, 355

Palmer, Th., vii, 356
Parmenter, M. M., 140, 142, 275, 344, 356
Passman, D. S., 142, 186, 356
Patterson, E. M., 262, 356
Pchelintsev, S. V., 316, 317, 356
Perlis, S., 25, 356
Pilz, G., 325, 327, 357
Plotkin, B. I., 28, 357
Preuß, G., vii, 357
Price, D. T., 279, 350
Procesi, C., 198, 357
Propes, R. E., 108, 257, 284, 357
Puczyłowski, E. R., 48, 65, 103, 105, 106, 107, 108, 136, 142, 144, 147, 149, 159, 208, 262, 273, 275, 284, 289, 291, 306, 340, 356, 357, 358

Raftery, J. G., 293, 358
Ramakotaiah, D., 327, 358
Rashid, M. A., 50, 358
Riley, D. M., 200, 358
Roberts, M. L., 134, 343
Robson, J. C., 16, 117, 353, 358
Roos, C., 40, 41, 45, 48, 76, 77, 78, 347, 352, 358

Rosa, B. de la, 28, 72, 98, 132, 134, 256, 260, 284, 358
Rossa, R. F., 48, 60, 358
Roux, H. J. le, 74, 75, 358
Rowen, L. H., 8, 16, 144, 208, 211, 218, 359
Rubin, R. A., 140, 359
Ryabukhin, Yu. M., vi, 47, 49, 75, 78, 84, 87, 88, 98, 100, 119, 120, 127, 128, 176, 185, 235, 299, 302, 303, 304, 338, 359, 360

Sakhajev, I., 144, 360
Salavová, Katarina, 88, 149, 164, 319, 341, 360
Sands, A. D., 40, 41, 42, 44, 71, 72, 101, 142, 149, 150, 151, 152, 155, 156, 159, 164, 166, 168, 170, 172, 261, 283, 284, 297, 348, 355, 360, 361
Sąsiada, E., 208, 227, 361
Schoeman, M. J., 36, 349
Shafarevich, I. M., 190, 346
Shestakov, I. P., 8, 316, 366
Shirshov, A. I., 8, 316, 366
Shulgeifer, E. G., 21, 361
Shum, Kar-Ping, 181, 340
Sidorov, A. V., 65, 353, 361
Sierpińska, Anna, 275, 361
Slin'ko, A. M., 8, 316, 318, 361, 366
Slover, Rebecca, 262, 361
Small, L., 289
Smoktunowicz, Agata, 235, 275, 287, 358, 361
Snider, R. L., 57, 256, 257, 361
Sokolskiĭ, A. G., 164, 361
Steinfeld, O., 216, 218, 362
Stewart, P. N., vii, 58, 59, 81, 88, 140, 142, 148, 174, 175, 176, 178, 302, 345, 353, 356, 360, 362
Stokes, T., 102, 354
Storrer, H. H., 249, 362
Struik, Ruth R., 190, 344

Suliński, A., 40, 45, 51, 56, 57, 60, 145, 146, 149, 208, 256, 338, 343, 361, 362
Szász, F. A., vi, 164, 205, 236, 256, 278, 282, 284, 350, 362
Szele, T., 278, 282, 362
Szendrei, J., 11, 362

Tang Ai-ping, 65, 352
Tangeman, R., 28, 49, 358, 362
Terlikowska-Osłowska, Barbara, 306, 350, 363
Thierrin, G., 87, 363
Tholen, W., viii, 348
Tumurbat, S., 88, 103, 134, 142, 146, 164, 256, 275, 354, 361, 363

Utumi, Y., 196, 218, 344, 363

Varadarajan, K., 288, 343
Veldsman, S., vii, 21, 28, 99, 100, 101, 131, 133, 134, 142, 262, 293, 306, 327, 333, 334, 335, 336, 342, 355, 356, 358, 363

Walker, R., 327, 347
Walt, A. P. J. van der, 293, 327, 345, 364
Wang, C. S., 149, 340
Wang, Xion Hau, 134, 188, 364
Watters, J. F., 49, 53, 59, 65, 89, 142, 158, 161, 162, 163, 275, 304, 351, 355, 364
Wedderburn, J. H. M., v, 7, 21, 365
Weinert, H. J., ix, 346, 365
Widiger, A., 278, 349, 365
Wiegandt, R., vi, vii, 21, 28, 34, 37, 40, 41, 45, 48, 50, 67, 69, 71, 73, 81, 84, 85, 88, 89, 98, 102, 103, 131, 134, 140, 146, 164, 174, 181, 186, 254, 256, 275, 278, 279, 283, 284, 287, 291, 293, 303, 319, 323, 325, 327, 334, 336, 338, 339, 340, 341, 342, 344, 346, 351, 352, 352,

Author Index

353, 355, 356, 358, 360, 363, 365
Wielandt, H., 325
Wong, E. T., 244, 349
Wu, Tong-suo, 181, 365
Wyck, L. van, 262, 346, 365

Xiao Jie, 65, 352
Xu Yong Hua, 218, 365

Zaidi, A. M., 258, 339
Zand, H., 142, 149, 164, 208, 256, 284, 358, 363
Zassenhaus, H., 325
Zel'manov, E. I., 188, 366
Zelmanowitz, J., 236, 243, 245, 249, 366
Zhevlakov, K. A., 8, 316, 366
Zhou Yiqiang, 89, 90, 366
Zorn, M., 1, 366

Subject Index

abstract property, 6
a.c.c., 196
ADS-property, 319
ADS-Theorem 40
A-lattice, 245
algebra
   K-, 2
   Weyl, 16
Amitsur Conjecture, 275
Amitsur–Levitzki Theorem, 6
Amitsur property, 267
Andrunakievich Lemma 11
Andrunakievich s-variety, 316
annihilator, 86, 118
   left, 87
   middle, 72
   right, 86
antisimple radical, 67
A-radical, 165
artinian ring, 211

Baer radical, 29
Behrens radical, 67
biideal, 164
   *-, 325
bijection, 3
Birkhoff's Theorem
   on subdirect decompositions, 5
   on varieties, 5
Brown–McCoy radical, 31

center, 17
centralizer, 209
class
   abstract, 6
   Brown–McCoy radical, 31
   — closed under essential extensions, 47
   — closed under extensions, 22
   — closed under subdirect sums, 4, 32
   hereditary, 4
   homomorphically closed, 4, 22
   Jacobson radical, 25
   left hereditary, 48
   left regular, 145
   Levitzki radical, 25
   lower, 102
   lower radical, 28
   matric-extensible, 257
   N-, 158
   nilpotent rings, 2
   nil radical, 24
   normal, 158
   radical semisimple, 173
   regular, 30
   right hereditary, 48
   semisimple, 31
   special class, 80
   strongly hereditary, 48
   torsion, 36
   torsionfree, 36

Subject Index 382

universal, 4
upper, 102
upper radical, 30
von Neumann regular radical, 25
weakly homomorphically closed, 68
weakly special, 78
weak radical, 235, 242
closure
  homomorphic, 28
  semisimple, 37, 44
completely prime ring, 287
condition
  ascending chain, 196
  descending chain, 211
  — (F), 66
constant part of a near-ring, 326
construction
  Kurosh lower radical, 51
  semisimple class, 31
  Tangeman–Kreiling lower radical, 28
  upper radical, 31
  Watters lower radical, 53
  Yu-lee Lee radical, 52
cover
  essential, 73

d.c.c., 211
degree
  of nilpotency, 2
  of solvability, 309
dense ring, 211
density theorem, 211
  generalized, 245
derivation, 14
  σ-, 14
direct limit, 218
direct product, 4
direct sum, 4
  complete, 4
  restricted, 4
Divinsky's lower radical, 279
divisible radical, 104

division ring, 6
domain, 6
  left Ore, 7
Dorroh extension, 10
dual radical, 96

element
  cancellable, 6
  divisible, 8
  F_1-regular, 253
  G-regular, 253
  λ-, 132
  $\mathbb{P}$-, 131
  quasi-regular, 202
embedding, 3
endomorphism ring, 10, 12
extension
  closed under essential, 47
  closed under, 22
  Dorroh, 10
  essential, 47
  split-null, 10, 68
  unital, 254

factor ring, 3
field, 6
  skew, 6
finite topology, 211
formal power series ring, 18

Gardner's Lemma, 171
generalized nil radical, 87
Goldie ring, 249
Groenewald–Heyman radical, 141
group
  divisible abelian, 8
  N-group, 327
  Prüfer, 8
  quasi-cyclic, 8
  reduced abelian, 104

heart, 5
hereditary class, 4
  left, 48
  strongly, 48

Subject Index

hereditary radical, 45
Hoehnke radical, 27
homomorphic closure, 28
homomorphically closed, 4, 22
    largest — subclass, 90
homomorphism, 3
homomorphism theorem, 3
Hopkins Lemma, 213
hull
    injective, 8
    quasi-injective, 242
hyperconstant radical, 335
hypernilpotent radical, 101, 330
hypoidempotent radical, 101

ideal, 2
    completely prime, 287
    distinguished, 306
    essential, 46
    F_1-regular, 254
    G-regular, 254
    left, 2
    maximal divisible, 8, 104
    maximal torsion, 8
    modular left, 122
    P-, 106
    prime, 12
    primitive, 203
    quasi-, 216
    quasi-semiprime, 72
    right, 2
    semiprime, 12, 79
    singular, 250
    small, 254
    uniform left, 237
    *-, 319
idealizer, 111
identity
    proper polynomial, 198
    standard polynomial, 6
image
    prime, 84
index of an Andrunakievich variety, 316

injection, 3
intersection property, 75
insulator, 140
    uniform, 290
involution, 318
    exchange, 319
isomorphism theorem
    first, 3
    second, 3

Jacobson radical, 25
Jacobson's Commutativity Theorem, 7
Jacobson's Conjecture, 208
Jenkins radical, 260

kernel, 3
Köthe's Problem, 144
Köthe's radical, 24
Krempa's Lemma, 55
Kurosh–Amitsur radical, 22

λ-radical, 132
Laurent polynomial ring, 18
van Leeuwen's Theorem, 48
left-ADS-radical, 166
left stable radical, 136
left strong radical, 142
Levitzki radical, 25
Levitzki's Problem, 235, 287
Litoff–Ánh Theorem, 216
lower radical, 28
    Kurosh, 51
    Tangeman–Kreiling, 28
    Watters', 53

matric equation, 257
matric-extensible class, 257
module, 7
    compressible, 236
    critically compressible, 236
    faithful, 118
    injective, 7
    irreducible, 121
    monoform, 242

noetherian, 238
prime, 124
quasi-injective, 7, 242
quasi-simple, 249
uniform, 126, 237
modularity law, 4
Morita context, 149
dual, 150
factor context, 158
S-faithful, 158
multiplication, 1
left, 111
scalar, 2
zero-, 2
multiplier, 110

near-ring, 325
constant, 326
constant part, 326
equiprime, 327
— ideal, 326
— left ideal, 326
0-symmetric, 326
0-symmetric part, 326
nilpotency
degree of, 2
nil radical, 24
nil ring, 24
nilpotent ring, 2
nilpotent subset, 186
noetherien ring, 196
normal radical, 150
N-radical, 156

Olson's radical, 290
operation
circle, 25
operator
essential cover, 73
hereditary closure, 77
largest homomorphically closed subclass, 90
lower radical, 28
semisimple, 32

upper radical, 31
Ore domain, 7

partial endomorphism, 237
PI-ring, 198
Plotkin radical, 28
polynomial
cyclotomic, 223
standard, 6
polynomial ring
Laurent, 18
skew, 15
prime radical, 29
prime ring, 29
primitivity, 203
ν-, 327
primitive ring, 203
principally left hereditary radical, 151
property
abstract, 6
ADS-, 319
Amitsur, 267
coinductive, 33
inductive, 23
intersection, 75
p-torsion radical, 104

quasi-ideal, 216
quasi-inverse, 202
quasi-regular element, 202
quasi-regular ring, 202

radical, 22
A-, 165
antisimple, 67
Baer, 29
Behrens, 67
Brown–McCoy, 31
complementary, 89
Divinsky's lower, 279
divisible, 104
dual, 96

Subject Index

equiprime, 328
generalized nil, 87
Groenewald–Heyman, 141
hereditary, 45
Hoehnke, 27
hyperconstant, 335
hypernilpotent, 101, 330
hypoidempotent, 101
Jacobson, 25
Jenkins, 260
Köthe's, 24
Kurosh–Amitsur, 22
λ-, 132
left-ADS-, 166
left stable, 136
left strong, 142
Levitzki, 25
lower, 28
matric-extensible, 257
N-, 156
N^*-, 186
nil, 24
normal, 150
Olson's, 290
overnilpotent, 336
pair of supplementing, 96
Plotkin, 28
prime, 29
principally left hereditary, 151
p-torsion, 104
right invariantly strong, 327
special, 80
strict, 148
strongly prime, 141
strongly regular, 296
subidempotent, 94
supernilpotent, 65
supplementing, 89, 96
Thierrin, 51
torsion, 95
uniform, 127
uniformly strongly prime, 290
von Neumann regular, 25
weak radical, 235, 242

radical class, 22
  hereditarily idempotent rings, 95, 98
  idempotent rings, 46
radical construction
  Kurosh lower, 51
  Tangeman–Kreiling, 28
  upper, 31
  Watters' lower, 53
  Yu-lee Lee, 52
reduced
  abelian group, 104
  ring, 176
regular class, 30
  left, 145
ring, 1
  artinian, 211
  Brown–McCoy, 275
  commutative, 2
  completely prime, 287
  division, 6
  endomorphism, 10, 12
  factor, 3
  f-regular, 99
  F_1-regular, 253
  formal power series, 18
  free, 6
  γ-radical, 22
  γ-semisimple, 32
  Goldie, 249
  G-regular, 253
  Jacobson, 275
  λ-, 132
  Laurent polynomial, 18
  locally nilpotent, 25
  m-, 286
  minimally embeddable, 114
  nil, 24
  nilpotent, 2
  noetherian, 196
  p-, 38, 104
  PI-, 198
  prime, 12
  (left) primitive, 203

Subject Index

P-simple, 106
quasi-regular, 202
quasi-semiprime, 72
reduced, 176
right primitive, 205
semiprime, 12, 79
semiprimitive, 205
simple, 2
∗-simple, 319
skew polynomial, 15
solvable, 309
strongly locally matrix, 216
strongly prime, 140
strongly regular, 296
subdirectly irreducible, 5
torsion, 9
torsionfree, 9
transitive, 306
trivial, 2
twisted Laurent polynomial, 114
twisted polynomial, 15
unequivocal, 103
uniformly strongly prime, 290
von Neumann regular, 25
weakly primitive, 235, 245
weakly semiprimitive, 245
Zassenhaus, 60
zero-, 2
ring of left quotients, 6
ring with involution, 318

Sands' Theorem, 42
Schur's Lemma, 209
semiprime ring, 12, 79
semiprimitive ring, 205
   weakly, 245
semisimple class, 31
semisimple closure, 37
simple ring, 2
   P-, 106
   ∗-, 319
skew polynomial ring, 15
solvable ring, 309
special class, 80

weakly special class, 78
special radical, 80
standard polynomial identity, 6
Stewart's Lemma, 58
strict radical, 148
strongly locally matrix ring, 216
strongly prime radical, 141
   uniformly, 290
strongly prime ring, 140
strongly regular radical, 296
strongly regular ring, 296
subdirectly irreducible ring, 5
subdirect sum, 4
subidempotent radical, 94
subnear-ring
   invariant, 326
subring, 2
   accessible, 43
   dense, 211
   fixed, 17
   trivial, 2
   ∗-, 319
Suliński–Anderson–Divinsky
Problem, 60
supernilpotent radical, 65
supplementing radical, 89
surjection, 3
Szász' Theorem, 278
Szendrei's Theorem, 11

Thierrin radical, 51
torsion class, 36
torsion ring, 9
torsion theory 36
torsionfree class, 36
torsionfree ring, 9
transitive ring, 306
trivial subring, 2
twisted Laurent polynomial ring, 19, 114
twisted polynomial ring, 15

unequivocal ring, 103
uniform radical, 127

Subject Index

uniformly strongly prime radical, 290
uniformly strongly prime ring, 290
unity element, 2
universal class, 4
    normal, 306
upper radical, 31

variety, 4
    Andrunakievich s-, 316
von Neumann regular radical, 25
von Neumann regular ring, 25

weakly primitive ring, 245
weak radical, 235, 245
Wedderburn's Theorem, 7
Wedderburn–Artin Structure Theorem
    first, 212
    second, 213
wreath-sum, 331

Zassenhaus ring, 60
Zorn's Lemma, 1